Introduction to Biometry

Introduction to Biometry

Pierre Jolicoeur

Department of Biological Science
University of Montreal

Forewords by

Richard Tomassone

Department of Mathematics and Computer Science
National Agronomical Institute, Paris

and

Jean-Marie Legay

Claude-Bernard Lyon I University, Villeurbanne

Kluwer Academic / Plenum Publishers
New York, Boston, Dordrecht, London, Moscow

This book has been translated by the author, with assistance from his wife, Veronika Meinow, from the fifth edition *Introduction à la biométrie*, which will be published in 1999 by Décarie Éditeur, Inc.

The front cover illustration is based on an original analysis of data published in 1905 by J. S. Haldane and J. G. Priestley, who pioneered the study of human respiration.

ISBN 0-306-46163-3

233 Spring Street, New York, N.Y. 10013

10 9 8 7 6 5 4 3 2 1

A C.I.P. record for this book is available from the Library of Congress

Printed in the United States of America

Foreword

The collaboration between biologists and mathematicians, mentioned by Jean-Marie Legay in his foreword to the first French edition of Pierre Jolicoeur's book, is of course as necessary as ever. The interactions between the two different fields of thinking must go on! The mathematician, on the one hand, may perhaps more easily function within his own universe without suffering too much from a lack of collaboration, although he will be intellectually frustrated to see that the tools he conceives are not readily used by other scientists, as he thinks they should be. The biologist, on the other hand, can hardly ignore mathematical and statistical methods: he must progress in his own field, developing his ideas and making decisions with tools often created by others but indispensable in order to extract all information available in experimental data.

Language is undoubtedly of prime importance to transmit knowledge and motivations from the mathematician to the biologist and vice versa, a fact too often neglected by many scientists. In order for the objectives of a study to be well understood, and for mathematical formulae to be applied correctly to life phenomena, communication through well-chosen language is as important as technical training, but this is often underestimated in teaching, especially at the university level. Biometry aims at being the melting pot of mathematical statistics and biology, which makes an aptitude for dialogue mandatory in the biometrician. This aptitude can only be developed through an intimate acquaintance with both mathematical (statistical) tools and a particular field of application, as illustrated by the case of Pierre Jolicoeur.

The author of this textbook belongs to a group of scientists who feel a strong yearning to fill this need. His treatment of the subject is simple and clear, yet highly rigorous: it should satisfy all kinds of readers. The mathematician will find the essentials of what must be learned and taught, and undoubtedly also a manner of presenting mathematical notions in such a way as to adapt them to biological contexts. The biologist will get even more. First, he will obtain tools which can be applied immediately to his own problems, something which he might consider as a strictly decent minimum... In addition to basic statistical formulae, however, the biologist will also discover a simple and intelligent manner of reasoning which can be used to adapt statistical models to complex biological situations. He will realize that biometry enables him to emphasize the salient features of an analysis while discarding the random peculiarities which are present in any experiment or observation. But many other categories of potential users, besides biologists, could profit from this book, and all those whose results are affected by random should be able to transpose to their own field the ideas and techniques discussed here.

Admittedly, contemporaneous biometry is a rapidly evolving science (or is it a craft?), and the new tools which are created every day are not all mentioned in this textbook, but this was not the author's purpose. However, once the reader has read the book and understood its contents, and has applied some of the methods to his own data, he will have reached a "point of no return" in his growing acquaintance with biometry. The biometrical way of thinking will have become anchored in his mind, and he will be able to read and understand more specialized and possibly more technical textbooks, if necessary. But he will already have acquired the essential basis needed to develop his own personal reflections.

I think that the readers of this book should be aware of their chance. As for myself, I can only regret not to have had Pierre Jolicoeur's book at hand when I began my career as a biometrician!

Richard Tomassone

Foreword to the first French edition (1991)

The necessity of multidisciplinary studies is widely acknowledged today. The training of mathematicians should therefore include some biology but, except in a few universities or other schools of higher learning, this is still generally not done on a satisfactory scale. Similarly, experimenters, including biologists, can no longer get along without some knowledge of mathematics, which they will need either in data analysis or in model building. Whether experimental scientists carry out such activities themselves or collaborate with mathematicians, they must understand quantitative techniques, and be aware of their possibilities and limitations.

In the various biological sciences, including medicine and agriculture, many students are therefore faced with a major problem since, although they do not wish to take advanced training in mathematics, their studies and their own research will require them to know and understand some mathematical concepts and techniques.

Fortunately, some biologists and some mathematicians have investigated questions involving biology and mathematics for decades (models of differential growth, population dynamics, spatial dispersion, prey-predator interactions, etc.). Their joint activities have given birth to an interface field: biometry. Not surprisingly, courses in biological mathematics or in mathematical biology are often taught by biometricians, who do research at the frontier of both fields.

As a biometrician, Pierre Jolicoeur is known throughout the world for his early studies of complex biological phenomena. He pioneered the use of multivariate statistical methods in morphometrics, in the case of bilateral symmetry for instance. More recently, he has developed original nonlinear models for somatic growth. Pierre Jolicoeur's interdisciplinary stand is clearly shown by the title of this book. Drawing upon his experience both as a research worker in biometry (see the list of his publications) and as a teacher, the author is undertaking the difficult task of making statistics palatable to biologists and biology students.

What makes this textbook markedly original is that elementary notions of probability and statistics are presented in an unabashedly simple and clear fashion, with helpful comments, and are illustrated using new and often personal data sets. Moreover, the text includes frequent hints at more elaborate developments, which makes careful reading very rewarding.

While the reader is generally spared detailed rigorous mathematical demonstrations, assumptions and hypotheses are always clearly stated, and the advantages and disadvantages of the various methods and approaches with respect to the aims pursued are always discussed. The author has evidently chosen to guide the reader's efforts toward a qualitative understanding of procedures rather than the technical derivation of already-known results.

Emphasis is placed on the distribution concept, which appears generally adequate for the description of biological variation. After presenting the statistical distributions most frequently useful in biology, the author discusses their utilization for testing hypotheses concerning means, variances, frequency tables, goodness of fit, and simple or partial and multiple correlations and regressions. While the total number of variates is limited to two or three[1] in order to make it easy for the (beginning) reader to visualize relationships, natural extensions are evoked whenever possible.

While the author generally does not treat his subject in a highly technical manner, his attempt to give the nonmathematical reader a deep and intuitive understanding of biological statistics is obviously a considerable challenge. Moreover, by letting beginning students of biometry glimpse at more advanced topics, this book may stimulate some of them to pursue their studies further. Let us wish the author all the success he deserves.

Jean-Marie Legay

[1] Editor's note: in the present edition, the simultaneous analysis of more than three variates is covered in chapters 24, 25 and 29 to 34.

Contents

The statistical tables most frequently used in biometry

Introduction

The word *biometry* comes from the Greek (βιος, *life* + μετρον, *measurement*) and means literally the quantitative study of life phenomena. Since living organisms generally differ from each other in size and shape as well as in their functions, the study of these differences requires statistical methods. Consequently, the word *biometry* is often considered as a synonym of *biostatistics*. However, biometricians generally pay more attention to biological aspects than statisticians would do. For many years, it has been realized that a knowledge of biometry is a must for practicing biologists as well as for graduate students. Therefore, most universities are now aware of the necessity of giving their biology students at least one introductory course in biometry.

The present textbook is aimed at university level biology students as well as at biologists wishing to improve their knowledge and understanding of biometry. The author's viewpoint is intermediate between *classical statistics* and *data analysis*. In addition to presenting a broad spectrum of statistical methods, the author has emphasized understanding as much as possible, in order to enable readers to become gradually self-reliant. However, since this book will presumably be read mostly by biologists, in some cases technical explanations are given in ordinary English rather than through formal mathematical demonstrations.

The coverage of the so-called nonparametric methods is limited to the analysis of frequency tables (chapter 15), tests of goodness of fit (chapter 16), and the sign test (section 17.11): the author believes that parametric methods provide better illustrations of the logical role which statistical methods can play in scientific research, the statistical parameters corresponding to the desired theoretical knowledge. Moreover, nonparametric methods somewhat duplicate their parametric counterparts and are redundant to some extent in this respect. Finally, nonparametric techniques are generally less sensitive ("powerful") than their parametric equivalents. The frequency distribution of biological data is often similar enough to the normal distribution (chapters 5, 19, 23, 29), or can be made similar enough through transformations, for methods based on the normal distribution to be used. When other nonparametric methods are truly needed, the reader should consult Sprent's thoughtful introduction (1993).

Biologists and biology students are more strongly motivated for the study of statistical methods when the latter are illustrated on genuine biological data. For this reason, most examples in this textbook are based on real data extracted from scientific publications. In the few cases where artificial data have been used, they have been made to simulate biological reality as closely as possible.

An effort has been made to organize the subject matter in the most efficient pedagogical order. Whenever there is overlap between chapters or between the sections into which they are subdivided, cross-references are given. Exercises (problems) are not included in the present edition but may be published separately in the future.

The five statistical tables which are most frequently used in biometry have been entirely recomputed using algorithms discussed by Abramowitz and Stegun (1968) and by Kennedy and Gentle (1980). These tables cover a particularly extensive range of numbers of degrees of freedom in order to make interpolation seldom necessary.

Tabulated values generally include four or five digits, which should all be accurate since computations were carried out in triple precision. Moreover, the risk of accidental errors has been eliminated by transferring numerical entries electronically. In order to reduce ambiguities and to promote methodical working habits, all tables are given in terms of the *cumulative probability,* i. e. the probability that the numerical value of a *variate* (random variable) is less than the tabulated value.

Some readers may wonder whether it is still pertinent to provide extensive statistical tables at a time when personal computers enable anyone to compute almost instantly the probability of an observed value or, inversely, the value corresponding to a specified probability. The decision to include such tables in this textbook was made for pedagogical reasons: the author is convinced that a university student's familiarity with statistical methods would remain superficial if his training were limited to using computer programs developed by other persons. Nevertheless, every biometrician should possess and use either a good scientific calculator or a personal computer.

The author hopes that this textbook will help the reader to become familiar with the basic notions of biometry and to discover the interesting challenge of applying mathematics and statistics to biology. Should any reader notice obscurities or errors, he is invited to contact the author, who would be grateful for constructive suggestions. Thanks are expressed to the readers of the first three French editions who suggested improvements.

Several generations of students incited me, by their questions and comments, to present many topics more clearly. Several research workers, as indicated within the text, gave permission to base examples on their interesting data. Many years ago, Dr. James E. Mosimann, then at the Department of Biological Sciences of the University of Montreal, stimulated my early interest in biometry, and Professor Stanley W. Nash, of the Department of Mathematics of the University of British Columbia, answered many questions when I began to develop an acquaintance with multivariate analysis. Professor Jean-Marie Legay, of Claude-Bernard Lyon I University, France, and Professor Richard Tomassone, of the Department of Mathematics and Computer Science of the National Agronomical Institute, Paris, kindly accepted to write forewords to the first French edition (1991) and to the present edition. Professor Jacques Pontier, also of Claude-Bernard University, made valuable suggestions. Professors William H. Kruskal and Stephen M. Stigler, of the Department of Statistics of the University of Chicago, provided historical information. My children, my wife, Veronika Meinow, Mr. André Décarie, Mrs. Edenise Garcia and Mrs. Anne-Marie Blais contributed to the improvement of the text and the illustrations. Finally, my wife helped checking the English translation. I sincerely thank all persons mentioned here, as well as many others whose names are not listed for conciseness.

I dedicate this book to the memory of my parents and to the happiness and success of my children, Lucie, Francine, and André.

Montreal, July 1, 1998 Pierre Jolicoeur

Author's address:

Département de Sciences biologiques, Université de Montréal, Case Postale 6128, Succursale Centre-Ville, Montréal, Québec H3C 3J7

or

1226, Rang Égypte, Case Postale 160, St-Valérien, Québec J0H 2B0

Chapter 1
Looking at quantitative biological data through scatter diagrams

Section 1.1: scatter diagrams

In applied statistics in general, and in biometry in particular, graphical representations and numerical techniques are complementary and often equally important. One of the most useful graphical methods in statistics is the *scatter diagram,* in which variable measurements are represented simply by dots dispersed on a surface overlaid by Cartesian coordinate axes (named after the French mathematician René Descartes, 1596-1650). Such coordinate axes are usually perpendicular to each other. While a scatter diagram may have a single coordinate axis (one-dimensional dispersion) or more than two coordinate axes (three-dimensional or multidimensional dispersion), the most commonly used version is the *bivariate scatter diagram,* which contains two coordinate axes (two-dimensional dispersion) and where dots represent pairs of measurements (see figures 1.2.1 and 1.2.2).

Section 1.2: an example of the graphical examination of quantitative data

The numerical measurements in table 1.2.1 were obtained on undissociated human skeletons following archeological excavations in England, and are represented in a bivariate scatter diagram (figure 1.2.1, next page). Data given on any single line, in table 1.2.1, correspond to measurements made on the skeleton of a single individual.

Table 1.2.1
Length measurements of the left humerus and right humerus of female human skeletons; data drawn from the study of Münter (1936)
(data reproduced with the permission of Oxford University Press on behalf of the Biometrika Trustees)

Number of skeleton (subscript)	Left humerus (mm)	Right humerus (mm)
1	$X_1 = 311$	$Y_1 = 315$
2	$X_2 = 302$	$Y_2 = 306$
3	$X_3 = 301$	$Y_3 = 311$
4	$X_4 = 322$	$Y_4 = 333$
5	$X_5 = 312$	$Y_5 = 316$
6	$X_6 = 285$	$Y_6 = 292$
7	$X_7 = 305$	$Y_7 = 308$
8	$X_8 = 310$	$Y_8 = 318$
9	$X_9 = 328$	$Y_9 = 326$
10	$X_{10} = 304$	$Y_{10} = 309$

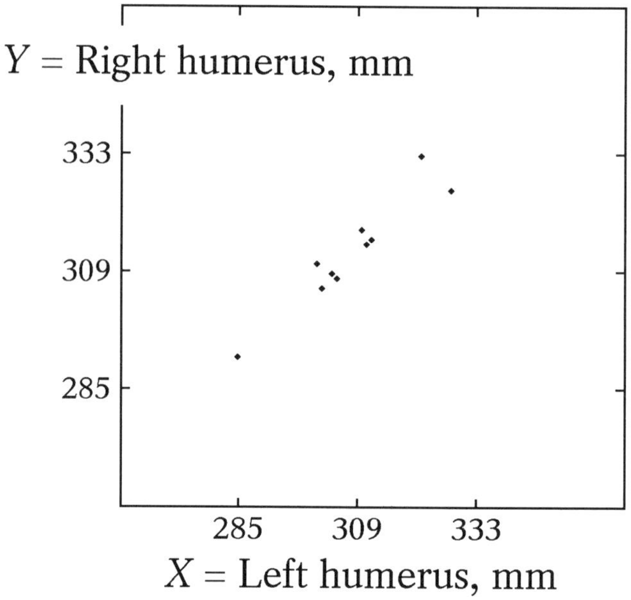

Figure 1.2.1
Scatter diagram of left humerus length *X* and right humerus length *Y* of ten female human skeletons; data from Münter (1936)

Similarly, in figure 1.2.1, each dot represents one individual (or, more exactly, its skeleton) whose left humerus length and right humerus length correspond to the coordinates on the abscissa (the *X*-axis) and on the ordinate (the *Y*-axis) respectively. A careful examination of this scatter diagram reveals one important feature of this small data set: dots are spread approximately along a hypothetical straight line going from the lower left corner to the upper right corner of the diagram. This suggests that a person having a particularly long left humerus tends to have also a particularly long right humerus; inversely, a person having a short left humerus also has a short right humerus. This diagram thus appears to show that, in the human skeleton, the left humerus and the right humerus tend to have closely similar lengths. In fact, this is an instance of the phenomenon of *bilateral symmetry,* which occurs in man, in many other animals, and even in certain structures of some plants.

If bilateral symmetry were perfect, one would expect all dots to lie exactly on a straight line passing through points having equal *X*- and *Y*-coordinates, of which the equation would be $Y = X$. Such a line has been drawn in figure 1.2.2. Bilateral symmetry is obviously not perfect, since dots do not all lie exactly on the line $Y = X$. Moreover, 9 out of the 10 individual dots, as well as the mean dot ($\overline{X}$, $\overline{Y}$) (see chapter 4), lie above the line $Y = X$, in the part of the scatter diagram where $Y > X$, that is where the right humerus is longer than the left humerus. As numerical methods will confirm in sections 9.6, 9.9 and 17.11, the right humerus thus seems slightly longer that the left humerus in most female human skeletons.

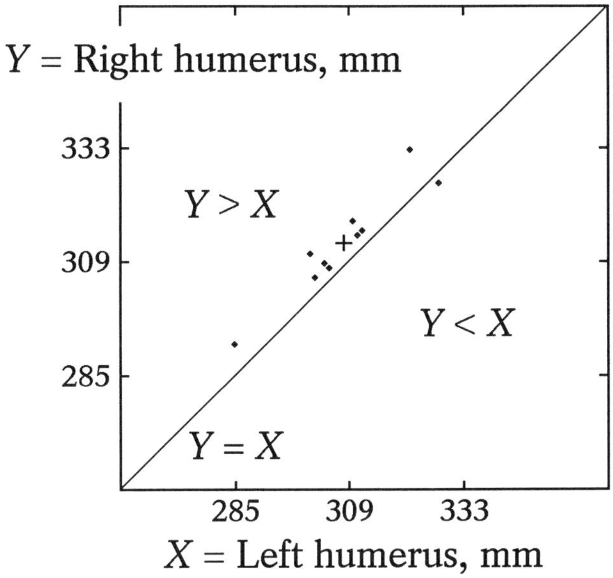

Figure 1.2.2
Scatter diagram of left humerus length X and right humerus length Y of ten female human skeletons; the diagonal line $Y = X$ represents the hypothesis of bilateral symmetry and the cross (+) represents the mean dot ($\bar{X}$, $\bar{Y}$)

One may wonder whether this slightly greater length of the right humerus might be related to the fact that, in all contemporaneous human populations, and even in earlier populations according to the analysis of prehistoric drawings (Coren and Porac, 1977), most individuals (from 90% to 95%) are right-handed. It is rather striking that the simple graphical examination of such a small data set should provide such interesting information and raise such fascinating questions.

As for bilateral symmetry, the functional asymmetry of the human brain has received considerable attention from neurobiologists during the last two decades, the question being reviewed by Bradshaw and Rogers in 1993. In a multivariate statistical analysis of limb bone asymmetry in man and in the North American Marten, Jolicoeur (1963c) has discussed the hypothesis that, in animals possessing a cephalic pole (a head), bilateral symmetry (with respect to a median plane) is an adaptation to rapid straight-line locomotion in an environment vertically stratified because of gravity. Contrariwise, *radial symmetry* (with respect to a single axis) often occurs in animals whose cephalo-caudal axis is weakly differentiated and generally vertical, and which are fixed or move slowly without a marked directional preference, like Coelenterates (sea anemones and jellyfish) and adult stages of many Echinoderms (sea urchins and sea stars). In plants, *radial symmetry* is most frequent, but bilateral symmetry does occur in some diatoms and in the so-called *zygomorphous* flowers of several angiosperms (orchids in particular), which have a slanting habit and of which many interact strongly with pollinizing insects.

Chapter 2
Samples and populations, estimates and parameters

Section 2.1: samples and populations

The words *sample* and *population* have different and more restricted meanings in statistics and in biometry than in common everyday language. In statistics, a *population* (also called a *universe*) is a set of beings (or the set of qualitative or quantitative observations which can be made on those beings) about which information is desired and which is studied through a subset, called a *sample*, drawn from it at *random,* that is in as impartial (unbiased) a manner as possible.

In the biological sciences, but not in statistics, the word population implies the existence of living organisms differing from each other with respect to age and sex and able to reproduce themselves sexually or asexually. Biometricians must therefore always watch out and carefully distinguish the cases where the word population is used in its biological as opposed to its statistical meaning. However, this does not rule out the possibility that a particular group of living organisms may be justifiably considered as a *biological population* by a biologist and as a *statistical population* by a statistician or a biometrician.

As for the word sample, its statistical meaning is that of a subset drawn from the statistical population in order to get information about it. This subset generally includes several beings or observations, and is only exceptionally reduced to a single unit. On the contrary, in everyday language, a sample is often a single specimen, like a piece of fabric in textile marketing, a piece of rock in mineralogy, or a small amount of blood or urine in medical biochemistry.

The number of beings or observations included in a statistical sample is known as the *sample size* and is often denoted by the symbol N. As for statistical populations, finite populations, containing limited numbers of beings or of possible observations, are occasionally considered (sometimes for didactic reasons, see section 9.5), but many populations are made up of so many units that they can be considered as practically infinite. Moreover, even if a population is thought to be finite, its size is seldom known exactly. Ecologists are frequently interested in estimating the size of finite natural populations.

A statistical population is a whole which one tries to know by studying one of its parts. The logical operation through which findings made on the part are extrapolated to the whole constitutes a *generalization* (an *inductive reasoning* or *inference*). Let us remember that the conclusion of an inductive inference is uncertain, except in the rather special case of *complete induction* (where the whole population is studied).

Would it not be simpler to study all members of the statistical population directly instead of taking a sample? – Theoretically yes, but this would not always be feasible, since some statistical populations are extremely large. Moreover, some biological studies require living organisms to be killed: in such cases, only part of the population is usually taken in order to prevent extermination. Finally, the direct study of a whole population would often require efforts disproportionate to the importance of the information sought. While the distinction between the statistical population and the sample may seem rather abstract at first, it is of the utmost importance for anyone who wishes to understand the logical role which statistical methods are called upon to play in scientific research (section 2.4).

Section 2.2: estimates and parameters

In biometry, a sample characteristic, such as the ordinary mean $\overline{X}$ (section 4.2) for instance, is the result of an attempt to know the corresponding characteristic of the statistical population from which the sample has come and which it represents. It is therefore important to distinguish a numerical population characteristic, called a *parameter*, from the corresponding sample characteristic, called an *estimate* (the word *estimator* is used to denote the algebraic formula through which the numerical estimate is obtained). In order to be easily recognized, statistical parameters are represented as often as possible by Greek letters. Thus the mean of a quantitative variate X in a statistical population is denoted by the symbol μ_X (or μ in short), while the mean of the same variate in a sample of size N is represented by the symbol $\overline{X}$. The sample mean $\overline{X}$ is therefore an estimate resulting from an attempt to know the parametric population mean μ_X . The Greek letter μ *(mu)* is used because it corresponds to the letter m of our alphabet, which is the initial letter of the word *mean* (section 4.2).

Section 2.3: major kinds of sampling

The sample is generally collected mostly at random (*random sampling*) in order to prevent the conclusions drawn by research workers from being affected by preconceived ideas. Thus, when an ichthyologist studies the length of fish belonging to a particular species in a lake, if he chose mostly the largest and nicest specimens because he considered them as more "representative", this would tend to artificially increase the sample mean $\overline{X}$ and to decrease the similarity of the latter with the mean μ of the population about which information is sought.

However, when a population is known in advance to be heterogeneous and to contain different subpopulations (known technically as strata), more precise information may be obtained from *stratified sampling*: the size of the subsample taken in each stratum is then proportional to the importance of the latter (defined by its relative frequency or probability, sections 3.1 and 3.2) within the global population. This is what is usually done when a survey is made to predict the result of a popular vote in a population which is heterogeneous from an ethnic, linguistic, political, religious or social viewpoint.

One may also distinguish *sampling with replacement* from *sampling without replacement* according to whether each individual is put back in the population or not after being measured and may be sampled more than once or only once. In biology, sampling is necessarily carried out without replacement when each individual studied must be sacrificed, but sampling with replacement may be done in other cases or for didactic reasons (see section 9.5).

Section 2.4: the logical role of statistical methods in scientific research

A diagram illustrating the logical role of statistical methods in scientific research is represented in figure 2.4.1. While all of our contacts with reality consist in sensory perceptions on samples of a limited number of concrete beings in particular, we are eager to know the general and abstract characteristics of the whole statistical population comprising all beings of the kind studied. Estimating the parameters of the population on the basis of the sample, using *point estimates* or *interval estimates* (chapters 9 and following), is therefore a quantitative form of inductive reasoning. As for the prediction of new particular observations using population parameters, it is a statistical form of deductive reasoning, carried out either (a) to test a hypothesis against facts (chapter 9) or (b) to predict an as yet unobserved particular case for pragmatic reasons.

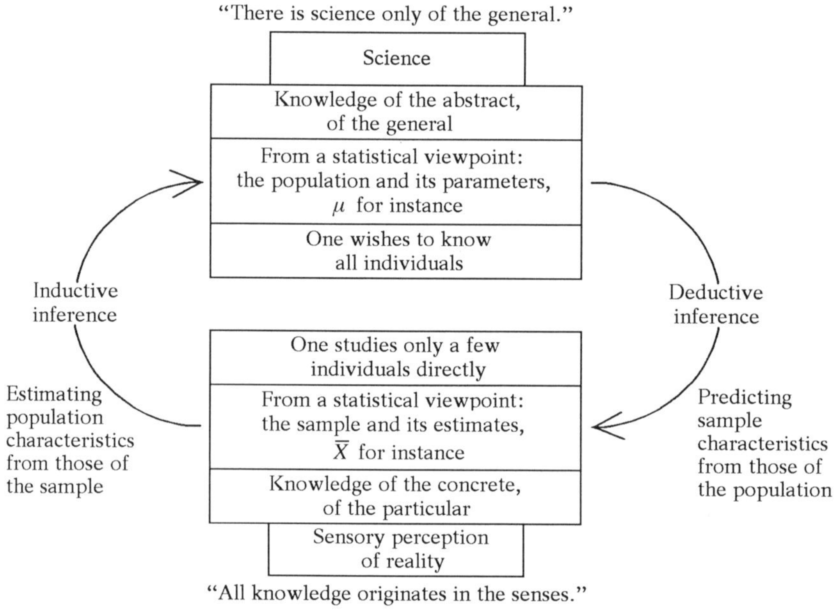

Figure 2.4.1
The logical role of statistical methods in scientific research

Chapter 3
Frequencies and probabilities

Section 3.1: absolute and relative frequencies

When a sample (chapter 2) of beings or of observations is classified with respect to a quantitative or qualitative variate, several observations may be located in the same class or have the same value. The number of beings or of observations "falling" in that class or possessing that value is then called the *frequency* (or *absolute frequency*) of that class or of that value. Since the sample is usually spread over several classes, the sum of the frequencies of all classes is called the *total frequency*. The total frequency of a sample is thus synonymous with sample size. Table 3.1.1 shows how a sample of 76 male North American martens (*Martes americana*) is spread, distributed, with respect to the length of the skull measured in mm. Such a table is called a *frequency table*.

Table 3.1.1
Frequency table of the skull length X in mm of male North American martens captured in 1955 in Montana (Jolicoeur, 1963b, 1963c)

Length classes (mm)	Absolute frequencies	Relative frequencies
68	0	0.0000
69	0	0.0000
70	0	0.0000
71	0	0.0000
72	0	0.0000
73	1	0.0132
74	0	0.0000
75	2	0.0263
76	3	0.0395
77	7	0.0921
78	13	0.1711
79	20	0.2632
80	12	0.1579
81	12	0.1579
82	6	0.0789
Sums of frequencies	76	1.0001

When two samples of unequal sizes must be compared, it may be desirable to compensate for sample size by considering the *relative frequency* of each class instead of its *absolute* (ordinary) *frequency*: the relative frequency of a class is obtained by dividing the absolute frequency of that class by the total frequency of all classes, that is by sample size. Conversely, the absolute frequency of a class may be recalculated by multiplying its relative frequency by the total frequency. The relative frequency may be expressed as a fraction of unity, but it may also be multiplied by 100 and expressed as a percentage. While the sum of the absolute frequencies of all classes is equal to the total frequency, the sum of relative frequencies is equal to 1.0 or 100% (within rounding errors).

Section 3.2: observed and expected relative frequencies, probabilities

The relative frequency of a particular class in a sample is an estimate of the relative frequency of the same class in the statistical population from which the sample has come. The relative frequencies of the various classes of a statistical population are fixed (somewhat theoretically) and are called *probabilities*. The probability of a class is therefore the relative frequency of that class in the statistical population from which a sample or an event (section 3.6) is about to come, and the relative frequency which will be observed in the sample is expected to be more or less similar to the corresponding probability. In short, probabilities may be said to be *relative frequencies in a statistical population* or *expected relative frequencies* predicted for a sample on the basis of the population, in opposition to *observed relative frequencies,* which are based on a sample.

Section 3.3: frequency and probability distributions

A frequency table like table 3.1.1 shows how the total frequency of a sample is spread or distributed over the various values or classes of a random variable (variate) X. In practice, the particular manner in which such a set of data is spread may therefore be called a *frequency distribution.* Similarly, the particular way in which a set of theoretical or expected relative frequencies is dispersed may be called a *probability distribution* (section 3.2). A frequency or probability distribution may be represented graphically in a Cartesian diagram by inscribing the successive values of the variate X on the abscissa and the corresponding frequencies or probabilities on the ordinate. The frequency of each class may then be shown as a vertical line segment in a *bar diagram* (figure 3.3.1.a) or as a rectangle in a *frequency histogram* (figure 3.3.1.b). The bar diagram is most appropriate for a variate X which is truly discrete, because it suggests discontinuity, while the frequency histogram is preferable for a variate which is basically continuous and in which a limited number of classes are distinguished mostly for the sake of concision. In the case of a continuous variate, the frequencies of consecutive classes may be represented simply by dots which are then joined by straight-line segments forming a frequency polygon (figure 3.3.1.c) or by curved segments constituting a frequency curve (figure 3.3.1.d). When the frequency or probability distribution of a continuous variate X is represented by a curve, the ordinate $f(X)$ of that curve must be interpreted as a frequency density or a probability density instead of a frequency or a probability (see sections 3.8, 3.10 and 5.2). The frequency distribution of the skull length of 76 male North American martens (table 3.1.1) is represented by a frequency histogram in figure 3.3.2.

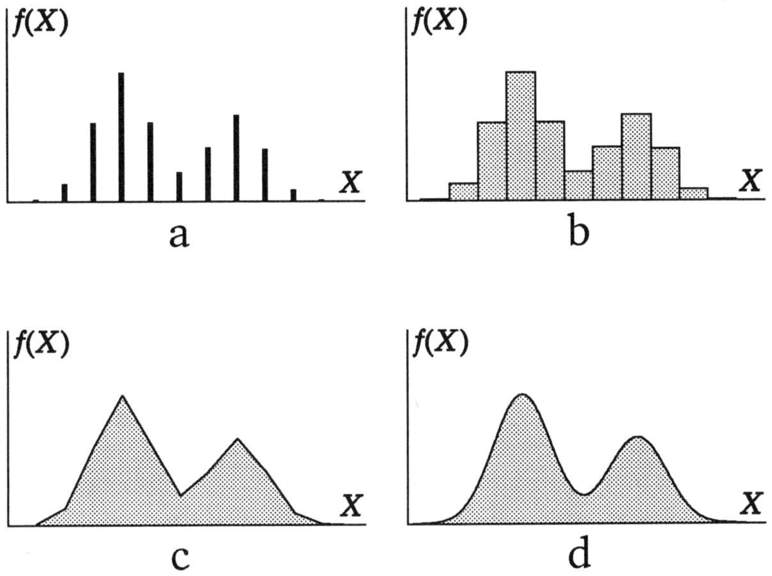

Figure 3.3.1
Frequency or probability distribution represented by a bar diagram (a), by a histogram (b), by a polygon (c), and by a curve (d)

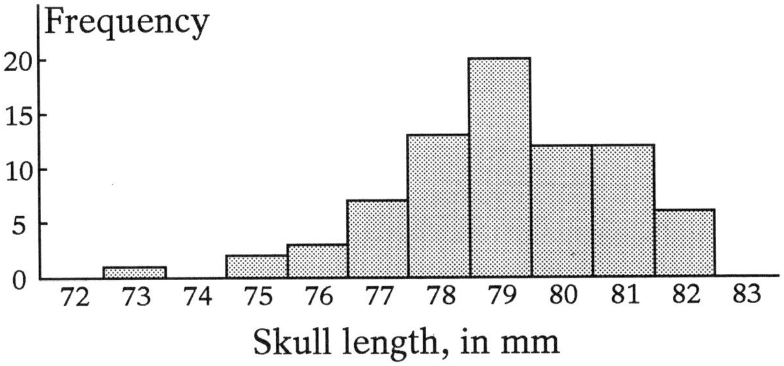

Figure 3.3.2
Histogram of the frequency distribution of the skull length in mm of male North American martens captured in 1955 in Montana (see table 3.1.1)

Section 3.4: unimodal and plurimodal distributions

A *mode* is a value or a class possessing a frequency or a probability markedly higher than the adjacent values or classes of the distribution. In the case of the skull length of 76 male North American martens (table 3.1.1 and figure 3.3.2), for instance, the value 79 mm, of which the frequency is 20, is a mode. A frequency or probability distribution is said to be unimodal, bimodal, trimodal or plurimodal insofar as it possesses one, two, three or several modes. When a distribution has more than one mode, this can often be interpreted as a symptom of heterogeneity. For instance, the frequency distribution of the skull length of 53 female in addition to 76 male North American martens is represented in table 3.4.1 and figure 3.4.1.

The frequency distributions of males and females are clearly shifted with respect to each other, the mode of males being located at 79 mm with a frequency of 20 while the mode of females is located at 71 mm with a frequency of 16. This shift reflects the marked sexual dimorphism of body size in Mustelidae (the family to which the North American marten belongs), males being much larger than females. If the sex of specimens is neglected, the overall frequency distribution appears strongly bimodal, because of the superposition of the two different distributions (table 3.4.1 and figure 3.4.1). Even when sexual dimorphism is less evident, it is always preferable to distinguish males from females carefully in a data analysis.

Table 3.4.1
Absolute frequencies of the skull length *X* in mm of male and female North American martens captured in 1955 in Montana (Jolicoeur, 1963b, 1963c)

Length (mm)	Males	Females	Both sexes
68	0	2	2
69	0	2	2
70	0	9	9
71	0	16	16
72	0	13	13
73	1	7	8
74	0	4	4
75	2	0	2
76	3	0	3
77	7	0	7
78	13	0	13
79	20	0	20
80	12	0	12
81	12	0	12
82	6	0	6
Total frequencies	76	53	129

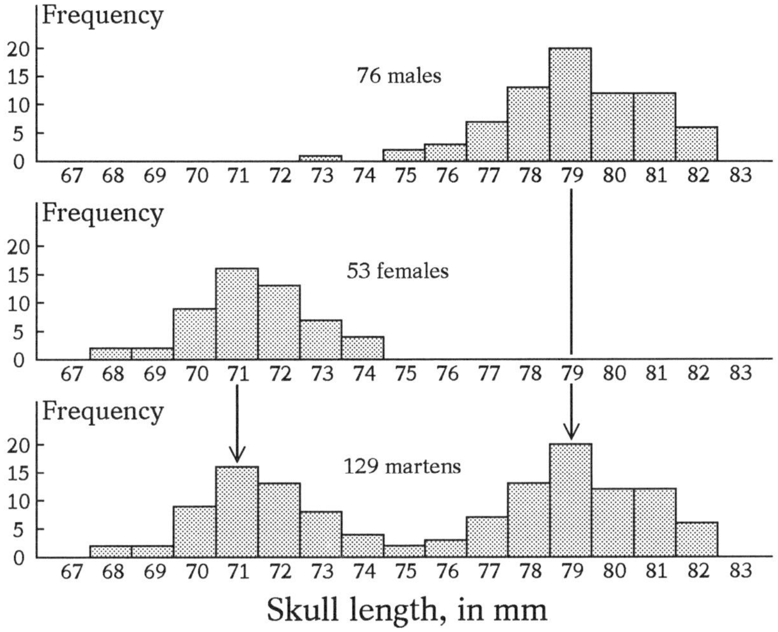

Figure 3.4.1
Frequency distributions of the skull length in mm of male and female North American martens captured in 1955 in Montana (see table 3.4.1)

As for the sexual dimorphism of body size in mammals, it is sometimes strikingly related to social behavior and to the structure of (biological) populations. In some species of Pinnipedia (the suborder of seals and sea lions), the elephant seal for instance (*Mirounga angustirostris* in the northern hemisphere and *Mirounga leonina* in the southern hemisphere), the adult male reaches a much larger body size than females, is socially dominant and polygamous, and chases other males aggressively away from its harem. In other species of pinnipeds, on the contrary, males reach approximately the same body size as females and are rather placid and generally monogamous.

Section 3.5: cumulative frequencies and probabilities

The preceding sections dealt with the number of beings or observations having a certain value or contained in a certain class; thus, table 3.1.1 and the histogram in figure 3.3.2 showed how the total frequency $N = 76$ was *distributed* among the various classes. That table and that histogram and the frequencies therein could therefore be said to be *distributive*. In other cases, however, the quantity of interest is the number of beings or observations of which the measurement is lower than a specified value, and the corresponding frequencies are then said to be *cumulative*. Finally, the object of a study may also be the number of beings of which the measurement is equal to or higher than a specified value, and the frequencies may then be said to be *decumulative* (table 3.5.1 and figure 3.5.1). In the particular case of the survival times of drosophilae, taken here as an example in table 3.5.1 and figure 3.5.1, the histogram of cumulative frequencies is an approximate description of a *death curve* while the histogram of decumulative frequencies is an approximate description of a *survival curve*.

Table 3.5.1
Survival times in minutes in dry air at 33.5 °C of female drosophilae raised at 15 °C; data digitized from diagrams published by Smith (1957)

Distributive frequencies		Cumulative frequencies		Decumulative frequencies	
Survival time	Dying flies	Survival time	Dead flies	Survival time	Live flies
$X = 50$	0	$X < 50$	0	$X \geq 50$	6
$X = 55$	1	$X < 55$	0	$X \geq 55$	6
$X = 60$	0	$X < 60$	1	$X \geq 60$	5
$X = 65$	1	$X < 65$	1	$X \geq 65$	5
$X = 70$	3	$X < 70$	2	$X \geq 70$	4
$X = 75$	1	$X < 75$	5	$X \geq 75$	1
$X = 80$	0	$X < 80$	6	$X \geq 80$	0
Total frequency = 6					

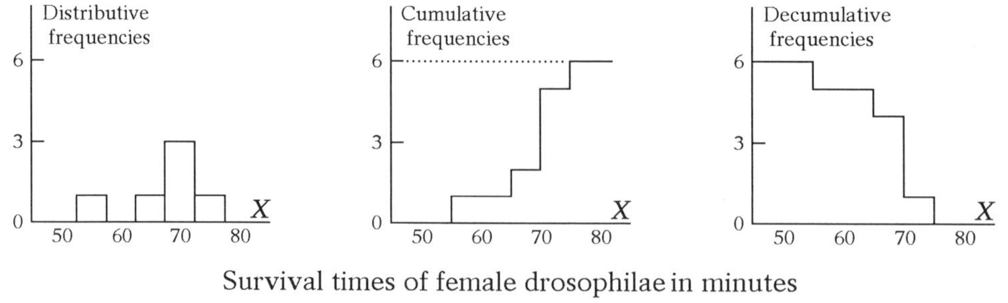

Figure 3.5.1
Distributive (left), cumulative (center), and decumulative (right) frequencies of the survival times X in minutes in dry air at 33.5°C of female drosophilae raised at 15°C (see table 3.5.1)

The concept of *cumulative frequency* or *probability* is important not only in practice but also in theory, because a theoretical probability distribution, such as the *normal distribution* (chapter 5) for instance, can be described efficiently by giving the cumulative probability $P(X_0) = P(X < X_0)$ of the variate X, that is the probability that X takes a value lower than a specified number X_0 (figure 3.5.2). While the cumulative probability is represented by the stippled surface left of the value X_0 below the distributive curve (figure 3.5.2, left), it is also equal to the ordinate of the cumulative curve above the abscissa X_0 (figure 3.5.2, right). The ordinate of the distributive curve (figure 3.5.2, left) is the derivative with respect to X of the ordinate of the cumulative curve (figure 3.5.2, right).

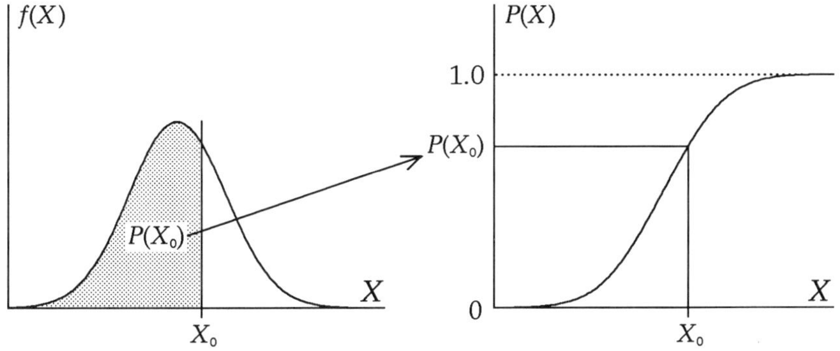

Figure 3.5.2
Distributive (left) and cumulative (right) curves of the normal probability distribution

When a table of cumulative probabilities is available, the probability that the corresponding variate X will fall within any specified region is very easily determined: the probability that a continuous random variable (variate) X will fall between two specified values $X_1 < X_2$ is indeed simply equal to the cumulative probability of the largest (X_2) less that of the smallest (X_1) of the two values (figure 3.5.3), the cumulative probability of $-\infty$ being 0 and that of $+\infty$ being 1.0.

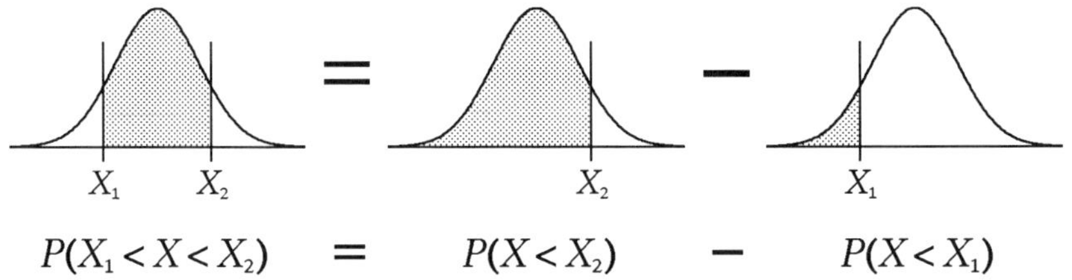

Figure 3.5.3
The probability that a continuous random variable (variate) X falls between two specified values $X_1 < X_2$ is equal simply to the cumulative probability of the largest (X_2) less that of the smallest (X_1) of the two values

In theoretical statistics, the cumulative probability $P(X_0) = P(X < X_0)$ is often denoted by $F(X)$ and is called the *cumulative distribution function* but, in this introductory textbook, the symbol $F(X)$ will rather be used in practice to denote an observed absolute frequency (section 4.7, chapters 15, 16, 17 and 18).

Section 3.6: events (simple, compound, joint) and occasions (trials)

In the study of probabilities, the word *event* is used to signify a phenomenon which may happen when an occasion arises, while the *occasion* (also called a *trial* in technical language) is a precise set of circumstances in which an event may occur. In practice, an occasion may be conceived as a small space-time portion within which an event may take place. The fact that each of these two concepts (*event* and *occasion*) is involved in the definition of the other one may seem circular at first, but the two ideas are essentially linked and form a pair.

A *simple event* is an event which cannot be broken down into more elementary events which could happen separately, while a *compound event* is made up of two or several simple events occurring jointly (simultaneously or successively). *Joint events* are simple events of which the simultaneous or successive realization makes up a compound event. In practice, it is scientifically useful to consider the joint realization of simple events as a compound event only if there is a certain relationship, real or hypothetical, between those simple events.

Section 3.7: compatible, exclusive and complementary events

Compatible events are events which can occur jointly, while *exclusive events* cannot happen jointly. However, events which are exclusive in a *simple occasion* may be compatible in a *compound occasion*. The birth of a boy and the birth of a girl are mutually exclusive if a woman is expecting only one child, for instance, but they are compatible if she is expecting two children (except if they are monozygotic twins). If two events, A and B, are exclusive, the probability that one or the other happens if an occasion arises is equal to the sum of the probabilities of separate occurrence:

$$P(\text{A or B}) = P(\text{A}) + P(\text{B}).$$

There may also be several exclusive events A, B, C, ... such that

$$P(\text{A or B or C or ...}) = P(\text{A}) + P(\text{B}) + P(\text{C}) + \ldots .$$

Complementary events are exclusive events of which one will necessarily happen (with a probability of 1.0) because the set of these exclusive events covers all possibilities and makes up a *complete system of events*. If events A, B, C, ... etc. are complementary,

$$P(\text{A or B or C or ...}) = P(\text{A}) + P(\text{B}) + P(\text{C}) + \ldots = 1 .$$

Thus, if a single individual is picked at random in a human population containing 51% men and 49% women, the probability of picking a man is $P(\text{A}) = 0.51$, the probability of picking a woman is $P(\text{B}) = 0.49$, and the probability of picking an individual of either sex is $P(\text{A}) + P(\text{B}) = 0.51 + 0.49 = 1.00$, because it is certain that the individual picked will be either a man or a woman.

Section 3.8: summing up or integrating probability distributions

In the case of a random variable (variate) X which can take only a limited number K of values $X_1, X_2, \ldots, X_K$ possessing probabilities $P(X_1), P(X_2), \ldots, P(X_K)$ respectively, the set of possible values $X_1, X_2, \ldots, X_K$ makes up a complete system of events (section 3.7) and the sum of their probabilities is equal to unity:

$$\sum_{i=1}^{K} P(X_i) = 1.0 \ .$$

The above equation is a summarized description of the probability distribution because, inasmuch as the value of each term of the sum is known, the latter indicates how the probabilities are distributed. In the case of a continuous variate, the probability distribution may be summarized by an integral where each of the infinite number of possible values is represented by the corresponding probability density $f(X)$ (see sections 3.3, 3.10 and 5.2):

$$\int_{-\infty}^{+\infty} f(X)\, dX = 1.0 \ .$$

Section 3.9: conditional probabilities; association and independence

Unlike the ordinary probability (also called *marginal probability*) P(A) of an event A, considered until now, which is the probability of that event without taking into account its possible relationships with other events B, C ... etc., the *conditional probability of event* A *given* B, $P(\mathrm{A}|\mathrm{B})$, is the probability that, if event B occurs, event A also occurs. The conditional probability $P(\mathrm{A}|\mathrm{B})$ is greater than the marginal probability $P(\mathrm{A})$ if there is a *positive association* but smaller than the marginal probability if there is a *negative association* between events A and B. When events A and B are independent, the conditional probability $P(\mathrm{A}|\mathrm{B})$ is simply equal to the marginal probability $P(\mathrm{A})$.

If a compound event is made up of the joint occurrence of two simple events A and B, the probability of that compound event is

$$P(\mathrm{AB}) = P(\mathrm{A\ and\ B}) = P(\mathrm{A}|\mathrm{B})\, P(\mathrm{B}) = P(\mathrm{B}|\mathrm{A})\, P(\mathrm{A}) \,.$$

In the particular case where two joint events A and B are independent, the probability of the compound event AB is simply equal to the product of marginal probabilities:

$$P(\mathrm{AB}) = P(\mathrm{A}) P(\mathrm{B}) \,.$$

Biological examples of independent and associated phenomena will be found in sections 15.3 and 15.5.

Section 3.10: expected values (parametric means)

The concepts of an *expected value* (or *expectation*) and of a *parametric mean* (sections 2.2, 4.2, 5.3) are basically equivalent. In the case of a quantitative random variable (variate) X which can take only a limited number K of values $X_1, X_2, \ldots, X_K$ possessing probabilities $P(X_1), P(X_2), \ldots, P(X_K)$ respectively, the expected value of X is

$$\mathscr{E}(X) = X_1 P(X_1) + X_2 P(X_2) + \ldots + X_K P(X_K) \,,$$

that is

$$\mathscr{E}(X) = \sum_{i=1}^{K} [X_i P(X_i)] = \mu_X .$$

The expected value is thus a *weighted sum* of all possible values of the variate X where the *weight* of each possible value X_i is its probability $P(X_i)$, that is its relative frequency in the statistical population. The above *weighted sum* is related to the formula for calculating the sample mean $\bar{X}$ in the case of grouped data (section 4.7). Because the weights of the possible values of variate X are their respective probabilities, the expected value $\mathscr{E}(X)$ is located approximately at the center of the set of possible values and constitutes a measurement of *central tendency* (chapter 4).

In the case of a continuous variate, the expected value is defined by an integral where each of the infinite number of possible values X is multiplied by the corresponding probability density $f(X)$ (sections 3.8, 5.2 and 5.3):

$$\mathscr{E}(X) = \int_{-\infty}^{+\infty} X f(X)\, dX = \mu_X .$$

However, since an integral is a limit of a sum of products, the expected value possesses in all cases a series of properties which follow from the associativity and distributivity of algebraic sums and products.

$\mathscr{E}(K) = K$ if K is a constant;

$\mathscr{E}(KX) = K\,\mathscr{E}(X)$;

$\mathscr{E}[\mathscr{E}(X)] = \mathscr{E}(X)$;

$\mathscr{E}(a + b\,X) = a + b\,\mathscr{E}(X)$;

$\mathscr{E}[X - \mathscr{E}(X)] = \mathscr{E}(X) - \mathscr{E}[\mathscr{E}(X)] = \mathscr{E}(X) - \mathscr{E}(X) = 0$;

$\mathscr{E}(X \pm Y \pm Z) = \mathscr{E}(X) \pm \mathscr{E}(Y) \pm \mathscr{E}(Z)$, that is $\mu_{(X \pm Y \pm Z)} = \mu_X \pm \mu_Y \pm \mu_Z$.

The properties of expected values are the basis of many theoretical relationships concerning probabilities and statistics (see section 12.4 for instance). Moreover, some important statistical quantities can be defined using the concept of expected value, such as

the variance (section 4.5),

$$\sigma_X^2 = \mathrm{var}(X) = \mathscr{E}\{[X - \mathscr{E}(X)]^2\} = \mathscr{E}(X^2) - [\mathscr{E}(X)]^2 = \mathscr{E}(X^2) - \mu_X^2 ,$$

and the covariance (chapter 19),

$$\sigma_{XY} = \mathrm{covar}(X, Y) = \mathscr{E}\{[X - \mathscr{E}(X)]\,[Y - \mathscr{E}(Y)]\} = \mathscr{E}(XY) - \mathscr{E}(X)\mathscr{E}(Y) = \mathscr{E}(XY) - \mu_X \mu_Y .$$

It may be shown that $\mathrm{var}(a + bX) = b^2 \mathrm{var}(X)$

and that $\mathrm{covar}(a + bX, c + dY) = bd\,\mathrm{covar}(X, Y)$.

Moreover,

$$\sigma^2_{(X\pm Y)} = \mathscr{E}\{[(X\pm Y)-\mathscr{E}(X\pm Y)]^2\} = \mathscr{E}\{[[X-\mathscr{E}(X)] \pm [Y-\mathscr{E}(Y)]]^2\},$$

whence

$$\sigma^2_{(X\pm Y)} = \mathscr{E}[X-\mathscr{E}(X)]^2 \pm 2\mathscr{E}\{[X-\mathscr{E}(X)][Y-\mathscr{E}(Y)]\} + \mathscr{E}[Y-\mathscr{E}(Y)]^2 = \sigma_X^2 \pm 2\sigma_{XY} + \sigma_Y^2 .$$

Finally, when two variates X and Y are statistically independent, $\mathscr{E}(XY) = \mathscr{E}(X)\mathscr{E}(Y)$, their covariance σ_{XY} is null, and $\sigma^2_{(X\pm Y)} = \sigma_X^2 + \sigma_Y^2$. However, even though the nullity of the covariance is a necessary condition of statistical independence, it is not sufficient: two variates may have a null covariance without being independent (see figure 19.10.1).

Section 3.11: permutations, combinations and factorials

Combinations are different arrangements of objects which may be obtained by changing only the nature of the objects taken into the arrangement, the order in which these objects are taken having no importance. As for *permutations,* they are the different linear arrangements which may be obtained by changing the order as well as the nature of the objects taken into the arrangement.

If the total number of objects available is K while the number of objects taken into the arrangement is X, the number of permutations is denoted by ${}_KP_X$ and is equal to

$${}_KP_X = K(K-1)(K-2)\ldots(K-X+1) = K!\,/\,(K-X)!\,,$$

where $K!$ denotes the factorial of K, that is the product of K by all integers smaller than K down to unity:

$$K! = K(K-1)(K-2)\ldots 3\times 2\times 1 .$$

The expression $(K-X)!$ retains its validity even in the case where $X=K$ since $0!=1$ by convention.

Because the number of combinations does not take into account the order in which objects are taken, it is necessarily smaller than the number of permutations and may be obtained by noting that each combination could give rise to $X!$ permutations. The number of combinations is denoted by ${}_KC_X$ and is therefore equal to

$${}_KC_X = {}_KP_X/X! = K!/[X!\,(K-X)!] .$$

The symmetry of the above expression shows that ${}_KC_X = {}_KC_{(K-X)}$ and reflects the fact that each combination of X objects taken among a total number of K objects corresponds to a combination of $(K-X)$ objects which are left out and vice versa. The number of combinations may also be denoted by

$$\binom{K}{X}.$$

The factorial function and the number of combinations appear in the successive terms of the binomial and of the Poisson distributions (chapters 17 and 18).

Chapter 4
Measures of central tendency and of dispersion

Section 4.1: introduction

Most frequency and probability distributions of quantitative variates studied in biology may be interpreted using two opposite but complementary principles: the tendency of observations toward a central value on the one hand, and the dispersion of the same observations about that central value on the other hand. This state of things may be summarized algebraically by a *model equation*:

$$X_h = \mu + e_h ,$$

where h is the subscript (order number) of the hth observation in a sample, X_h is the numerical value of that h^{th} observation, μ is a constant value corresponding to the *central tendency,* and e_h is a random value which varies from each observation to the next one and accounts for *dispersion*. While the letter i is often used as a subscript, the letter h is preferred here because the letter i is saved for later use. Subtracting μ from both members of the above model equation yields $e_h = (X_h - \mu)$, which shows that the random variable e_h corresponds geometrically to the distance (or *deviation* or *deviate* in statistical terminology) between the hth observation X_h and the central tendency μ. Variation may be represented visually as a back-and-forth motion of an observed point X_h about an equilibrium point μ when an observation is succeeded by the next one (figure 4.1.1).

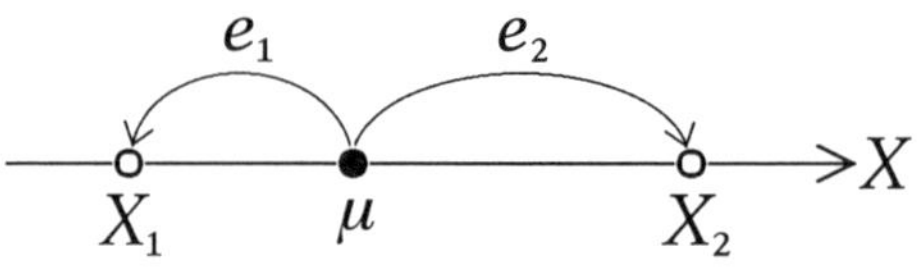

Figure 4.1.1
Variation may be visualized as a back-and-forth motion of an observed point X_h about an equilibrium point μ

Section 4.2: the ordinary or arithmetic mean (average)

The most usual measure of central tendency is the ordinary *mean* (or *average*), also called *arithmetic mean*; the arithmetic mean of a sample is denoted by the symbol $\overline{X}$ and is calculated by dividing the sum of the numerical values of the observations by sample size N, that is by the number of observations in the sample:

$$\overline{X} = \frac{X_1 + X_2 + X_3 + \ldots + X_N}{N} = \sum_{h=1}^{N} X_h / N .$$

When two variates are analyzed simultaneously and when pairs of observations (X, Y) are represented by dots in a scatter diagram, for instance, the mean dot $(\overline{X}, \overline{Y})$, of which the coordinates are the mean of the X's on the abscissa and the mean of the Y's on the ordinate, is located approximately at the center of the group of individual observation dots (see figure 1.2.2). The concept of a mean is thus related to the notions of *center* and of *central tendency,* and an analogy is sometimes drawn between the mean of a group of observation dots and the *center of gravity* of the solid which would be constituted by those dots if each of the latter had a mass equal to that of the others. The numerical value of the arithmetic mean $\overline{X}$ of a sample is an estimate of the arithmetic mean of the population, while the latter is a parameter (section 2.2) and is denoted by the symbol μ_X or μ in short.

The geometrical meaning of the (arithmetic) mean may be made explicit by comparing two statistical populations which follow the normal probability distribution (chapter 5) and which differ from each other only with respect to their means. In figure 4.2.1, for instance, populations **A** and **B** have unequal means $\mu_A = 5$ and $\mu_B = 15$ but equal standard deviations $\sigma_A = \sigma_B = 1$ (section 4.5). Population **B**, which has the highest mean, is located to the right of population **A** on the coordinate axis of variate X. Increasing or decreasing the mean of a normal distribution thus moves its center and shifts the whole distribution toward the right or toward the left. The mean is therefore considered not only as a measure of *central tendency* but also as a measure of *location*.

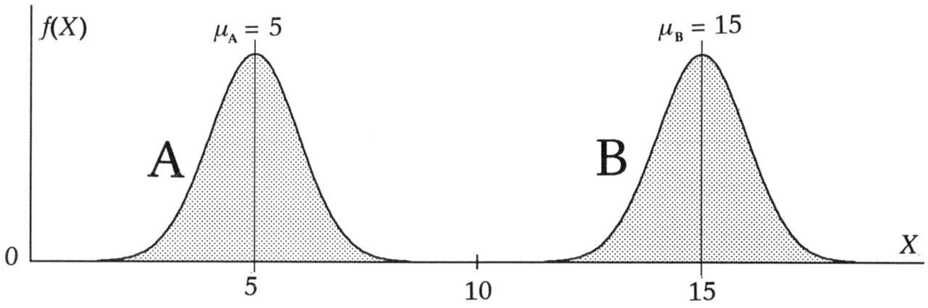

Figure 4.2.1
Geometrically, if two normal distributions differ only with respect to their means, the distribution having the largest mean is located to the right

The sum $\Sigma(X_h - \overline{X})$ of the deviations with respect to the arithmetic mean is null, since

$$\Sigma(X_h - \overline{X}) = \Sigma X_h - \Sigma\overline{X} = N\overline{X} - N\overline{X} \equiv 0 .$$

Because of this, it will be impossible to use the sum or the mean of the algebraical (signed) values of the deviations $(X_h - \overline{X})$ as a measure of dispersion (sections 4.5 and 4.8). Moreover, the arithmetic mean satisfies the important *principle of least squares.* It can indeed be shown that the sum of the squares of the deviations from the arithmetic mean $\overline{X}$ is never greater than the sum of squares of the deviations with respect to any other value K:

$$\sum_{h=1}^{N} (X_h - \overline{X})^2 \leq \sum_{h=1}^{N} (X_h - K)^2 ,$$

for $\Sigma(X_h - K)^2 = \Sigma[(X_h - \overline{X}) + (\overline{X} - K)]^2 = \Sigma(X_h - \overline{X})^2 + 2\Sigma(X_h - \overline{X})(\overline{X} - K) + \Sigma(\overline{X} - K)^2$

$$= \Sigma(X_h - \overline{X})^2 + 2(\overline{X} - K)\Sigma(X_h - \overline{X}) + \Sigma(\overline{X} - K)^2 = \Sigma(X_h - \overline{X})^2 + 0 + N(\overline{X} - K)^2,$$

where $N(\overline{X} - K)^2 \geq 0$. The fact that the arithmetic mean satisfies the principle of least squares may also be confirmed by showing that the derivative of the sum of the squares of the deviations between the observations and a number K, $d[\Sigma(X_h - K)^2]/dK$, is null when $K = \overline{X}$. While the sample mean $\overline{X}$ satisfies the least-squares principle at the level of the sample, the population mean μ satisfies the same principle at the population level (section 5.3). The least-squares principle ensures that, in some sense, the arithmetic mean $\overline{X}$ (or μ in the case of the population) is located as close as possible to the center of the group of dots representing the observations in a scatter diagram or on a graduated axis (see figures 1.2.2 and 4.1.1).

Section 4.3: the geometric mean

While the arithmetic mean is often satisfactory, it is not the only measure of central tendency used in biometry; the *geometric mean* (or *geometric average*) G_X is often desirable:

$$G_X = \sqrt[N]{[X_1\, X_2\, X_3\, \ldots\, X_N]} = [\prod_{h=1}^{N} X_h]^{1/N}.$$

In fact, many biological variates, such as the concentration of a chemical substance in an organism, the length or the weight of an individual, or the number of individuals in a biological population, cannot take negative numerical values. Moreover, the probability distributions of those variates are often positively skewed (chapter 13): they have a longer tail on the right than on the left and are more closely similar to the *lognormal distribution* (chapter 14) than to the *normal distribution* (chapter 5). Consequently, it is often desirable to carry out the *logarithmic transformation* of the data, that is to replace each observation X_h by its natural (base e) logarithm $Y_h = \log_e(X_h)$ before calculating the arithmetic mean $\overline{Y}$ of logarithms:

$$\overline{Y} = \sum_{h=1}^{N} Y_h / N = \sum_{h=1}^{N} \log_e(X_h) / N.$$

In the end, the above result may be reexpressed on the arithmetic scale on which the original observations X_h have been made by taking the antilogarithm of the arithmetic mean $\overline{Y}$ of logarithms, which yields just the geometric mean G_X of original variates: $G_X = \exp(\overline{Y})$. The geometric mean is therefore particularly useful when the logarithmic transformation must be applied to data or when the lognormal distribution must be used. The geometric mean of a set of positive values is never greater than their arithmetic mean. The geometric mean is null if some of the observations are null, and it is not mathematically defined if some of the observations are negative.

Section 4.4: other measures of central tendency: the mode, the median

Two other measures of central tendency are also used in some cases: the *mode* and the *median*. The mode of a unimodal distribution (section 3.4) is defined as the value (or the class in the case of grouped data; see section 4.7) which has the highest frequency or probability. In the case of the skull length of 76 male North American martens for instance, the mode is the value 79 mm, of which the frequency is 20. In the case of the survival time of 6 female drosophilae (table 3.5.1 and figure 3.5.1), the mode is the value 70 minutes, of which the frequency is 3. When the distribution is markedly bimodal or plurimodal (figure 3.4.1), however, the utilization of the principal mode (of which the frequency is the highest) as a measure of central tendency would be difficult to justify.

The median of a distribution is the value on each side of which lies one half of the total frequency (or a probability of 1/2 or 50% in the case of a probability distribution). The median of a sample is often denoted by the symbol $\tilde{X}$ and is defined as the value located at the center of the set of observations when the latter are arranged by order of increasing magnitude and sample size is odd. When sample size is even, the median is defined as the arithmetic mean (and thus the midpoint) of the two observations located at the center of the set of observations.

Theoretically, the parametric mode and median coincide with the arithmetic mean in a normal distribution (chapter 5), and the median coincides with the geometric mean in a lognormal distribution (chapter 14). The various measures of central tendency are often rather close to each other, but they may differ markedly in skewed distributions.

Section 4.5: measures of dispersion: the variance and the standard deviation

The most widely used measures of dispersion are the *variance* as well as its square root the *standard deviation* (also called *root-mean-square*). While the mean absolute deviation (section 4.8) would seem at first easier to calculate, the variance and the standard deviation are more widely used because they appear explicitly or implicitly in the definition of several theoretical distributions, such as the normal distribution (chapter 5), the lognormal distribution (chapter 14), the chi-squared (χ^2) distribution (chapter 7), the variance ratio (F) distribution (chapter 8), the binomial distribution (chapter 17), the Poisson distribution (chapter 18), etc. . The sample variance is usually denoted by the symbol S_X^2 while the population variance is represented by σ_X^2, the lower case Greek letter σ *(sigma)* being the equivalent of our "*s*". The corresponding standard deviations are denoted by S_X and σ_X.

It may be shown (sections 4.10 and 12.4) that, in order to obtain *unbiased estimates* (which are exact on the average) of the parametric variance σ_X^2, one must define the sample variance as follows:

$$S_X^2 = \sum_{h=1}^{N} (X_h - \overline{X})^2/(N-1) .$$

The denominator $(N-1)$ is called the *number of degrees of freedom* and is often denoted by the lower case Greek letter ν *(nu)*, which corresponds to the "n" of our alphabet (see also sections 6.2 and 7.1). The standard deviation S_X is obtained simply by taking the square root of the variance S_X^2.

Geometrically speaking, while the mean describes the location of the center of a group of observations, the standard deviation (or its square, the variance) indicates whether the dots representing the data are weakly or strongly dispersed in a scatter diagram. As for that graphical dispersion, it shows whether the numerical data differ little or much from each other as well as from their mean $\overline{X}$. If all numerical data were equal to each other and to their mean, all observation dots would coincide with the mean dot and the standard deviation (and variance) would be null. On the contrary, when numerical variation and graphical dispersion are large, the standard deviation is also large. Figure 4.5.1 represents the probability distributions of two normal statistical populations **C** and **D** (chapter 5) of which the means are equal ($\mu_C = \mu_D = 10$) but the standard deviations are different: $\sigma_C = 1$ and $\sigma_D = 2$. Distribution **D**, which has the largest standard deviation, is more widely spread.

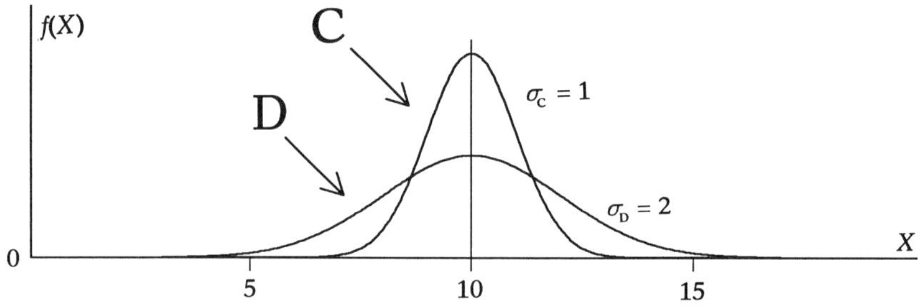

Figure 4.5.1
Geometrically, if two normal distributions differ only by their standard deviations, the distribution having the largest standard deviation is more widely spread

Section 4.6: calculating the mean, the variance, and the standard deviation

In order to reduce the risk of accidental numerical errors, the calculation of the mean, of the variance and of the standard deviation should be done in a methodical manner. Unless a computer is used, calculations should be arranged as in table 4.6.1. Each quantity like $(X_h - \overline{X})$ is known as a *deviation* (or a *deviation from the mean,* or a *deviate*), and the *sum of the squares of the deviations from the mean,*

$$\sum_{h=1}^{N} (X_h - \overline{X})^2,$$

is also known as a *centered sum*. Once columns h and X_h are filled (column h is optional), the X_h values are summed up. The sample mean $\overline{X}$ is then obtained by dividing the sum of the X_h values by the sample size N. The columns of deviations and of their squares can next be filled and summed up. If calculations are correct and have been done with enough accuracy, the sum of deviations should be very close to zero. Dividing the sum of squared deviations by $(N-1)$ then yields the variance, and the standard deviation is finally obtained by taking the square root of the latter.

Table 4.6.1
Detailed steps of calculations for the arithmetic mean, the variance, and the standard deviation of the length of ten right humeri (mm); (see table 1.2.1)

individual h	observation X_h	deviation $(X_h - \overline{X})$	squared deviation $(X_h - \overline{X})^2$
1	315	+1.6	+2.56
2	306	−7.4	+54.76
3	311	−2.4	+5.76
4	333	+19.6	+384.16
5	316	+2.6	+6.76
6	292	−21.4	+457.96
7	308	−5.4	+29.16
8	318	+4.6	+21.16
9	326	+12.6	+158.76
10	309	−4.4	+19.36
Sums	3134	0.0	+1140.4

the mean $\overline{X} = \Sigma X/N = 3134/10 = 313.4$ mm ,

the centered sum $\Sigma(X - \overline{X})^2 = 1140.4$ mm^2,

the variance $S_X^2 = 1140.4/(10-1) = 126.711$ mm^2,

and the standard deviation $S_X = \sqrt{126.711} = 11.2566$ mm .

When a computer or a scientific calculator is available, the first quantities calculated are usually the *raw sums*,

$$N, \quad \sum_{h=1}^{N} X_h , \quad \text{and} \quad \sum_{h=1}^{N} X_h^2 ,$$

from which the mean and the centered sum are then obtained through the formulae

$$\overline{X} = (\Sigma X)/N \qquad \text{and} \qquad \Sigma(X-\overline{X})^2 = \Sigma X^2 - (\Sigma X)^2/N = \Sigma X^2 - N\overline{X}^2;$$

(the summation subscript h is omitted here because there is no ambiguity). The quantity $(\Sigma X)^2/N = N\overline{X}^2$ is called a *centering term* (also known as a *correction factor*) because its subtraction corresponds to the fact of using the *center of observations* (the mean $\overline{X}$) as an origin instead of point 0. However, while the calculation of the centered sum by using the centering term is algebraically equivalent to the summation of the squares of the deviations from the mean (see section 13.2), it may yield numerical results which are inaccurate or even completely wrong when computing registers are short and the numerical variation of data is slight relatively to their order of magnitude. Even when a computer or a calculator is used, numerical accuracy is therefore better ensured if the mean $\overline{X}$ is calculated at first and the centered sum $\Sigma(X_h - \overline{X})^2$ is obtained next by subtracting the mean $\overline{X}$ from each individual observation X_h before adding up squared deviations. Analogous numerical difficulties may occur when the sums of the third and fourth powers of deviations are computed in order to obtain skewness and peakedness indices (section 13.6). Centering terms are nevertheless useful in several kinds of analyses of variance because they make the geometrical meaning of calculations more obvious (see sections 12.3 and 28.4 for instance).

Section 4.7: computing methods for grouped data

The methods outlined in the preceding section are particularly suitable for continuous data which have not been grouped into classes. Thanks to the power of present-day computers and calculators, continuous data are indeed no longer grouped into classes for the only sake of simplifying computations as done a few decades ago. Such arbitrary grouping operations did entail some losses of accuracy and would no longer save much computing time at present. Nevertheless, there are still today some cases where data are naturally grouped, such as in the case of discrete variates (chapters 17 and 18), and continuous data must still be arranged into groups when tests of goodness of fit must be carried out (chapter 16). Computing methods for grouped data speed up the calculation of raw sums and of centered sums because they prevent similar terms from being repeated. For instance, the part of the sum of the squares of the observations belonging to the same (i^{th}) class is obtained by multiplying the square of the value or of the midpoint X_i of that class by its absolute frequency $F(X_i)$. Computing formulae adapted to grouped and to ungrouped data are listed in parallel in table 4.7.1. The total number of classes is denoted here by K.

Section 4.8: other measures of dispersion: mean absolute deviation, range

Two other measures of dispersion are occasionally used: the *mean absolute deviation* and the *range*. The mean absolute deviation is obtained by dividing the sum of the absolute values of the deviations from the mean $\overline{X}$ by sample size:

$$\sum_{h=1}^{N} |X_h - \overline{X}| \,/ N .$$

The mean absolute deviation is an instance of the ordinary (arithmetic) mean and may seem initially simpler and more natural than the variance or the standard deviation (section 4.5) as a measure of dispersion. As indicated in section 4.5, however, the variance or its square root, the standard deviation, are often preferred because they appear in several theoretical probability distributions. The mean absolute deviation is nevertheless sometimes preferable because it is less affected than the variance and the standard deviation by extreme or atypical observations (also called *outliers*).

The *range* is even simpler than the mean absolute deviation, for it is the difference between the largest and the smallest observations of a sample:

$$\text{Range} = \text{maximum value} - \text{minimum value} .$$

Table 4.7.1
Computing formulae for ungrouped and grouped data; the summation subscripts *h* and *i* are omitted when there is no ambiguity

quantity	ungrouped data	grouped data
sample size	N	$\sum_{i=1}^{K} F(X_i)$
sum of observations	$\sum_{h=1}^{N} X_h$	$\sum_{i=1}^{K} [X_i\, F(X_i)]$
sum of squares of observations	$\sum_{h=1}^{N} X_h^2$	$\sum_{i=1}^{K} [X_i^2\, F(X_i)]$
sample mean	$\overline{X} = (\Sigma X)/N$	$\overline{X} = \{\Sigma[X\, F(X)]\}/[\Sigma F(X)]$
sum of squares of deviations	$\sum_{h=1}^{N} (X_h - \overline{X})^2$	$\sum_{i=1}^{K} [(X_i - \overline{X})^2\, F(X_i)]$
number of degrees of freedom	$\nu = (N - 1)$	$\nu = [\Sigma F(X)] - 1$
sample variance	$S_X^2 = (1/\nu)\ \Sigma(X_h - \overline{X})^2$	$S_X^2 = (1/\nu)\ \Sigma[(X_i - \overline{X})^2\, F(X_i)]$

Thus the range of the frequency distribution of the skull length of 76 male North American martens (table 3.1.1 and figure 3.3.2) is (82–73) = 9 mm. The range is interesting because it depicts the extent of individual variation, but it extracts little of the information present in the sample because it is based only on the two most extreme observations.

Section 4.9: measures of relative dispersion

All of the measures of dispersion mentioned up to this point, the variance and the standard deviation (section 4.5), the mean absolute deviation and the range (section 4.8), involve the unit of measurement in which individual observations are expressed: in the case of the skull length of 76 male martens, for instance, the variance is expressed in mm^2 while the standard deviation, the mean absolute deviation and the range are expressed in mm. All of those quantities are therefore said to be measures of *absolute dispersion.*

In practice, the measures of absolute dispersion of most biological data are not independent from their central tendency: when several samples of values of the same variate X are compared, those having large means also tend to have large standard deviations. Means and standard deviations are given in table 4.9.1, for instance, for the stature (height) of 20 Pygmy men of the Éfé tribe and for the length of 20 male blue whales. The standard deviation S_X of the length of blue whales (115 cm) is almost nineteen times larger than the standard deviation (6 cm) of the stature of Pygmies. At first, this might seem to indicate that blue whales vary much more than Pygmies. However, the average length $\overline{X}$ of blue whales (2399 cm) is also approximately seventeen times greater than the average stature of Pygmies (142 cm), and an individual difference of one meter in length or stature would obviously not have the same relative importance in whales as in Pygmies. In order to take into account the order of magnitude of organisms or of characters to be compared, one must therefore use a measure of *relative dispersion* rather than a measure of absolute dispersion.

Table 4.9.1
Absolute and relative dispersion of the height (stature) of 20 Pygmy men of the Éfé tribe and of the length of 20 male blue whales

Variate	Stature of 20 male Éfé Pygmies	Length of 20 male blue whales
Source of data	Schebesta & Lebzelter, 1933	MacKintosh & Wheeler, 1929
$\overline{X}$	142.4 cm	2399.25 cm
S_X	6.08190 cm	114.780 cm
$V = S_X / \overline{X}$	0.0427100	0.0478399
$S_{\log_e(X)}$	0.0432670	0.0495337

The best-known measure of relative dispersion is the coefficient of variation, V, obtained by dividing the standard deviation by the arithmetic mean. The coefficient of variation may also be expressed as a percentage, in which case the standard deviation is multiplied by 100 before being divided by the mean:

$$V = S_X/\overline{X} = 100\,(S_X/\overline{X})\% \,.$$

Because the numerator and the denominator of the coefficient of variation possess the same unit of measurement (the cm in the above examples), that unit cancels out upon division and the coefficient of variation by itself is a *dimensionless number* (also called a *pure number*). The coefficients of variation of the 20 Pygmies and of the 20 blue whales (table 4.9.1) are approximately equal. This shows that, if their considerably greater length is taken into account, blue whales do not appear to vary more than Pygmies.

While the coefficient of variation V is biologically interesting, it suffers from a serious statistical drawback: because it is estimated by a ratio of two other estimators, S_X and $\overline{X}$, its probability distribution is complicated. Fortunately, the coefficient of variation may often be conveniently replaced by another measure of relative variation, the standard deviation of logarithms $S_{\log_e(X)}$.

The latter may be obtained by transforming the data into logarithms (section 4.3) before calculating their standard deviation. As noted by Kermack and Haldane (1950), the standard deviation of (natural) logarithms often has approximately the same numerical value as the coefficient of variation of the original (untransformed) data.

In quantitative genetics and taxonomy, measures of relative dispersion tend to be large for heterogeneous groups of individuals, such as natural or experimental hybrids. In physiology, measures of relative dispersion tend to be small for quantitative characters which are closely regulated, such as the blood concentration of glucose (glycemia), but large for variable characters, such as the liver or muscle concentration of glycogen, a reserve substance of which the fluctuations help to keep glycemia constant.

Unfortunately, measures of relative dispersion possess no clear meaning for some variates of which the scale of measurement has an arbitrary origin, which makes their average also arbitrary. From a physiological point of view, for instance, it would be interesting to get a measure of relative dispersion for body temperature, which varies widely in poikilothermal organisms but is rather constant in homeothermal organisms (including man and most mammals and birds). But measures of relative dispersion of body temperature would differ considerably depending on whether the Celsius zero or the absolute (Kelvin) zero is used as an origin, and there are no fully convincing reasons to choose one over the other. In the case of such variates, it is undoubtedly preferable to use measures of absolute dispersion.

Section 4.10: principles of estimation

As an estimator of the mean μ_X of a statistical population, the sample mean $\overline{X} = \sigma_{g_1}(\Sigma X)/N$ satisfies the *least-squares principle* (section 4.2). Moreover, the mean $\overline{X}$ of a sample drawn from a normal statistical population (chapter 5) also satisfies another current principle of estimation, the *principle of maximum likelihood,* according to which the estimate of a parameter is the value which would have maximized the probability density of drawing the sample obtained in fact. The *likelihood* concept is also suitable to describe a scientist's uncertainty concerning a statistical conclusion (section 9.1).

In the search for satisfactory estimators, other qualities are also desirable, such as *unbiasedness* and *efficiency.* The expected value (section 3.10) of an unbiased estimator is equal to the corresponding parameter, while the sampling variance of an efficient estimator is as small as possible. A particular estimator may not necessarily possess all desirable qualitics. For instance, the maximum-likelihood estimator of the parametric variance σ_X^2 of a normal distribution (chapter 5) is $\hat{\sigma}_X^2 = \Sigma(X-\overline{X})^2/N$, but that estimator is biased because its expected value is smaller than the corresponding parameter (sections 4.5 and 12.4):

$$\mathscr{E}[\Sigma(X-\overline{X})^2/N] = [(N-1)/N]\,\sigma_X^2 < \sigma_X^2 .$$

In order to correct the bias of the maximum-likelihood estimator, the latter is therefore replaced by $S_X^2 = \Sigma(X-\overline{X})^2/(N-1)$. As for the sample median $\tilde{X}$, while it could be used as an estimator of the mean μ_X of a normal statistical population (which coincides with its parametric median, section 5.3), that would be inefficient because the sampling variance of the median is larger than that of the sample mean $\overline{X}$ (section 9.5).

Chapter 5
The normal distribution

Section 5.1: introduction

The so-called *normal distribution* is a continuous, unimodal and symmetrical probability distribution of which the distributive curve (section 3.5) is bell-shaped (figure 5.1.1). This distribution was discovered independently by the Frenchman Pierre Simon de Laplace (1749-1827) and the German Karl Friedrich Gauss (1777-1855), who both studied mathematics, physics and astronomy. The normal distribution is expected according to probability theory when the variation of a continuous random variable (variate) is due to causes which are very numerous and independent from each other, which act additively, and which all have small and approximately equal effects. The normal distribution is theoretical and is almost never perfectly observed in reality. For instance, even though the normal distribution extends in principle from $-\infty$ to $+\infty$, most biological variates, such as the concentration of a chemical substance in an organism, the length or the weight of an individual organism, or the number of individuals in a biological population, cannot take negative numerical values (section 4.3). In practice, the normal distribution is nevertheless obtained approximately in many fields, such as physics, biology, psychology, etc., when the variation is due largely to objective (real) factors as well as when it is due mostly to inaccuracies or accidental measuring errors. Moreover, even when the normal distribution by itself is not suitable, it can often be satisfactorily replaced by a related distribution, the lognormal distribution (chapter 14).

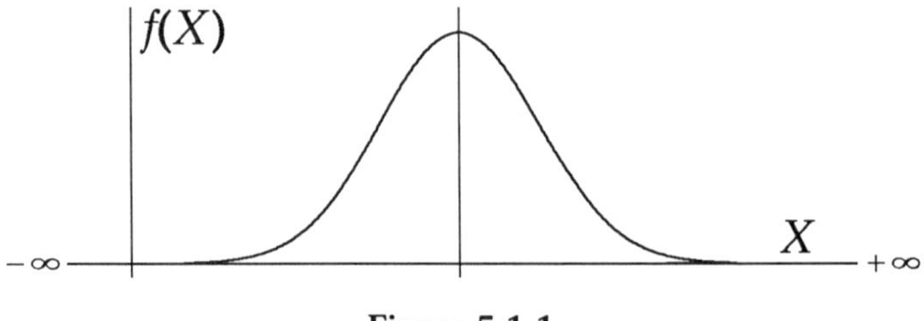

Figure 5.1.1
The distributive normal curve is bell-shaped

The normal distribution is very important in statistics because several other theoretical distributions are related to it, such as the chi-squared (χ^2) distribution (chapter 7), or become increasingly similar to it when sample size or a parameter increases, such as Student's t distribution (chapter 6), the binomial distribution (chapter 17) or the Poisson distribution (chapter 18).

Section 5.2: the normal probability density

Like all other probability distributions of continuous variates, the normal distribution involves a paradox: even though a continuous variate X may take any integer or noninteger numerical value, including a perfectly exact value like $X_0 = 13.000...$ (with an infinite number of zeros), the probability that a continuous variate X takes an infinitely exact value is null: $P(X = X_0) = 0$. That probability is indeed equal to the area of an infinitely narrow rectangle in a frequency histogram where the number of classes would tend toward infinity while class breadth would tend toward zero. Nevertheless, the occurrence of a perfectly exact value must be considered as possible in principle and the ordinate of the distributive curve above the value $X = X_0$ is not null (figure 5.2.1). Therefore, as already mentionned in sections 3.3, 3.8 and 3.10, the ordinate $f(X)$ of the distributive curve of a continuous probability distribution must be defined not as a probability but as a *probability density,* that is as the amount of *probability per unit of the X variate.* In practice, moreover, a statistician cannot consider the probability that a continuous variate takes an exact value: he is compelled to consider the probability that a continuous variate takes a value smaller or greater than a specified number or included within a certain interval. The probability density $f(X)$ of the normal distribution is defined by the equation

$$f(X) = \frac{1}{\sigma_X \sqrt{2\pi}} \, e^{-\frac{1}{2}\left(\frac{X-\mu}{\sigma_X}\right)^2} ,$$

where $X = X_0$ = the value of the variate of which the probability density is evaluated,

$\mu = \mu_X$ = the parametric mean of the distribution,

$\sigma^2 = \sigma_X^2$ = the parametric variance of the distribution (the square root of the variance being the standard deviation σ_X),

π = 3.141592653590 ... , a numerical constant,

and e = 2.718281828459 ... , the numerical constant used as the base of natural logarithms.

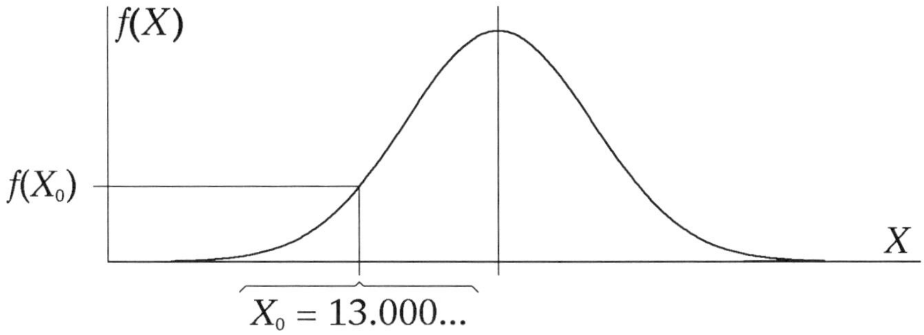

Figure 5.2.1
The ordinate of the distributive curve of a continuous distribution such as the normal distribution is a probability density

As indicated in section 3.5, the probability density $f(X)$ is the ordinate of the distributive curve (figure 3.5.2, left, and figure 5.2.1) and the derivative of the cumulative probability $P(X < X_0)$, which is the ordinate of the cumulative curve (figure 3.5.2, right). In the particular case of the normal distribution, the integral of the probability density $f(X)$ has not been found to be expressible in terms of elementary functions, and the cumulative probability can be evaluated only by using numerical algorithms (computational procedures).

Section 5.3: the two parameters of the normal distribution, μ and σ^2

If a continuous variate X has (or is assumed to have) a normal probability distribution, only two quantities need to be known for the distribution to be completely determined: the mean μ_X and the variance σ_X^2. These are indeed the only two quantities, besides the variate X, which are liable to change in the equation of the probability density $f(X)$ of section 5.2. These are therefore the only two quantities with respect to which various normal distributions may differ from each other. Like in the case of other continuous distributions, the parametric mean and variance of the normal distribution are defined by integrals (section 3.10):

$$\mu_X = \int_{-\infty}^{+\infty} X\, f(X)\, dX \quad \text{and} \quad \sigma_X^2 = \int_{-\infty}^{+\infty} (X-\mu)^2 f(X)\, dX\,.$$

The above integrals, as well as the probability density in section 5.2, are mostly of theoretical interest. In practice, the mean and the variance of a normal statistical population are usually estimated from a sample by using the formulae given in chapter 4. In order to be able to use statistical hypotheses and confidence intervals in a clear and logical manner (chapters 9 and 10), however, consumers of statistical methods must have a good understanding of the concepts of parametric mean and variance.

Because of the symmetry and unimodality of the normal distribution, the mean of that distribution coincides with its parametric mode and median (section 4.4). Moreover, at the population level, the parametric mean μ possesses properties similar to those of the sample mean $\overline{X}$ (section 4.2):

$$\int_{-\infty}^{+\infty} (X-\mu)\, f(X)\, dX \equiv 0\,.$$

and

$$\int_{-\infty}^{+\infty} (X-\mu)^2 f(X)\, dX \leq \int_{-\infty}^{+\infty} (X-K)^2 f(X)\, dX\,.$$

Finally, as illustrated in chapter 4, the parametric mean reflects the location of the center of the distribution (figure 4.2.1) while the variance (or its square root, the standard deviation) reflects the spread of the distribution (figure 4.5.1). The normal density curve has two inflection points located one standard deviation from the parametric mean on the left and on the right (figure 5.3.1); the curve is convex upwards within the interval between these two inflection points and concave upwards outside of the interval.

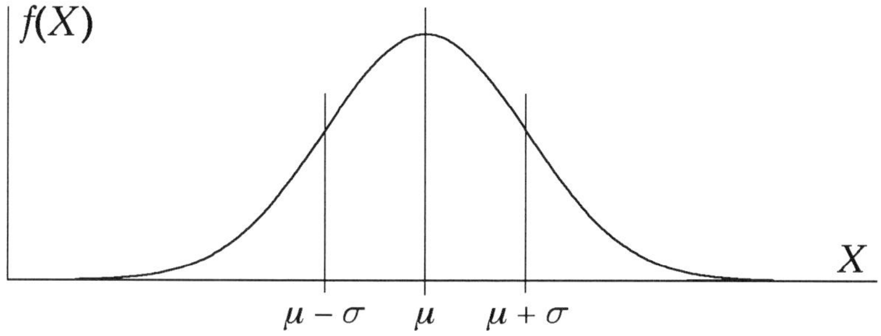

Figure 5.3.1
The normal density curve has two inflection points located one standard deviation from the parametric mean on the left and on the right

To indicate that a continuous variate X possesses a normal probability distribution of which the mean is μ_X and the variance σ_X^2, the following notation may be used:

$$X \leftarrow \mathcal{N}(\mu_X, \sigma_X^2) \ .$$

Section 5.4: the standardized normal distribution

In practice, most normal distributions differ from each other with respect to their means and variances. In order to save the labor of reevaluating the cumulative probability (section 3.5) of a normal variate through numerical algorithms in each particular case, all normal distributions can be made similar through a transformation called a *standardization*. Since normal distributions can differ from each other only with respect to two parameters (section 5.3), the standardized variate is obtained by subtracting the mean μ from each observation X and dividing the difference $(X-\mu)$ by the standard deviation σ_X. This transformed variate may be denoted either by the lower case x or by the lower case t letter and is equal to

$$x = t = (X - \mu)/\sigma \ .$$

Following this transformation, the new variate x may be positive, null or negative, its parametric mean is null, and its variance and standard deviation are equal to unity:

$$\mu_x = 0 \quad \text{and} \quad \sigma_x^2 = \sigma_x = 1 \ .$$

One may therefore write $x \leftarrow \mathcal{N}(0, 1)$. The cumulative probability of the normal distribution can thus be given once and for all in a single table (section 5.5). Once statistical calculations have been done on the standardized variate x, results can be reexpressed with respect to the original variate X through the inverse transformation

$$X = \mu + x\sigma \ .$$

Geometrically speaking, replacing the original variate X by the standardized variate x and coming back later to the original variate may be said to be simply a temporary regraduation of the coordinate axis (figure 5.4.1):

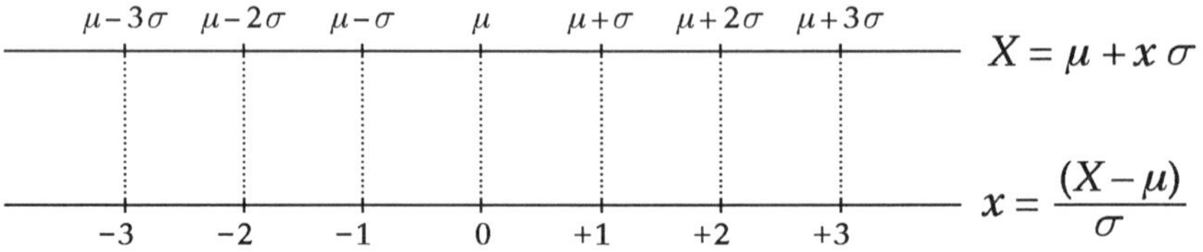

Figure 5.4.1
Correspondence between the values of the original variate X and those of the standardized variate $x = (X - \mu)/\sigma$

The equation of the normal probability density (section 5.2) is markedly simpler for the standardized variate x than for the original variate X:

$$f(x) = \frac{1}{\sqrt{2\pi}} e^{-x^2/2} .$$

The standardized normal probability density curve is illustrated in figure 5.4.2.

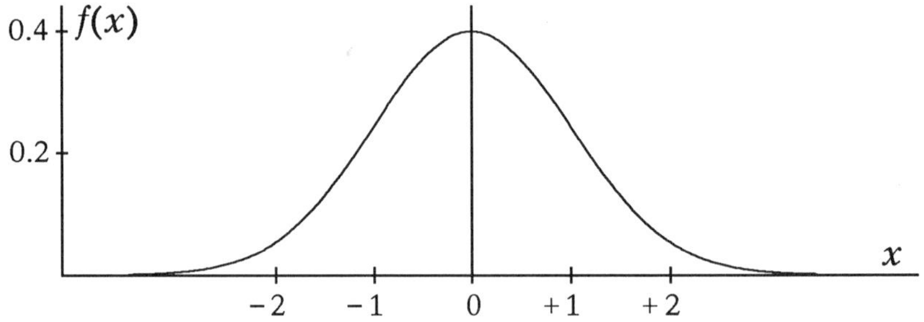

Figure 5.4.2
Probability density curve of the standardized normal distribution

In practice, the parametric values of the mean μ_X and of the standard deviation σ_X are often unknown, and the standardization of the original variate X and the inverse transformation are then done using the sample mean $\overline{X}$ and standard deviation S_X:

$$x' = (X - \overline{X})/S_X \qquad \text{and} \qquad X = \overline{X} + x' S_X .$$

In this case, the mean and standard deviation of the standardized variate x' are 0 and 1 respectively at the sample level, but generally not at the population level:

$$\overline{x}' = 0 \qquad \text{and} \qquad S_{x'} = 1 ,$$

but

$$\mu_{x'} \neq 0 \qquad \text{and} \qquad \sigma_{x'} \neq 1 .$$

Section 5.5: numerical tables of the normal distribution

A table of the cumulative probabilities of the standardized form of the normal distribution is given at the end of this book. The probability density is not included because it is used less often in practice and, unlike its integral, it can easily be evaluated through the expressions given in sections 5.2 and 5.4. The normal distribution may be considered as the limiting case, when the sample size N becomes infinitely large, of the Student's t distribution discussed in chapter 6. While many authors use the letter z to denote a tabulated value of a standardized normal variate corresponding to a specified cumulative probability P, the present writer thus prefers the letter t bearing a subscript parenthesis containing the cumulative probability P followed by a semicolon and the symbol ∞ : $t_{(P;\,\infty)}$. The symbol ∞ is a reminder that the normal distribution is similar to Student's t distribution when the number ν of degrees of freedom of the latter becomes infinitely large.

For instance, the equation $t_{(0.95;\,\infty)} = +1.644854$ means that, in a standardized normal distribution, the probability that $t < +1.644854$ is 0.95, that is 95%. If the subscript parenthesis does not contain a cumulative probability, this implies that the corresponding t value is an observed value rather than a tabulated value: thus, $t_{(\infty)} = 2.87$ would denote an observed value of a continuous variate known or assumed to follow a standardized normal distribution. Finally, when there is no ambiguity concerning the fact that the normal distribution is used, the notation may be simplified by omitting the symbol ∞ .

Because the standardized normal distribution is symmetrical and centered on 0, the table at the end of this book contains only positive values of t, but the cumulative probability P of a negative value $t_0 < 0$ may be obtained from the relationship $P[t < t_0] = 1 - P[t < -t_0]$ and the t values corresponding to cumulative probabilities smaller than 0.5 can be obtained from $t_{(P;\,\infty)} = -t_{(1-P;\,\infty)}$. The values of the standardized normal variates corresponding exactly to the most frequently used cumulative probabilities (0.95, 0.975, 0.99, 0.995, 0.999 and 0.9995) are given on the last line (corresponding to $\nu = \infty$) of the Student's t table.

Section 5.6: normal probability paper

The cumulative curve of a normal distribution is sigmoid, which means that it has the shape of a drawn-out S (figure 3.5.2, right). However, this sigmoid curve can be transformed into a straight line by modifying the graduation of the ordinate, on which the cumulative frequencies or probabilities are inscribed. This can be done simply by spacing the graduations of the ordinate according to the inverse of the distribution (cumulative probability) function (section 3.5). The result is known as *normal probability paper* and can be used to check whether the frequency distribution of a set of data resembles the normal distribution, for it is visually easier to detect departures from a straight line than from a sigmoid curve.

Chapter 6
The distribution of Student's t

Section 6.1: introduction

The distribution of Student's t was discovered by the British biometrician William Sealy Gossett (1876-1937), who published a paper on this topic in *Biometrika* in 1908 under the pseudonym *Student*. When a sample of observations must be used to test hypotheses concerning the mean μ of a normal statistical population, Student's t distribution yields exact probabilities even though the standard deviation σ of that population is unknown. Before the discovery of Student's t, the normal distribution was used as an approximation in that context, but that procedure was satisfactory only if the sample studied was large ($N \geq 30$). Student's t distribution is therefore more accurate in the case of the small samples ($N < 30$) with which scientists often have to work in many fields, such as experimental biology, ecology, paleontology, etc. . Present-day tables of Student's t distribution reach rather large sample sizes, and the normal distribution seldom has to be substituted for it as an approximation.

Section 6.2: the probability density

The probability density of Student's t distribution with ν *degrees of freedom* (see sections 4.5, 7.1 and 12.4) is defined by the following equation:

$$f(t_{(\nu)}) = \frac{(1 + t_{(\nu)}^2/\nu)^{-(\nu+1)/2}}{\sqrt{\nu}\ B(1/2, \nu/2)} ,$$

$$\text{where } B(1/2, \nu/2) = \frac{\Gamma(1/2)\,\Gamma(\nu/2)}{\Gamma(1/2 + \nu/2)} .$$

As indicated above, the Beta function appearing in the probability density can be evaluated using Gamma functions. As for the latter, they can be calculated using Stirling's approximate method or various algorithms discussed by Abramowitz and Stegun (1968). The Gamma function of a positive integer $(X+1)$ is related to the factorial (section 3.11), since $\Gamma(X+1) = X!$, but it is defined also for noninteger arguments. In the case of Student's t distribution like in those of other theoretical distributions (chapters 5, 7 and 8), the probability density is interesting mostly because it shows what quantities affect the shape of the distribution. What users rather need in practice, however, is the cumulative probability, which is obtained by numerical integration of the probability density (section 6.3). The probability density of Student's t distribution is represented by a few typical curves in figure 6.2.1. The higher the number ν of degrees of freedom, the greater the similarity between Student's t distribution and the normal distribution, with the result that, in the limiting case where $\nu = \infty$, these two distributions become numerically and geometrically identical.

When sample sizes and numbers of degrees of freedom are small, however, Student's t distribution is flatter and broader than the normal distribution. In order to include the same probability within a symmetrical interval centered on the origin 0, a broader interval must therefore be taken in the case of Student's t than in that of the normal distribution. This compensates for the uncertainty due to the fact that the parametric standard deviation σ is unknown and estimated on the basis of a sample of finite size.

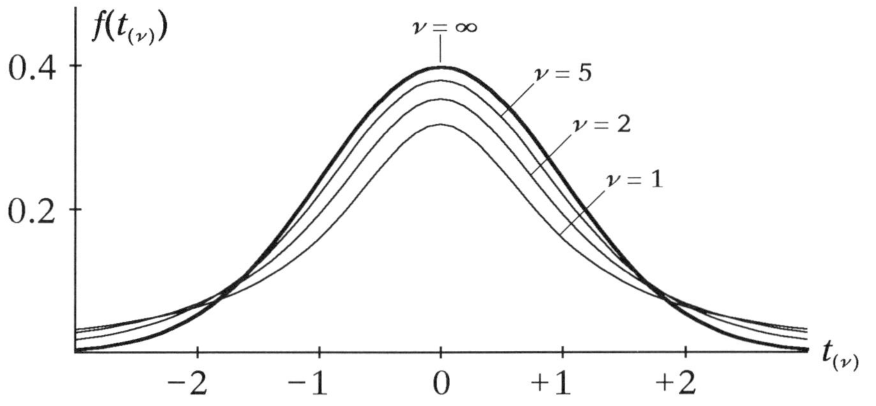

Figure 6.2.1
Probability density curve of Student's *t*

Section 6.3: the numerical table of Student's *t* distribution

A table of the cumulative probabilities of Student's t distribution will be found at the end of this book. The notation is very similar to the one used for the normal distribution (section 5.5): if a variate t follows Student's t distribution with ν degrees of freedom, $t_{(P;\ \nu)}$ denotes a tabulated value of which the cumulative probability is P. This means that

$$\text{Probability}\ [t < t_{(P;\ \nu)}] = P\ .$$

If the parenthesis does not contain a cumulative probability (which is generally a fractional value), this means that the corresponding t value is an observed rather than a tabulated value: thus, $t_{(9)} = 2.87$ would denote an observed value of a continuous variate following Student's t distribution with $\nu = 9$ degrees of freedom. Like the standardized normal distribution, Student's t distribution is symmetrical and centered on the origin 0, and the table at the end of this book contains only positive values of t which correspond to cumulative probabilities equal to or greater than 1/2, but a negative value of t corresponding to a cumulative probability smaller than 1/2 may easily be obtained from the relationship

$$t_{(P;\ \nu)} = -t_{(1-P;\ \nu)}\ .$$

The utilization of Student's t distribution to test hypotheses or to determine confidence intervals concerning one or two means will be discussed in chapter 9.

Chapter 7
The distribution of χ^2 (chi squared)

Section 7.1: introduction

The distribution of χ^2 (chi squared) is a continuous and asymmetrical distribution which ranges from 0 to $+\infty$ and is followed by a sum of squares of independent standardized normal variates. The χ^2 distribution and its application to frequency tables (chapter 15) were discovered by the British biometrician Karl Pearson (1857-1936), who was considered as the "father of biometry" and was one of the founders of the periodical *Biometrika*. The χ^2 distribution is exact when hypotheses must be tested or confidence intervals must be determined (chapter 10) concerning the parametric variance σ^2 of a continuous variate X which follows a normal distribution (chapter 5). Moreover, the χ^2 distribution may be used as an approximation in many cases, including Bartlett's test of the homogeneity of variance (section 12.7), tests of hypotheses concerning frequency tables (chapter 15), tests of goodness of fit (chapters 16, 17 and 18), the binomiality (chapter 17) and Poissonianity (chapter 18) tests, etc. . Paradoxically, the approximate utilizations of the χ^2 distribution are perhaps better known than the exact ones.

If each of a set of N independent observations, X_h, $h = 1, 2, \dots N$, follows a normal distribution $\mathcal{N}(\mu, \sigma^2)$, the sum of the squares of the standardized variates $(X_h - \mu)/\sigma$ follows the χ^2 distribution with *N degrees of freedom* (see sections 4.5, 6.2 and 12.4):

$$\sum_{h=1}^{N} [(X_h - \mu)/\sigma]^2 \leftarrow \chi^2_{(N)} \, .$$

However, the real number of completely independent observations is often smaller then the apparent number N of observations because of various constraints, and it is this real number of completely independent observations which is called the number of degrees of freedom and is denoted by the lower case Greek letter ν (*nu*). For instance, when the population mean μ is unknown and is replaced by the sample mean $\overline{X}$, there is a loss of one degree of freedom:

$$\sum_{h=1}^{N} [(X_h - \overline{X})/\sigma]^2 \leftarrow \chi^2_{(\nu = N-1)} \, .$$

This results from the fact that, unlike the deviations from the population mean $(X_1 - \mu)$, $(X_2 - \mu)$, ... $(X_N - \mu)$, which are independent from each other, the deviations from the sample mean $(X_1 - \overline{X})$, $(X_2 - \overline{X})$, ... $(X_N - \overline{X})$ are not fully independent because the relationship $\Sigma(X_h - \overline{X}) \equiv 0$ (section 4.2) can be used to express one of these deviations as a function of the others.

Section 7.2: the probability density

The probability density of the χ^2 distribution with ν degrees of freedom is defined by the following equation:

$$f(\chi^2_{(\nu)}) = \frac{\chi^{2\,(\nu/2-1)}_{(\nu)}\, e^{-\chi^2_{(\nu)}/2}}{2^{\nu/2}\;\Gamma(\nu/2)} \,.$$

The above probability density is represented by a few typical curves in figure 7.2.1. The χ^2 distribution is strongly skewed when ν is small, but it shifts toward the right and becomes less asymmetrical when the number of degrees of freedom increases, which makes it finally more closely similar to the normal distribution.

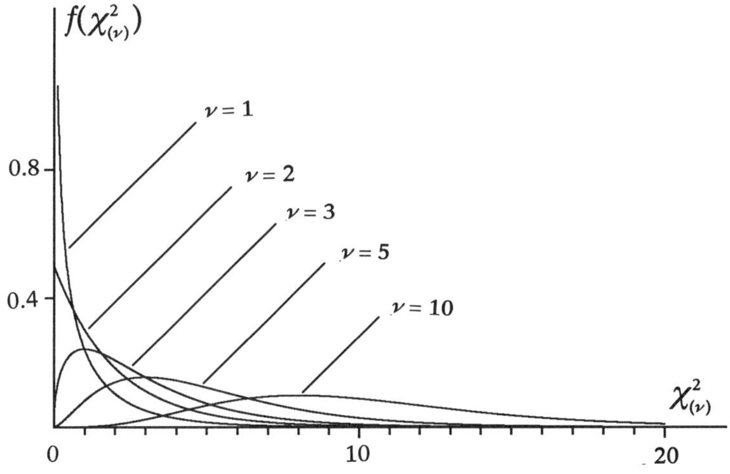

Figure 7.2.1
Probability density curve of the χ^2 distribution

Section 7.3: the numerical table of the χ^2 distribution

A table of the cumulative probabilities of the χ^2 distribution is given at the end of this book. If $\chi^2_{(\nu)}$ denotes a variate which follows the chi-squared distribution with ν degrees of freedom, $\chi^2_{(P;\,\nu)}$ denotes a tabulated value of which the cumulative probability is P, which means that

$$\text{Probability}\,[\,\chi^2_{(\nu)} < \chi^2_{(P;\,\nu)}\,] = P\,.$$

When the number ν of degrees of freedom is greater than 200, the value of χ^2 can generally be evaluated with at least four accurate digits by using the very good approximation of Wilson and Hilferty (1931)[1] :

$$\chi^2_{(P;\,\nu)} \doteq \nu\{1 - 2/(9\nu) + t_{(P;\,\infty)}\sqrt{[2/(9\nu)]}\,\}^3\,,$$

where $t_{(P\,;\,\infty)}$ denotes the algebraical (signed) value of a standardized normal variate having a cumulative probability of P.

---

[1] In this book, the symbol $\doteq$ indicates an approximate equality.

Chapter 8
The distribution of the variance ratio, $F = S_1^2 / S_2^2$

Section 8.1: introduction

The theoretical foundations of the distribution of the variance ratio were established by the British biometrician Ronald Aylmer Fisher (1890-1962), but it is the American statistician George W. Snedecor (1881-1974) who expressed that distribution with respect to the ratio of two variances and paved the way for the convenient tables available today. Snedecor denoted the ratio of two observed variances, S_1^2/S_2^2, by an upper case F in honor of Fisher. The F distribution is the probability distribution of the ratio of two independent variance estimates, of which each is obtained by dividing a sum of squares of standardized normal deviations, which follows a χ^2 distribution, by the corresponding number of degrees of freedom. The variance ratio distribution is continuous and asymmetrical and ranges from 0 to $+\infty$ like the χ^2 distribution, but it differs from the latter by having two numbers of degrees of freedom instead of only one. The variance ratio distribution is useful not only to test the hypothesis that two normal statistical populations have equal individual variabilities (chapter 11) but also in many kinds of analyses of variance (chapters 12, 21, 26, 27, 28, etc.) where the numerator of F may reflect the possible presence of differences between the means or the estimation (regression) lines of several groups of data.

Section 8.2: the probability density

The probability density of the F distribution with ν_1 and ν_2 *degrees of freedom* is defined by the following equation:

$$f(F_{(\nu_1,\,\nu_2)}) = \frac{\nu_1^{\nu_1/2}\ \nu_2^{\nu_2/2}}{B(\nu_1/2,\, \nu_2/2)}\ F_{(\nu_1,\,\nu_2)}^{(\nu_1/2\,-\,1)}\ (\nu_2 + \nu_1 F_{(\nu_1,\,\nu_2)})^{(-\nu_1/2\,-\,\nu_2/2)}\ .$$

ν_1 denotes the number of degrees of freedom of the variance estimate S_1^2 appearing in the numerator of the variance ratio F while ν_2 is the number of degrees of freedom of the denominator S_2^2. As in section 6.2, the Beta function may be evaluated using Gamma functions which may themselves be calculated using Stirling's approximate method or various algorithms discussed by Abramowitz and Stegun (1968).

As in chapters 5, 6 and 7, what is most useful in practice is the cumulative probability, which is the integral of the above probability density. The computation of the cumulative probability is generally rather tedious and usually done on computers.

The probability density of the variance ratio distribution is illustrated by a few typical curves in figure 8.2.1. When the numbers of degrees of freedom ν_1 and ν_2 are large, the distribution tends to concentrate about the value $F = 1$.

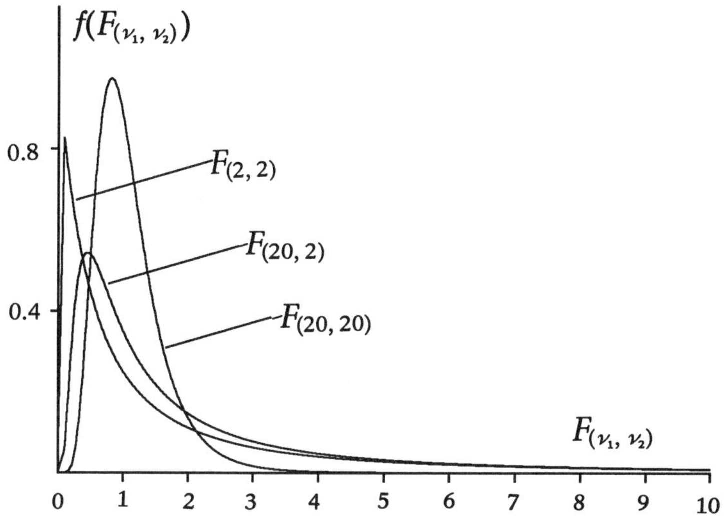

Figure 8.2.1
Probability density curve of the F distribution

Section 8.3: the numerical table of the *F* distribution

A table of the cumulative probabilities of the F distribution is given at the end of this book. $F_{(P;\ \nu_1, \nu_2)}$ denotes a tabulated value of the variance ratio $F_{(\nu_1, \nu_2)}$ having ν_1 and ν_2 degrees of freedom and possessing a cumulative probability equal to P, which means that

$$\text{Probability}\ [F_{(\nu_1, \nu_2)} < F_{(P;\ \nu_1, \nu_2)}] = P.$$

Although the variance ratio distribution is not symmetrical, it possesses an analogous property:

$$F_{(1-P;\ \nu_2, \nu_1)} = 1/F_{(P;\ \nu_1, \nu_2)}\ .$$

Because the table at the end of this book includes only probabilities equal to or greater than 0.95, the preceding property is used to evaluate variance ratio values corresponding to smaller cumulative probabilities (0.05, 0.025, 0.01, 0.005, 0.001 and 0.0005). The variance ratio (F) distribution is related to Student's t distribution as follows:

$$F_{(P;\ 1, \nu)} = t^2_{[(1-P)/2;\ \nu]} = t^2_{[(1+P)/2;\ \nu]}\ .$$

In the case of the variance ratio (F) as well as in that of the χ^2 and of Student's t distributions, interpolation is more accurate when it is carried out with the inverse of the numbers of degrees of freedom instead of the numbers of degrees of freedom themselves.

Chapter 9
Hypothesis testing and confidence intervals concerning one or two means

Section 9.1: introduction

In this chapter, the continuous variate X under study will be assumed to follow a normal probability distribution (chapter 5) of which the mean is μ_X and the variance σ_X^2: $X \leftarrow \mathcal{N}(\mu_X, \sigma_X^2)$. However, the methods discussed here will retain their validity in many other cases where the variate X is not normally distributed but can be transformed into another variate $Y = g(X)$ having a normal distribution $\mathcal{N}(\mu_Y, \sigma_Y^2)$. In such cases, once the statistical analysis of the transformed variate Y has been completed through methods based on normal theory, results can be reexpressed with respect to the original (untransformed) variate X by using the inverse transformation $X = g^{-1}(Y)$. For instance, if the original variate X follows a lognormal distribution (chapter 14), the transformed variate $Y = \log_e(X)$ has a normal distribution and the results of the statistical analysis of Y can be reexpressed with respect to X thanks to the transformation $X = \exp(Y)$.

A most important kind of question in statistics is whether one of the parameters of a population is equal or not to a particular numerical value. For instance, the mean μ of a normal distribution may be assumed to have a particular value $\mu_0 = 15$ while being suspected of having a different value: $\mu \neq 15$. The two opposing statements, $\mu = \mu_0 = 15$ on the one hand and $\mu \neq 15$ on the other hand, are said to be hypotheses. In practice, attempts are often made to check whether a hypothesis such as $\mu = \mu_0 = 15$ is acceptable by confronting it with a sample of data drawn from the population studied. Such a validation attempt is called a *hypothesis test* or a *significance test*.

When a parameter does not have any known or hypothetical numerical value, attempts are usually made to estimate that value on the basis of a sample taken in the population. There are two major kinds of estimates: point and interval estimates. The simplest approach is to take a characteristic of a sample (a sample mean such as $\overline{X} = 14.7$, for instance) as an estimate of the corresponding characteristic of the population (the mean μ of a normal distribution). This procedure is summarized by the equation $\hat{\mu} = \overline{X} = 14.7$, where the circumflex accent ("hat") over the Greek letter μ (*mu*) indicates that $\hat{\mu}$ is an estimated value. Because the estimate $\hat{\mu} = \overline{X} = 14.7$ is a single value, it can be represented by a single point on the coordinate axis of the X's, and $\overline{X} = 14.7$ is said to be a *point estimate*. While a point estimate has the advantage of simplicity, it has the drawback that it does not indicate by itself the uncertainty entailed by the fact that it is obtained from a sample of finite size. That uncertainty can be reflected by taking as an estimate of μ a whole range of numerical values, such as

$$12.7 \leq \mu \leq 16.7 ,$$

which constitutes an *interval estimate*.

The most commonly used interval estimate is the *confidence interval* (section 9.7), of which the limits are based on sample data and called *confidence limits*. Determining a confidence interval and testing a hypothesis are mutually inverse approaches and, in principle, the conclusions reached through either must agree.

The theory of hypothesis testing and confidence intervals outlined here was developed by the statisticians J. Neyman (1894-1981) and E. S. Pearson (1895-1980). The latter was the son of Karl Pearson (section 7.1). According to Fisher (1970, pages 10-11) as well as Neyman and Pearson, the *probability* concept should be used only to describe the uncertain occurrence of an objective process (external to the observer), and should be replaced by the *likelihood* concept (see section 9.7) for the description of a scientist's subjective uncertainty regarding a conclusion. While there are other schools of thought, the Bayesian school in particular, the most currently used statistical methods are based on Neyman and Pearson's theory.

The methods of this chapter are suitable to test hypotheses or to determine confidence intervals regarding one or two parametric means. When the means of more than two groups have to be compared, the analysis of variance (chapter 12) should be used.

Section 9.2: hypothesis testing: a simple example

In order to introduce the problem as simply as possible, a fictitious example will be considered at first. Assume that, after studying a very large sample, an anthropologist knows that the stature (body height) X of the adult men of a homogeneous local population **A** follows exactly a normal probability distribution of which the mean and the standard deviation are $\mu_A = 1700$ mm and $\sigma_A = 60$ mm. The same anthropologist has also measured the stature $X = 1500$ mm of another adult man whose origin is doubtful and who may or may not belong to the same population (**A**) or to another population **B** of which the parametric mean would be different: $\mu_B \neq 1700$ mm (figure 9.2.1).

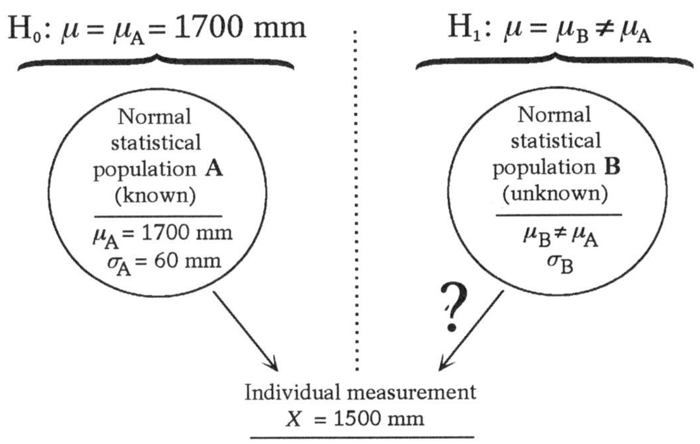

Figure 9.2.1
The preliminary ("null") hypothesis H_0 and the alternative hypothesis H_1

Even if the individual $X = 1500$ mm is suspected of belonging to an unknown population **B**, the assumption that it belongs to the known population **A** is logically simpler unless there is evidence to the contrary. If the hypothesis that the individual X belongs to population **A** is plausible, there is indeed no necessity to postulate more than a single statistical population **A** (or two identical populations, which is mathematically equivalent); this prevents multiplying entities beyond need, as would be done if a second population **B** were postulated without adequate justification. Assuming at first that the new individual X comes from the known population is in accord with the *principle of parsimony,* also known as *Occam's razor* and attributed to the English philosopher William of Occam (approximately 1285-1349); according to this principle, among several possible explanations of a phenomenon, the simplest one should be preferred if it is as satisfactory as the others. Because individual stature varies even within a single homogeneous population, the fact that the stature of an individual differs slightly from the mean of a known population does not warrant the conclusion that such an individual comes from a different population. If an individual is suspected of coming from another population **B** and if his stature differs markedly from the mean of population **A**, however, the initial hypothesis that he belongs to population **A** should possibly be rejected.

The initial hypothesis, which is logically simpler and according to which the individual $X = 1500$ mm comes from the known population **A**, will be called a *preliminary hypothesis* and denoted by the symbol H_0. The preliminary hypothesis is called a *null hypothesis* by most writers, but the expression *preliminary hypothesis* is more appropriate because the expression *null hypothesis* should be reserved for the cases where a parameter is really assumed to be null (sections 9.8 and 9.9). As for the opposite hypothesis according to which the individual X would come from an unknown population **B** having a different mean, it is called an *alternative hypothesis* and denoted by the symbol H_1.

It remains to be decided whether, considering the stature of the individual $X = 1500$, the preliminary hypothesis H_0 should be accepted, or whether it should be rejected, in which case the alternative hypothesis H_1 would be accepted. It would seem natural to accept H_0 if the individual lies close enough to the center of population **A** but to reject H_0 and accept H_1 if the individual lies too far from the center of population **A** (figure 9.2.2). In the present case, the alternative hypothesis, $\mu_B \neq \mu_A$, comprises two possibilities: $\mu'_B < \mu_A$ and $\mu''_B > \mu_A$. The preliminary hypothesis H_0 will therefore be accepted if the individual X lies within a central region called an *acceptance region*; on the contrary, the preliminary hypothesis H_0 will be rejected and the alternative hypothesis H_1 will be accepted if the individual X lies either to the left or to the right of the acceptance region, in what will be called a *rejection region*.

The limits of the acceptance and rejection regions of the preliminary hypothesis H_0 are determined in such a way that, when an individual observation X is randomly drawn from population **A**, that observation has a large probability of falling within the acceptance region but a small probability of falling in the rejection region. A table of the standardized form of the normal distribution (or the last line of Student's t distribution, corresponding to $\nu = \infty$) indicates that the standardized variate $(X - \mu)/\sigma$ has a probability of 0.025 of being smaller than -1.96 and of 0.025 of being larger than $+1.96$.

If $\mu_A - 1.96\,\sigma_A = 1582.4$ is used as a lower limit and $\mu_A + 1.96\sigma_A = 1817.6$ as an upper limit, there is therefore a probability of $0.95 = 95\%$ that an individual value of X randomly drawn from population **A** falls within the interval [1582.4, 1817.6] and a probability of $(0.025 + 0.025) = 0.05 = 5\%$ that it falls outside of the interval. The individual measurement $X = 1500$ falls outside of the interval, in the rejection region, and an individual measurement differing from μ_A as much as $X = 1500$ toward the left or toward the right is unlikely to have come at random from population **A**. The preliminary hypothesis should be rejected and the alternative hypothesis should be accepted.

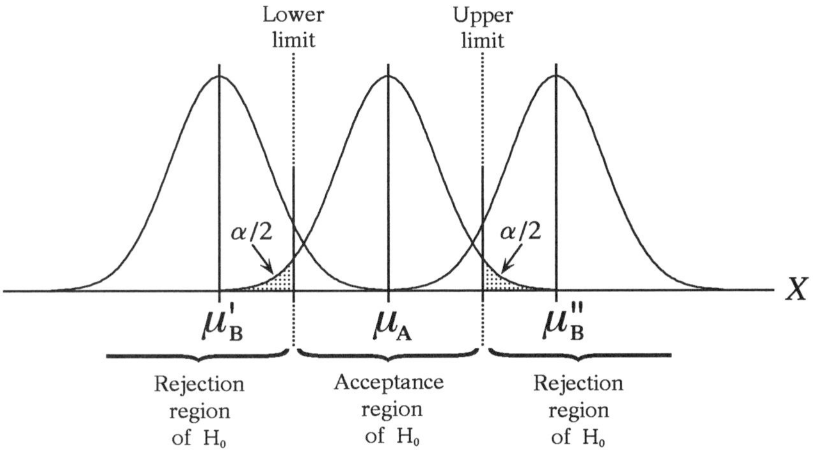

Figure 9.2.2
Acceptance and rejection regions of the preliminary hypothesis H_0; the sum of the two small stippled areas corresponds to the significance level α (section 9.4)

In statistical parlance, the difference $(X - \mu_A)$ is said to be *significant*, which means that it is worthy of attention, conclusive. It must be emphasized that, even if a difference is significant from a statistical point of view, it does not necessarily have a major biological significance. To prevent any ambiguity, the difference $(X - \mu_A)$ could therefore be said to be *statistically significant*. If, on the contrary, the individual measurement fell within the acceptance region, the difference $(X - \mu_A)$ would be said to be nonsignificant from a statistical viewpoint; it would be attributed to individual variation, and the preliminary hypothesis would be retained.

Section 9.3: deductive statistical inference: a reduction to the absurd

The manner in which a statistician goes about answering the kind of question considered in the preceding section may seem uselessly complicated at first: why indeed begin with the preliminary hypothesis H_0 while it is in fact the alternative hypothesis H_1 which corresponds to what is suspected, to what one is trying to detect or to discover? – It is because the kind of inference done in this context is a *reduction to the absurd*.

What reveals something new in scientific research is the observation of a difference which was not expected on the basis of previous knowledge. Because of the variability of biological phenomena such as human stature, however, an individual observation may be expected to differ somewhat from the parametric mean of a known population even if it does belong to that population. A scientist must therefore sift observed differences and retain only those that are important enough.

A statistical analysis thus begins with the preliminary hypothesis, according to which the observation would come from an already-known population even if it differs from the parametric mean of that population. The observed difference $(X - \mu_A)$ would then be due strictly to individual variation. If the probability of obtaining such an observation turns out to be too small according to the preliminary hypothesis, the latter is considered unlikely (*"absurd"*) and rejected, but that conclusion is motivated by reality itself, by way of the data analyzed. It must be emphasized that, when the preliminary hypothesis is accepted, it cannot be considered to have been proven, but only not to have been disproven. The contemporary Austrian philosopher Karl Popper (1902–1994) has stressed the importance of considering hypotheses and theories capable of being refuted (an idea translated somewhat ambiguously by the adjective *falsifiable*).

Section 9.4: type I and type II errors and the statistical significance level α

Because of the variability of biological phenomena, the conclusions drawn in biometry do not always correspond to reality (figure 9.4.1). Even when the preliminary hypothesis H_0 is true, it will nevertheless be rejected with a relative frequency or probability α *(alpha)* which depends on the locations chosen for the limits of the acceptance and rejection regions (figure 9.2.2). This erroneous rejection of the preliminary hypothesis H_0 is called a *type I error,* and its probability α corresponds to the sum of the two small stippled areas at the left and right ends of the central curve (of population **A**) in figure 9.2.2.

When the preliminary hypothesis H_0 is false, it will nevertheless be accepted with a probability β *(beta)* which depends not only on the locations of the limits of the acceptance and rejection regions but also on the distance between the mean μ'_B or μ''_B on the one hand and the mean μ_A on the other hand. This erroneous acceptance of the preliminary hypothesis H_0 is called a *type II error*.The Greek letters α and β correspond respectively to the lower case *a* and *b* of the Roman alphabet. The probability α of erroneously rejecting the preliminary hypothesis H_0 is called the *significance level,* while the probability $(1 - \beta)$ of rejecting it correctly is called the *power of the test*. The power of the test is a desirable characteristic, because a scientist would like to detect the alternative hypothesis H_1 when it is true. Ideally, it would be desirable to decrease the probabilities α and β of both types of errors as much as possible, but a compromise has to be made because shifting the limits of acceptance and rejection regions in such a way as to decrease α would increase β and vice versa. The significance levels most currently used in practice are $\alpha = 0.05$ (already used in section 9.2), $\alpha = 0.01$, and $\alpha = 0.001$. Statistical conclusions drawn at the $\alpha = 0.05$ significance level may be considered as provisional indications, while those drawn at $\alpha = 0.01$ or $\alpha = 0.001$ may be considered as well or very well established.

Conclusion drawn	Reality: H_0 true	Reality: H_0 false
H_0 accepted	Correct conclusion $P = (1 - \alpha)$	Type II error $P = \beta$
H_0 rejected	Type I error $P = \alpha$ (significance level)	Correct conclusion $P = (1 - \beta)$ (power)

Figure 9.4.1
Type I and type II errors, and the statistical significance level α

Section 9.5: variation of sample means within a single population

In this section, the variation of the means $\overline{X}$ of samples drawn from a single statistical population is illustrated through a *sampling experiment* based on the lengths in meters of 200 male blue whales [of the species *Sibbaldus* (*Balaenoptera*) *musculus*] captured in 1925-1926 in the southern hemisphere (MacKintosh and Wheeler, 1929). These data are reproduced in table 9.5.1 (next page) and have been arranged here at random (using random numbers on a computer) in order to prevent a possible lack of independence between successive values.

For didactic purposes, the 200 whales of table 9.5.1 are used here as a statistical population of which the parametric mean is estimated by drawing random samples. However, contrary to what happens in practice, where population parameters are usually unknown, the experimental population sampled here is finite and its mean and standard deviation can be calculated ($\mu = 20.6971$ and $\sigma = 2.50095$). The extent to which sample means deviate from the parametric mean of which they are estimates will therefore be easy to see.

In order to simulate *sampling without replacement* (section 2.3) in an infinite population, which is often closely approximated in biology, experimental samples have been drawn from the finite population of table 9.5.1 *with replacement*, each whale being possibly "captured" more than once. The results of the sampling experiment are illustrated in figure 9.5.1. Ten samples of each size, ranging from $N = 1$ to $N = 30$, have been drawn successively at random, and the mean of each has been calculated and represented in figure 9.5.1 by a short horizontal dash of which the abscissa indicates sample size N while the ordinate indicates the mean value $\overline{X}$. In the particular case where $N = 1$, each sample contains only one whale and the sample mean is exceptionally equal to the length of that whale.

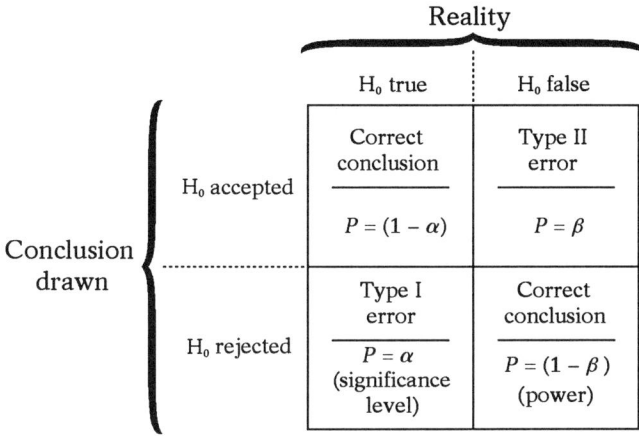

Table 9.5.1
Length measurements of 200 male blue whales in meters (MacKintosh and Wheeler, 1929)
(data reproduced with the permission of Cambridge University Press)

Columns									
1	2	3	4	5	6	7	8	9	10
22.74	17.25	16.95	18.97	18.00	19.40	23.40	18.45	17.60	21.30
19.80	21.65	24.50	22.10	22.05	21.80	18.45	25.83	17.50	20.15
21.50	17.80	21.70	24.40	21.70	19.64	19.45	23.50	20.30	18.60
18.70	20.00	19.20	22.45	24.80	19.60	24.90	17.53	19.00	25.20
18.60	19.20	18.20	18.00	21.60	17.90	18.10	19.00	20.35	19.90
21.50	18.60	16.90	18.95	21.90	18.55	24.40	22.55	20.30	18.90
19.80	25.45	18.05	18.40	18.80	23.80	17.85	20.50	19.10	18.50
19.85	22.90	21.40	24.55	24.40	22.60	24.50	18.70	19.10	19.00
25.50	19.35	18.95	17.75	22.70	18.80	17.80	17.50	20.55	21.70
24.90	26.30	18.35	24.45	19.30	18.07	22.70	18.30	19.40	23.30
21.85	23.20	17.75	17.70	24.45	26.00	24.10	18.40	22.90	25.10
24.65	17.30	19.85	19.10	20.50	16.85	20.40	24.10	17.75	21.30
23.90	23.10	18.50	19.90	18.25	20.85	19.20	20.37	24.80	18.80
19.05	19.20	22.20	22.25	18.60	19.40	17.00	22.35	18.40	24.10
24.70	15.83	18.40	21.97	20.90	23.30	18.40	17.85	18.30	24.25
17.25	18.80	17.25	21.65	17.50	18.75	23.20	23.40	21.75	18.10
18.10	19.55	25.25	22.60	23.80	18.80	19.30	25.50	20.80	23.80
24.80	20.65	20.10	17.40	18.30	23.40	19.50	21.20	20.07	23.05
23.00	22.00	21.25	20.80	21.50	24.40	19.50	20.00	18.05	19.30
22.00	24.15	24.00	20.63	19.40	20.10	18.17	24.20	20.20	21.60

To the extent that sample size N increases, the vertical dispersion of the dashes representing sample means decreases (figure 9.5.1). Therefore, the larger the samples are, the less their means differ from each other as well as from the parametric mean $\mu = 20.70$ of which they are estimates. The decrease in the dispersion of sample means $\bar{X}$ as sample size N increases could be described numerically by calculating the standard deviation of each successive group of 10 means: first, the standard deviation of the means of the 10 samples of size $N = 1$ (which are simply 10 individual measurements); second, the standard deviation of the means of the 10 samples of size $N = 2$; and so on up to $N = 30$. In fact, the standard deviation of the sample mean $\bar{X}$ is currently used in statistics and is denoted by the symbol $\sigma_{\bar{X}}$ in the case of the parameter and by $S_{\bar{X}}$ in the case of the estimate.

However, the circumstances in which the standard deviation of the mean is used in practice are very different from those of the present sampling experiment. Here, for didactic reasons, many samples of increasing size are drawn from a known population in order to study the way in which sample means vary according to sample size. In this context, the standard deviation of sample means can be estimated by direct calculations on the means of several samples of equal size. In practice, on the contrary, a statistician is often studying an unknown population on the basis of a single sample, and the standard deviation of the mean, $\sigma_{\bar{X}}$, is then estimated from that single sample by way of a theoretically derived formula:

$$S_{\bar{X}} = S_X/\sqrt{N}\,.$$

Provided the individual measurements X follow a normal distribution of which the mean is $\mu = \mu_X$ and the variance is $\sigma^2 = \sigma_X^2$, the means $\overline{X}$ of samples of size N can indeed be shown to follow a normal distribution of which the mean is also μ but the variance is N times smaller: $\sigma_{\overline{X}}^2 = \sigma_{(\Sigma X/N)}^2 = (1/N^2)\sigma_{(\Sigma X)}^2 = (1/N^2)N\sigma_X^2 = \sigma_X^2/N$. Therefore,

$$\overline{X} \leftarrow \mathcal{N}(\mu_X, \sigma_X^2/N)\,.$$

Moreover, statistical theory (the *"central limit theorem"*) indicates that, even when the distribution of individual measurements X is not normal, that of $\overline{X}$ resembles the normal distribution if the sample size N is large enough.

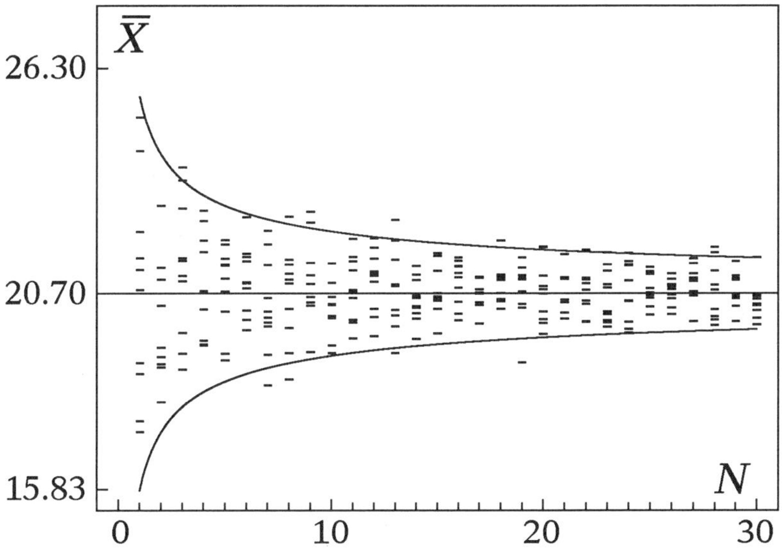

Figure 9.5.1
Decrease in the dispersion of sample means with the increase of sample size N; the horizontal line $\mu = 20.70$ represents the parametric mean of the population, and the two curves delimit the region within which 95% of sample means $\overline{X}$ are expected theoretically to fall

The standard deviation of the mean is generally used in practice to provide a numerical description of the uncertainty of the sample mean $\overline{X}$ as an estimate of the parametric mean μ of the population (sections 9.6 and 9.7). The decrease of the standard deviation of the mean with the increase of sample size is one particular consequence of the law of large numbers and has a fundamental importance in statistics: the greater the size N of a sample, the greater the chances that the mean $\overline{X}$ of that sample lies close to the parametric mean μ of the population and is a good estimate of it. Therefore, the larger the sample, the more information it provides about the population from which it has come.

It should be noted that the standard deviation of the mean is called the *standard error of the mean* by most writers. However, the expression *standard error* has a misleading connotation in biometry because, unlike in other fields, most of the variations observed in biology are due to objective factors (genetical, physiological, environmental, etc.) rather than to inaccuracies or accidental measuring errors. The expression *standard deviation of the mean* will therefore be preferred in this book.

Section 9.6: hypotheses concerning a population mean, one-sided and two-sided alternative hypotheses

Thanks to the concepts discussed in earlier sections of this chapter, it is possible to test a hypothesis concerning the parametric mean μ of a normal statistical population from which a sample has been drawn. In the case of the length measurements of left and right humeri of ten female human skeletons (table 1.2.1), for instance, if the question of bilateral symmetry is of particular interest, the ratio

$$X = \text{(length of right humerus)/(length of left humerus)}$$

can be calculated for each of these ten skeletons. The ratio of each skeleton is larger or smaller than 1.0 to the extent that the right humerus is longer or shorter than the corresponding left humerus. These ten ratios are given with 6-digit accuracy in table 9.6.1.

Table 9.6.1
Ratios of right humerus length to left humerus length for ten female human skeletons (see table 1.2.1)

1.01286	1.01325	1.03322	1.03416	1.01282
1.02456	1.00984	1.02581	0.993902	1.01645

The ratio X is assumed here to be normally distributed, the mean and standard deviation of the sample are $\overline{X} = 1.01769$ and $S_X = 0.0121176$, and the standard deviation of the mean is $S_{\overline{X}} = 0.00383192$. The preliminary hypothesis H_0: $\mu = \mu_0 = 1.0$, according to which the bilateral symmetry of humerus length would be perfect on the average, may be tested. If that preliminary hypothesis were true, the sample mean $\overline{X}$ would belong to a normal distribution of which the mean would be $\mu = \mu_0 = 1.0$ and the standard deviation would be $\sigma_{\overline{X}}$. If the parametric value of the standard deviation of the mean, $\sigma_{\overline{X}}$, were known, the normal distribution could be used, for the standardized variate $(\overline{X} - \mu_0)/\sigma_{\overline{X}}$ would follow the standardized normal distribution $\mathcal{N}(0, 1)$. Since the value of $\sigma_{\overline{X}}$ is unknown, however, it is replaced by its estimate $S_{\overline{X}}$ and the variate $(\overline{X} - \mu_0)/S_{\overline{X}}$ (also called a *criterion*) follows Student's t distribution with $\nu = (N - 1)$ degrees of freedom.

Before the limits of acceptance and rejection regions can be calculated, the appropriate kind of alternative hypothesis must be determined. Although only one kind of alternative hypothesis (H_1: $\mu = \mu_1 \neq \mu_0$) was mentioned in section 9.2 in order to keep things simple, two other kinds are also currently used: H_1: $\mu = \mu_1 < \mu_0$ and H_1 : $\mu = \mu_1 > \mu_0$. The acceptance and rejection regions corresponding to the three most common kinds of alternative hypotheses are illustrated in figure 9.6.1. These acceptance and rejection regions may also be reexpressed (figure 9.6.2) with respect to the criterion $t_{(N-1)} = (\overline{X} - \mu_0)/S_{\overline{X}}$.

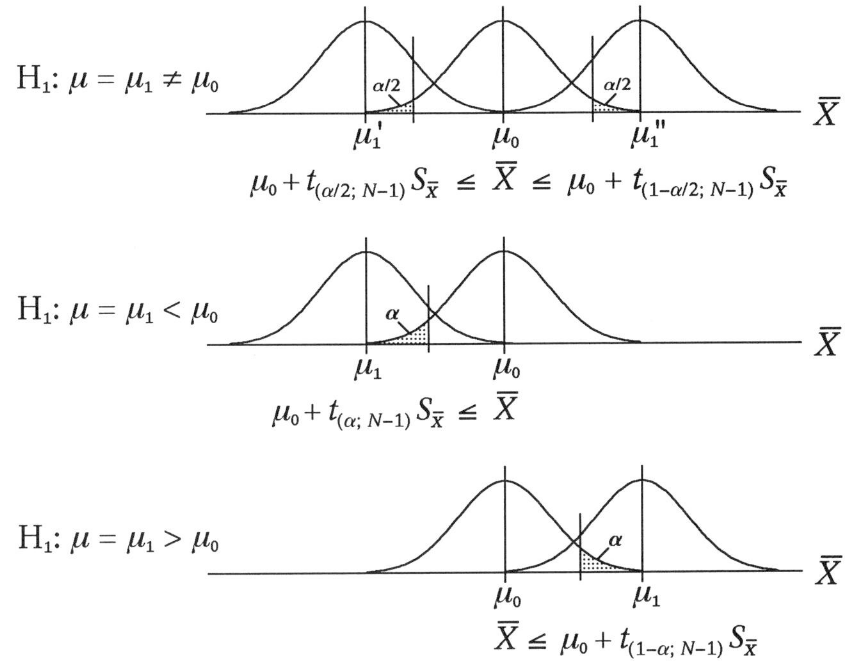

Figure 9.6.1
Acceptance and rejection regions of the preliminary hypothesis $H_0: \mu = \mu_0$ corresponding to the three current kinds of alternative hypotheses

It must be emphasized without delay that the kind of alternative hypothesis should be chosen for *a priori* reasons (based on earlier knowledge) and not suggested by the examination of the data. All methods in this chapter, as well as most current statistical methods, are indeed designed to test hypotheses set up independently of the data (see section 26.4, however, concerning *a posteriori* hypotheses).

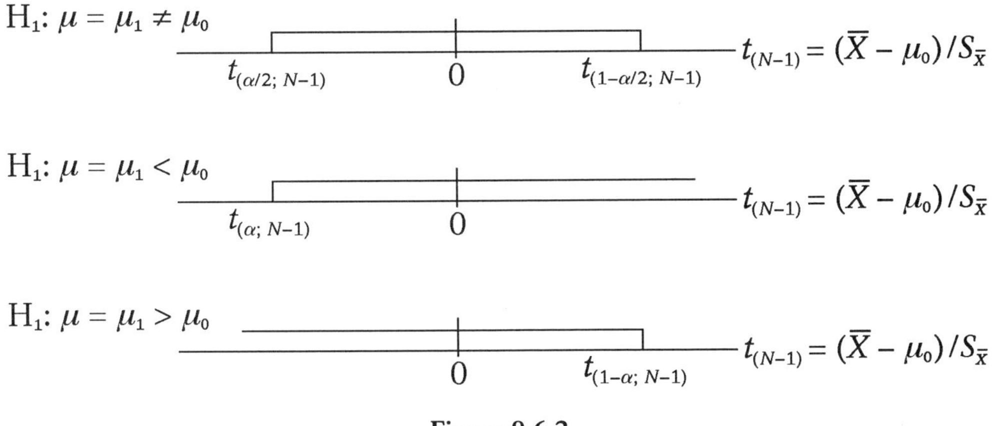

Figure 9.6.2
Acceptance and rejection regions reexpressed with respect to the criterion $t_{(N-1)}$

The alternative hypothesis H_1: $\mu = \mu_1 \neq \mu_0$ is said to be *two-sided* because it comprises two possibilities: $\mu = \mu_1 < \mu_0$ and $\mu = \mu_1 > \mu_0$. This kind of alternative hypothesis, already discussed in section 9.2, is appropriate whenever there are no more reasons to expect the mean $\mu = \mu_1$ to be small than to expect it to be large. The two other kinds of alternative hypotheses, H_1: $\mu = \mu_1 < \mu_0$ and H_1: $\mu = \mu_1 > \mu_0$, are said to be *one-sided* because each one includes a single possibility, and they are appropriate when there are *a priori* reasons to suspect that the mean $\mu = \mu_1$ is only smaller than μ_0 (in the case of H_1: $\mu = \mu_1 < \mu_0$) or only larger than μ_0 (in the case of H_1: $\mu = \mu_1 > \mu_0$). If attempts are being made to develop a kind of fertilizer suitable in agriculture, for instance, the only interesting substances would be those that increase production, and a one-sided alternative hypothesis like H_1: $\mu = \mu_1 > \mu_0$ might be appropriate. Similarly, in medicine, only substances which speed up healing might be desirable. When it is based on *a priori* reasons, a one-sided alternative hypothesis is worthwile because it increases the chances that it will be detected when it is true and that "a discovery will be made".

In the case of the ratios of humerus lengths in table 9.6.1, because the majority of individuals are right-handed in all known human populations, the one-sided alternative H_1: $\mu = \mu_1 > \mu_0 = 1.0$ would perhaps be appropriate. If this hypothesis were true, this would tend to increase the value of the sample mean $\overline{X}$ as well as the criterion

$$t_{(N-1)} = (\overline{X} - 1.0)/S_{\overline{X}}.$$

The rejection region should therefore be placed to the right of the acceptance region (bottom of figures 9.6.1 and 9.6.2). The sample mean $\overline{X}$ thus falls in the acceptance region if

$$\overline{X} \leq \mu_0 + t_{(1-\alpha;\,N-1)}\, S_{\overline{X}}$$

that is if the *criterion* $\quad t_{(N-1)} = (\overline{X} - \mu_0)/S_{\overline{X}} \leq t_{(1-\alpha;\,N-1)}$.

In the case of humerus length ratios,

$$t_{(9)} = (1.01769 - 1.0) / 0.00383192 = 4.6165 > 4.2968 = t_{(0.999\,;\,9)},$$

and the criterion falls in the rejection region at the $\alpha = 0.001$ significance level. H_0 must therefore be rejected and the parametric mean of the ratio of right and left humerus lengths is concluded not to be equal to 1.0. The difference $(\overline{X} - \mu_0)$ may be said to be *statistically significant at significance level* $\alpha = 0.001$ and the probability of getting a difference at least as large as $(\overline{X} - \mu_0)$ at random is $P < 0.001$. Therefore, this numerical analysis confirms the graphical indications given by figure 1.2.2, according to which the bilateral symmetry of humerus length is not perfect.

If the alternative hypothesis were H_1: $\mu = \mu_1 \neq \mu_0$, the acceptance region of H_0 would be

$$t_{(\alpha/2\,;\,N-1)} \leq (\overline{X} - \mu_0)/S_{\overline{X}} \leq t_{(1-\alpha/2\,;\,N-1)},$$

and, if the alternative hypothesis were H_1: $\mu = \mu_1 < \mu_0$, the acceptance region of H_0 would be

$$t_{(\alpha;\,N-1)} \leq (\overline{X} - \mu_0)/S_{\overline{X}}.$$

When a difference is found to be statistically significant, the finest and most rigorous significance level (corresponding to the smallest α) at which that difference remains significant is usually stated, because this indicates at once all less rigorous levels at which the same difference would also be significant. In the case of the ten skeletons above, for instance, the statement $P < 0.001$ would be made because the reader would thus know that $P < 0.001 < 0.01 < 0.05$ and that the observed difference is worthy of attention at all usual levels from $\alpha = 0.05$ to $\alpha = 0.001$. Conversely, when an observed difference does not reach statistical significance, the least rigorous level (corresponding to the largest α) at which that difference remains non significant is usually stated, because this simultaneously indicates all more rigorous levels at which the same difference would also fail to reach statistical significance. Thus, when it is known that $P > 0.05$, it is also known that $P > 0.05 > 0.01 > 0.001$. As for the statement $0.01 < P < 0.05$, it would indicate that an observed difference may be considered as a provisional indication ($P < 0.05$) but not as a well-established conclusion ($P > 0.01$; section 9.4).

Section 9.7: confidence intervals of the mean of a population

The hypothesis testing methods outlined in the preceding section are useful when a hypothetical mean value is already known and must be confronted with a sample of data. When the mean value μ of a population is completely unknown, the determination of a *confidence interval* is more appropriate. The confidence interval is the most current kind of interval estimate (section 9.1).

In practice, when a confidence interval is determined, the mean $\overline{X}$ of the sample is known and the mean μ of the population is unknown. Theoretically, however, in order to set up the problem correctly, the assumption that the population mean μ is known and that the sample mean $\overline{X}$ is unknown must be made temporarily and the question must be framed initially as a case of hypothesis testing. Let us consider, for instance, the case where the mean $\overline{X}$ of a sample is going to be used to determine the confidence interval of the mean μ of a population and where, if a hypothesis were being tested, the alternative hypothesis would be two-sided:

$$H_0\colon\ \mu = \mu_0 \text{ and } H_1\colon\ \mu = \mu_1 \neq \mu_0 .$$

If the parametric standard deviation σ_0 is assumed to be known, the parametric standard deviation of the mean $\overline{X}$ of a sample of size N is $\sigma_0/\sqrt{N}$ and the probability that the mean $\overline{X}$ of the sample falls in the acceptance region is

$$P\,[\mu_0 + t_{(\alpha/2;\,\infty)}\,\sigma_0/\sqrt{N} \le \overline{X} \le \mu_0 + t_{(1-\alpha/2;\,\infty)}\,\sigma_0/\sqrt{N}] = (1-\alpha)\,.$$

The normal distribution is suitable here because the parametric standard deviation is assumed to be known. The preceding equation is correct provided the preliminary hypothesis H_0 is true, but that condition can be removed if μ_0 and σ_0 are replaced by μ and σ, for μ and σ without subscripts are defined theoretically as the mean and the standard deviation of the population from which the sample will be drawn in reality. Therefore, the new equation

$$P\,[\mu + t_{(\alpha/2;\,\infty)}\,\sigma/\sqrt{N} \le \overline{X} \le \mu + t_{(1-\alpha/2;\,\infty)}\,\sigma/\sqrt{N}] = (1-\alpha)\,,$$

is always true, whether the preliminary hypothesis is true or false. Because the parametric standard deviation σ is generally unknown in practice, it is replaced at this stage by its estimate S_X, and the normal distribution must then be replaced by Student's t distribution with $(N-1)$ degrees of freedom:

$$P[\mu + t_{(\alpha/2\,;\,N-1)}\, S_X/\sqrt{N} \leq \overline{X} \leq \mu + t_{(1-\alpha/2\,;\,N-1)}\, S_X/\sqrt{N}] = (1-\alpha)\,.$$

The double inequality of which the probability is expressed above may next be modified by a sequence of simple algebraic operations: the means μ and $\overline{X}$ are subtracted from the three members, the signs of the three members are changed and the inequality signs are reversed, the three members are rewritten by order of increasing magnitude, and the values $-\,t_{(\alpha/2;\,N-1)}$ and $-\,t_{(1-\alpha/2;\,N-1)}$ are replaced respectively by $t_{(1-\alpha/2;\,N-1)}$ and $t_{(\alpha/2;\,N-1)}$ thanks to the symmetry of Student's t distribution. This yields:

$$P[\overline{X} + t_{(\alpha/2\,;\,N-1)}\, S_X/\sqrt{N} \leq \mu \leq \overline{X} + t_{(1-\alpha/2\,;\,N-1)}\, S_X/\sqrt{N}] = (1-\alpha)\,.$$

As long as the sample has not yet been drawn from the population, the above expression corresponds to the probability that the random interval

$$[\overline{X} + t_{(\alpha/2\,;\,N-1)}\, S_X/\sqrt{N}\,,\quad \overline{X} + t_{(1-\alpha/2\,;\,N-1)}\, S_X/\sqrt{N}]$$

falls in such a way on the coordinate axis of the possible values of the parametric mean that it *contains* (or *covers* in statistical language) the true value μ of the mean of the sampled population. Once the sample has been drawn from the population, however, the probability concept becomes unsuitable because each of the two limits of the interval now has a fixed value and the parametric mean μ too (even though the latter is unknown). The *confidence interval,*

$$\overline{X} + t_{(\alpha/2\,;\,N-1)}\, S_X/\sqrt{N} \leq \mu \leq \overline{X} + t_{(1-\alpha/2\,;\,N-1)}\, S_X/\sqrt{N}\,,$$

of which the limits have been evaluated by this time, will therefore be said *to be likely to contain the mean μ of the population with a confidence coefficient* $(1-\alpha)$.

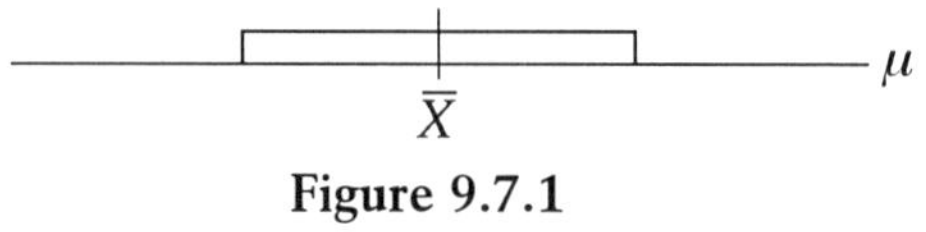

Figure 9.7.1
Graphical representation of a two-sided confidence interval

In short, this interval may also be said to be the *confidence interval of μ at the α significance level.* A confidence interval should usually be represented graphically as in figure 9.7.1, for this makes its meaning more concrete.

The determination of one-sided confidence intervals requires some care because, if it were done in a purely intuitive manner, it might well lead to erroneous results. Let us consider, for instance, the case where, if a hypothesis were being tested, the alternative hypothesis would be H_1: $\mu = \mu_1 < \mu_0$ (figure 9.7.2). The probability that the mean $\overline{X}$ of the sample (not yet drawn) would fall in the acceptance region would be:

$$P\,[\mu_0 + t_{(\alpha\,;\,\infty)}\, \sigma_0/\sqrt{N} \leq \overline{X}] = (1-\alpha)\,.$$

Carrying out the same kinds of operations as in the preceding case shows that

$$P[\,\mu \leq \overline{X} + t_{(1-\alpha;\,N-1)}\, S_X/\sqrt{N}\,] = (1-\alpha)\,.$$

Once the sample has been drawn, the one-sided confidence interval,

$$\mu \leq \overline{X} + t_{(1-\alpha;\,N-1)}\, S_X/\sqrt{N}\,,$$

is thus likely to contain the mean μ of the population with a confidence coefficient $(1-\alpha)$. The (upper) limit of this confidence interval may be interpreted as the largest likely value for the mean μ of the population.

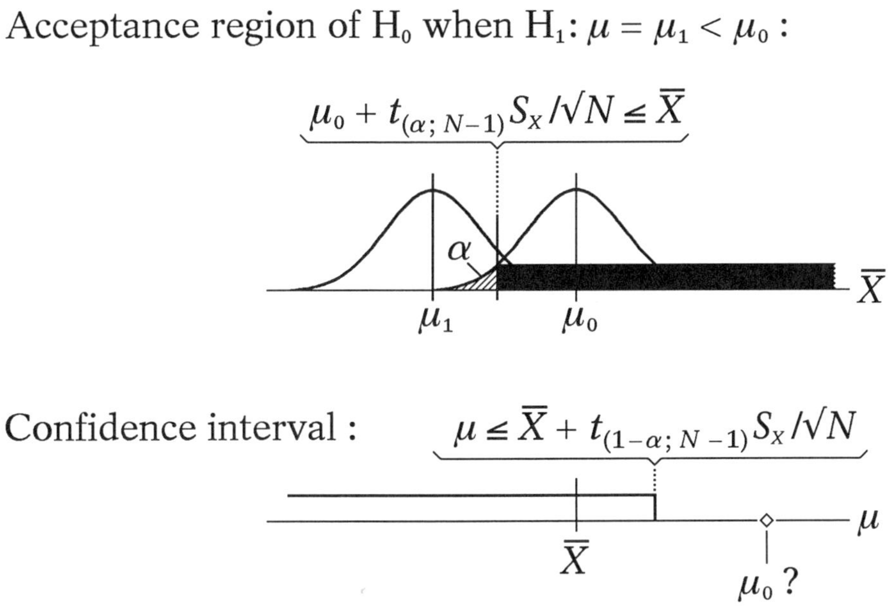

Figure 9.7.2
In the case where, if a hypothesis were tested, the alternative hypothesis would be H$_1$: $\mu = \mu_1 < \mu_0$, while the acceptance region would reach $+\infty$ toward the right, the confidence interval reaches $-\infty$ toward the left

It is disturbing at first to note in this case that, while the acceptance region ranges up to $+\infty$ toward the right, the confidence interval ranges down to $-\infty$ toward the left (figure 9.7.2). However, further reflection shows that this is clearly correct: for the sample to give evidence that it comes from a population of which the mean μ is lower than a specified value μ_0, the sample mean $\overline{X}$ must be so much lower than μ_0 that even the upper limit of the confidence interval, which is above $\overline{X}$, is below μ_0. Intuitively, it would seem natural to set up a confidence interval extending toward the right like the acceptance region but, with such an erroneous interval, possible indications (given by the sample mean $\overline{X}$) that the population mean μ is small could not be detected!

In the opposite case where, if a hypothesis were tested, the alternative hypothesis would be H_1: $\mu = \mu_1 > \mu_0$, the confidence interval would be

$$\overline{X} + t_{(\alpha;\, N-1)}\, S_X / \sqrt{N} \leq \mu \, .$$

The (lower) limit of this interval could be interpreted as the lowest likely value for μ .

Section 9.8: do the means of two populations differ?

When a single sample is in hand, the methods discussed up to this section can be used to confront a hypothetical mean with data or to determine the confidence interval of a theoretical mean μ. It will now be seen that a simple extension of those methods may be used to compare the means $\overline{X}_A$ and $\overline{X}_B$ of two samples which are assumed to come from the same normal statistical population (or from two identical normal populations **A** and **B**, which is mathematically equivalent), but are suspected of coming from populations having different means μ_A and μ_B.

When the individual values of a continuous random variate X follow a normal distribution $X \leftarrow \mathcal{N}(\mu, \sigma_X^2)$, the mean values $\overline{X}$ of samples of size N drawn from the same population also follow a normal distribution of which the mean is μ but the variance is $\sigma_{\overline{X}}^2 = \sigma_X^2 / N$ (section 9.5): $\overline{X} \leftarrow \mathcal{N}(\mu, \sigma_X^2/N)$. In this section, the process will be pushed one step further, and a new random variate Y will be considered which is equal to the difference $(\overline{X}_B - \overline{X}_A)$ between the mean values $\overline{X}_A$ and $\overline{X}_B$ of two samples coming from normal statistical populations of which the means are μ_A and μ_B respectively. In practice, the methods of this section are usually applied when the means $\overline{X}_A$ and $\overline{X}_B$ of a single pair of samples must be compared but, in theory, in order to set up the problem correctly, it must be assumed that there is a large (infinite?) number of pairs of samples, the two samples of each pair coming one from population **A** and the other from population **B**. Theoretically, there is thus a large number of differences such as $Y = (\overline{X}_B - \overline{X}_A)$, and those differences can be shown to follow also a normal distribution of which the mean is $\mu_Y = (\mu_B - \mu_A)$ and the variance is

$$\sigma_Y^2 = \sigma_{(\overline{X}_B - \overline{X}_A)}^2 = \sigma_{\overline{X}_A}^2 + \sigma_{\overline{X}_B}^2 = \sigma_X^2/N_A + \sigma_X^2/N_B = \sigma_X^2 (1/N_A + 1/N_B) \, .$$

Therefore, $$Y = (\overline{X}_B - \overline{X}_A) \leftarrow \mathcal{N}\big(\mu_B - \mu_A\, ,\, \sigma_X^2 (1/N_A + 1/N_B)\big) \, .$$

The preliminary hypothesis here is generally a *null hypothesis*:

$$H_0\colon\ \mu_Y = (\mu_B - \mu_A) = 0\, , \quad \text{that is} \quad H_0\colon\ \mu_A = \mu_B\, ,$$

and the alternative hypothesis usually belongs to one of the three kinds discussed in section 9.6 and illustrated in figure 9.6.1:

$$H_1\colon\ (\mu_B - \mu_A) \neq 0\, , \quad \text{that is} \quad H_1\colon\ \mu_A \neq \mu_B\, ,$$

$$H_1\colon\ (\mu_B - \mu_A) < 0\, , \quad \text{that is} \quad H_1\colon\ \mu_A > \mu_B\, ,$$

or $$H_1\colon\ (\mu_B - \mu_A) > 0\, , \quad \text{that is} \quad H_1\colon\ \mu_A < \mu_B\, .$$

In order to keep the problem of comparing means simple, the variances of the two populations are usually assumed to be equal, $\sigma_A^2 = \sigma_B^2 = \sigma^2$, and a pooled estimate of the variance (and of its square root the standard deviation) is obtained by adding up the sums of squared deviations (centered sums) of both samples before dividing them by the sum of their numbers of degrees of freedom: $(N_A - 1) + (N_B - 1) = (N_A + N_B - 2)$. If the assumption of variance equality is true, the criterion

$$t_{(N_A + N_B - 2)} = \frac{[(\overline{X}_B - \overline{X}_A) - (\mu_B - \mu_A)]}{S_{x\,\text{pooled}} \sqrt{(1/N_A) + (1/N_B)}} \, ,$$

$$\text{where } S_{x\,\text{pooled}} = \sqrt{\frac{\Sigma(X - \overline{X})_A^2 + \Sigma(X - \overline{X})_B^2}{N_A + N_B - 2}} \, ,$$

follows Student's t distribution with $(N_A + N_B - 2)$ degrees of freedom and that distribution can be used to test the preliminary hypothesis as in section 9.6, the nature of the alternative hypothesis being taken into account. It should be noted that, when the preliminary hypothesis H_0: $\mu_A = \mu_B$ is being tested, the numerator of the above criterion becomes simpler since the difference $(\mu_B - \mu_A)$ is replaced by 0. If the assumption of variance equality were not true, the above criterion would not follow Student's t distribution exactly, which could make inaccurate the significance level α (sections 9.4 and 9.6) selected for comparing means. The assumption that variances are equal, $\sigma_A^2 = \sigma_B^2$, is therefore a necessary condition for using Student's t distribution validly when testing the hypothesis that means are equal, H_0: $\mu_A = \mu_B$. If the assumption of variance equality is strongly suspected of being untrue, it can also be treated as a hypothesis and tested with the variance ratio (F) distribution, as indicated in section 11.2. However, the assumption of variance equality is usually not tested unless there are strong reasons for doing so, because the comparison of sample means through Student's t distribution can be shown to be a *robust* statistical method, which is little affected when its validity conditions are not completely fulfilled.

As an example of the utilization of Student's t distribution to test the hypothesis that the means of two statistical populations are equal, the mean survival times in minutes in dry air at 33.5°C of male and female drosophilae raised at 15°C are compared (table 9.8.1, next page). Since there are no *a priori* reasons to suspect that males would survive shorter than females, the alternative hypothesis is

$$H_1: (\mu_B - \mu_A) \neq 0 \, .$$

The criterion falls in the rejection region at significance level $\alpha = 0.001$, and the observed difference of survival time between males and females is strongly conclusive despite small sample sizes.

Table 9.8.1
Comparison of the means of two samples with Student's t distribution: survival times in minutes in dry air at 33.5°C of male and female drosophilae raised at 15°C (Smith, 1957)

Sample	N	$\overline{X}$	$\Sigma(X-\overline{X})^2$	S_X^2
A : Males	6	44.1667	470.833	94.1666
B : Females	6	67.5	237.5	47.5
	$t_{(10)} = +4.80196 > +4.5869 = t_{(0.9995;\,10)}$			

When the alternative hypothesis is one-sided, attention must be paid to the order of the sample means $\overline{X}_A$ et $\overline{X}_B$ in the numerator of Student's t. If H_0: $\mu_A = \mu_B$ and H_1: $\mu_A < \mu_B$ and if one uses the formula

$$t_{(N_A+N_B+2)} = [(\overline{X}_B - \overline{X}_A) - 0] \,/\, \{S_{X_{\text{pooled}}} \sqrt{[(1/N_A) + (1/N_B)]}\},$$

the difference $(\overline{X}_B - \overline{X}_A)$ then follows a distribution of which the mean is $(\mu_B - \mu_A) = 0$ if H_0 is true but $(\mu_B - \mu_A) > 0$ if H_1 is true (figure 9.8.1).

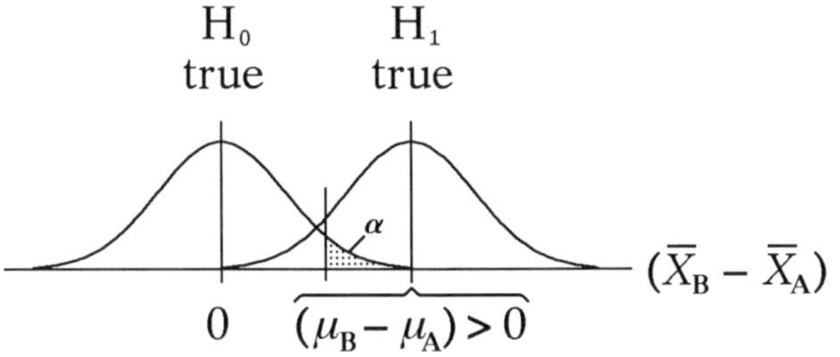

Figure 9.8.1
Acceptance and rejection regions of the preliminary hypothesis H_0: $(\mu_B - \mu_A) = 0$ in the case of a one-sided alternative hypothesis H_1: $(\mu_B - \mu_A) > 0$, expressed with respect to the difference $(\overline{X}_B - \overline{X}_A)$

Expressed with respect to the difference $(\overline{X}_B - \overline{X}_A)$, the rejection region of H_0 is thus located on the right, and the acceptance region of H_0 is

$$t \leq t_{(1-\alpha;\, N_A+N_B-2)}\,.$$

If, however, in the same context, one uses the formula

$$t_{(N_A+N_B+2)} = [(\overline{X}_A - \overline{X}_B) - 0] \,/\, \{S_{X_{\text{pooled}}} \sqrt{[(1/N_A) + (1/N_B)]}\},$$

the difference $(\overline{X}_A - \overline{X}_B)$ follows a distribution of which the mean is $(\mu_A - \mu_B) = 0$ if H_0 is true but $(\mu_A - \mu_B) < 0$ if H_1 is true. The situation illustrated in figure 9.8.1 is then reversed from left to right and the rejection region of H_0 is located on the left while the acceptance region of H_0 is

$$t_{(\alpha;\, N_A+N_B-2)} \leq t\,.$$

Section 9.9: a particularly favorable context: paired measurements

The method discussed in section 9.8 is suitable for an overall comparison of two independent samples, coming from two different populations **A** and **B**, and of which all individual observations are unrelated to each other. In such a case, the sizes N_A and N_B of the two samples are not necessarily equal. However, it happens frequently in biology that a comparison must be made between two groups of data **A** and **B** in which there is a special relationship between each observation of the first group (**A**) and one (and only one) of the observations of the second group (**B**) and vice versa. Such data, between which there is a one-to-one relationship, are called *paired data,* and the sizes of two samples of paired data are necessarily equal: $N_A = N_B$. The ten pairs of length measurements of left and right humeri given in table 1.2.1 are a good example of paired data. The special relationship between the measurements of each pair correspond here to the fact that they are the two humeri (left and right) of the same skeleton. While other procedures are possible, such as the use of ratios (section 9.6), such relationships may be taken into account by calculating the algebraical (signed) difference between the measurements of each pair and by replacing the two samples of measurements by a single sample of differences between paired measurements (table 9.9.1).

Table 9.9.1
Differences between paired length measurements in mm of left and right humeri of female human skeletons (from table 1.2.1)

Number (subscript) of skeleton	Left humerus X_A	Right humerus X_B	Difference $Y=(X_B-X_A)$
1	311	315	+4
2	302	306	+4
3	301	311	+10
4	322	333	+11
5	312	316	+4
6	285	292	+7
7	305	308	+3
8	310	318	+8
9	328	326	−2
10	304	309	+5

If each of the two variates X_A and X_B follows a normal distribution, the difference $Y=(X_B-X_A)$ too follows a normal distribution of which the parametric mean μ_Y and variance σ_Y^2 may be estimated by the mean $\overline{Y}$ and variance S_Y^2 of the sample of observed differences. If bilateral symmetry were perfect on the average, the mean length difference μ_Y between left and right humeri would be null: H_0: $\mu_Y=0$. This preliminary hypothesis (which is a true *null hypothesis* here as in section 9.8) may be tested by using the methods discussed in section 9.6 and calculating the criterion

$$t_{(N-1)} = (\overline{Y} - \mu_Y)/S_{\overline{Y}}, \text{ where } N = N_A = N_B .$$

In the case of the humerus lengths of table 9.9.1, if the preliminary hypothesis H_0: $\mu_Y = 0$ is tested against the alternative hypothesis H_1: $\mu_Y > 0$,

$$t_{(9)} = +5.4/1.1944 = +4.5210 > +4.2968 = t_{(0.999;\,9)}\,,$$

and the criterion falls in the rejection region at significance level $\alpha = 0.001$. The preliminary hypothesis H_0 must therefore be rejected and the parametric mean of paired length differences between left and right humeri is concluded not to be null. This result confirms the conclusion reached using ratios in section 9.6 and using the sign test in section 17.11.

An overall comparison of two samples of paired data based on the criterion discussed in section 9.8 would be inefficient because the relationship between the two measurements of each pair would be neglected. For instance, if the methods of section 9.8 were applied to the humerus length data, they would yield

$$t_{(18)} = +5.4/5.1687 = +1.0448 < +1.7341 = t_{(0.95\,;\,18)}\,.$$

The criterion would fall in the acceptance region at the level $\alpha = 0.05$ and the null hypothesis would be retained. The overall comparison would thus be much less sensitive here than the method of paired measurements. In cases where there are no *a priori* reasons to set up a null hypothesis and where what is desired is an interval estimate of the parametric mean difference, the confidence interval of μ_Y may be determined using the methods of section 9.7.

Pairing increases the sensitivity of a comparison because it reduces the influence of factors of variation other than the one at which the study is aimed. For the method of paired measurements to be worthwile, however, the special relationship between the measurements of each pair must be strong enough, since the number of degrees of freedom is cut by half when two samples of N measurements are replaced by a single sample of N differences. The strength of the tendency of the measurements of each pair to be similar can be evaluated by using the correlation coefficient (chapter 19). A particularly efficient graphical method consists in plotting paired data in a scatter diagram where a straight line represents the equality hypothesis (figure 1.2.2).

In spite of its simplicity, the method of paired measurements is very powerful and can be used not only in quantitative morphology but also in many other fields. In physiology, for instance, the two measurements of each pair may be taken on organisms which are as similar as possible genetically and treated in the same way, except that one is submitted to an experimental treatment while the other one is used as a *control.* The control organism thus gives as good an idea as possible of what the experimental organism would have been like if it had not been submitted to the experimental treatment. True (monozygotic) twins are sometimes used as paired organisms because of their practically perfect genetic similarity. Another possible design consists in taking two paired measurements successively on the same organism, the control measurement before and the experimental measurement after the treatment.

In ecology, the method of paired measurements can be used to compare the abundance or the activity of a particular species in two different but neighboring habitats. For instance, the number of small mammals captured could be compared in a wood **A** and in an adjacent prairie **B** during several successive nights, the number of captures of a same pair being observed on the same night. Because the nocturnal activity and the chances of capture of small mammals are strongly affected by climatic conditions (wind, rain, temperature, lighting, etc.), the method of paired measurements would reduce the influence of climatic conditions and bring out the influence of the habitat difference.

Section 9.10: comparing an individual observation and a sample mean

Perhaps more often in biology than in other fields, the comparison of an individual observation with a group of observations may have a considerable interest. In taxonomy, for instance, a specimen found at the boundary of the geographical range of one species may be suspected of being a natural hybrid with a neighboring species. In physiology, an isolated individual may have a stronger or a weaker reaction than the members of a group because of a *"social effect"*. From a statistical point of view, the comparison of an individual observation with a group may be considered as a particular case of the comparison of two groups, one of the two groups being reduced to a single observation. The criterion used in section 9.8 may then be simplified as follows:

$N_A = N$ is the size of the sample which contains several observations,

$N_B = 1$ is the size of the sample reduced to a single observation,

$\overline{X}_A = \overline{X}$ is the mean of the first sample (of size N),

$\overline{X}_B = X$ is the single observation making up the second sample,

$\Sigma(X-\overline{X})^2_A = \Sigma(X-\overline{X})^2$ is the sum of the squared deviations of the first sample,

and $\Sigma(X-\overline{X})^2_B = 0$ is the (null) sum of the squared deviations of the second sample.

The criterion used to compare the means of two samples in section 9.8 then takes the following form:

$$t_{(N-1)} = \frac{[(X-\overline{X})-(\mu_B-\mu_A)]}{S_X}\sqrt{\frac{N}{N+1}}\ , \quad \text{where} \quad S_X = \sqrt{\frac{\Sigma(X-\overline{X})^2}{N-1}}\ .$$

During the test of the preliminary hypothesis H_0: $\mu_A = \mu_B$, according to which the individual observation X would come from the same population as sample **A** (or from a population having an equal mean), the above formula becomes even simpler since $(\mu_B - \mu_A) = 0$. The limits of the acceptance and rejection regions of H_0 are determined by taking into account the nature of the alternative hypothesis H_1 as in section 9.8, except that the number of degrees of freedom is $(N_A - 1) = (N-1)$ here instead of $(N_A + N_B - 2)$.

Section 9.11: prediction intervals and variation intervals

When an individual observation X and a sample having the mean $\overline{X}$ are available but the parametric means μ_A and μ_B are unknown, the significance test discussed in section 9.10 could be inverted in order to get a confidence interval for the difference $(\mu_B - \mu_A)$. For instance, if the context implied a two-sided alternative hypothesis, the confidence interval would be

$$(X-\overline{X}) + t_{(\alpha/2;\, N-1)} S_X \sqrt{[(N+1)/N]} \leq (\mu_B - \mu_A) \leq (X-\overline{X}) + t_{(1-\alpha/2;\, N-1)} S_X \sqrt{[(N+1)/N]}.$$

Multiplying the standard deviation S_X of individual observations by the factor $\sqrt{[(N+1)/N]}$ yields an estimate $S_{(X-\overline{X})}$ of the standard deviation of the difference $(X-\overline{X})$ between an individual observation X and a sample mean $\overline{X}$, since $\sigma^2_{(X-\overline{X})} = \sigma^2_X + \sigma^2_{\overline{X}} = \sigma^2_X + \sigma^2_X/N = \sigma^2_X\,[1+(1/N)] = \sigma^2_X\,[(N+1)/N]$. While the above confidence interval is seldom used in biology, it can be transformed into another kind of interval which is often useful. If the individual observation X has not yet been obtained but is going to come from the same population as the sample of which the mean is $\overline{X}$, the above double inequality may be rewritten as

$$(X-\overline{X}) + t_{(\alpha/2;\,N-1)}\,S_X\sqrt{[(N+1)/N]} \;\le\; 0 \;\le\; (X-\overline{X}) + t_{(1-\alpha/2;\,N-1)}\,S_X\sqrt{[(N+1)/N]}\,,$$

and, after terms are rearranged,

$$\overline{X} + t_{(\alpha/2;\,N-1)}\,S_X\sqrt{[(N+1)/N]} \;\le\; X \;\le\; \overline{X} + t_{(1-\alpha/2;\,N-1)}\,S_X\sqrt{[(N+1)/N]}\,.$$

The latter interval indicates between which limits a new observation X is likely to fall with a confidence coefficient $(1-\alpha)$ if it comes from the same normal statistical population as the sample of size N of which the mean is $\overline{X}$. This kind of interval is called a *prediction interval* in statistical language. The breadth of a prediction interval is approximately proportional to the standard deviation S_X of individual observations and is generally much greater than the breadth of the confidence interval of the mean (section 9.7) based on the same sample, because the latter is proportional to the standard deviation of the mean $S_{\overline{X}} = S_X/\sqrt{N}$. Prediction intervals are used fairly often in animal and plant taxonomy and paleontology because they reflect approximately the extent of individual variation. In the case of small samples, however, even though the prediction interval remains suitable for statistical inference, it becomes much broader than the observed or estimated extent of individual variation because, in addition to the latter, it is affected by the uncertainty of sample estimates $\overline{X}$ and S_X.

When what is needed is an interval which reflects strictly the extent of individual variation, the normal distribution should be used and the sample estimates $\overline{X}$ and S_X should be substituted for the population parameters μ and σ_X:

$$\overline{X} + t_{(\alpha/2;\,\infty)}\,S_X \;\le\; X \;\le\; \overline{X} + t_{(1-\alpha/2;\,\infty)}\,S_X\,.$$

This kind of interval may be called a *variation interval*.

Variation intervals are equivalent to the approximate procedures which were formerly used to determine prediction intervals in the case of large samples. Jolicoeur and Heusner (1986) have proposed that variation intervals be used to represent individual variation above and below growth curves, but the approach could be applied to all cases where what is needed is a description of the observed or estimated extent of individual variation rather than a prediction of the region where a new observation will fall. Other interval estimates known as tolerance intervals are used in industrial quality control.

Chapter 10
Hypothesis testing and confidence intervals concerning one variance

Section 10.1: introduction

As indicated in section 7.1, when a set of N completely independent observations $X_1, X_2, X_3, \dots X_N$ follows a normal distribution $\mathcal{N}(\mu_X, \sigma_X^2)$, the sum of the squares of the standardized variates $(X_h - \mu_X)/\sigma_X$ follows a χ^2 distribution with N *degrees of freedom*:

$$\sum_{h=1}^{N} (X_h - \mu_X)^2/\sigma_X^2 \leftarrow \chi^2_{(N)} \ .$$

However, when the parametric mean μ is unknown and is replaced by the sample mean $\overline{X}$, the number of degrees of freedom ν is decreased from N to $(N-1)$ and

$$\sum_{h=1}^{N} (X_h - \overline{X})^2/\sigma_X^2 \leftarrow \chi^2_{(\nu\,=\,N-1)} \ .$$

The cumulative probability (section 3.5) of a particular value of $\Sigma(X_h - \overline{X})^2/\sigma_X^2$ is therefore (section 7.3 and figure 10.1.1)

$$\text{Probability}\,[\,\chi^2_{(N-1)} = \Sigma(X_h - \overline{X})^2/\sigma_X^2 < \chi^2_{(P;\,N-1)}\,] = P.$$

The preceding equation may be used to test hypotheses or to determine confidence intervals concerning the variance σ_X^2 or the standard deviation σ_X of the population from which the sample has been drawn. Unlike most other uses of the χ^2 distribution, which are approximate, the methods of the present chapter are exact if the observations $X_1, X_2, X_3, \dots X_N$ really follow a normal distribution.

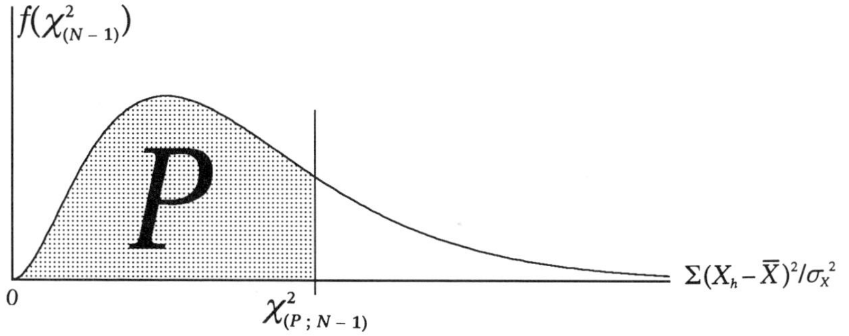

Figure 10.1.1
Cumulative probability of a particular value of $\Sigma\,(X_h - \overline{X})^2/\sigma_X^2$

Section 10.2: hypotheses concerning the parametric variance σ^2

In order to test the preliminary hypothesis H_0: $\sigma^2 = \sigma_0^2$, the first step consists in the calculation of the value of the criterion

$$\chi^2_{(N-1)} = \Sigma(X_h - \overline{X})^2/\sigma_0^2 = (N-1)\ S_X^2/\sigma_0^2.$$

This criterion follows the χ^2 distribution with $(N-1)$ degrees of freedom exactly when the preliminary hypothesis is true. The acceptance and rejection regions of the preliminary hypothesis are illustrated in figure 10.2.1 for the three current kinds of alternative hypotheses.

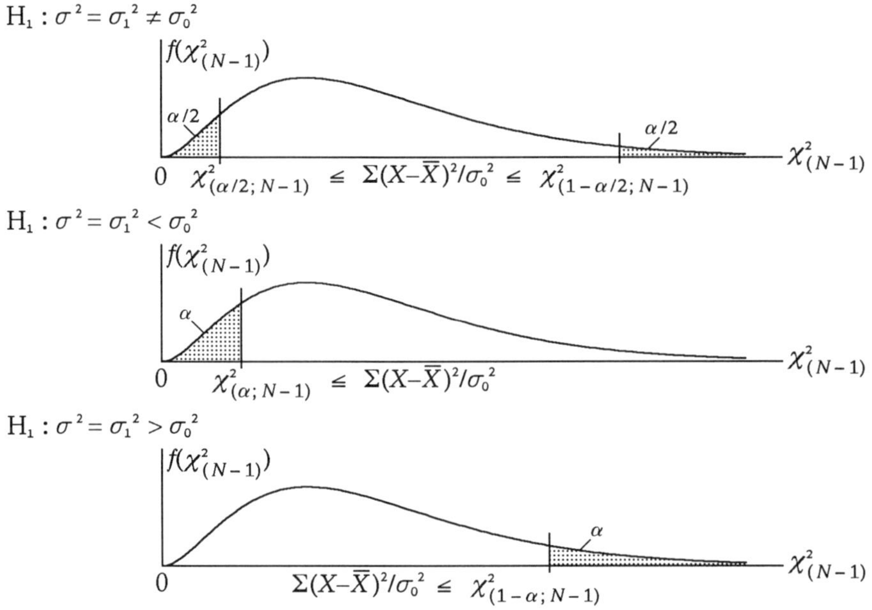

Figure 10.2.1
Acceptance and rejection regions of the preliminary hypothesis H_0: $\sigma^2 = \sigma_0^2$ for the three current kinds of alternative hypotheses

The two-sided alternative hypothesis, H_1: $\sigma^2 = \sigma_1^2 \neq \sigma_0^2$, is appropriate when there are no more reasons to expect the variance σ^2 to be large than to be small. The alternative hypothesis H_1: $\sigma^2 = \sigma_1^2 < \sigma_0^2$ is appropriate if the variance σ^2 is suspected of being small, for instance for a quantitative character in a closely related *(inbred)* group of individuals. Finally, the alternative hypothesis H_1: $\sigma^2 = \sigma_1^2 > \sigma_0^2$ is appropriate if the variance σ^2 is suspected of being large, as in a genetically heterogeneous group, for instance the F_2 generation resulting from a cross of two very different races or species. In practice, hypotheses are seldom tested about the variance σ^2 in biology because theoretical reasons of expecting a particular numerical value σ_0^2 are usually not available. Therefore, it is more often useful to determine confidence limits for σ^2 (section 10.3).

Section 10.3: confidence intervals of the parametric variance σ^2

In practice, when a confidence interval of the variance σ_X^2 is needed, the sample variance S_X^2 is known and the population variance σ_X^2 is unknown. Theoretically, however, in order to set up the problem correctly, the assumption must temporarily be made that σ_X^2 is known and S_X^2 is unknown and the question must initially be considered as a case of hypothesis testing, as for the mean (section 9.7). Consider first the case where, if a hypothesis were to be tested, the alternative hypothesis would be two-sided: H_1: $\sigma^2 = \sigma_1^2 \neq \sigma_0^2$. The probability that the sample criterion (assumed to be still unknown) falls in the acceptance region of H_0 would be (figure 10.2.1, top):

$$P[\chi^2_{(\alpha/2;\, N-1)} \leq \Sigma(X-\overline{X})^2/\sigma_0^2 \leq \chi^2_{(1-\alpha/2;\, N-1)}] = (1-\alpha)\,.$$

The preceding equation is valid provided the preliminary hypothesis H_0: $\sigma^2 = \sigma_0^2$ is true, but this condition may be lifted if σ_0^2 is replaced by σ^2 without subscript, for the latter is defined as the variance of the population from which the sample will come in reality:

$$P[\chi^2_{(\alpha/2;\, N-1)} \leq \Sigma(X-\overline{X})^2/\sigma^2 \leq \chi^2_{(1-\alpha/2;\, N-1)}] = (1-\alpha)\,.$$

This new equation is always valid, whether the preliminary hypothesis H_0 is true or not, and it may be reexpressed as follows if its three members are inverted, the inequality signs are reversed, etc.:

$$P[\Sigma(X-\overline{X})^2/\chi^2_{(1-\alpha/2;\, N-1)} \leq \sigma^2 \leq \Sigma(X-\overline{X})^2/\chi^2_{(\alpha/2;\, N-1)}] = (1-\alpha)\,.$$

As long as the sample has not yet been drawn, the above equation gives the probability that the random interval

$$[\Sigma(X-\overline{X})^2/\chi^2_{(1-\alpha/2;\, N-1)}\,,\quad \Sigma(X-\overline{X})^2/\chi^2_{(\alpha/2;\, N-1)}]$$

will contain *(cover)* the value of the parametric variance σ^2. Once the sample has been drawn, the above interval, of which the limits are determined by now, is the *confidence interval of* σ^2 with a confidence coefficient $(1-\alpha)$. This confidence interval is represented graphically in figure 10.3.1. Unlike the two-sided confidence interval of the mean, which was symmetrical with respect to $\overline{X}$ (figure 9.7.1), the two-sided confidence interval of the variance is slightly asymmetrical with respect to S_X^2 : this follows from the skewness of the χ^2 distribution.

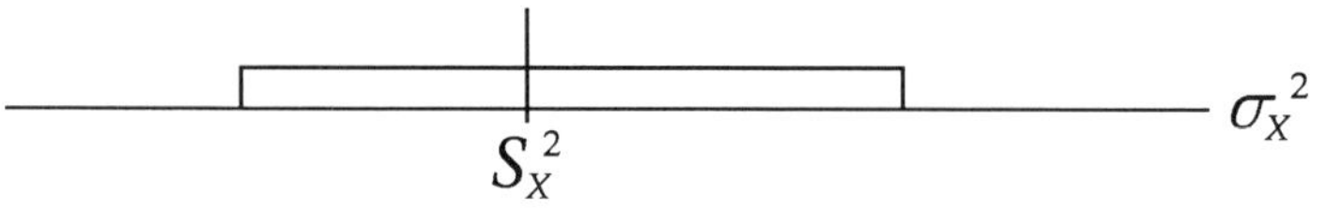

Figure 10.3.1
The two-sided confidence interval of σ_X^2 is asymmetrical with respect to S_X^2

In the case where the alternative hypothesis is H_1: $\sigma^2 = \sigma_1^2 < \sigma_0^2$, the probability that the criterion falls in the acceptance region of H_0 is

$$P[\chi^2_{(\alpha;\,N-1)} \le \Sigma(X-\overline{X})^2/\sigma_0^2 \le +\infty] = (1-\alpha)\,,$$

that is

$$P[0 \le \sigma^2 \le \Sigma(X-\overline{X})^2/\chi^2_{(\alpha;\,N-1)}] = (1-\alpha)\,,$$

and, unlike the one-sided acceptance region which ranged up to $+\infty$ toward the right, this one-sided confidence interval ranges down to 0 toward the left. The (upper) limit of this confidence interval is the largest numerical value which is likely for the parametric variance of the population.

Finally, in the case where the alternative hypothesis is H_1: $\sigma^2 = \sigma_1^2 > \sigma_0^2$, the acceptance region of H_0 is

$$0 \le \Sigma(X-\overline{X})^2/\sigma_0^2 \le \chi^2_{(1-\alpha;\,N-1)}$$

and the confidence interval is

$$\Sigma(X-\overline{X})^2/\chi^2_{(1-\alpha;\,N-1)} \le \sigma^2 \le +\infty\,.$$

Section 10.4: hypotheses and confidence intervals concerning σ

Hypotheses can be tested or confidence intervals can be determined concerning the parametric standard deviation σ simply by taking into account the fact that the latter is equal to the square root of the variance σ^2. For instance, if the preliminary hypothesis H_0: $\sigma = \sigma_0$ must be tested, it can be reexpressed with respect to the variance: H_0: $\sigma^2 = \sigma_0^2$. Conversely, if a confidence interval is required for the parametric standard deviation σ, the square roots of the confidence limits of the variance σ^2 are extracted. Thus, while the two-sided confidence interval of the variance is

$$\Sigma(X-\overline{X})^2/\chi^2_{(1-\alpha/2;\,N-1)} \le \sigma^2 \le \Sigma(X-\overline{X})^2/\chi^2_{(\alpha/2;\,N-1)}\,,$$

the corresponding confidence interval of the standard deviation is

$$\sqrt{[\,\Sigma(X-\overline{X})^2/\chi^2_{(1-\alpha/2;\,N-1)}\,]} \le \sigma \le \sqrt{[\,\Sigma(X-\overline{X})^2/\chi^2_{(\alpha/2;\,N-1)}\,]}\,.$$

Section 10.5: an example from anthropometry

A sample of height measurements in cm of 20 Pygmy men of the Éfé tribe extracted from the data of Schebesta and Lebzelter (1933) yields the following values: $\Sigma(X-\overline{X})^2 = 702.80$ and $S_X = 6.08190$. The two-sided confidence interval of the standard deviation at significance level $\alpha = 0.01$ is

$$\sqrt{(702.80/38.582)} = 4.26798 \text{ cm} \le \sigma \le 10.1336 \text{ cm} = \sqrt{(702.80/6.8440)}\,.$$

The parametric standard deviation of the statistical population from which the sample has been drawn is thus likely to be between approximately 4.3 and 10.1 cm.

Chapter 11
Hypothesis testing and confidence intervals concerning a variance ratio

Section 11.1: introduction

In this chapter, the distribution of the variance ratio $F = S_1^2/S_2^2$, already mentioned in chapter 8, will be used to test hypotheses or to determine confidence intervals concerning the ratio of the parametric variances σ_1^2 and σ_2^2 of two normal statistical populations. The distribution of the variance ratio can also be used to do an overall comparison of the means or of the estimation (regression) lines of several samples, but those other uses will be covered in later chapters (chapters 12, 21, 26, 27, 28, etc.).

Section 11.2: testing the preliminary hypothesis H_0: $\sigma_1^2/\sigma_2^2 = 1$

In the case of a variance ratio σ_1^2/σ_2^2 as in those of the mean μ (chapter 9) and of the variance σ^2 (chapter 10), the locations of the limits of the acceptance and rejection regions of the preliminary hypothesis H_0 depend on the nature of the alternative hypothesis. Besides the two-sided alternative hypothesis H_1: $\sigma_1^2/\sigma_2^2 \neq 1$, one-sided alternative hypotheses like H_1: $\sigma_1^2/\sigma_2^2 < 1$ or H_1: $\sigma_1^2/\sigma_2^2 > 1$ are occasionally considered, depending on the biological context. In practice, however, the case of the two-sided alternative hypothesis H_1: $\sigma_1^2/\sigma_2^2 \neq 1$ has a particular technical importance because the parametric variances σ_1^2 and σ_2^2 must be assumed to be equal for Student's t distribution to be valid in the comparison of two means (section 9.8), and that assumption is unfulfilled if $\sigma_1^2/\sigma_2^2 < 1$ as well as if $\sigma_1^2/\sigma_2^2 > 1$. Figure 11.2.1 (on next page) illustrates two equivalent manners of testing the preliminary hypothesis H_0: $\sigma_1^2/\sigma_2^2 = 1$ when the alternative hypothesis is H_1: $\sigma_1^2/\sigma_2^2 \neq 1$. If the preliminary hypothesis is true, the probability that the ratio of observed variances, S_1^2/S_2^2, falls in the acceptance region (figure 11.2.1, top) is

$$P\,[\,F_{(\alpha/2\,;\,\nu_1,\,\nu_2)} \leq S_1^2/S_2^2 \leq F_{(1-\alpha/2\,;\,\nu_1,\,\nu_2)}\,] = (1-\alpha)\,,$$

where $\nu_1 = (N_1 - 1)$ and $\nu_2 = (N_2 - 1)$. However, if the criterion S_1^2/S_2^2 is inverted, one should also have (figure 11.2.1, bottom)

$$P\,[\,F_{(\alpha/2\,;\,\nu_2,\,\nu_1)} \leq S_2^2/S_1^2 \leq F_{(1-\alpha/2\,;\,\nu_2,\,\nu_1)}\,] = (1-\alpha)\,.$$

It may easily be shown that the two above probabilities are in fact equal by inverting the three members of the double inequality

$$F_{(\alpha/2\,;\,\nu_2,\,\nu_1)} \leq S_2^2/S_1^2 \leq F_{(1-\alpha/2\,;\,\nu_2,\,\nu_1)}\,,$$

by reversing inequality signs, and rewriting the three members in order of increasing magnitude, which yields

$$1/F_{(1-\alpha/2\,;\,\nu_2,\,\nu_1)} \leq S_1^2/S_2^2 \leq 1/F_{(\alpha/2\,;\,\nu_2,\,\nu_1)}\,.$$

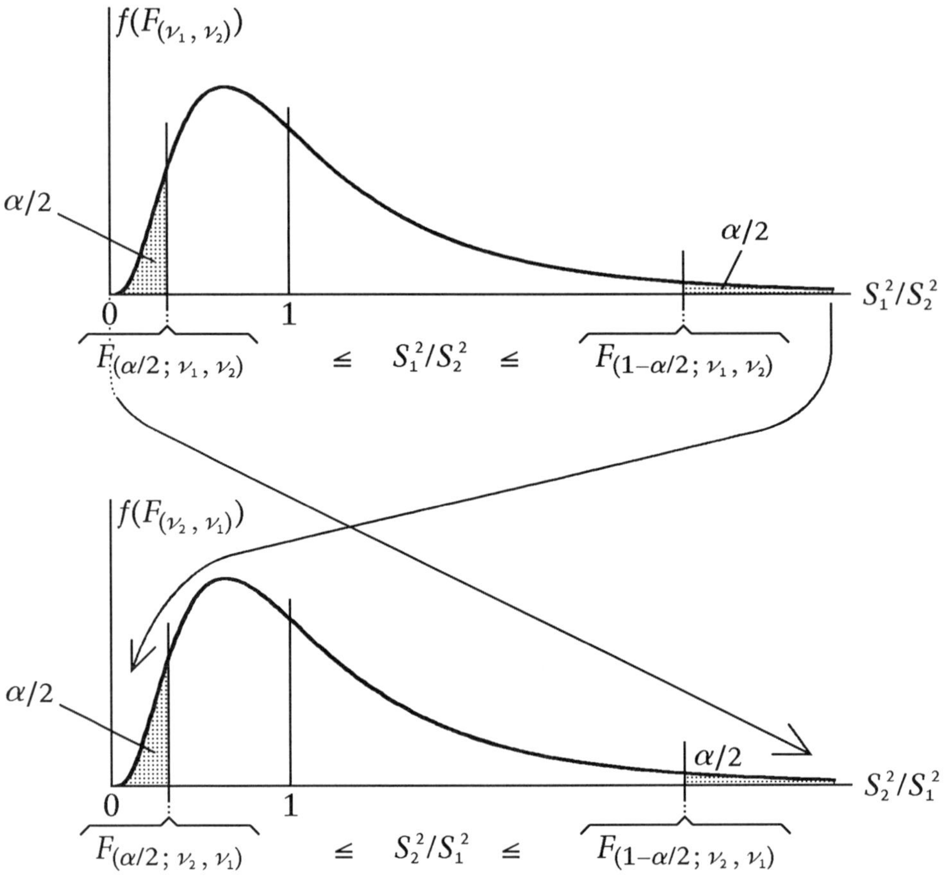

Figure 11.2.1
Two equivalent manners of testing the preliminary hypothesis H_0: $\sigma_1^2/\sigma_2^2 = 1$ when the alternative hypothesis is H_1: $\sigma_1^2/\sigma_2^2 \neq 1$

The fact that $F_{(\alpha/2;\, \nu_1,\, \nu_2)} = 1/F_{(1-\alpha/2;\, \nu_2,\, \nu_1)}$ and $F_{(1-\alpha/2;\, \nu_1,\, \nu_2)} = 1/F_{(\alpha/2;\, \nu_2,\, \nu_1)}$ follows from the existence of the relationship $F_{(1-P;\, \nu_2,\, \nu_1)} = 1/F_{(P;\, \nu_1,\, \nu_2)}$ (section 8.3). Geometrically, the lower diagram of figure 11.2.1 may be derived from the upper diagram by interchanging the subscripts 1 and 2 and inverting everything from left to right and vice versa, the origin 0 of the upper diagram corresponding to $+\infty$ in the lower diagram and the origin 0 of the lower diagram corresponding to $+\infty$ in the upper diagram. Whether S_1^2/S_2^2 (figure 11.2.1, top) or S_2^2/S_1^2 (figure 11.2.1, bottom) is used, the F distribution thus gives consistent results (fortunately!).

In principle, either the upper or the lower procedure illustrated in figure 11.2.1 could thus be used to test the hypothesis that the parametric variances are equal (that is to say that their ratio is equal to unity). In practice, however, both of these methods have the drawback that they require two F values to be looked up in the variance ratio table.

A third method, which requires only one F value, is therefore often used in practice: the largest observed variance (let us say S_i^2) is conventionally placed in the numerator of the variance ratio and the acceptance region of H_0 is defined as

$$1 \leq S_i^2 / S_j^2 \leq F_{(1-\alpha/2;\ \nu_i,\ \nu_j)} \ .$$

This is equivalent to using the right side of the upper diagram in figure 11.2.1 when $S_1^2 > S_2^2$ and the right side of the lower diagram (which corresponds to the left side of the upper diagram) when $S_1^2 < S_2^2$. This explanation is important for a clear understanding of the reasons for which an F value possessing a cumulative probability $(1-\alpha/2)$ is used here instead of $(1-\alpha)$ even though there seems superficially to be a rejection region only on the right. This misleading appearance is due to the fact that placing the largest observed variance conventionally in the numerator flips to the right the second rejection region which would otherwise be on the left. In spite of appearances, the alternative hypothesis is therefore truly two-sided in this case. A different situation will be found in the analysis of variance, where the alternative hypothesis will really be one-sided and where the *between-groups* estimate of variance will always be placed in the numerator of the variance ratio, even when it is smaller than the *within-groups* estimate (section 12.6).

In the case of the survival times of male and female drosophilae (table 9.8.1), for instance,

$$S_A^2 / S_B^2 = S_1^2 / S_2^2 = F_{(5,\ 5)} = 1.98246 < 7.1464 = F_{(0.975;\ 5,\ 5)} \ ,$$

and the ratio of the largest observed variance (of males) over the smallest variance (of females) falls in the acceptance region of the hypothesis that parametric variances are equal even at the $\alpha = 0.05$ significance level. Using Student's t distribution to compare means (section 9.8) was therefore justifiable.

Section 11.3: testing the hypothesis that $\sigma_1^2/\sigma_2^2 = \Phi$

The variance ratio distribution may also be used even when the parametric variances are not equal. Assume for instance that, because of theoretical reasons or of other previous knowledge, the variance ratio σ_1^2/σ_2^2 is suspected of being equal not to unity but to another positive number denoted by Φ (the upper case Greek letter *phi*, which corresponds to the Roman F). Even though the usual variance ratio table is based on the assumption that $\sigma_1^2/\sigma_2^2 = 1$, it can still be used here if the F values drawn from that table are multiplied by Φ or if the ratio of observed variances S_1^2/S_2^2 is divided by Φ. Thus, if the preliminary hypothesis H_0: $\sigma_1^2/\sigma_2^2 = \Phi$ must be tested against the alternative hypothesis H_1: $\sigma_1^2/\sigma_2^2 \neq \Phi$, the acceptance region of H_0 is

$$(\sigma_1^2/\sigma_2^2)\, F_{(\alpha/2;\ \nu_1,\ \nu_2)} \leq S_1^2/S_2^2 \leq (\sigma_1^2/\sigma_2^2)\, F_{(1-\alpha/2;\ \nu_1,\ \nu_2)} \ ,$$

that is

$$F_{(\alpha/2;\ \nu_1,\ \nu_2)} \leq (S_1^2/S_2^2)\,/(\sigma_1^2/\sigma_2^2) \leq F_{(1-\alpha/2;\ \nu_1,\ \nu_2)} \ ,$$

where the ratio σ_1^2/σ_2^2 is given the value Φ which it possesses according to H_0.

Section 11.4: confidence intervals of the ratio σ_1^2/σ_2^2

The double inequality of the preceding section may also be used to obtain a confidence interval for the ratio of parametric variances, σ_1^2/σ_2^2, when the latter is unknown. The parametric ratio σ_1^2/σ_2^2 is placed alone in the central member while the left and right members involve only the observed variance ratio S_1^2/S_2^2 as well as the F values drawn from the table, which yields

$$(S_1^2/S_2^2)/F_{(1-\alpha/2;\ \nu_1,\ \nu_2)} \leq (\sigma_1^2/\sigma_2^2) \leq (S_1^2/S_2^2)/F_{(\alpha/2;\ \nu_1,\ \nu_2)} ,$$

or, equivalently,

$$(S_1^2/S_2^2)\ F_{(\alpha/2;\ \nu_2,\ \nu_1)} \leq (\sigma_1^2/\sigma_2^2) \leq (S_1^2/S_2^2)\ F_{(1-\alpha/2;\ \nu_2,\ \nu_1)} .$$

It must be carefully noted that, in the second (and last) inequality above, the first number of degrees of freedom (ν_2) of the F values drawn from the table corresponds to the denominator instead of the numerator of the ratio (S_1^2/S_2^2) of observed variances. This is the reverse of what one usually has in significance tests.

Section 11.5: hypotheses and confidence intervals for σ_1/σ_2

As in chapter 10 (section 10.4), if a hypothesis must be tested concerning the ratio of parametric standard deviations, σ_1/σ_2, that hypothesis can be reexpressed with respect to the variance ratio by squaring the hypothetical value of the ratio of standard deviations. Conversely, a confidence interval may be obtained for the ratio of parametric standard deviations by extracting the square roots of the confidence limits of the corresponding variance ratio.

Section 11.6: an example from plant genetics

The mean and standard deviation in cm of the height of maize plants obtained by crossing the short *Tom Thumb* and the tall *Missouri Dent* varieties were extracted from the work of earlier workers by the American geneticist Sewall Wright (1968, page 376). Values concerning the first (F_1) and second (F_2) hybrid generations are reproduced in table 11.6.1.

Table 11.6.1
Mean and standard deviation of the height of hybrid maize plants according to data compiled by Wright (1968)

Generation	Number of plants (sample size N)	Mean height (cm)	Standard deviation (cm)
F_1	20	175	11.18
F_2	223	159	26.41

Theoretically, F_2 hybrids are expected to be more variable than F_1 hybrids because of the segregation phenomenon. The two-sided confidence interval of the ratio of standard deviations (σ_2/σ_1) at significance level $\alpha = 0.001$ is

$$(26.41/11.18)/\sqrt{3.9719} = 1.1853 \leq \sigma_2/\sigma_1 \leq 3.7955 = (26.41/11.18)/\sqrt{0.38736},$$

which confirms the greater variability of F_2 individuals.

Chapter 12
The analysis of variance or "ANOVA" (one-way, type I)

Section 12.1: introduction

If the means of several groups were compared pairwise using Student's t (section 9.8), the number of comparisons would rapidly become larger than the number of groups (figure 12.1.1). Besides breaking up and drawing out the analysis, this would also increase the probability of rejecting erroneously at least once the hypothesis that parametric means are equal. In fact, if ten statistical conclusions are drawn at significance level α, the probability α that the preliminary hypothesis is rejected at least once when it is true is multiplied by ten. If ten conclusions are drawn at level $\alpha = 0.05$, there is thus a probability of 1/2 that the preliminary hypothesis is rejected erroneously at least once. Even though it is possible to compensate for such a risk (section 26.2), it is often desirable to carry out a single overall comparison of the means of several samples.

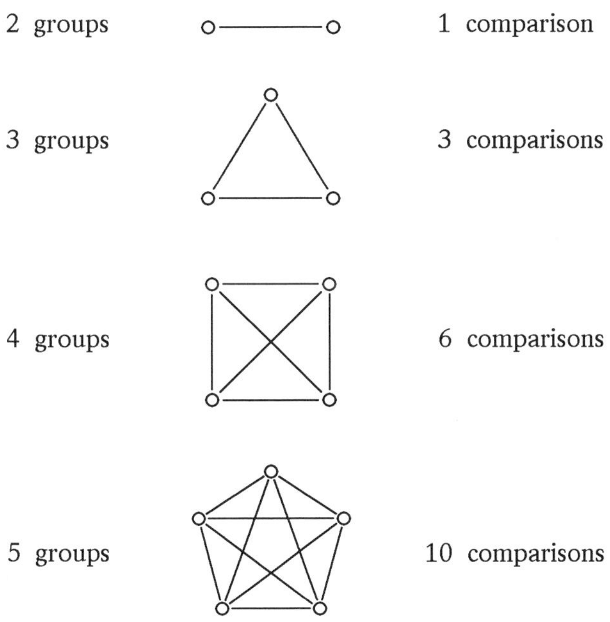

Figure 12.1.1
Increase in the number of two-by-two comparisons according to the number of groups

The method for doing such a comparison is called the *analysis of variance* (also designated by the acronym ANOVA). The present chapter deals with the simplest kind of analysis of variance, in which the groups to be compared differ in only one way, such as taxonomy, rather than with respect to several criteria simultaneously, such as taxonomy and geographical origin (chapter 28). Moreover, the present discussion is limited to type I analysis of variance, where information is sought about population means and their differences. Type II analysis of variance, where the relative importance of random variation at several levels is investigated, will be considered in chapter 27.

Section 12.2: a notation with double subscripts

Each individual observation X_{gh} is denoted by the letter X with a double subscript of which the first one, g, indicates the group of origin while the second one, h, is the order number of the observation within the sample. Because the data of a one-way analysis of variance may be arranged in a table where each group is placed in its own column, the total number of groups of data is usually denoted by C (the first letter of the word *column*). The group subscript g (the first letter of the word *group*) thus varies from 1 to C. The parametric means of the C populations being studied are μ_1, ... μ_g, ... μ_C and the preliminary hypothesis is

$$H_0: \ \mu_1 = \ldots = \mu_g = \ldots = \mu_C ,$$

while the alternative hypothesis is

$$H_1: \ \mu_i \neq \mu_j .$$

It should be noted that, while the preliminary hypothesis H_0 indicates that all population means are equal to each other, the alternative hypothesis H_1 only states that one of the means (of the ith group) differs from another one (of the jth group). In fact, the alternative hypothesis is false as soon as one of the C means differs from the others, even if all other means are equal to each other. The usual notation for the quantities relative to the C samples is given in table 12.2.1. The quantities pertaining to the aggregate of all data are denoted by the same kinds of symbols as those of each sample, except that the group subscript is replaced by a dot (last column of table 12.2.1). Those quantities are obtained by the formulae given at the top of the next page (the notation is occasionally simplified when there is no ambiguity).

Table 12.2.1
Notation for the quantities relative to the C samples and to the aggregate of all data

Quantity	Sample					All data
	1	...	g	...	C	
Sample size	N_1	...	N_g	...	N_C	$N.$
Sum of observations	$(\Sigma X)_1$	...	$(\Sigma X)_g$	...	$(\Sigma X)_C$	$(\Sigma X).$
Sum of squares of observations	$(\Sigma X^2)_1$	...	$(\Sigma X^2)_g$	...	$(\Sigma X^2)_C$	$(\Sigma X^2).$
Centering term	$(\Sigma X)_1^2/N_1$	...	$(\Sigma X)_g^2/N_g$	...	$(\Sigma X)_C^2/N_C$	$(\Sigma X)_.^2/N.$
Mean	$\overline{X}_1$	...	$\overline{X}_g$	...	$\overline{X}_C$	$\overline{X}.$
Sum of squares of deviations	$\Sigma(X-\overline{X})_1^2$	...	$\Sigma(X-\overline{X})_g^2$	...	$\Sigma(X-\overline{X})_C^2$	$\Sigma(X-\overline{X})_.^2$
Variance	S_1^2	...	S_g^2	...	S_C^2	$S_.^2$

Computing formulae

$$\overline{X}_g = (\Sigma X)_g / N_g = \sum_{h=1}^{N_g} X_{gh} / N_g$$

$$\Sigma(X-\overline{X})_g^2 = (\Sigma X^2)_g - (\Sigma X)_g^2/N_g\,, \text{ where } (\Sigma X)_g^2/N_g = N_g\overline{X}_g^2 \text{ is the } \textit{centering term}$$

$$S_g^2 = \Sigma(X-\overline{X})_g^2/(N_g-1)\,, \text{ where } (N_g-1) \text{ is the } \textit{number of degrees of freedom}$$

$$N. = N_1 + \ldots + N_g + \ldots + N_C = \Sigma N_g$$

$$(\Sigma X). = (\Sigma X)_1 + \ldots + (\Sigma X)_g + \ldots + (\Sigma X)_C$$

$$(\Sigma X^2). = (\Sigma X^2)_1 + \ldots + (\Sigma X^2)_g + \ldots + (\Sigma X^2)_C$$

Section 12.3: the *within-groups* and *between-groups* sums of squares

The analysis of variance is based on the fact that, when all samples are combined into a single aggregate before computations, the resulting sum of squared deviations (called the *total sum of squares* and denoted by the acronym SS_{total}) is generally greater than if sums of squares are calculated for each sample separately and summed up afterwards (which yields the so-called *within-groups* SS). The difference between the *total* SS and the *within-groups* SS is called the *between-groups* SS and reflects the differences between the sample means $\overline{X}_1, \ldots \overline{X}_g, \ldots \overline{X}_C$. It may indeed be shown algebraically that

$$SS_{total} = SS_{within\text{-}groups} + SS_{between\text{-}groups}\,,$$

where

$$SS_{total} = \Sigma\Sigma(X-\overline{X}.)^2 = \sum_{g=1}^{C}\sum_{h=1}^{N_g}(X_{gh}-\overline{X}.)^2\,,$$

$$SS_{within\text{-}groups} = \Sigma\Sigma(X-\overline{X})_g^2 = \sum_{g=1}^{C}\sum_{h=1}^{N_g}(X_{gh}-\overline{X}_g)^2\,,$$

and

$$SS_{between\text{-}groups} = SS_{total} - SS_{within\text{-}groups}$$

$$= N_1(\overline{X}_1-\overline{X}.)^2 + \ldots + N_g(\overline{X}_g-\overline{X}.)^2 + \ldots + N_C(\overline{X}_C-\overline{X}.)^2$$

$$= \sum_{g=1}^{C} N_g(\overline{X}_g-\overline{X}.)^2\,.$$

The between-groups sum of squares is thus an increasing function of the differences between the sample means, $\overline{X}_g$, and the general mean $\overline{X}.$, and that function would be null if all sample means coincided with each other and with the general mean. The between-groups SS may also be expressed algebraically with respect to centering terms, since

$$SS_{\text{between-groups}} = N_1\overline{X}_1^2 + \ldots + N_g\overline{X}_g^2 + \ldots + N_C\overline{X}_C^2 - N.\overline{X}.^2 .$$

While that would sometimes be numerically inaccurate, the between-groups sum of squares could thus be evaluated by subtracting the general centering term, $N.\overline{X}.^2$, from the sum of sample centering terms taken separately. Since each centering term reflects the value of the mean of the corresponding group, the between-groups sum of squares reflects the dispersion of sample means about the general mean $\overline{X}.$. The decomposition of the total sum of squares into a within-groups portion and a between-groups portion is illustrated in figure 12.3.1.

$$SS_{\text{total}} = SS_{\text{within-groups}} + SS_{\text{between-groups}}$$

$$\sum_g\sum_h(X_{gh}-\overline{X}.)^2 = \sum_g\sum_h(X_{gh}-\overline{X}_g)^2 + \sum_g\sum_h(\overline{X}_g-\overline{X}.)^2$$

$$(N.-1) = (N.-C) + (C-1)$$

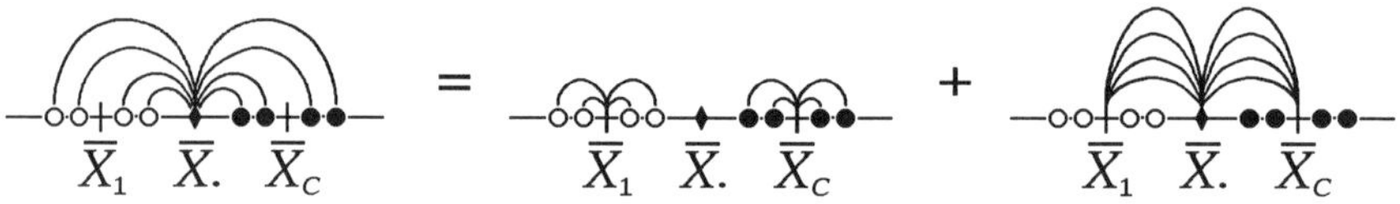

Figure 12.3.1
Breaking up the total sum of squares into a within-groups portion and a between-groups portion

Section 12.4: the numbers of degrees of freedom of sums of squares

The number of degrees of freedom of the total aggregate of all data, $(N.-1)$, may also be broken up, like the total sum of squares, into two portions of which one,

$$(N_1-1) + \ldots + (N_g-1) + \ldots + (N_C-1) = \sum_{g=1}^{C}(N_g-1) = (N.-C),$$

corresponds to the within-groups sum of squares while the other one, $(C-1)$, corresponds to the between-groups sum of squares. The proof that the within-groups SS has $(N.-C)$ degrees of freedom and that the between-groups SS has $(C-1)$ degrees of freedom requires the expected values (section 3.10) of these sums of squares to be evaluated. The manner in which such problems are solved will now be illustrated by determining the expected value of the sum of squares $\Sigma(X-\overline{X})^2$ in the simpler case where there is a single sample. The reader will be spared the detailed manner in which the expected values of the within-groups and between-groups sums of squares would be determined because that would be more complicated, but the procedures involved would be basically similar. How about proving then that, in order to get an unbiased estimate of the population variance σ_x^2 the sample sum of squares $\Sigma(X-\overline{X})^2$ must be divided by the number of degrees of freedom $(N-1)$ instead of sample size N as would seem to be required by the concept of variance (which is defined in principle as the mean value of a squared deviation)?

The processes which bring about the hth individual observation are thought to be summarized by the *model equation* (section 4.1):

$$X_h = \mu + e_h \ ,$$

where μ = the parametric mean, that is the expected value $\mathscr{E}(X_h)$ of variate X in the normal statistical population from which the sample has been drawn,

and e_h = the random deviation which has a normal distribution $\mathscr{N}(0, \sigma^2)$ following which $\mathscr{E}(e_h) = 0$ and $\mathscr{E}\{[e_h - \mathscr{E}(e_h)]^2\} = \mathscr{E}[(e_h - 0)^2] = \mathscr{E}(e_h^2) = \sigma^2$; moreover, since successive observations (let us say the hth and i^{th}) are assumed to be independent, their deviations from μ too are independent from each other and $\mathscr{E}[e_h e_i - \mathscr{E}(e_h)\mathscr{E}(e_i)] = \mathscr{E}(e_h e_i - 0) = \mathscr{E}(e_h e_i) = 0$.

Let us then attempt to determine the expected value of the sample sum of squares

$$\sum_{h=1}^{N} (X_h - \overline{X})^2 = \sum_{h=1}^{N} X_h^2 - N\overline{X}^2 \ .$$

Since $\mathscr{E}(X \pm Y) = \mathscr{E}(X) \pm \mathscr{E}(Y)$, let us first evaluate $\mathscr{E}(\Sigma X_h^2)$ and then $\mathscr{E}(N\overline{X}^2)$ which will be subtracted from it later.

Note that
$$\sum_{h=1}^{N} X_h^2 = \sum_{h=1}^{N} (\mu + e_h)^2 = N\mu^2 + 2\,\mu \sum_{h=1}^{N} e_h + \sum_{h=1}^{N} e_h^2 \ ;$$

therefore,
$$\mathscr{E}\Big(\sum_{h=1}^{N} X_h^2\Big) = N\mu^2 + 2\,\mu \sum_{h=1}^{N} \mathscr{E}(e_h) + \sum_{h=1}^{N} \mathscr{E}(e_h^2)$$

$$= N\mu^2 + 2\mu(0) + N\sigma^2 = N\mu^2 + N\sigma^2 \ .$$

As for the centering term
$$N\overline{X}^2 = (1/N)\Big(\sum_{h=1}^{N} X_h\Big)^2 ,$$

note that
$$\sum_{h=1}^{N} X_h = \sum_{h=1}^{N} \mu + \sum_{h=1}^{N} e_h = N\mu + \sum_{h=1}^{N} e_h \ ;$$

therefore,
$$\Big(\sum_{h=1}^{N} X_h\Big)^2 = N^2\mu^2 + 2\,N\mu \sum_{h=1}^{N} e_h + \Big(\sum_{h=1}^{N} e_h\Big)^2 \ ;$$

therefore, $(1/N)\,(\sum_{h=1}^{N} X_h)^2 = N\mu^2 + 2\mu \sum_{h=1}^{N} e_h + (1/N)\,(\sum_{h=1}^{N} e_h)^2$;

therefore, $\mathscr{E}[(1/N)\,(\sum_{h=1}^{N} X_h)^2] = N\mu^2 + 2\mu \sum_{h=1}^{N} \mathscr{E}(e_h) + (1/N)\,\mathscr{E}[(\sum_{h=1}^{N} e_h)^2]$

$$= N\mu^2 + 2\mu(0) + (1/N)\,\mathscr{E}(\sum_{h=1}^{N} e_h^2 + \sum_{h=1}^{N} \sum_{\substack{i=1 \\ h\neq i}}^{N} e_h e_i)$$

$$= N\mu^2 + (1/N)\,[\sum_{h=1}^{N} \mathscr{E}(e_h^2) + \sum_{h=1}^{N} \sum_{\substack{i=1 \\ h\neq i}}^{N} \mathscr{E}(e_h e_i)]$$

$$= N\mu^2 + (1/N)\,[N\sigma^2 + 0] = N\mu^2 + \sigma^2 .$$

Subtracting the above result from the one of the preceding page shows that

$$\mathscr{E}[\sum_{h=1}^{N} (X_h - \bar{X})^2] = (N\mu^2 + N\sigma^2) - (N\mu^2 + \sigma^2) = (N-1)\,\sigma^2 .$$

Consequently, for the estimate of the parametric variance σ^2 to be equal to the latter on the average, the sample sum of squares $\Sigma(X_h - \bar{X})^2$ must be divided by $(N-1)$: if it were divided by sample size N, the resulting estimate would be too small on the average in the ratio of $(N-1)/N$. If the parametric mean μ were known, however, $\Sigma(X_h - \mu)^2$ should be divided by N for the expected value of $\Sigma(X_h - \mu)^2 = \Sigma\, e_h^2$ is $N\sigma^2$. Replacing the parametric mean μ by its sample estimate $\bar{X}$ thus entails a loss of one degree of freedom. The fact that $\Sigma(X_h - \bar{X})^2$ possesses $(N-1)$ degrees of freedom may also be shown more concisely, since

$$\Sigma(X_h - \mu)^2 = \Sigma[(X_h - \bar{X}) + (\bar{X} - \mu)]^2 = \Sigma[(X_h - \bar{X})^2 + 2(X_h - \bar{X})(\bar{X} - \mu) + (\bar{X} - \mu)^2]$$

$$= \Sigma(X_h - \bar{X})^2 + 2(\bar{X} - \mu)\Sigma(X_h - \bar{X}) + N(\bar{X} - \mu)^2 = \Sigma(X_h - \bar{X})^2 + 2(\bar{X} - \mu)0 + N(\bar{X} - \mu)^2 .$$

Thus, $\Sigma(X_h - \mu)^2 = \Sigma(X_h - \bar{X})^2 + N(\bar{X} - \mu)^2$, whence $\Sigma(X_h - \bar{X})^2 = \Sigma(X_h - \mu)^2 - N(\bar{X} - \mu)^2$,

and $\mathscr{E}[\Sigma\,(X_h - \bar{X})^2] = \Sigma\mathscr{E}[(X_h - \mu)^2] - N\mathscr{E}[(\bar{X} - \mu)^2] = N\sigma^2 - N\sigma_{\bar{X}}^2 = N\sigma^2 - \sigma^2 = (N-1)\sigma^2$.

By procedures similar to those used above, it could be shown that

$$\mathscr{E}(\text{SS}_{\text{within-groups}}) = (N. - C)\,\sigma^2$$

and that

$$\mathscr{E}(\text{SS}_{\text{between-groups}}) = (C-1)\,\sigma^2 + \sum_{g=1}^{C} N_g(\mu_g - \bar{\mu})^2 - [\sum_{g=1}^{C} N_g(\mu_g - \bar{\mu})]^2 / \sum_{g=1}^{C} N_g$$

where $\bar{\mu} = \sum_{g=1}^{C} \mu_g / C$.

When the sizes of all samples are equal to a common value denoted by N without subscript,

$$N_1 = \dots = N_g = \dots = N_C = N ,$$

the expected value of the between-groups sum of squares takes a simpler form:

$$\mathscr{E}(\text{SS}_{\text{between-groups}}) = (C-1)\,\sigma^2 + N \sum_{g=1}^{C} (\mu_g - \bar{\mu})^2 .$$

In conclusion, when the hypothesis that means are equal is true,

$$\text{H}_0\text{: } \mu_1 = \dots = \mu_g = \dots = \mu_C ,$$

the differences $(\mu_g - \bar{\mu})$ are all null and the expected value of the between-groups SS is simply $(C-1)\,\sigma^2$. On the contrary, if some parametric means are unequal, the expected value of the between-groups SS is increased by the presence of some nonnull squared deviations such as $(\mu_g - \bar{\mu})^2$. In order to estimate σ^2 when H_0 is true, the between-groups SS must therefore be divided by $(C-1)$ degrees of freedom. As for the within-groups SS, its expected value is $(N. - C)\,\sigma^2$ whether the preliminary hypothesis is true or false, and its number of degrees of freedom is therefore $(N. - C)$.

Section 12.5: the analysis of variance table

The many quantities which appear in an analysis of variance are usually presented in an orderly manner as in table 12.5.1.

Table 12.5.1
One-way type I analysis of variance

Variation	Sum of squares (SS)	D. of F.	Observed variance	Expected variance
group 1	$\Sigma(X-\bar{X}_1)^2$	(N_1-1)	S_1^2	σ^2
...	...	...	...	...
group g	$\Sigma(X-\bar{X}_g)^2$	(N_g-1)	S_g^2	σ^2
...	...	...	...	...
group C	$\Sigma(X-\bar{X}_C)^2$	(N_C-1)	S_C^2	σ^2
within-groups	"$\text{SS}_{\text{within}}$" = $\sum_g \sum_h (X_{gh}-\bar{X}_g)^2$	$(N. - C)$	$S^2_{\text{within-groups}}$ = S^2_{pooled} = $\text{SS}_{\text{within}}/(N. - C)$	σ^2
between-groups	"$\text{SS}_{\text{between}}$" = $\text{SS}_{\text{total}} - \text{SS}_{\text{within}}$	$(C-1)$	$S^2_{\text{between-groups}}$ = $\text{SS}_{\text{between}}/(C-1)$	$\sigma^2 + \Delta^2$
total	"SS_{total}" = $\sum_g \sum_h (X_{gh}-\bar{X}.)^2$	$(N. - 1)$	S^2_{total} = $\text{SS}_{\text{total}}/(N. - 1)$	(generally not used)

On the line pertaining to between-groups variation, in table 12.5.1 (preceding page), the symbol Δ^2 stands for the expression

$$\{ \sum_{g=1}^{C} N_g(\mu_g - \bar{\mu})^2 - [\sum_{g=1}^{C} N_g(\mu_g - \bar{\mu})]^2 / \sum_{g=1}^{C} N_g \} / (C-1)$$

in the expected variance, for the sake of conciseness. In addition, the reader should note that, wherever algebraical expressions appear in table 12.5.1, they are replaced by the corresponding numerical values during the analysis of variance of a particular set of data. Moreover, the lines relative to groups 1 to C and the "Expected variance" column are often omitted when a numerical analysis is summarized as briefly as possible. Finally, the adjectives *within-groups* and *between-groups* are often reduced to the prefixes "*within*" and "*between*". An example of analysis of variance based on entomological data extracted from Brower (1959) is given in table 12.5.2.

Table 12.5.2
Example of a one-way type I analysis of variance: length in mm of the right forewing of male butterflies belonging to four species of genus *Papilio* (Brower, 1959)

Variation	Sum of squares (SS)	Number of degrees of freedom	Observed variance (estimated variance)
species 1 (*Papilio glaucus*)	42.5	9	4.722222
species 2 (*Papilio rutulus*)	50.9	9	5.655555
species 3 (*Papilio eurymedon*)	108.5	9	12.05555
species 4 (*Papilio multicaudatus*)	108.1	9	12.01111
within-groups	310.0	36	8.611111
between-groups	1053.9	3	351.3
total	1363.9	39	–

A variance estimate (or "*observed variance*") may be obtained by dividing the sum of squares (SS) on each line of the table by the corresponding number of degrees of freedom (D. of F.).

Section 12.6: the overall test of equality of means

The most important step, in a one-way type I analysis of variance, is the comparison of the between-groups with the within-groups estimate of variance. If the preliminary hypothesis that means are equal is true,

$$H_0: \mu_1 = \dots = \mu_g = \dots = \mu_C ,$$

and if the assumption that population variances are equal corresponds to reality,

$$\sigma_1^2 = \dots = \sigma_g^2 = \dots = \sigma_C^2 = \sigma^2 ,$$

the observed variances $S^2_{\text{between-groups}}$ and $S^2_{\text{within-groups}}$ are indeed estimates of the same parameter, σ^2, and the ratio $S^2_{\text{between-groups}} / S^2_{\text{within-groups}}$ follows the F distribution with

$(C-1)$ and $(N. - C)$ degrees of freedom. On the contrary, if the hypothesis of mean equality is false, the parameter $\sigma^2 + \Delta^2$ estimated by $S^2_{\text{between-groups}}$ is greater than σ^2_i and the ratio $S^2_{\text{between}} / S^2_{\text{within}}$ tends to reflect this by being often large.

Because the presence of differences between parametric means may only increase (and may not decrease) the parameter $\sigma^2 + \Delta^2$, the rejection region of H_0 should be placed completely on the right side, and the acceptance region of H_0 is defined by the inequality

$$S^2_{\text{between-groups}} / S^2_{\text{within-groups}} \leq F_{(1-\alpha;\, C-1,\, N.-C)} \;.$$

In the present context, the $S^2_{\text{between-groups}}$ variance estimate is always placed in the numerator, even when it is smaller than $S^2_{\text{within-groups}}$, unlike what was done in section 11.2. In the case of the length of the right forewing of male butterflies of four species of genus *Papilio* (table 12.5.2), for instance,

$$S^2_{\text{between-groups}} / S^2_{\text{within-groups}} = 40.7961 > 6.7436 = F_{(0.999;\, 3,\, 36)} \;.$$

The observed variance ratio thus falls in the rejection region at the $\alpha = 0.001$ significance level. Consequently, the conclusion that the average length of the right forewing differs in the male butterflies of these four species may be considered as well established.

Section 12.7: Bartlett's test of the homogeneity of variances

As indicated in the preceding section, the variances of the populations being compared must be equal (*homogeneous* in statistical language; unequal variances are said to be *heterogeneous*) for the ratio $S^2_{\text{between-groups}} / S^2_{\text{within-groups}}$ to follow the F distribution exactly. An analysis of variance thus implies that $\sigma_1^2 = \ldots = \sigma_g^2 = \ldots = \sigma_C^2 = \sigma^2$, where σ^2 without subscript denotes the common value to which all parametric variances are equal. Like Student's t (section 9.8), the analysis of variance is a *robust* statistical method which is not much affected when its validity conditions are not completely fulfilled. In practice the assumption that variances are equal is therefore not questioned unless there are strong reasons to do so. Nevertheless, when variances are suspected of being unequal, the hypothesis

$$H_0:\; \sigma_1^2 = \ldots = \sigma_g^2 = \ldots = \sigma_C^2 = \sigma^2$$

may be tested by an approximate method due to the British statistician Bartlett (1954), who showed that the distribution of the criterion

$$(1/K) \sum_{g=1}^{C} (N_g - 1) \left[\log_e(S^2_{\text{within-groups}} / S_g^2)\right],$$

where

$$K = 1 + \left\{ \sum_{g=1}^{C} [1/(N_g - 1)] - [1/(N. - C)] \right\} / [3\,(C-1)],$$

is approximately similar to the χ^2 distribution with $(C - 1)$ degrees of freedom.

If one of the samples (let us say the gth) contains only one observation, it may still be included in the analysis of variance properly speaking but it must be excluded from the calculations leading to Bartlett's test because it would contribute no information about the variance of the population from which it comes; moreover, if that sample were included, coefficient K could not be determined since $1/(N_g - 1) = 1/0 = \infty$. Bartlett's criterion would be null if the variances of all samples were equal to each other as well as to the within-groups variance estimate since the ratios $S^2_{\text{within-groups}} / S^2_g$ would all have a unit value and their logarithms would be null. On the contrary, Bartlett's criterion is positive and its numerical value increases to the extent that sample variances differ from each other. The rejection region must therefore be placed completely on the right side, and the acceptance region of H_0 is defined by the inequality

$$(1/K)\,\Sigma(N_g - 1)\,[\log_e(S^2_{\text{within-groups}} / S^2_g)] \leq \chi^2_{(1-\alpha\,;\,C-1)}\,.$$

In the case of the length of the right forewing of male butterflies of four species of genus *Papilio* (table 12.5.2), Bartlett's criterion is equal to

$$3.02729 < 7.8147 = \chi^2_{(0.95;\,3)}$$

and falls in the acceptance region even at the $\alpha = 0.05$ level. The hypothesis that variances are equal may therefore be accepted (and the earlier utilization of the analysis of variance to compare means may be considered as justifiable). Bartlett's test is recommended only if each sample has at least four degrees of freedom; otherwise, the χ^2-approximation might be unsatisfactory. Moreover, Bartlett's criterion is sensitive to possible departures of the variate studied from normality, and its utilization is advisable only if parametric variances are suspected of differing strongly enough to make the analysis of variance invalid (in spite of the robustness of the latter with respect to variance heterogeneity).

There is a simpler procedure for checking the assumption that variances are equal, but the answer provided by that procedure may be trusted only when the equality hypothesis is acceptable. The criterion is obtained by dividing the largest (let us say S_i^2) by the smallest (let us say S_j^2) of the variances of the C samples (see section 11.2). When that ratio falls in the acceptance region,

$$S_i^2/S_j^2 \leq F_{(1-\alpha/2;\,N_i-1,\,N_j-1)}\,,$$

the hypothesis $H_0: \sigma_1^2 = \ldots = \sigma_g^2 = \ldots = \sigma_C^2 = \sigma^2$ may be accepted conclusively. When the ratio S_i^2/S_j^2 falls in the rejection region, however, the hypothesis H_0 cannot be rejected for sure because the variances S_i^2 and S_j^2 are not picked completely at random, since S_i^2 is the largest and S_j^2 is the smallest among more than two variance estimates. A truly valid comparison of the largest with the smallest among several variance estimates should be based on a table designed specifically for that purpose (Pearson and Hartley, 1966, table 31).

When there are strong differences between sample variances S_1^2 , ... S_g^2 , ... S_C^2 and when population variances σ_1^2 , ..., σ_g^2, ..., σ_C^2 cannot be assumed to be equal, the validity of the analysis of variance for testing the hypothesis that means are equal,

$$H_0:\ \mu_1 = \ldots = \mu_g = \ldots = \mu_C\,,$$

becomes doubtful, but that technical problem may often be circumvented by finding a transformation of the original variate X which makes sample variances approximately equal. The logarithmic transformation $Y = \log_e(X)$ or $Y = \log_{10}(X)$ is often useful in that respect (sections 4.3, 4.9, 9.1 and chapter 14). Once the analysis of the transformed variate Y is completed, its results may be reexpressed with respect to the original variate X by using the inverse transformation [for instance, $X = \exp(Y)$ if $Y = \log_e(X)$, $X = 10^Y$ if $Y = \log_{10}(X)$, or $X = Y^2$ if $Y = \sqrt{X}$ (section 14.4; see also sections 19.8 and 30.9)].

Section 12.8: comparing confidence intervals of means

The analysis of variance reveals whether the means of several populations differ, but it indicates neither the nature nor the magnitude of those differences. An analysis of variance should therefore always be completed by the calculation and the graphical representation of the confidence intervals of means. Figure 12.8.1 illustrates two-sided confidence intervals at the $\alpha = 0.05$ significance level for the length in mm of the right forewing of male butterflies of four species of genus *Papilio* according to data from Brower (1959) analyzed in table 12.5.2. The first species appears to have the shortest forewing while the fourth species appears to have the longest forewing. Moreover, forewing length appears to be approximately equal in species 2 and 3. If the second and third species turned out to be closely similar with respect to other characters too, this might suggest that they are closely related. The comparison of the confidence intervals of means is thus a useful complement of the analysis of variance, for the latter only indicated that there were unspecified differences between means. While the answer provided by the analysis of variance is indispensable in order to know whether the conclusion is statistically well established, the indications obtained from the examination of confidence intervals are biologically more meaningful. More advanced aspects of the analysis of variance will be considered in chapters 26, 27 and 28.

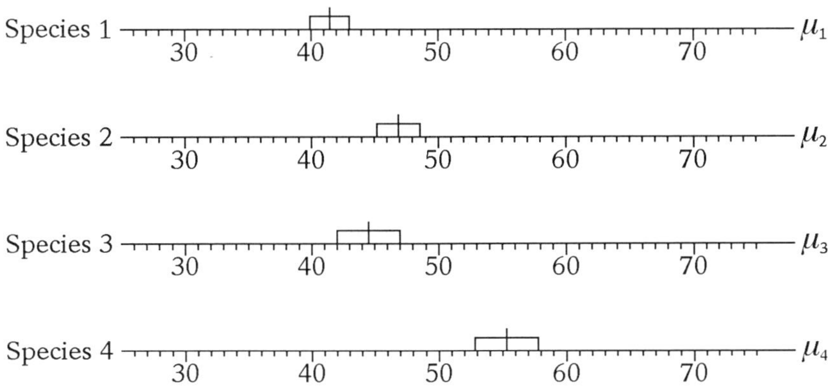

Figure 12.8.1
Two-sided confidence intervals at the $\alpha = 0.05$ significance level for the average length in mm of the right forewing of male butterflies of four species of genus *Papilio* (Brower, 1959)

Chapter 13

The skewness and peakedness indices, g_1 and g_2

Section 13.1: introduction

The probability distribution of a continuous variate is usually assumed to be normal in a majority of statistical methods. The *skewness* and *peakedness* indices, g_1 and g_2, may be used to test that normality assumption. While the agreement between the frequency distribution of a set of data and a theoretical distribution (such as the normal distribution) can also be checked through tests of goodness of fit (chapter 16), the latter require a large sample (containing preferably 100 or more observations). The g_1 and g_2 indices can be used even in the case of a small sample and also have the advantage of providing explicit answers concerning the *skewness* (also called *asymmetry*) and the *peakedness* (also called *kurtosis*) of an observed distribution (figures 13.2.1 and 13.2.2). Peaked distributions generally have longer tails than flat distributions, but peakedness is a complex property which cannot be reduced to dispersion (section 4.5) and cannot be measured like the latter simply by the variance or the standard deviation.

Section 13.2: moments

The skewness and peakedness indices are derived from a family of statistical quantities which are known as the *moments* of a probability distribution and which also include the mean and the variance (chapter 4). The moments of the probability distribution of a continuous variate may be expressed as in table 13.2.1 and are classified with respect to two criteria: the order r and the origin, 0, μ or $\overline{X}$. The parametric moments μ_r and μ_r' are defined by integrals and concern the statistical population (table 13.2.1), but the corresponding sample moments m_r and m_r' are defined by sums (table 13.2.2). The moments about the mean μ_r and m_r may be obtained from the moments about the origin μ_r' and m_r' and vice versa through formulae derived from the expansion of the binomial $(X-\mu)^r$ for the population and $(X-\overline{X})^r$ for the sample. In the case of m_2, for instance, it may be shown (see also section 4.6) that

$$\sum_{h=1}^{N} (X_h-\overline{X})^2 \;=\; \sum_{h=1}^{N} (X_h^2-2\overline{X}X_h+\overline{X}^2) \;=\; \sum_{h=1}^{N} X_h^2-2\overline{X}\sum_{h=1}^{N}X_h+\sum_{h=1}^{N}\overline{X}^2 \;=\; \sum_{h=1}^{N}X_h^2-2N\overline{X}^2+N\overline{X}^2,$$

whence

$$\sum_{h=1}^{N} (X_h-\overline{X})^2 \;=\; \sum_{h=1}^{N} X_h^2 \;-\; N\overline{X}^2 .$$

Consequently,

$$m_2 \;=\; \sum_{h=1}^{N} (X_h-\overline{X})^2/N \;=\; (1/N)\sum_{h=1}^{N} X_h^2 \;-\; \overline{X}^2 \;=\; m_2'-m_1'^2 .$$

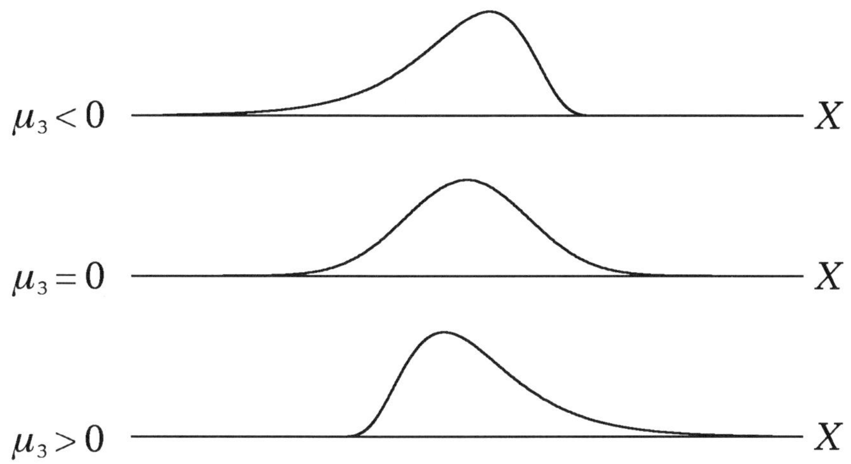

Figure 13.2.1
Geometrical meaning of the third moment about the mean, μ_3: the distribution is negatively skewed at the top, symmetrical at the center, and positively skewed at the bottom

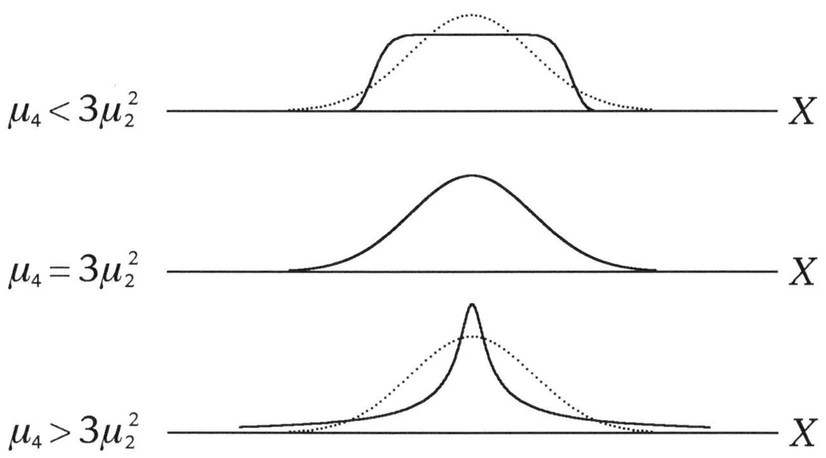

Figure 13.2.2
Geometrical meaning of the fourth moment about the mean, μ_4: the distribution is flat at the top, normal at the center, and peaked at the bottom

It may be shown in the same way that

$$\Sigma(X-\overline{X})^3 = \Sigma X^3 - 3\overline{X}\,\Sigma X^2 + 2N\overline{X}^3,$$

and that

$$\Sigma(X-\overline{X})^4 = \Sigma X^4 - 4\overline{X}\,\Sigma X^3 + 6\overline{X}^2\,\Sigma X^2 - 3N\overline{X}^4,$$

and the relationships between m_3 and m_3' as well as m_4 and m_4' may be deduced. Moreover, since integrals are limits of sums, there are analogous relationships between μ_2 and μ_2', μ_3 and μ_3', μ_4 and μ_4', etc. .

In order to understand the reasons for which the various moments are useful to describe the shape of a frequency distribution, the reader should note the following facts: (1) in the case of the moments about the sample mean of which the order r is odd, the contribution $(X_h-\overline{X})^r$ of the hth observation has the same sign as the deviation $(X_h-\overline{X})$; this contribution is thus positive if the observation X is greater than the mean $\overline{X}$ but negative if the observation is smaller than the mean; (2) as for the moments about the mean of which the order r is even, the contribution of an observation is positive whatever the sign of the deviation (whether the observation lies at the left or at the right of the mean); (3) the moments about the mean of which the order r is high are more sensitive to extreme observations than those of which the order is low, because the relative importance of the deviations $(X_h-\overline{X})$ which have a large absolute value is increased when the latter are raised to a high power r.

The geometrical meaning of the first two moments has already been discussed in the case of the measures of central tendency and dispersion (chapter 4), since the first parametric moment about the origin μ_1' is equal to the mean μ of the population and describes the central tendency, while the second moment about the mean μ_2 is equal to the variance σ^2 and describes the dispersion, the spread of a distribution. As for the third moment about the mean, μ_3, it reflects the skewness of a distribution (figure 13.2.1), since it is negative if individual observations fall farther from the mean on the left than on the right (figure 13.2.1, top), but positive if observations fall closer to the mean on the left than on the right (figure 13.2.1, bottom). The third moment about the mean is null if the distribution is symmetrical. Finally, the fourth moment about the mean μ_4 reflects the peakedness (figure 13.2.2) of the distribution: $\mu_4 = 3\mu_2^2$ in the case of the normal curve (chapter 5), but $\mu_4 < 3\mu_2^2$ if the distribution is flat (also called *platykurtic*; short-tailed; figure 13.2.2, top) and $\mu_4 > 3\mu_2^2$ if the distribution is peaked (also called *leptokurtic*; long-tailed; figure 13.2.2, bottom). It must be emphasized again that the degree of peakedness reflected by the fourth moment about the mean is a more complex property that the dispersion measured by the variance σ^2. Thus, a peaked distribution has a higher peak and higher tails but lower flanks than a normal distribution, while a flat distribution has a lower top and lower tails but higher flanks than a normal distribution.

The moments have the advantage of being relatively simple, but they have the drawback that, except for the first order moment about the origin of the sample m_1', which provides an unbiased estimate of the population mean μ, the sample moments are biased estimators of the corresponding parameters. For instance, the second moment about the mean of the sample m_2, would underestimate the second moment about the mean of the population, $\mu_2 = \sigma_X^2$, on the average because its expected value is $(N-1)\sigma_X^2/N$ (see section 12.4).

Table 13.2.1
Moments of the probability distribution of a continuous variate classified with respect to the order and the origin

		Origin	
		moments about the origin	moments about the mean
Order	1	$\mu_1' = \int_{-\infty}^{+\infty} X\, f(X)\, dX = \mu$	$\mu_1 = \int_{-\infty}^{+\infty} (X-\mu)\, f(X)\, dX = 0$
	2	$\mu_2' = \int_{-\infty}^{+\infty} X^2\, f(X)\, dX$	$\mu_2 = \int_{-\infty}^{+\infty} (X-\mu)^2\, f(X)\, dX = \sigma^2$
	3	$\mu_3' = \int_{-\infty}^{+\infty} X^3\, f(X)\, dX$	$\mu_3 = \int_{-\infty}^{+\infty} (X-\mu)^3\, f(X)\, dX$
	4	$\mu_4' = \int_{-\infty}^{+\infty} X^4\, f(X)\, dX$	$\mu_4 = \int_{-\infty}^{+\infty} (X-\mu)^4\, f(X)\, dX$
	...	...	...
	r	$\mu_r' = \int_{-\infty}^{+\infty} X^r\, f(X)\, dX$	$\mu_r = \int_{-\infty}^{+\infty} (X-\mu)^r\, f(X)\, dX$

Table 13.2.2
Sample moments classified with respect to the order and the origin

		Origin	
		moments about the origin	moments about the mean
Order	1	$m_1' = (1/N) \sum_{h=1}^{N} X_h = \overline{X}$	$m_1 = (1/N) \sum_{h=1}^{N} (X_h - \overline{X}) = 0$
	2	$m_2' = (1/N) \sum_{h=1}^{N} X_h^2$	$m_2 = (1/N) \sum_{h=1}^{N} (X_h - \overline{X})^2 = (N-1)S_X^2/N$
	3	$m_3' = (1/N) \sum_{h=1}^{N} X_h^3$	$m_3 = (1/N) \sum_{h=1}^{N} (X_h - \overline{X})^3$
	4	$m_4' = (1/N) \sum_{h=1}^{N} X_h^4$	$m_4 = (1/N) \sum_{h=1}^{N} (X_h - \overline{X})^4$
	...	...	...
	r	$m_r' = (1/N) \sum_{h=1}^{N} X_h^r$	$m_r = (1/N) \sum_{h=1}^{N} (X_h - \overline{X})^r$

Section 13.3: cumulants (*k-statistics*)

Fortunately, there is another related family of quantities, the *cumulants,* which may be used to characterize probability distributions and for which unbiased estimators are available (table 13.3.1). The sample cumulants are denoted by a lower case k letter and are also called *k–statistics* (Fisher, 1970). The parametric (population) cumulants are denoted by the lower case Greek letter $\varkappa$ (*kappa*) which corresponds to the k of the Roman alphabet. The upper case Roman K and Greek *kappa* letters are not used because they are similar and could not be distinguished from each other. The cumulants $\varkappa_3$ and $\varkappa_4$ are null in the case of a normal distribution. While they are more satisfactory than moments, the cumulants too possess a drawback, namely that they are not *pure numbers* (also called *dimensionless numbers*): their numerical value depends on the unit in which variate X is measured. Thus, if X is measured in cm, $\varkappa_1$ is expressed in cm, $\varkappa_2$ is expressed in cm^2, $\varkappa_3$ is expressed in cm^3, and $\varkappa_4$ is expressed in cm^4.

Table 13.3.1
Population and sample cumulants

		population	sample
Order	1	$\varkappa_1 = \mu_1' = \mu_X$	$k_1 = (1/N) \sum_{h=1}^{N} X_h = \overline{X}$
	2	$\varkappa_2 = \mu_2 = \sigma_X^2$	$k_2 = [1/(N-1)] \sum_{h=1}^{N} (X_h - \overline{X})^2 = S_X^2$
	3	$\varkappa_3 = \mu_3$	$k_3 = \{N/[(N-1)(N-2)]\} \sum_{h=1}^{N} (X_h - \overline{X})^3$
	4	$\varkappa_4 = \mu_4 - 3\mu_2^2$	$k_4 = N \Delta_4 / [(N-1)(N-2)(N-3)]$ where $\Delta_4 = (N+1) \sum_{h=1}^{N} (X_h - \overline{X})^4 - 3[(N-1)/N] \{ \sum_{h=1}^{N} (X_h - \overline{X})^2 \}^2$

Section 13.4: The skewness and peakedness indices, g_1 and g_2

In order to compare various kinds of distributions with each other, cumulants may be used to construct a *skewness index* and a *peakedness index* which are pure numbers and do not depend on the unit in which variate X is measured (table 13.4.1). The skewness and peakedness indices are denoted by the lower case g letter (with subscript 1 or 2) in the case of the sample but by the corresponding lower case Greek letter γ (*gamma*) in the case of the population. In the case of a variate X following a normal distribution, the parametric indices γ_1 and γ_2 are null and the estimators g_1 and g_2 have approximate normal distributions of which the means are γ_1 and γ_2 respectively and the standard deviations are

$$\sigma_{g_1} = \sqrt{\{[6\, N\, (N-1)] / [(N-2)(N+1)(N+3)]\}}\ ,$$

and
$$\sigma_{g_2} = \sqrt{\{[24\, N\, (N-1)^2] / [(N-3)(N-2)(N+3)(N+5)]\}}\ .$$

The lower case Greek letter σ is used for the standard deviations (or "standard errors" as they are usually called in this context) of g_1 and g_2 like for parameters because the values of these standard deviations depend only on sample size N and do not have to be estimated from the data.

Table 13.4.1
Skewness and peakedness indices

	population	sample
skewness	$\gamma_1 = \kappa_3/\kappa_2^{1.5} = \kappa_3/\sigma_X^3$	$g_1 = k_3/k_2^{1.5} = k_3/S_X^3$
peakedness	$\gamma_2 = \kappa_4/\kappa_2^2 = \kappa_4/\sigma_X^4$	$g_2 = k_4/k_2^2 = k_4/S_X^4$

Section 13.5: tests of normality

The skewness and peakedness indices can be used to test the hypothesis that the variate studied, X, follows a normal distribution,

$$X \leftarrow \mathcal{N}(\mu_X, \sigma_X^2) ,$$

or, more exactly, that the distribution of X in the population is symmetrical ($\gamma_1 = 0$) and possesses the same degree of peakedness ($\gamma_2 = 0$) as a normal distribution. Since the skewness index g_1 and the peakedness index g_2 also have distributions which are (approximately) normal,

$$g_1 \leftarrow \mathcal{N}(\gamma_1 , \sigma_{g_1}^2) \quad \text{and} \quad g_2 \leftarrow \mathcal{N}(\gamma_2 , \sigma_{g_2}^2) ,$$

the observed values g_1 and g_2 may be compared with the hypothetical values γ_1 and γ_2 by using the normal distribution (chapter 5): there is no need to use Student's t distribution here even if sample size is small, because the formulae given for σ_{g_1} and σ_{g_2} at the bottom of the preceding page already take sample size into account. It must nevertheless be noted that the normal approximation is somewhat less satisfactory for the peakedness index than for the skewness index, particularly in the case of small samples.

The hypothesis that the distribution of variate X is symmetrical (like the normal distribution) is thus H_0: $\gamma_1 = 0$ and, if the alternative hypothesis is two-sided, H_1: $\gamma_1 \neq 0$, the acceptance region of the preliminary hypothesis is

$$t_{(\alpha/2;\, \infty)} \leq t_1 = (g_1 - 0)/\sigma_{g_1} \leq t_{(1-\alpha/2;\, \infty)} \cdot$$

As for the hypothesis that the distribution of X has the same degree of peakedness as the normal distribution, H_0: $\gamma_2 = 0$, if the alternative hypothesis is two-sided here again, H_1: $\gamma_2 \neq 0$, the acceptance region of H_0 is

$$t_{(\alpha/2\, ;\, \infty)} \leq t_2 = (g_2 - 0)/\sigma_{g_2} \leq t_{(1-\alpha/2\, ;\, \infty)} \cdot$$

The preceding significance tests provide separate answers to the questions about skewness and peakedness, and it may occasionally be preferable to combine the results of these two tests into a single answer. This can be done by using the χ^2 distribution: the acceptance region of the double preliminary hypothesis,

$$H_0: \gamma_1 = 0 , \gamma_2 = 0 ,$$

is then

$$t_1^2 + t_2^2 \leq \chi^2_{(1-\alpha;\,2)} ,$$

where

$$t_1 = g_1/\sigma_{g_1} \quad \text{and} \quad t_2 = g_2/\sigma_{g_2} .$$

Section 13.6: numerical accuracy of computations

Theoretically, if the calculator or computer used for computations had very long registers (containing over 20 and preferably 30 decimal digits!), the raw sums N, ΣX, ΣX^2, ΣX^3 and ΣX^4 could be calculated at first and the centered sums $\Sigma(X-\overline{X})^2$, $\Sigma(X-\overline{X})^3$ and $\Sigma(X-\overline{X})^4$ could then be obtained through the formulae discussed in section 13.2. In practice, however, if the calculator or computer has rather short registers, the sample mean $\overline{X}$ should be calculated at first and the centered sums should then be calculated directly, by subtracting $\overline{X}$ from each individual observation, without passing through the raw sums. Otherwise, the numerical values of centered sums and of all other quantities subsequently derived from them might be affected by unacceptably large rounding errors, because high order centered sums would be obtained by subtracting from each other large numbers which short computing registers could not contain completely. Analogous numerical inaccuracies sometimes occur in the calculation of the variance (section 4.6).

Section 13.7: is the body weight of adult men normally distributed?

Body weight is an important quantitative character which is often used in human biology and in medicine. It may be wondered whether the probability distribution of body weight resembles the normal distribution or differs from it. In an attempt to answer that question, the skewness and peakedness indices have been calculated for a sample of 100 body weights of adult men deceased at the London Hospital, England (Greenwood and Brown, 1913). The indices g_1 and g_2 , their standard deviations σ_{g_1} and σ_{g_2} , and the criteria t_1 and t_2 have the following numerical values:

skewness	$g_1 = +0.931752$	$\sigma_{g_1} = 0.241380$	$t_1 = +3.86011$
peakedness	$g_2 = +1.08346$	$\sigma_{g_2} = 0.478331$	$t_2 = +2.26507$

The hypothesis that the probability distribution of the body weight of adult men is symmetrical ($P < 0.001$) and the hypothesis that it has the same degree of peakedness as the normal distribution ($P < 0.025$) can therefore both be rejected. If a conclusion based on a single criterion is preferred, $\chi^2_{(2)} = t_1^2 + t_2^2 = 20.031$ ($P < 0.0005$). It would therefore not be advisable to apply the normal distribution to the body weight measurements of Greenwood and Brown (1913). The analysis of these interesting data will be carried on in chapters 14 and 16.

Chapter 14
The lognormal distribution

Section 14.1: introduction

As already mentioned in sections 4.3, 5.1, 9.1 and 12.7, the normal distribution is not always completely suitable in biology because many biological variates cannot take negative numerical values and have positively skewed frequency distributions (sections 13.2 and 13.7). The lognormal distribution is often better adapted to biological data. While the normal distribution extends theoretically from $-\infty$ to $+\infty$ and covers negative values (figure 14.1.1, top and bottom), the lognormal distribution extends from 0 to $+\infty$ and is limited to positive values (figure 14.1.1, center).

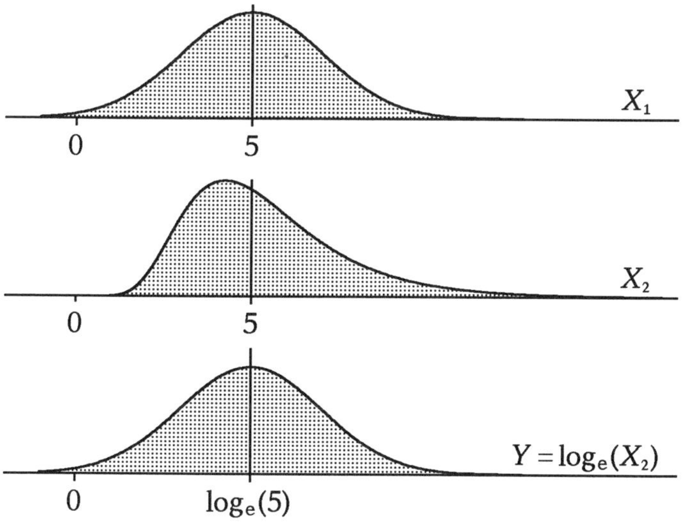

Figure 14.1.1
Top: the variate X_1 follows a normal distribution of which the arithmetic mean is 5; center : the variate X_2 has a lognormal distribution of which the geometric mean is 5; bottom: the logarithm $Y = \log_e(X_2)$ has a normal distribution of which the arithmetic mean is $\log_e(5)$

The vast array of tools available for the normal distribution can be applied indirectly to lognormal data since the logarithmic transform $Y = \log_e(X_2)$ of a lognormal variate X_2 (figure 14.1.1, center) is normally distributed (figure 14.1.1, bottom). Once the statistical analysis of the transformed variate Y has been completed, its results can be reexpressed with respect to the original variate X_2 through the antilogarithmic transformation $X_2 = \exp(Y)$ (see also the bivariate case in figures 19.8.1 and 30.9.1).

Section 14.2: additive and multiplicative variation

Variation is said to be *additive* in the case of the normal distribution (chapter 5) because the *model equation*

$$X_h = \mu_X + e_h , \quad \text{where} \quad e_h \leftarrow \mathcal{N}(0, \sigma_X^2) ,$$

represents the individual observation X_h as the result of the addition of the (parametric) arithmetic mean μ_X and of the random deviation e_h (sections 4.1, 12.4). Since the normal distribution extends theoretically from $-\infty$ to $+\infty$,

$$-\infty < e_h < +\infty \quad \text{and} \quad -\infty < X_h < +\infty .$$

As for the lognormal distribution, variation is additive in the case of the logarithm $Y_h = \log_e(X_h)$ because the latter follows the normal distribution and corresponds to the model equation

$$Y_h = \mu_Y + e_h, \quad \text{where} \quad e_h \leftarrow \mathcal{N}(0, \sigma_Y^2) ,$$

but variation is said to be *multiplicative* in the case of the original variate X because, when the model equation is reexpressed with respect to the latter,

$$X_h = \Gamma_X \exp(e_h) ,$$

where $\Gamma_X = \exp(\mu_Y)$ is the (parametric) geometric mean of X . The upper case Greek letter Γ (*gamma*) corresponds to letter G of the Roman alphabet. The lognormal variate X_h is thus pictured as the product of the geometric mean Γ_X by a random deviation $\exp(e_h)$ of which the value fluctuates about unity, since $\exp(0) = 1$, but is never negative, since its minimum value is $\exp(-\infty) = 0$. Consequently,

$$-\infty < e_h < +\infty \quad \text{but} \quad 0 < \exp(e_h) < +\infty \quad \text{and} \quad 0 < X_h < +\infty .$$

As stated above, multiplicative variation models and the lognormal distribution are often suitable in biology because many biological variates, such as the bodily dimensions of an individual organism, the concentration of a chemical within an organism, or the number of organisms in a biological population, cannot have negative numerical values.

Section 14.3: positive skewness and heteroscedasticity

The logarithmic transformation $Y = \log_e(X)$ often eliminates positive skewness within statistical populations and *heteroscedasticity* (the inequality of variances; the word heteroscedasticity comes from the Greek ετερος, *other* + σκεδαστος, *dispersed;* the equality of variances is called *homoscedasticity*) among populations (sections 4.3, 5.1, 9.1 and 12.7); conversely, if a variate Y has a symmetrical distribution and if the variances of several populations are equal in spite of the fact that their means differ, the antilogarithmic transformation $X = \exp(Y)$ will generally yield a positively skewed and heteroscedastic variate. The lognormal distribution thus provides a satisfactory descriptive model of the positive skewness and heteroscedasticity which often occur in biology.

Consider for instance a fictitious population which follows a lognormal distribution of which the geometric mean and the variance are $\Gamma_X = 5$ and $\sigma_Y^2 = [\log_e(2)/2]^2$ (figure 14.3.1). In a normal distribution like that of $Y = \log_e(X)$,

$$t_{(0.025;\,\infty)} = -1.96 \doteq -2 \quad \text{and} \quad t_{(0.975;\,\infty)} = +1.96 \doteq +2 .$$

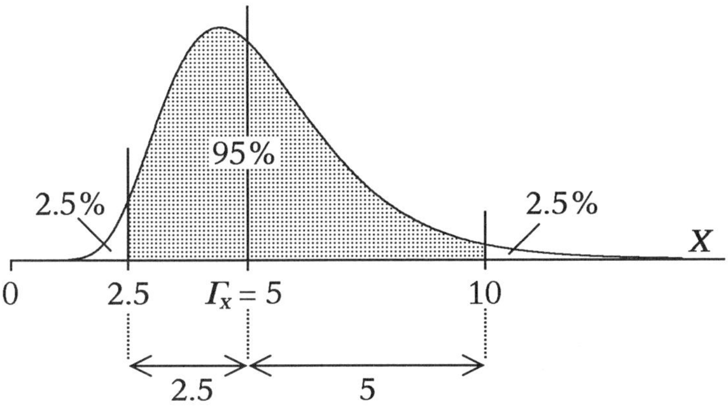

Figure 14.3.1
Positive skewness of the two-sided interval containing 95% of the individual observations in a lognormal distribution having a standard deviation of $\log_e(2)/2$

Approximately 95% of the individual values of Y thus lie in the interval

$$\mu_Y - 2\sigma_Y < Y < \mu_Y + 2\sigma_Y ,$$

that is

$$\log_e(5) - \log_e(2) < Y < \log_e(5) + \log_e(2) ,$$

and approximately 95% of the individual values of X therefore lie in the interval

$$2.5 < X < 10 .$$

The interval of the original variate X extends twice as far from the geometric mean $\Gamma_X = 5$ toward the right $(10-5=5)$ as toward the left $(5-2.5=2.5)$ and reflects the positive skewness of the lognormal probability distribution of X (figure 14.3.1).

In a second example (figure 14.3.2, next page), compare the preceding lognormal population with another lognormal population which has the same variance $\sigma_Y^2 = [\log_e(2)/2]^2$ but a geometric mean three times larger $(\Gamma_X = 15)$. The two-sided interval containing approximately 95% of the individual values of this second population is

$$\log_e(15) - \log_e(2) < Y < \log_e(15) + \log_e(2) ,$$

that is

$$7.5 < X < 30 .$$

The interval of the second population is therefore three times broader $(30-7.5=22.5)$ than the interval of the first population $(10-2.5=7.5)$, which reflects the heteroscedasticity of the lognormal distribution (figure 14.3.2).

According to the lognormal distribution, the breadth of the two-sided interval containing approximately 95% of the individual observations is proportional to the geometric mean Γ_X. This situation agrees rather well with the approximate constancy of the coefficient of variation generally observed in biology (section 4.9) and differs drastically from what would be expected in the case of the normal distribution, where the variance is independent from the mean (figures 4.2.1 and 4.5.1).

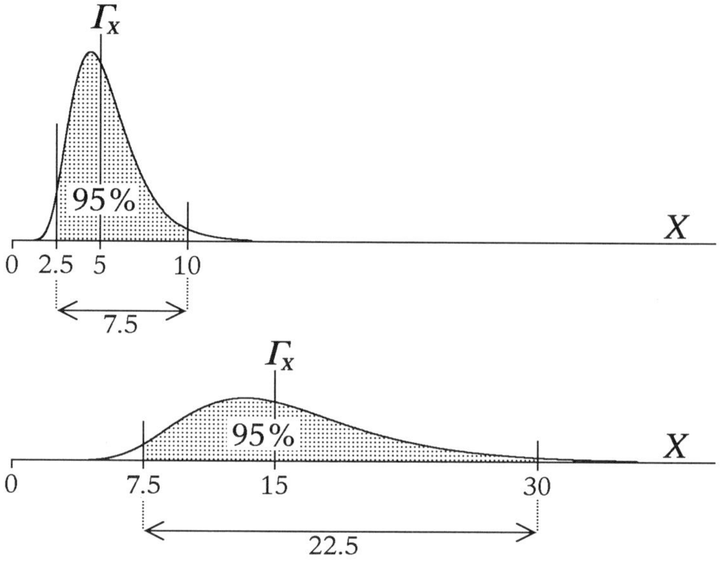

Figure 14.3.2
Heteroscedasticity of the lognormal distribution: even if the logarithmic standard deviations of two lognormal distributions are equal [$\log_e(2)/2$ here], the variation is greater in the population which has the largest geometric mean

When two or several variates are analyzed simultaneously, in addition to eliminating positive skewness and heteroscedasticity, the logarithmic transformation may help to straighten relationships which would otherwise be curved. This will be discussed in chapters 19, 20, 22, 31 and 35.

Section 14.4: lognormal approximation of discrete distributions

Positive skewness and heteroscedasticity are often particularly strong when numbers of organisms are determined in one or several biological populations, but this kind of variate is basically discrete (discontinuous) because it can take only integer values, such as 0, 1, 2, 3, ... 99, ...999, ...9999, etc.. Since the lognormal distribution itself is a continuous distribution, where the variate X can take noninteger as well as integer values, it can be applied to numbers of organisms only as an approximation.

Moreover, while the probability that a (continuous) lognormal variate is exactly equal to zero is theoretically null (section 5.2), the probability that a nonnegative discrete variate is null cannot be neglected, because the value zero then corresponds to a class of finite width. A positive constant should therefore be added to a discrete variate before submitting it to the logarithmic transformation, in order to prevent the possible occurrence of the value $\log_e(0) = -\infty$, for the latter would make impossible the calculation of the mean and of the variance. The logarithmic transformation of a discrete variate is thus done through the formula $Y = \log_e(X + a)$, where $a > 0$.

But, if $Y \leftarrow \mathcal{N}(\mu_Y, \sigma_Y^2)$, $-\infty < Y < +\infty$, $\exp(-\infty) < (X+a) < \exp(+\infty)$, and $-a < X < +\infty$. The value $a = 1/2$ seems particularly suitable since the point $X = -1/2$ may be interpreted as the lower limit of a one-unit-wide class centered on zero. Many statisticians treat the constant a as a third parameter of the lognormal distribution and attempt to estimate it from the data, while others assign various arbitrary values to it, such as 1/4, 1/2, 1.0, 1.5, etc., apparently with the aim of adjusting the degree of skewness of the lognormal distribution, but these modifications of the numerical value of the constant a have little effect. Moreover, if values of a higher than 1/2 were used, part of the probability distribution would correspond to impossible values of the original variate X. The transformation $Y = \log_e(X + 1/2)$ is thus perhaps most suitable. In spite of its approximate nature, the utilization of the lognormal distribution for the analysis of numbers of organisms often gives excellent results in practice. In some cases, however, the square root transformation, $Y = \sqrt{X}$, stabilizes the variance of numbers of organisms better than the logarithmic transformation (section 12.7).

Section 14.5: is the body weight of adult men lognormally distributed?

At the end of the preceding chapter, the skewness and peakedness indices of the data of Greenwood and Brown (1913) led to a rejection of the hypothesis that the body weight of adult men was normally distributed. Those indices have been recalculated here for the logarithmically transformed data, and the values of g_1 and g_2 are clearly lower than for the original data. The values of the standard deviations σ_{g_1} and σ_{g_2} are unchanged, however, because they depend only on sample size.

skewness	$g_1 = +0.397302$	$\sigma_{g_1} = 0.241380$	$t_1 = +1.64596$
peakedness	$g_2 = +0.0979443$	$\sigma_{g_2} = 0.478331$	$t_2 = +0.204762$

The criteria t_1 and t_2, as well as the combined criterion $\chi^2_{(2)} = t_1^2 + t_2^2 = 2.75112$, all fall in the acceptance region ($P > 0.05$), and the application of the lognormal distribution to the human body weight data of Greenwood and Brown (1913) would be justifiable.

Section 14.6: using natural or decimal (common) logarithms?

From a mathematical viewpoint, many expressions take a simpler form when they are expressed with natural (Napierian) logarithms, of which the base is the constant $e = 2.718281828459\ldots$: thus, the inverse function of the natural logarithm $Y = \log_e(X)$ is simply the exponential of that logarithm: $X = \exp(Y) = e^Y$. In practice, however, many users are more familiar with decimal (common) logarithms, $Z = \log_{10}(X)$, of which the base is the number 10 and the inverse function is $X = 10^Z$. When the positive skewness and heteroscedasticity of statistical data are reduced or eliminated by the logarithmic transformation, this remains true whatever the kind of logarithms used. The natural logarithm of a number X is indeed equal simply to the product of the decimal logarithm of that number by the natural logarithm of the number 10: $\log_e(X) = \log_e(10) \times \log_{10}(X)$,

since $$X = e^{\log_e(X)} = 10^{\log_{10}(X)} = [e^{\log_e(10)}]^{\log_{10}(X)} = e^{\log_e(10) \times \log_{10}(X)}.$$

Moreover, the numerical values of the skewness and peakedness indices and of their standard deviations (chapter 13) are not affected by changing the base of logarithms.

Chapter 15
Testing hypotheses concerning frequency tables using the χ^2 distribution

Section 15.1: introduction

The distribution of χ^2 (chapter 7) has many uses in statistics. In the case where a continuous variate X follows a normal distribution (chapter 5), the χ^2 distribution may be used to test hypotheses or to determine confidence intervals exactly about the parametric variance σ_X^2 of the population (chapter 10). When the hypothesis that within-groups variances are equal must be tested in an analysis of variance, Bartlett's criterion follows the χ^2 distribution approximately (section 12.7). Moreover, the hypotheses that the distribution of a set of data has the same skewness index ($\gamma_1 = 0$) and the same peakedness index ($\gamma_2 = 0$) as a normal distribution can be tested jointly by using the χ^2 distribution (section 13.5). But the most common application of the χ^2 distribution, which was discovered by Karl Pearson (1900), is perhaps its approximate use to test hypotheses about *frequency tables* (also called *contingency tables*), one of the oldest among the so-called *nonparametric* methods.

Section 15.2: one-way frequency tables

The simplest kind of frequency table is the one-way table, in which data are classified with respect to only one criterion (tables 15.2.1 and 15.2.2). For instance, consider the number of shrimps of the species *Mysis mixta* captured in the Gulf of St. Lawrence by the marine ecologist Pierre Brunel (Jolicoeur and Brunel, 1966) on six successive dates during 1961 (table 15.2.1). Each observed absolute frequency $F(X_i) = F_i$ reflects the result of a similar *fishing effort,* for it is the total number of individuals captured by towing four shrimp nets during one hour (1 upper net and 1 lower net at 9 a.m. and at 1.00 p.m.). The environment sampled was relatively uniform and constant, since the bottom was muddy, the depth varied only between 110 and 117 meters, and water temperature was close to 0°C throughout the year. Under such conditions, the hypothesis that the population of *Mysis mixta* would not be affected by seasonal changes and would not fluctuate from one part of the year to another may be considered. According to that hypothesis, the probability $P(X_i)$ that each capture may have been made at any of the six dates is $1/6 = 0.16666667$. The expected frequency on the ith date is then equal to the product $E_i = N\,P(X_i)$ of the probability $P(X_i)$, which is an expected relative frequency (section 3.2), by the total frequency or sample size $N = 4568$, since the expected frequency considered here is an absolute frequency (section 3.1). The scrutiny of the fifth column of table 15.2.1 and the comparison of columns 2 and 4 show that observed frequencies differ markedly from the frequencies expected according to the hypothesis. Pearson showed, almost a century ago, that the criterion

$$\sum_{i=1}^{R} (F_i - E_i)^2/E_i \;=\; \Big[\sum_{i=1}^{R} (F_i^2/E_i)\Big] - N$$

follows the χ^2 distribution approximately (sections 17.12 and 29.9). The letter "R" (the first letter of the word *row*) denotes the number of observed categories. The number of degrees of freedom ν is equal to $(R-1)$ if the probabilities $P(X_1)$, ...$P(X_i)$, ...$P(X_R)$ are obtained independently from the data, as in the case of the present analysis. The loss of one degree of freedom is due to the fact that the sum of expected frequencies is equal to the sum of observed frequencies,

$$\sum_{i=1}^{R} E_i = \sum_{i=1}^{R} F(X_i) = N .$$

It follows that the sum of deviations $\Sigma(F_i - E_i) \equiv 0$ and that these deviations are not completely independent from each other since any one of them can be expressed as a function of the others. A careful examination of Pearson's criterion shows that it is null if observed frequencies agree exactly with expected frequencies and that its value increases when observed frequencies differ from expected frequencies. The preliminary hypothesis H_0, according to which the probabilities of the R classes are $P(X_1)$, ...$P(X_i)$, ...$P(X_R)$ should therefore be accepted if

$$\text{Pearson's criterion} \leq \chi^2_{(1-\alpha;\, \nu)} ,$$

where $\nu = (R-1)$ and α denotes the significance level (section 9.4) selected. The preliminary hypothesis should be rejected otherwise. In the case of the data of table 15.2.1,

$$1967.124 > \chi^2_{(0.9995;\, 5)} = 22.105 ,$$

and the hypothesis that the population of *Mysis mixta* sampled by Pierre Brunel is constant should be rejected.

Table 15.2.1
Numbers of shrimps of the species *Mysis mixta* captured on various dates in the Gulf of St. Lawrence by using shrimp nets (Jolicoeur and Brunel, 1966)

Date (1961)	Observed frequency $F(X)$	Hypothetical probability $P(X)$	Expected frequency $NP(X)$	Deviation $F(X) - NP(X)$	Contribution to Pearson's criterion
July 25	324	0.1666667	761.333	− 437.333	251.218
August 14	1369	0.1666667	761.333	607.667	485.016
August 25	709	0.1666667	761.333	− 52.3333	3.597344
September 8	347	0.1666667	761.333	− 414.333	225.489
September 25	1509	0.1666667	761.333	747.667	734.245
October 12	310	0.1666667	761.333	− 451.333	267.559
Sums	$N = 4568$	1.000000	4568	0.000	1967.124

Table 15.2.2
One-way frequency table

Class	Observed frequency $F(X)$	Hypothetical probability $P(X)$	Expected frequency $NP(X)$	Deviation $F(X) - NP(X)$	Contribution to Pearson's criterion
X_1	$F_1 = F(X_1)$	$P(X_1)$	$E_1 = NP(X_1)$	$F(X_1) - NP(X_1)$	$[F(X_1) - NP(X_1)]^2/E_1$
...	...	...	...	...	...
X_i	$F_i = F(X_i)$	$P(X_i)$	$E_i = NP(X_i)$	$F(X_i) - NP(X_i)$	$[F(X_i) - NP(X_i)]^2/E_i$
...	...	...	...	...	...
X_R	$F_R = F(X_R)$	$P(X_R)$	$E_R = NP(X_R)$	$F(X_R) - NP(X_R)$	$[F(X_R) - NP(X_R)]^2/E_R$
Sums	N	1.0	N	0.0	Pearson's criterion

The notation used in a one-way frequency table is shown in a more detailed manner in table 15.2.2. When a set of data is classified into several consecutive classes, the labels X_1, ..., X_i, ... etc. denote the successive states of the qualitative variate (or the central value of each class if the variate is quantitative and the data are grouped). Each row of the table contains, in successive columns, the class label X_i , the observed frequency $F(X_i)$, the probability $P(X_i)$, the expected frequency $E_i = NP(X_i)$, the deviation $F(X_i) - NP(X_i)$ and the contribution of the class to Pearson's criterion, $[F(X_i) - NP(X_i)]^2/E_i$ .The probabilities sum up to unity because the sequence of R classes contains all possibilities and constitutes a *complete system of events* (section 3.7). One-way frequency tables are often used in *tests of goodness of fit,* which will be discussed and illustrated in chapter 16.

Section 15.3: independence and association of qualitative variates

In order to prepare the discussion of two-way frequency tables given in section 15.4, the concepts of marginal probability, conditional probability, independence and association (section 3.9) will be illustrated here by two concrete examples. While the frequencies are fictitious, they have been made to simulate biological reality as closely as possible. Consider first a finite population of 200 persons (table 15.3.1) of whom half are men and half are women, and of whom half have light-colored hair and half have dark-colored hair (natural shade). The essential part of the table contains the frequencies of *compound events* (section 3.6): light-haired men (frequency = 50), dark-haired men (frequency = 50), light-haired women (frequency = 50), dark-haired women (frequency = 50).

Table 15.3.1
Independent variates (independent events)

		Y = hair shade		Right margin
		Y_1: light	Y_2: dark	
X = sex	X_1: men	50	50	100
	X_2: women	50	50	100
Bottom margin		100	100	$N = 200$

In addition to the frequencies of compound events, the so-called marginal frequencies are given in the right margin and the bottom margin of the table; each marginal frequency is equal to the sum of the frequencies of the compound events located in the same row or in the same column: men ($50+50=100$), women ($50+50=100$), light-haired individuals ($50+50=100$), dark-haired individuals ($50+50=100$). Finally, the total frequency ($N=200$) appears in the lower right corner of the table. If an individual is about to be picked at random in this population, the ordinary or marginal probability of picking a man without taking hair shade into account is determined by the frequencies entered in the right margin of the table. Since there are 100 men over a total of 200 individuals, this probability is

$$P(\text{man}) = 100/200 = 1/2 = 0.5 = 50\% .$$

If only dark-haired individuals are taken into account, the conditional probability of picking a man whose hair is dark is determined by the frequencies in the second column of the table (Y_2: dark) and is equal to

$$P(\text{man}|\text{dark hair}) = 50/100 = 1/2 = 0.5 = 50\% .$$

The conditional probability of picking a man whose hair is dark is thus equal to the marginal probability of picking a man without taking hair shade into account. This reflects the fact that, in this finite population, sex and hair shade are independent events.

Table 15.3.2
Associated variates (associated events)

		Y = stature		Right margin
		Y_1: low	Y_2: high	
X = sex	X_1 : men	25	75	100
	X_2 : women	75	25	100
Bottom margin		100	100	$N = 200$

Consider now a second finite population of 200 persons (table 15.3.2) of whom half are men and half are women, and of whom half are short and half are tall. This time the conditional probability of picking a man if the individual is tall,

$$P(\text{man}|\text{high stature}) = 75/100 = 3/4 = 0.75 = 75\% ,$$

is greater than the marginal probability of picking a man without taking stature into account,

$$P(\text{man}) = 100/200 = 1/2 = 0.5 = 50\% .$$

This reflects the fact that, in this second population, there is a positive association between the male sex and tallness.

When two events are independent, such as sex and hair shade (table 15.3.1), the frequencies of compound events (dark-haired men, etc.) are equal to the product of the corresponding marginal probabilities multiplied by the total frequency. Thus the frequency of dark-haired men (50) is equal to the product by the total frequency (200) of the product of the marginal probability of picking a man (100/200) by the marginal probability of picking a dark-haired individual (100/200). This boils down to the quotient by the total frequency (200) of the product of the marginal frequency of men (100) by the marginal frequency of dark-haired individuals (100), a relationship which does not exist in the case of associated events, such as sex and stature (table 15.3.2).

It should also be noted that, when two events are independent (table 15.3.1), the sequence of frequencies of the compound events in any row (or column) is equal or proportional to the sequence of frequencies in another row (or column). On the contrary, when two events are associated (table 15.3.2), the sequence of frequencies of the compound events changes in a nonproportional manner from one row (or column) to another one.

It should finally be emphasized that, because the fictitious examples of the present section involve finite populations which are assumed to be completely known, there is no need here to distinguish observed frequencies from expected frequencies as done elsewhere in this chapter.

Section 15.4: two-way frequency tables

The preliminary hypothesis which must be tested in a two-way frequency table is that the two classification criteria or variates X and Y are independent. The observed frequency of the compound event X_iY_j is denoted by $F(X_iY_j)$ or more simply by F_{ij} . Since observed frequencies already fill a two-way table comprising R rows and C columns (table 15.4.1), the expected frequencies E_{ij} are entered into another table (table 15.4.2). The letter C is used to denote the number of classes of the second classification criterion Y because it is the first letter of the word *column*.

As in the preceding section (tables 15.3.1 and 15.3.2), each marginal frequency in the right margin is equal to the sum of the frequencies of the compound events located in the same row, while each marginal frequency in the bottom margin is equal to the sum of the frequencies of the compound events located in the same column:

$$F_i. = \sum_{j=1}^{C} F_{ij} \quad \text{and} \quad F._j = \sum_{i=1}^{R} F_{ij} \ .$$

In the present notation, as in the analysis of variance (section 12.2), a subscript with respect to which the summation has been carried out is replaced by a dot. The total frequency $F..$ is equal to sample size N and may be calculated by summing up the marginal frequencies either in the right margin or in the bottom margin:

$$F.. = \sum_{i=1}^{R} \sum_{j=1}^{C} F_{ij} = \sum_{i=1}^{R} F_i. = \sum_{j=1}^{C} F._j \ .$$

As for the frequencies expected under the independence hypothesis (table 15.4.2), like in the case of the frequencies of a finite population (section 15.3), each one of them is obtained by multiplying the total frequency by the product of the marginal probabilities of the corresponding row and column, replaced here by their estimates $\hat{P}(X_i)$ and $\hat{P}(Y_j)$:

$$E_{ij} = F..\, \hat{P}(X_i)\, \hat{P}(Y_j) = F..\, (F_i./F..)\, (F._j/F..) = F_i.\, F._j/F.. \ .$$

Table 15.4.1
Two-way frequency table: observed frequencies

		Y					Right margin
		Y_1	...	Y_j	...	Y_C	
X	X_1	F_{11}	...	F_{1j}	...	F_{1C}	$F_{1\cdot}$
	...	...	...	...	...	...	...
	X_i	F_{i1}	...	F_{ij}	...	F_{iC}	$F_{i\cdot}$
	...	...	...	...	...	...	...
	X_R	F_{R1}	...	F_{Rj}	...	F_{RC}	$F_{R\cdot}$
Bottom margin		$F_{\cdot 1}$	...	$F_{\cdot j}$	...	$F_{\cdot C}$	$F_{\cdot\cdot}$

The marginal frequencies of the table of expected frequencies (table 15.4.2) are equal to those of the table of observed frequencies (table 15.4.1) within rounding errors. There are thus R relationships such as

$$\sum_{j=1}^{C} (F_{ij} - E_{ij}) \equiv 0$$

and C relationships such as

$$\sum_{i=1}^{R} (F_{ij} - E_{ij}) \equiv 0$$

and the deviations between observed and expected frequencies are not completely independent from each other. However, it may be shown that these $(R+C)$ relationships are not fully independent either, that there is in fact the equivalent of $(R+C-1)$ independent relationships, and that Pearson's criterion,

$$\sum_{i=1}^{R} \sum_{j=1}^{C} (F_{ij} - E_{ij})^2/E_{ij} = [\sum_{i=1}^{R} \sum_{j=1}^{C} (F_{ij}^2/E_{ij})] - N,$$

possesses $RC-(R+C-1) = (R-1)(C-1)$ degrees of freedom in this case.

Table 15.4.2
Two-way frequency table: frequencies expected according to the independence hypothesis

		Y					Right margin
		Y_1	...	Y_j	...	Y_C	
X	X_1	E_{11}	...	E_{1j}	...	E_{1C}	$F_{1\cdot}$
	...	...	...	...	...	...	...
	X_i	E_{i1}	...	E_{ij}	...	E_{iC}	$F_{i\cdot}$
	...	...	...	...	...	...	...
	X_R	E_{R1}	...	E_{Rj}	...	E_{RC}	$F_{R\cdot}$
Bottom margin		$F_{\cdot 1}$	...	$F_{\cdot j}$	...	$F_{\cdot C}$	$F_{\cdot\cdot}$

As in the case of one-way frequency tables (section 15.2), Pearson's criterion would be null if observed and expected frequencies were exactly equal, and the numerical value of that criterion is an increasing function of the magnitude of differences between observed and expected frequencies. The preliminary hypothesis H_0, according to which the two classification criteria X and Y would be independent from each other, must therefore be accepted if

$$\sum_{i=1}^{R} \sum_{j=1}^{C} (F_{ij} - E_{ij})^2 / E_{ij} \leq \chi^2_{(1-\alpha;\, \nu)},$$

where $\nu = (R-1)(C-1)$ and α is the significance level. The hypothesis H_0 must be rejected if Pearson's criterion is greater than $\chi^2_{(1-\alpha;\, \nu)}$.

Section 15.5: are eye colors of husbands and wives related?

Karl Pearson and Alice Lee (1900) have studied data collected earlier by Francis Galton on the color of human eyes. In order to make the present analysis simpler and to avoid the occurrence of expected frequencies which would be too small (section 15.6), the shades of eye color distinguished by Galton have been regrouped here as follows:

Class	Shades of eye color
a	light blue, blue-green, medium gray
b	medium blue, dark blue
c	dark gray, hazel
d	light brown, medium brown
e	dark brown, black

The absolute frequencies observed for the various combinations of eye colors of husbands and wives in 774 human couples are entered in table 15.5.1. The absolute frequencies expected according to the independence hypothesis, calculated as indicated in section 15.4, are given in table 15.5.2. The shades of eye colors of husbands and wives tend to resemble each other perceptibly in classes "a" (observed frequency of 105, expected 85.1486), "b" (observed 87, expected 64.3204) and "e" (observed 24, expected 11.8256), but not clearly in other classes.

Table 15.5.1
Observed frequencies of various combinations of the eye colors of husbands and wives in 774 human couples

		Eye color of husband					Right margin
		a	b	c	d	e	
Eye color of wife	a	105	69	34	18	19	245
	b	48	87	26	16	19	196
	c	39	32	18	16	7	112
	d	34	38	11	13	12	108
	e	43	28	9	9	24	113
Bottom margin		269	254	98	72	81	774

Table 15.5.2
Expected frequencies of various combinations of eye colors according to the independence hypothesis

		Eye color of husband					Right margin
		a	b	c	d	e	
Eye color of wife	a	85.1486	80.4005	31.0207	22.7907	25.6395	245
	b	68.1189	64.3204	24.8165	18.2326	20.5116	196
	c	38.9251	36.7545	14.1809	10.4186	11.7209	112
	d	37.5349	35.4419	13.6744	10.0465	11.3023	108
	e	39.2726	37.0827	14.3075	10.5116	11.8256	113
Bottom margin		269	254	98	72	81	774

Pearson's criterion is equal to 49.4228 and has $(R-1)(C-1)=16$ degrees of freedom. Since $\chi^2_{(0.999;\ 16)} = 39.252$, the independence hypothesis must be rejected even at level $\alpha = 0.001$, and the eye colors of husbands and wives appear to be related.

However, that statistical association must be interpreted with caution. Even though Pearson and Lee (1900, page 113) expressed the opinion that Galton's data indicated that a human being tends to select a resembling spouse, the apparent association shown by Galton's data might indeed be due to the fact that the data came from several different regions which were more or less isolated from each other. For instance, if part of the 774 couples studied came from a region where most individuals have eye color "a" while others came from another region where most individuals have eye color "b", and if there were very few population exchanges between the two regions, the similarity of eye colors of husbands and wives detected while analyzing the pooled heterogeneous sample could reflect mostly the fact that an individual had a high probability of taking a spouse coming from the same region, the same social stratum, etc., a phenomenon known as *homogamy*. In order to check the possible influence of individual preferences (*sexual selection* in genetical terminology), the data studied should be as homogeneous as possible. Each person included in the analysis should nevertheless have had enough freedom to choose a spouse among several individuals on the basis of many varying characteristics, a condition which is perhaps seldom completely fulfilled in practice.

In medicine and epidemiology, the detection of associations between the occurrence of a disease and the existence of antecedents is of considerable importance.

Section 15.6: conditions of validity

Pearson's criterion follows the χ^2 distribution closely only if expected frequencies are large enough. Before Pearson's criterion is calculated, it is therefore advisable to pool neighboring classes if necessary in such a way that no expected frequency is smaller than unity and that the percentage of expected frequencies which are smaller than 5 does not exceed 20% (Cochran, 1954). This is done by combining subsequent rows in a one-way frequency table (section 15.2) or subsequent rows and columns in a two-way frequency table (section 15.4). When frequencies are low, the continuity correction proposed by Yates (1934; see section 17.6) may also be advisable.

Chapter 16
Tests of goodness of fit

Section 16.1: introduction

Tests of goodness of fit are one particular instance of the methods applied to the analysis of frequency tables in chapter 15. Tests of goodness of fit may be used to check the hypothesis that a continuous or discontinuous variate X follows a specified theoretical probability distribution such as the normal distribution (chapter 5), the lognormal distribution (chapter 14), the binomial distribution (chapter 17), or the Poisson distribution (chapter 18). The frequency table analyzed in a test of goodness of fit is a one-way table if the distribution is univariate, but it becomes a two-way or a multi-way table if the distribution is bivariate or multivariate, such as the bivariate (chapter 19) or the trivariate (chapter 23) normal distribution.

Section 16.2: goodness of fit of the normal distribution

In order to check whether the frequency distribution of a sample resembles the normal distribution, the observed range of variation may be subdivided into a sequence of classes of equal width (classes of unequal width could also be used but will not be considered in this textbook). It is often convenient to obtain the limits of those classes by adding a multiple of the sample standard deviation S_X to the sample mean $\overline{X}$. For instance, making successive class limits equal to $\overline{X} + (k/4)\,S_X$, where k is a positive or negative integer, yields the sequence $-\infty, \ldots, \overline{X} - 3\,S_X, \overline{X} - 2.75\,S_X, \ldots, \overline{X} - 0.25\,S_X, \overline{X}, \overline{X} + 0.25\,S_X, \ldots, \overline{X} + 2.75\,S_X, \overline{X} + 3\,S_X, \ldots, +\infty$. The width of each class is thus $(S_X/4)$ except in the case of the end classes which reach $\pm\infty$. The possible coincidence of individual observations with class limits can usually be prevented by calculating the latter with an accuracy of several digits.

The probability that an individual observation X falls in a class is evaluated by subtracting the cumulative probability of the lower limit from the cumulative probability of the upper limit of that class (figure 3.5.3). In the case of the normal distribution, if the sequence of limits mentioned in the preceding paragraph is used, their cumulative probabilities are obtained by entering the table of the standardized normal distribution at the values $\ldots, -3, -2.75, \ldots, -0.25, 0, +0.25, \ldots, +2.75, +3, \ldots$, etc. . The absolute frequency expected for the ith class of variate X is then equal to the product $N\,P(X_i) = E_i$ of the probability $P(X_i)$ of that class by the total frequency N (sample size). The observed absolute frequency $F(X_i) = F_i$ and the expected absolute frequency $N\,P(X_i)$ are compared by calculating the deviation $[F(X_i) - N\,P(X_i)]$. A positive (or negative) value of the deviation $[F(X_i) - N\,P(X_i)]$ indicates that the observed frequency is higher (or lower) than expected. The test of goodness of fit of a univariate normal distribution is a particular case of the analysis of a one-way frequency table (section 15.2). Pearson's criterion is then equal to the expression

$$\sum_{i=1}^{R} (F_i - E_i)^2/A_i \;=\; \left[\,\sum_{i=1}^{R} (F_i^2/E_i)\right] - N\,,$$

which is distributed approximately like χ^2 with ν degrees of freedom and where R denotes the number of classes of the variate X.

For the χ^2 approximation to be satisfactory, however, none of the expected frequencies must be smaller than unity and the percentage of expected frequencies which are smaller than 5 must not exceed 20%. As indicated in section 15.6, these conditions may be met by pooling neighboring classes if necessary before calculating Pearson's criterion. In a symmetrical distribution like the normal distribution, the lowest expected frequencies are found at the left and right ends, and it is therefore the lowest and the highest classes which may have to be pooled. In the case of the normal distribution, moreover, the estimates $\overline{X}$ and S_X of parameters μ and σ must be obtained from the sample in order to set up class limits, which creates two additional constraints besides the one due to the fact that the sums of observed and expected frequencies are equal (section 15.2). The number of degrees of freedom of Pearson's criterion is therefore

$$\nu = (R' - 1 - 2) = (R' - 3) ,$$

where R' denotes the number of classes after pooling, and the hypothesis that the population sampled follows the normal distribution may be accepted if

$$\text{Pearson's criterion} \leq \chi^2_{(1-\alpha;\,\nu)} ,$$

where α is the significance level (section 9.4) selected for drawing conclusions.

The various quantities calculated in a test of goodness of fit are usually summarized in a numerical table organized like table 16.2.1. Possible differences between the theoretical distribution and the observed distribution may be brought out by superposing the histogram of expected frequencies (stippled, thin outline) and the histogram of observed frequencies (clear, heavy outline) as in figures 16.4.1 and 16.4.2.

Table 16.2.1
Test of goodness of fit of a univariate distribution

Class	Hypothetical probability	Observed frequency	Expected frequency	Deviation $[F(X) - NP(X)]$	χ^2 contribution
X_1	$P(X_1)$	$F_1 = F(X_1)$	$E_1 = NP(X_1)$	$(F_1 - E_1)$	$(F_1 - E_1)^2/E_1$
...	...	...	...	...	...
X_i	$P(X_i)$	$F_i = F(X_i)$	$E_i = NP(X_i)$	$(F_i - E_i)$	$(F_i - E_i)^2/E_i$
...	...	...	...	...	...
X_R	$P(X_R)$	$F_R = F(X_R)$	$E_R = NP(X_R)$	$(F_R - E_R)$	$(F_R - E_R)^2/E_R$
Sums	1.0	N	N	0.0	Pearson's criterion

Section 16.3: goodness of fit of other distributions

The goodness of fit of the lognormal distribution (chapter 14) to a set of data may be tested in the same way as that of the normal distribution (section 16.2) provided the original variate X is submitted beforehand to the logarithmic transformation $Y = \log_e(X)$ or $Y = \log_{10}(X)$. If the distribution of the transformed variate Y appears to be normal but that of the original variate X does not, the latter may be thought to be lognormally distributed (see section 16.4). The tests of goodness of fit of the binomial and of the Poisson distributions will be discussed in chapters 17 and 18.

Section 16.4: the body weight of adult men: normal or lognormal distribution?

In earlier chapters, skewness and peakedness indices have indicated that the frequency distribution of the body weight measurements of adult men (in pounds; 1 pound = 454 grams) published by Greenwood and Brown (1913) differs from the normal distribution (section 13.7) but resembles the lognormal distribution (section 14.5). In this section, an attempt will be made to check the preceding indications through tests of goodness of fit. Table 16.4.1 summarizes the test of goodness of fit of the normal distribution to the original (untransformed) body weight measurements of which the mean is $\bar{X} = 128.89$ and the standard deviation is $S_x = 24.4895$. The observed frequencies $F(X)$ are clearly lower than the expected frequencies $NP(X)$ in the first seven classes (before pooling) of the table but clearly higher in 5 of the last seven classes. The number of classes after pooling is $R' = 14$. Pearson's criterion = 28.6839 with $(R' - 3) = 11$ degrees of freedom and the normality hypothesis is rejected at the level $\alpha = 0.01$, since $\chi^2_{(0.99;\,11)} = 24.725$.

Table 16.4.1
Normal distribution fitted to untransformed body weight measurements

X in pounds (454 g)	$x' = (X - \bar{X})/S_x$	$P(X)$	$F(X)$	$NP(X)$	$F(X) - NP(X)$	χ^2
$-\infty < X \leq 55.42$	$-\infty < x' \leq -3.00$	0.00134990	0	0.134990		
$55.42 < X \leq 61.54$	$-3.00 < x' \leq -2.75$	0.00162987	0	0.162987	-1.22245	1.22245
$61.54 < X \leq 67.67$	$-2.75 < x' \leq -2.50$	0.00322990	0	0.322990		
$67.67 < X \leq 73.79$	$-2.50 < x' \leq -2.25$	0.00601481	0	0.601481		
$73.79 < X \leq 79.91$	$-2.25 < x' \leq -2.00$	0.0105257	0	1.05257		
$79.91 < X \leq 86.03$	$-2.00 < x' \leq -1.75$	0.0173090	1	1.73090	-4.45827	3.64148
$86.03 < X \leq 92.16$	$-1.75 < x' \leq -1.50$	0.0267480	0	2.67480		
$92.16 < X \leq 98.28$	$-1.50 < x' \leq -1.25$	0.0388426	5	3.88426	0.815195	0.0723524
$98.28 < X \leq 104.40$	$-1.25 < x' \leq -1.00$	0.0530055	5	5.30055		
$104.40 < X \leq 110.52$	$-1.00 < x' \leq -0.75$	0.0679721	16	6.79721	9.20279	12.4597
$110.52 < X \leq 116.65$	$-0.75 < x' \leq -0.50$	0.0819102	9	8.19102	0.808981	0.0798986
$116.65 < X \leq 122.77$	$-0.50 < x' \leq -0.25$	0.0927561	8	9.27561	-1.27561	0.175427
$122.77 < X \leq 128.89$	$-0.25 < x' \leq 0.00$	0.0987063	13	9.87063	3.12937	0.992129
$128.89 < X \leq 135.01$	$0.00 < x' \leq +0.25$	0.0987063	10	9.87063	0.129367	0.00169553
$135.01 < X \leq 141.14$	$+0.25 < x' \leq +0.50$	0.0927561	8	9.27561	-1.27561	0.175427
$141.14 < X \leq 147.26$	$+0.50 < x' \leq +0.75$	0.0819102	5	8.19102	-3.19102	1.24314
$147.26 < X \leq 153.38$	$+0.75 < x' \leq +1.00$	0.0679721	4	6.79721	-2.79721	1.15112
$153.38 < X \leq 159.50$	$+1.00 < x' \leq +1.25$	0.0530055	4	5.30055	-3.18481	1.10432
$159.50 < X \leq 165.62$	$+1.25 < x' \leq +1.50$	0.0388426	2	3.88426		
$165.62 < X \leq 171.75$	$+1.50 < x' \leq +1.75$	0.0267480	4	2.67480		
$171.75 < X \leq 177.87$	$+1.75 < x' \leq +2.00$	0.0173090	2	1.73090	0.541727	0.0537658
$177.87 < X \leq 183.99$	$+2.00 < x' \leq +2.25$	0.0105257	0	1.05257		
$183.99 < X \leq 190.11$	$+2.25 < x' \leq +2.50$	0.00601481	2	0.601481		
$190.11 < X \leq 196.24$	$+2.50 < x' \leq +2.75$	0.00322990	1	0.322990	2.77755	6.31095
$196.24 < X \leq 202.36$	$+2.75 < x' \leq +3.00$	0.00162987	0	0.162987		
$202.36 < X < +\infty$	$+3.00 < x' < +\infty$	0.00134990	1	0.134990		
Sums		1.00000000	100	100.000000	0.000000	28.6839

Table 16.4.2 summarizes the test of goodness of fit of the normal distribution to the logarithmically transformed body weight measurements. Decimal (base 10) logarithms, also called common logarithms, have been used here as often done in practice. The average is $\overline{Y} = 2.10292$ and the standard deviation is $S_Y = 0.0790786$. Pearson's criterion is much smaller (8.50553) than in the case of the original data and the normality hypothesis can be accepted even at the level $\alpha = 0.05$, for $\chi^2_{(0.95;\,11)} = 19.675$. While the normality hypothesis must be rejected in the case of the original variate X, it can therefore be accepted in the case of the transformed variate $Y = \log_{10}(X)$, which indicates that the original variate X may follow a lognormal distribution (see chapter 14).

Table 16.4.2
Normal distribution fitted to body weight measurements transformed into decimal logarithms

$Y = \log_{10}(X)$	$y' = (Y - \overline{Y})/S_Y$	$P(Y)$	$F(Y)$	$NP(Y)$	$F(Y) - NP(Y)$	χ^2
$-\infty < Y \leq 1.8657$	$-\infty < y' \leq -3.00$	0.00134990	0	0.134990		
$1.8657 < Y \leq 1.8855$	$-3.00 < y' \leq -2.75$	0.00162987	0	0.162987	−0.222447	0.0404785
$1.8855 < Y \leq 1.9052$	$-2.75 < y' \leq -2.50$	0.00322990	1	0.322990		
$1.9052 < Y \leq 1.9250$	$-2.50 < y' \leq -2.25$	0.00601481	0	0.601481		
$1.9250 < Y \leq 1.9448$	$-2.25 < y' \leq -2.00$	0.0105257	0	1.05257		
$1.9448 < Y \leq 1.9645$	$-2.00 < y' \leq -1.75$	0.0173090	0	1.73090	−1.45827	0.389603
$1.9645 < Y \leq 1.9843$	$-1.75 < y' \leq -1.50$	0.0267480	4	2.67480		
$1.9843 < Y \leq 2.0041$	$-1.50 < y' \leq -1.25$	0.0388426	2	3.88426	2.815195	0.862873
$2.0041 < Y \leq 2.0238$	$-1.25 < y' \leq -1.00$	0.0530055	10	5.30055		
$2.0238 < Y \leq 2.0436$	$-1.00 < y' \leq -0.75$	0.0679721	10	6.79721	3.20279	1.50913
$2.0436 < Y \leq 2.0634$	$-0.75 < y' \leq -0.50$	0.0819102	7	8.19102	−1.19102	0.173181
$2.0634 < Y \leq 2.0832$	$-0.50 < y' \leq -0.25$	0.0927561	8	9.27561	−1.27561	0.175427
$2.0832 < Y \leq 2.1029$	$-0.25 < y' \leq 0.00$	0.0987063	8	9.87063	−1.87063	0.354513
$2.1029 < Y \leq 2.1227$	$0.00 < y' \leq +0.25$	0.0987063	13	9.87063	3.129367	0.992129
$2.1227 < Y \leq 2.1425$	$+0.25 < y' \leq +0.50$	0.0927561	11	9.27561	1.72439	0.320573
$2.1425 < Y \leq 2.1622$	$+0.50 < y' \leq +0.75$	0.0819102	4	8.19102	−4.19102	2.14438
$2.1622 < Y \leq 2.1820$	$+0.75 < y' \leq +1.00$	0.0679721	6	6.79721	−0.79721	0.0935006
$2.1820 < Y \leq 2.2018$	$+1.00 < y' \leq +1.25$	0.0530055	4	5.30055	−2.18481	0.519703
$2.2018 < Y \leq 2.2215$	$+1.25 < y' \leq +1.50$	0.0388426	3	3.88426		
$2.2215 < Y \leq 2.2413$	$+1.50 < y' \leq +1.75$	0.0267480	4	2.67480		
$2.2413 < Y \leq 2.2611$	$+1.75 < y' \leq +2.00$	0.0173090	1	1.73090	1.541727	0.435472
$2.2611 < Y \leq 2.2809$	$+2.00 < y' \leq +2.25$	0.0105257	2	1.05257		
$2.2809 < Y \leq 2.3006$	$+2.25 < y' \leq +2.50$	0.00601481	1	0.601481		
$2.3006 < Y \leq 2.3204$	$+2.50 < y' \leq +2.75$	0.00322990	0	0.322990	0.77755	0.494572
$2.3204 < Y \leq 2.3402$	$+2.75 < y' \leq +3.00$	0.00162987	1	0.162987		
$2.3402 < Y < +\infty$	$+3.00 < y' < +\infty$	0.00134990	0	0.134990		
Sums		1.00000000	100	100.000000	0.000000	8.50553

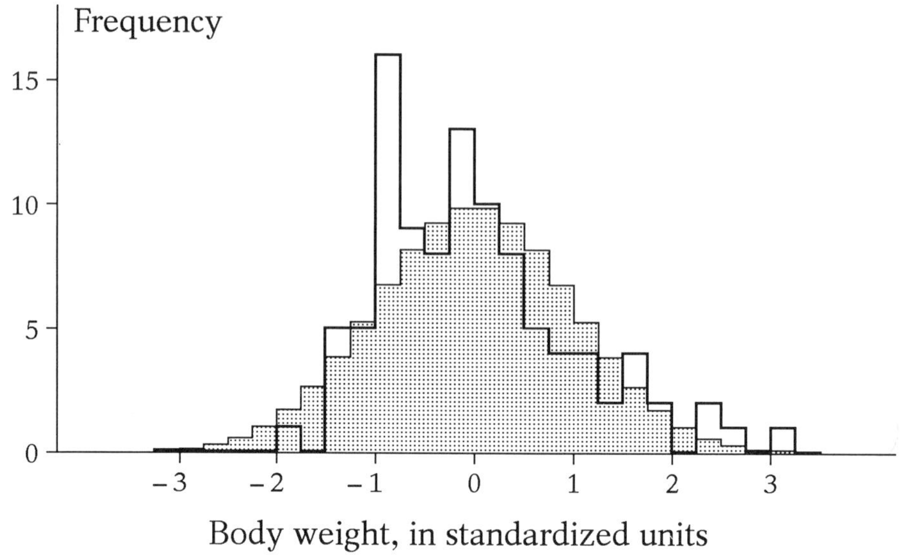

Figure 16.4.1
Histograms of observed frequencies (clear, heavy outline) and of frequencies expected according to the normal distribution (stippled, thin outline) in the case of the body weight of 100 adult men deceased at the London Hospital (Greenwood and Brown, 1913)

The comparison of the histograms of observed and expected frequencies before (figure 16.4.1) and after the logarithmic transformation (figure 16.4.2) is particularly revealing. Observed frequencies are visibly higher than expected frequencies at the right end and lower at the left end of the distribution in the case of body weight X, but not in the case of its logarithm Y. The logarithmic transformation thus appears to correct the positive skewness which can be perceived in the case of body weight.

It may be wondered whether the positive skewness of the frequency distribution of adult human body weight is a general phenomenon in human populations, since it has been brought out here on the basis of a sample which is not very large ($N = 100$) and which was obtained in a medical environment. The present conclusion appears to be confirmed by several other studies, including the analysis of a very large set of data obtained on recruits of the U. S. army during the second world war (Karpinos, 1958). In the case of adult stature, however, the logarithmic transformation does not appear to be generally necessary, since the normal distribution often fits data in an acceptable manner.

In the case of some variates, of which the relative dispersion is slight and for which only small samples are available, it may happen that the normality hypothesis can be rejected neither for the original variate nor for its logarithm. It may then be decided to apply the logarithmic transformation for other motives, for instance in order to avoid the possible occurrence of negative confidence limits on the original variate or in order to straighten exponential (section 20.6) or allometric (section 20.7) relationships.

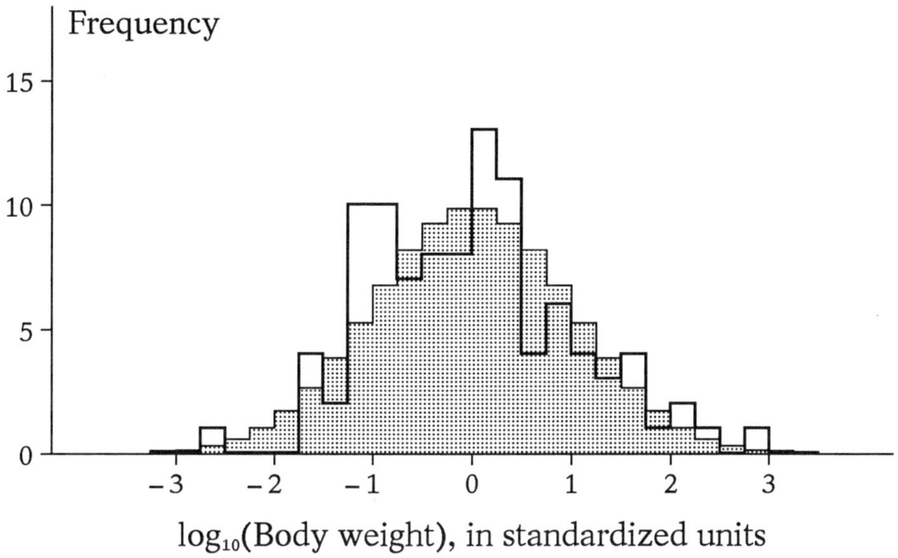

Figure 16.4.2
Histograms of observed frequencies (clear, heavy outline) and of frequencies expected according to the normal distribution (stippled, thin outline) in the case of the decimal logarithm of the body weight of 100 adult men deceased at the London Hospital (Greenwood and Brown, 1913)

Chapter 17
The binomial distribution

Section 17.1: introduction

The binomial distribution is the probability distribution of compound events (section 3.6) consisting of the joint occurrence of independent simple events (sections 3.9 and 15.3). Each simple event may take two complementary forms (section 3.7) of which the probabilities are p and q respectively. Because the two forms which each simple event may take are complementary, $(p+q)=1$. Moreover, since the simple events are independent from each other, the probabilities of the various kinds of compound events may be obtained by adding up the probabilities of the two forms and by multiplying the probabilities of the simple events, which yields

$$\underbrace{\underset{1^{\text{st}}}{(p+q)}\;\underset{2^{\text{nd}}}{(p+q)}\;\underset{\ldots\ldots}{\ldots\ldots}\;\underset{k^{\text{th}}}{(p+q)}}_{\text{simple events}} \;=\; (p+q)^k \;=\; 1\;.$$

In order to make the discussion of the binomial distribution concrete and easy to follow, the example of the frequency of male and female children in human families is introduced without delay. Each simple event here is the birth of a child and, since children of ill-determined sex are very rare, it may be considered in practice that the birth of a boy, of which the probability is p, and the birth of a girl, of which the probability is q, are complementary events, whence $(p+q)=1$. The compound event is the occurrence of a family of k children of whom X are male and $(k-X)$ are female. Except in the case of true (monozygotic) twins, who always have the same sex, the sex of simultaneous or successive children is thought to be independent and the probability of each type of family may be determined by multiplying each term of the binomial corresponding to the first birth by each term of the binomial corresponding to the second birth and so on. Because of the current knowledge about the mechanism of sex determination, and because the relative frequencies of male and female newborns are usually very close, it is generally considered that $p=q=0.5$, but the analysis of large samples may give indications that the probability of the birth of a boy differs slightly from that of a girl. The relative frequency of male and female children in human families is an excellent example of the binomial distribution and will be used through this chapter.

Section 17.2: expansion of the binomial $(p+q)^k$

If the kth power of the binomial $(p+q)$ is expanded, which means that all operations implied in the expression $(p+q)^k$ are carried out, and if similar terms are regrouped, a sum of $(k+1)$ terms is obtained. Thus, if $k = 2$, $(p+q)^2 = p^2 + 2pq + q^2$,

if $k = 3$, $(p+q)^3 = p^3 + 3p^2q + 3pq^2 + q^3$,

if $k = 4$, $(p+q)^4 = p^4 + 4p^3q + 6p^2q^2 + 4pq^3 + q^4$,

and so on. When the terms obtained by expanding a binomial such as $(p+q)^4$ are examined, in the case of the example of the frequencies of male and female children for instance, each term corresponds to a kind of family of k children:

the term p^4 is the probability of having a family comprising 4 boys and 0 girls, for it is the probability p of having a boy which has been taken in each of the 4 factors of $(p+q)^4$;

the term $4p^3q$ is the probability of having a family comprising 3 boys and 1 girl, for the probability p of having a boy has been taken in 3 factors and the probability q of having a girl has been taken in 1 factor;

the term $6p^2q^2$ is the probability of having a family comprising 2 boys and 2 girls, for the probability p of having a boy has been taken in 2 factors and the probability q of having a girl has been taken in 2 factors;

the term $4pq^3$ is the probability of having a family comprising 1 boy and 3 girls, for the probability p of having a boy has been taken in 1 factor and the probability q of having a girl has been taken in 3 factors;

the term q^4 is the probability of having a family comprising 0 boys and 4 girls, for it is the probability q of having a girl which has been taken in each of the 4 factors of $(p+q)^4$.

Since the total number of children in a family is equal to the exponent k and the number of boys is X, the number of girls is obviously equal to $(k-X)$, and each term of the expanded binomial may be obtained by multiplying the expression $p^X q^{(k-X)}$ by a *binomial coefficient* of which the evaluation will now be discussed. When the exponent k is somewhat higher than above, the direct expansion of the binomial would be tedious, and the coefficients of the successive terms may be determined (table 17.2.1) by using Pascal's triangle (named after the French mathematician Blaise Pascal, 1623-1662). This triangular table is constructed row by row, the exponent k being given the integer values 1, 2, 3, 4, 5, 6, 7, 8, etc. successively. The coefficients on each row fall between those of the preceding row, each coefficient being equal to the sum of the coefficients located immediately to its left and to its right on the preceding row.

Table 17.2.1
Determination of binomial coefficients with Pascal's triangle

exponent k	binomial coefficients										
1					1		1				
2				1		2		1			
3			1		3		3		1		
4		1		4		6		4		1	
5	1		5		10		10		5		1
etc.											

Pascal's triangle provides an easy way to determine binomial coefficients when the exponent k is not too high, but an algebraical expression of those coefficients is obviously desirable. This may be obtained by having recourse to the factorial and combination functions discussed in section 3.11, for the number of times where products such as $p^X q^{(k-X)}$ appear during the expansion of the binomial is equal to the number of combinations of k objects taken X at a time, that is to

$$k!/[X!\,(k-X)!]\,.$$

The *general term* of the binomial expansion is thus

$$\{k!/[X!\,(k-X)!]\}\,p^X q^{(k-X)}\,,$$

and the reader should make sure that each successive term of the expansions written out at the beginning of this section may also be found by attributing a suitable value to k and letting X vary from 0 to k and $(k-X)$ vary from k to 0 in the *general term*. It follows that the binomial distribution may be summarized by the following equation:

$$(p+q)^k = \sum_{X=0}^{k} \{k!/[X!\,(k-X)!]\}\,p^X q^{(k-X)} = 1.0\,.$$

Section 17.3: geometrical aspect of the binomial distribution

The geometrical aspect of the binomial distribution differs according to the value of the probability p or, what amounts to the same thing, of $q=(1-p)$, as well as of the exponent k. Figure 17.3.1 illustrates a binomial distribution of which the exponent $k=12$ and the probability p equals 0.9 at the top, 0.5 at the center and 0.1 at the bottom. The distribution is negatively skewed and squeezed to the right when the probability p is large, symmetrical about the center when $p=q=1/2$, and positively skewed and compressed to the left when the probability p is small.

The effect of the exponent k is illustrated in figure 17.3.2. A normal density curve (section 5.2) is superposed on a bar diagram representing a binomial distribution in which the probability $p=q=0.5$ and the exponent $k=50$. Even though the binomial distribution remains discontinuous, it becomes closely similar to the normal distribution when its exponent k is large, particularly if the probabilities p and q are approximately equal. The increasing similarity of the binomial to the normal distribution with the increase of exponent k has been demonstrated analytically by the French mathematician Abraham de Moivre (1667-1754), and it is the main reason for which the normal distribution is often used as an approximation for the binomial distribution (section 17.6).

Section 17.4: parametric mean and variance of the binomial distribution

The exponent k and the probability p of a theoretical binomial distribution are the parameters of that distribution (section 2.2). However, because of the practical usefulness of measures of central tendency and of dispersion (chapter 4), it is interesting to evaluate the parametric mean μ_X and variance σ_X^2 of the binomial variate X which, in the example of the frequencies of male and female children, denotes the number of boys per family.

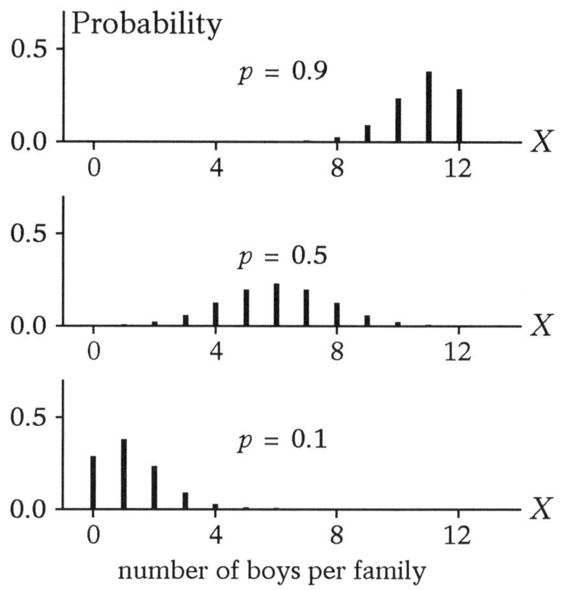

Figure 17.3.1
Geometrical aspect, according to the value of the probability p, of a binomial distribution in which the exponent $k = 12$

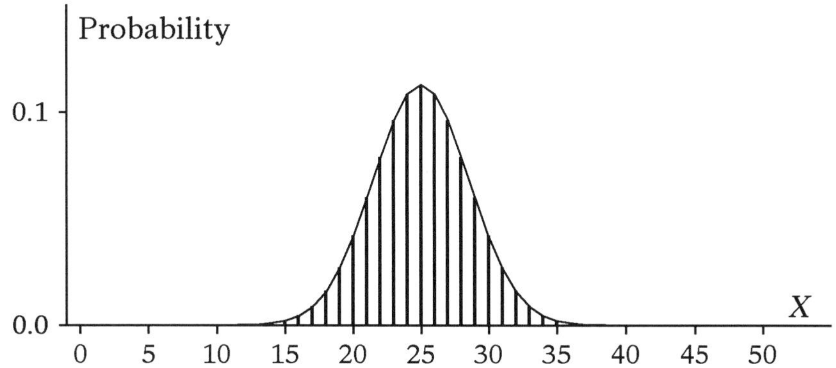

Figure 17.3.2
Similarity of the binomial to the normal distribution when the exponent k is large: normal density curve superposed on a bar diagram representing a binomial distribution in which the exponent $k = 50$ and the probability $p = 0.5$

The easiest way of determining the value of the parametric mean and variance of the binomial distribution is based on the fact that the binomial variate $X_{(k)}$ in a distribution having an exponent equal to k can be expressed as the sum of k binomial variates $X_{(1)}$ in distributions having exponents equal to unity:

$$X_{(k)} = \sum_{i=1}^{k} X_{(1)_i} .$$

In the case of the frequencies of boys and girls, this is equivalent to considering a family of 10 children, for instance, as the sum of 10 successive families comprising a single child each. The parametric mean and variance of a binomial distribution of which the exponent is unity are easily determined using the definitions given in section 3.10, if the binomial variate X itself is used as a summation index varying from 0 to k:

$$\mu_X = \sum_{X=0}^{k} [X\,P(X)]$$

and

$$\sigma_X^2 = \sum_{X=0}^{k} [(X-\mu_X)^2\,P(X)] = \sum_{X=0}^{k} [X^2\,P(X)] - \mu_X^2 .$$

When the exponent $k=1$, the binomial variate X indeed takes only two successive values, 0 and 1, of which the probabilities are q and p respectively. Therefore,

$$\mu_{X(1)} = 0 \times q + 1 \times p = p ,$$

and

$$\sigma_{X(1)}^2 = (0-p)^2\,q + (1-p)^2\,p = p^2q + q^2p = pq(p+q) = pq \times 1 = pq .$$

The mean $\mu_{X(k)}$ and the variance $\sigma_{X(k)}^2$ of a binomial distribution having an exponent equal to k may next be deduced from $\mu_{X(1)}$ and $\sigma_{X(1)}^2$ by using two properties (section 3.10):

(1) $\mu_{(X \pm Y \pm \ldots \pm Z)} = \mu_X \pm \mu_Y \pm \ldots \pm \mu_Z$, and (2) $\sigma^2_{(X \pm Y \pm \ldots \pm Z)} = \sigma_X^2 + \sigma_Y^2 + \ldots + \sigma_Z^2$;

The reader should note that the second of these two properties requires variates X, Y and Z to be statistically independent.

One then has

$$\mu_{X(k)} = \sum_{i=1}^{k} \mu_{X(1)i} = k\,\mu_{X(1)} = kp ,$$

and

$$\sigma_{X(k)}^2 = \sum_{i=1}^{k} \sigma_{X(1)i}^2 = k\,\sigma_{X(1)}^2 = kpq .$$

The fact that a binomial distribution of exponent k has a mean equal to kp and a variance equal to kpq may also be shown without recourse to the two properties used on the preceding page, but the demonstration is more laborious and will only be outlined. Recall that the binomial distribution may be expressed (section 17.2) by the equation:

$$(p+q)^k = \sum_{X=0}^{k} \{k!/[X!\,(k-X)!]\}\, p^X q^{(k-X)} = 1.0 \,.$$

But each of the $(k+1)$ terms of the above sum is the probability $P(X)$ of one of the successive values 0, 1, 2, ... k of the binomial variate $X_{(k)}$. The parametric mean may therefore be evaluated using its definition

$$\mu_X = \sum_{X=0}^{k} [X\,P(X)] = \sum_{X=0}^{k} X\{k!/[X!\,(k-X)!]\}\, p^X q^{(k-X)} ,$$

but the first term of the latter sum cancels out since it contains a null value of X . The change of variable $X' = (X-1)$ may then be carried out, the upper limit of the new variate being $k' = (k-1)$, and it may be shown that

$$\mu_{X(k)} = kp(p+q)^{k'} = kp \times 1 = kp \,.$$

In the case of the variance, the same change of variable leads to

$$\sigma^2_{X(k)} = \sum_{X=0}^{k} [X^2\,P(X)] - \mu^2_{X(k)} = kp[\,k'p + (p+q)^{k'}\,] - k^2p^2 = kp[\,(k-1)p + 1] - k^2p^2 = kpq \,.$$

Section 17.5: testing hypotheses concerning p with a single family

Hypotheses are often made about the probability p of a binomial distribution. In the case of the frequency of male and female children (section 17.1), for instance, on the basis of current knowledge about the mechanism of sex determination, the probability of the birth of a boy is expected theoretically to be $p = 0.5$. However, it might be wondered whether, for genetical or physiological reasons, the probability of a male birth might not be greater in some couples than in others. The preliminary hypothesis H_0: $p = p_0 = 0.5$ might then require to be tested. If the probability of the birth of a boy were suspected of being greater than 1/2 in a particular family, the alternative hypothesis would be H_1: $p = p_1 > p_0$. Figure 17.5.1 (next page) illustrates two binomial distributions having exponents equal to $k = 10$ and corresponding to the preliminary hypothesis H_0: $p = p_0 = 0.5$ (at the top) and to the alternative hypothesis H_1: $p = p_1 > p_0$ (at the bottom). The reader may note that, in order to picture the situation clearly, the probability selected here for the alternative hypothesis ($p_1 = 0.9$) is much greater than anything expected in practice in the case of the frequencies of male and female children.

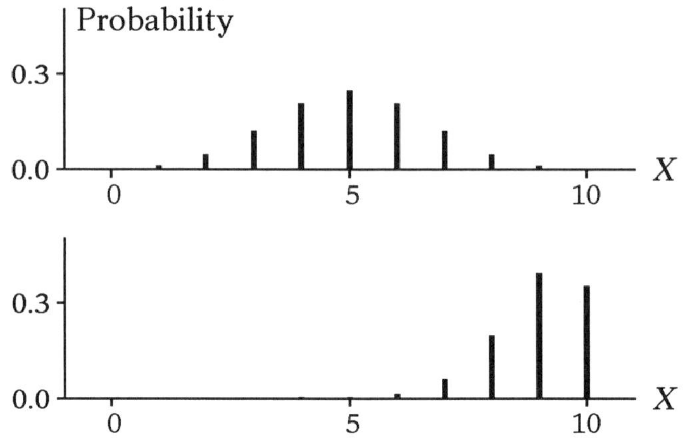

Figure 17.5.1
Binomial distributions having exponents equal to $k = 10$ and corresponding to the preliminary hypothesis H_0: $p = p_0 = 0.5$ (at the top) and to the alternative hypothesis H_1: $p = p_1 > p_0$ (at the bottom)

The probability of having a family of 10 children who would all be boys is very small according to the preliminary hypothesis ($p_0^{10} = 0.5^{10} = 1/1024$) but much larger according to the alternative hypothesis ($p_1^{10} = 0.9^{10} = 0.348678$). Therefore, if a family comprising 10 boys and 0 girls is observed and if there are *a priori* reasons to suspect that the probability of a male birth is greater than 0.5 in that family, rejecting the preliminary hypothesis would be justifiable. It must be emphasized, however, that the alternative hypothesis must not be inspired by the examination of the data because, as already noted in section 9.6, most current statistical methods are designed to test *a priori* rather than *a posteriori* hypotheses.

In the case of a discontinuous distribution like the binomial distribution, the concepts of significance level and of acceptance and rejection regions (sections 9.2 and 9.4) may be used but must be mitigated, because the limits of acceptance and rejection regions cannot be placed as exactly as in the case of continuous variates. In the preliminary hypothesis above ($k = 10$, H_0: $p = p_0 = 0.5$), for instance, the last 3 classes ($X = 8$, 9, 10) possess the probabilities $P(8) = 0.0439453$, $P(9) = 0.0097656$ and $P(10) = 0.0009766$. For the significance level to be exactly 0.01, the rejection region, which is completely on the right, should contain the class $X = 10$ plus part (but only part) of the class $X = 9$. If the rejection region contained both classes completely, the significance level would indeed be somewhat larger than desired [$P(9) + P(10) = 0.0107422$] while, if the rejection region contained only the class $X = 10$, the significance level would be much smaller than desired [$P(10) = 0.0009766$]. However, if the class $X = 9$ is placed in part in the acceptance and in part in the rejection region, it would not be clear whether the preliminary hypothesis should be accepted or rejected in a particular case where an observation falls in that class. In a discontinuous distribution, the acceptance region is therefore usually defined in practice in such a way that its probability of containing a random observation under the preliminary hypothesis is not smaller than the complement of the significance level [$P(H_0$ accepted if true$) \geq (1 - \alpha)$]. This is equivalent to granting a certain priority to the preliminary hypothesis.

Section 17.6: normal approximation and continuity correction

As already mentioned in section 17.3, the binomial distribution becomes more and more similar to the normal distribution when the exponent k increases ($k \rightarrow \infty$), particularly if $p \doteq q$. At the same time, the binomial distribution then has more numerous terms and requires heavier calculations. In practice, when k is large and $p \doteq q$, it is therefore advisable to use the normal distribution instead of the binomial distribution. This substitution is called the *normal approximation of the binomial distribution*.

The approximation of the binomial distribution by the normal distribution may be improved by applying the so-called *continuity correction*. The continuity correction was proposed in 1934 by the British statistician Frank Yates for the analysis of qualitative data with the χ^2 distribution, but it is appropriate also in the case of the normal approximation of the binomial distribution. The purpose of the continuity correction is to make the correspondence between a continuous and a discontinuous distribution as adequate as possible. In the binomial distribution, probabilities are concentrated at integer values of the binomial variate X but, in order to get a good agreement with the normal distribution, it is better to consider integer values as the midpoints of one-unit-wide classes and to represent the binomial distribution by a frequency histogram instead of a bar diagram (figure 17.6.1).

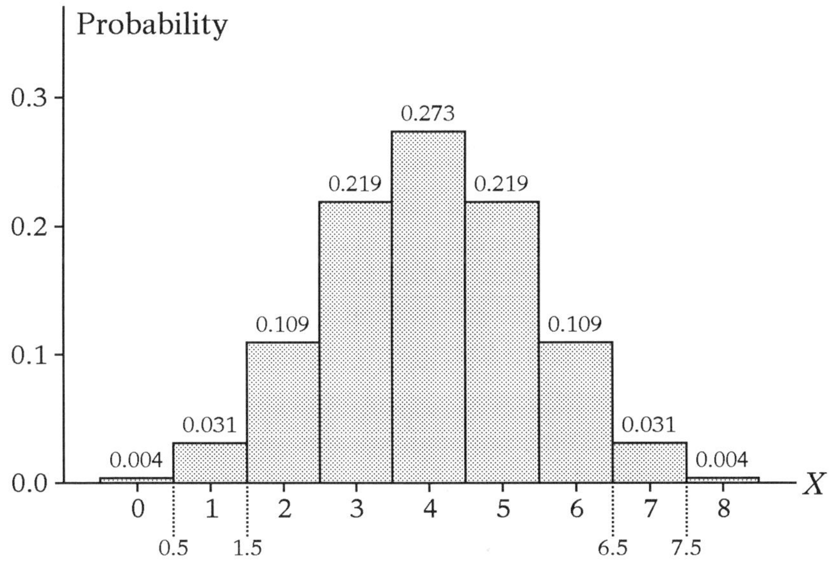

Figure 17.6.1
Continuous interpretation of a binomial distribution
($k = 8$, $p = 0{,}5$, and $\mu = kp = 4$)

Even if, theoretically, the binomial variate X can take only integer values, class limits are thus interpreted in practice as nonintegers. According to the nature of the questions asked, a particular class must then either be completely included or completely excluded:

(1) as a first example, assume that, in the binomial distribution represented in figure 17.6.1 (on the preceding page), the preliminary hypothesis $H_0: p=0.5$ (or equivalently $H_0: \mu=kp=4$) must be tested, the alternative hypothesis being $H_1: p>0.5$ or $H_1: \mu=kp>4$. Is the preliminary hypothesis acceptable if the value $X=7$ is observed? – Because a priority is granted to the preliminary hypothesis (section 17.5), the latter will be accepted if the class $X=7$ penetrates into the acceptance region (that is if its lower limit $X=6.5$ is located within it). Before a table of the normal distribution is consulted, the following standardized deviation must therefore be calculated (noting that $\mu=kp=4$ and that $\sigma=\sqrt{(kpq)}=\sqrt{2}$):

$$t = (6.5-\mu)/\sigma = [(7-\mu)-0.5]/\sigma = +[|7-4|-0.5]/\sqrt{2}\,;$$

since $$t_{(0.95;\,\infty)} = +1.645 < t = +1.768 < +1.960 = t_{(0.975;\,\infty)},$$

the preliminary hypothesis can be rejected at the 0.05 but accepted at the 0.025 level;

(2) in a second example, assume that the preliminary hypothesis $H_0: \mu=kp=4$ must be tested, the alternative hypothesis being $H_1: \mu=kp<4$. Is the preliminary hypothesis acceptable if the value $X=1$ is observed? – The preliminary hypothesis will be accepted if the observed class $X=1$ (that is its upper limit 1.5) penetrates into the acceptance region (figure 17.6.1, on the preceding page), and the standardized deviation is

$$t = (1.5-\mu)/\sigma = [(1-\mu)+0.5]/\sigma = -[|1-4|-0.5]/\sqrt{2}\,;$$

since $$t_{(0.025;\,\infty)} = -1.960 < t = -1.768 < -1.645 = t_{(0.05;\,\infty)},$$

the preliminary hypothesis can be rejected at the 0.05 but accepted at the 0.025 level.

The reader may note that, in both examples above, the standardized deviation must be calculated with the *most central* limit of the observed class (that is with the limit which is located closest to μ), which is equivalent to subtracting 0.5 from the absolute value $|X-\mu|$ of the deviation, and then giving back the sign of $(X-\mu)$ to the criterion:

$$t = [\text{sign}(X-\mu)]\left[\frac{|X-\mu|-0.5}{\sigma}\right] .$$

The function "sign of$(X-\mu)$" or "sign$(X-\mu)$" is available in many computer programming languages and takes the value -1, 0 or $+1$ depending on whether the difference $(X-\mu)$ is <0, $=0$, or >0.

Unless the exponent k is somewhat larger than in the examples above (let us say $k \geq 10$ or preferably $k \geq 20$), the direct utilization of the binomial distribution would be appreciably more accurate than the normal approximation. One particular use of the above criterion is the comparison of paired measurements through the sign test (discussed in section 17.11).

Section 17.7: estimating the probability *p*, the mean, and the variance

When observations are available about a large number of families of k children rather than on a single family, the probability p, the mean μ and the variance σ^2 can be estimated through computing methods adapted to grouped data (section 4.7). As in section 17.2, however, the binomial variate X is used itself as a summation index here and varies from 0 to k. The data may then be presented in a one-way frequency table. Table 17.7.1 gives the number X of boys per family in 53680 families of 8 children. The frequency $F(X)$ of each X value is the number of families of each type, that is 215 families of 0 boys (and 8 girls), 1485 families of 1 boy (and 7 girls), etc. .

Table 17.7.1
Number *X* of boys per family in families of eight children; data from Geissler (1889)

X	$F(X)$
0	215
1	1485
2	5331
3	10649
4	14959
5	11929
6	6678
7	2092
8	342
Total frequency (sample size)	53680

The total number of boys in the 53680 families is calculated by multiplying the frequency $F(X)$ of each type of family by the corresponding number X of boys and by summing up all of those products:

$$\text{Total number of boys} = \sum_{X=0}^{k} [X\,F(X)] = 221023\,;$$

the average number $\overline{X}$ of boys per family is estimated by dividing the total number of boys by the number of families ($N = \Sigma F(X) = 53680$):

$$\overline{X} = \sum_{X=0}^{k} [X\,F(X)] \Big/ \sum_{X=0}^{k} F(X) = 4.117418\,;$$

the estimate $\hat{p}$ of the probability of the birth of a boy is obtained by dividing the average number $\overline{X}$ of boys per family by the number k of children per family:

$$\hat{p} = \overline{X}/k = 0.514677\,;$$

there is therefore the same kind of relationship, $\overline{X} = k\hat{p}$, between the observed mean and probability, $\overline{X}$ and $\hat{p}$, as between the parametric mean and probability, $\mu = kp$.

Finally, the variance may be estimated by dividing the centered sum of squares by the equivalent $[\Sigma F(X) - 1]$ of the number of degrees of freedom:

$$S_X^2 = \sum_{X=0}^{k} [(X-\overline{X})^2 F(X)] \,/\, [\sum_{X=0}^{k} F(X) - 1] = 110979/53679 = 2.06745 .$$

Section 17.8: testing hypotheses concerning p with several families

When data are available about a large number N of families of k children, as in section 17.7, and a hypothesis has to be tested about the probability p, as in section 17.6, the binomial distribution and its normal approximation may be used, but by considering that all families put together constitute a *superfamily* of Nk children. The binomial distribution then takes the form

$$(p+q)^{Nk} = 1$$

and has the mean $\mu = Nkp$ and the variance $\sigma^2 = Nkpq$, the binomial variate here being the total number of boys

$$\sum_{X=0}^{k} [X\,F(X)] .$$

Because the exponent Nk is generally very large, the approximation of the binomial distribution of the superfamily by the normal distribution is particularly satisfactory. When the formula given for the standardized deviation in section 17.6 is adapted to the present case, it becomes

$$t = \text{sign}[\Sigma[X\,F(X)] - Nkp] \left[\frac{|[\Sigma[X\,F(X)] - Nkp]| - 0.5}{\sqrt{Nkpq}}\right] .$$

However, if the numerator and denominator of this standardized deviation are both divided by Nk, one gets

$$t = \text{sign}(\hat{p} - p) \left[\frac{|\hat{p} - p| - \dfrac{0.5}{Nk}}{\sqrt{\dfrac{pq}{Nk}}}\right] ,$$

which shows that, by comparing the total observed number of boys $\Sigma[X\,F(X)]$ with the total expected number of boys Nkp, one is comparing in fact the estimated probability $\hat{p}$ with the hypothetical probability p. Moreover, the fact that the continuity correction 0.5 is divided by Nk shows that the relative importance of the continuity correction decreases when the sample size or total frequency N increases. In the case of Geissler's data (table 17.7.1), if the hypothesis H_0: $p = 0.5$ is tested, $t = 19.235$ ($P < 0.001$). The conclusion must therefore be drawn that the probability of having a boy in a family of eight children is not exactly 1/2.

Section 17.9: tests of binomiality using the variance

While, in practice, the variance of a binomial variate X may be estimated directly from data (section 17.7),

$$S_X^2 = \sum_{X=0}^{k} [(X-\bar{X})^2 F(X)] / [\sum_{X=0}^{k} F(X) - 1],$$

in theory, the variance of a binomial distribution is $\sigma_X^2 = kpq$, unless the estimated probability $\hat{p}$ differs significantly from the hypothetical probability p and is substituted for it:

$$\hat{\sigma}_X^2 = k\hat{p}\hat{q} = k\hat{p}(1-\hat{p}).$$

The two values above, S_X^2 and $\hat{\sigma}_X^2 = k\hat{p}\hat{q}$ (or $\sigma_X^2 = kpq$), are practically never exactly equal because of sampling variation, but they should be approximately equal if the shape of the observed distribution is closely similar to that of the binomial distribution. The comparison of these two values is indeed the basis of a method for testing the *binomiality* hypothesis, H_0: $\sigma_X^2 = kpq$, according to which the distribution from which the sample has come is binomial in reality.

Recall that, in the case of the normal distribution, the ratio of the centered sum of squares to the parametric variance follows the χ^2 distribution (chapters 7 and 10):

$$\sum_{h=1}^{N} (X_h - \bar{X})^2 / \sigma_X^2 \leftarrow \chi^2_{(N-1)}.$$

Since the binomial distribution resembles the normal distribution, particularly when $p \doteq q$ and k is large (sections 17.3 and 17.6), the binomiality hypothesis (H_0: $\sigma_X^2 = kpq$) may be tested in an approximate manner by using a criterion analogous to the preceding expression:

$$\nu S_X^2 / (k\hat{p}\hat{q}) = \sum_{X=0}^{k} [(X-\bar{X})^2 F(X)] / (k\hat{p}\hat{q}) \doteq \chi^2_{(\nu)}, \quad \text{where } \nu = \Sigma F(X) - 1.$$

Because it is the numerator of the above criterion which reflects the shape of the distribution observed in reality, the acceptance region of H_0 is

$$\nu S_X^2 / (k\hat{p}\hat{q}) \leq \chi^2_{(1-\alpha;\, \nu)} \quad \text{if the alternative hypothesis is } H_1\text{: } \sigma_X^2 > kpq,$$

$$\chi^2_{(\alpha;\, \nu)} \leq \nu S_X^2 / (k\hat{p}\hat{q}) \quad \text{if the alternative hypothesis is } H_1\text{: } \sigma_X^2 < kpq,$$

and

$$\chi^2_{(\alpha/2;\, \nu)} \leq \nu S_X^2 / (k\hat{p}\hat{q}) \leq \chi^2_{(1-\alpha/2;\, \nu)} \quad \text{if } H_1\text{: } \sigma_X^2 \neq kpq.$$

If the conclusion is that $\sigma_X^2 < kpq$, this indicates that the data are less dispersed than what could be expected according to the binomial distribution. Successive births are then not fully independent, and there are fewer families than expected with many boys or with many girls. This may indicate a tendency to compensation, to regulation. The murder of female newborns, regrettably carried out in some countries, may tend to regularize the proportion of the number of boys per family. Nowadays, the fact that some parents would like to have children of both sexes may tend to decrease the variance of the number of boys (or of girls) in families of two or three children. The parents who have two or three children of the same sex may indeed try having additional children in an attempt to have at least one child of the other sex and, by doing so, they leave the categories of two-children or three-children families. Moreover, in some countries where abortion has been legalized and may be easily obtained, some parents have begun requesting an abortion when the fetus is not of the desired sex ... . Such kinds of behaviors are liable to decrease the variance of the number of boys (or of girls) per family.

If the conclusion were that $\sigma_X^2 > kpq$, this would indicate that the data would be more dispersed than according to the binomial distribution. Successive births would also not be not fully independent in this case, but there would be more families than expected with many boys or with many girls. There would then be a positive association (section 15.3) between births of the same sex. This could suggest that, possibly for genetical reasons, some couples would tend to have mostly boys and other couples would tend to have mostly girls.

The number of degrees of freedom occurring in binomiality tests are often much larger that those currently found in χ^2 tables. The required χ^2 values may usually be obtained with a satisfactory accuracy by using Wilson and Hilferty's approximation (section 7.3).

In the case of Geissler's data (table 17.7.1), the observed variance, $S_X^2 = 2.06745$, is slightly larger than the variance expected under the binomiality hypothesis, $\hat{\sigma}_X^2 = k\,\hat{p}\,\hat{q} = 1.99828$, and the binomiality criterion has 53679 degrees of freedom and takes the value 55537.3. The hypothesis that the data of Geissler (1889) follow the binomial distribution exactly must be rejected.

Section 17.10: tests of goodness of fit

In addition to the binomiality test discussed in section 17.9, one may also use a test of goodness of fit (chapter 16) to check the hypothesis that data come from a binomial distribution. The test of goodness of fit of a binomial distribution is similar to that of a normal distribution (section 16.2) except that the successive classes of the binomial variate X correspond simply to the integers 0, 1, 2, ... k and that the maximum number of classes is $R = k + 1$. Extreme classes must be pooled here also when the corresponding expected absolute frequencies are too low, in order to satisfy the validity conditions mentioned in sections 15.6 and 16.2, and the number of classes after pooling is denoted by R' instead of R. The number of degrees of freedom of Pearson's criterion is $(R-2)$ or $(R'-2)$ if the probabilities of the successive classes are based on the probability $\hat{p}$ estimated from the data. On the contrary, if the hypothetical probability p is acceptable and can be used to evaluate the probabilities of the successive classes, then the number of degrees of freedom of Pearson's criterion is $(R-1)$ or $(R'-1)$.

The results yielded by the binomiality test (section 17.9) and by the test of goodness of fit usually do not agree completely because the binomiality test is affected mostly by the strength of dispersion while the test of goodness of fit reflects the general shape of the distribution. While the binomiality test may be used even when there is little data, the test of goodness of fit requires a total frequency which is large enough: otherwise, too many classes might have to be pooled, which would reduce the sensitivity (power) of the test.

In the case of Geissler's data (table 17.7.1), thanks to the large number of observations, no classes have to be pooled and Pearson's criterion takes the value

$$\chi^2_{(R-2)} = \chi^2_{(7)} = 91.8731 > 26.018 = \chi^2_{(0.9995;\,7)}\ ;$$

the test of goodness of fit thus confirms that the binomial distribution does not fit the data of Geissler (1889) satisfactorily.

Section 17.11: an application of the binomial distribution: the sign test

Even though nonparametric methods are treated only briefly in this textbook, the *sign test* is discussed in this section because of its simplicity and because it is an interesting example of the use of the binomial distribution. The sign test may be used to compare paired measurements (see section 9.9) without assuming that the data follow the normal distribution. Like the analysis of frequency tables using χ^2 (chapters 15 and 16), the sign test belongs to the category of the so-called *nonparametric* methods, which should perhaps rather be called *distribution-free* methods although that name would not be completely appropriate either. Nonparametric methods have the advantage of not requiring precise assumptions concerning the shape of the probability distribution of the variates studied, but they have the drawback that they are generally less sensitive (less "powerful") than classical (parametric) statistical methods.

When paired measurements are compared (section 9.9), the preliminary hypothesis states that the two measurements X_A and X_B of each pair come from identical populations, and their difference $Y=(X_B-X_A)$ then has equal probabilities of being positive $(p=0.5)$ or negative $(q=0.5)$. However, because the accuracy of observational or experimental measurements is usually limited, it may happen that the measurements X_A and X_B of some pairs are equal and that their difference Y is null: such pairs do not provide any information concerning the true sign of the difference and are usually excluded from the sign test.

In the case of the data of table 9.9.1, on the length of the left and right humeri of female skeletons, $X=9$ of the $k=10$ differences are positive and none are null. According to the preliminary hypothesis, the number X of positive differences possesses the binomial distribution

$$(p+q)^k = (0.5+0.5)^{10} = 1$$

of which the mean and the standard deviation are

$$\mu = kp = 10 \times 0.5 = 5$$

and

$$\sigma_X = \sqrt{(kpq)} = \sqrt{(10 \times 0.5 \times 0.5)} = \sqrt{2.5}\ .$$

The compatibility of the observed number $X=9$ of positive differences with the preliminary hypothesis may therefore be tested by using the criterion described in section 17.6:

$$t_{(\infty)} = [\text{sign}(9-5)]\left[\frac{|9-5|-0.5}{\sqrt{2.5}}\right] = +2.2136 > +1.96 = t_{(0.975;\,\infty)}\ .$$

If, like in section 9.9, the alternative hypothesis states that the second measurement X_B of each pair comes from a population of which the central tendency is higher than that of the population from which comes the first measurement X_A, the difference $Y = (X_B - X_A)$ should tend to take positive values more often than negative values, and the acceptance region of the preliminary hypothesis is

$$t_{(\infty)} \leq t_{(1-\alpha;\,\infty)} ;$$

the preliminary hypothesis may thus be rejected here at the significance level $\alpha = 0.025$.

It should be noted that the length difference of left and right humeri is statistically less significant with the sign test ($P < 0.025$) than with the paired Student's t test (section 9.9, $P < 0.001$). This reflects the fact that, in general, nonparametric methods are less sensitive (less "powerful") than the corresponding parametric methods.

Section 17.12: a generalization, the polynomial (multinomial) distribution

When the k independent simple events which jointly make up a compound event may happen in more than two complementary manners, the binomial distribution is replaced by the *polynomial distribution* (also called *multinomial distribution*). In palynology, for instance, if grains of pollen of three plant genera are present in a peat bog, and if the relative frequencies (probabilities) of these three genera are p_1, p_2 and p_3 respectively, and satisfy the condition $p_1 + p_2 + p_3 = 1.0$, the probability of getting a peat sample containing k pollen grains of which X_1 are of genus #1, X_2 of genus #2 and X_3 of genus #3 is

$$\{k!/[X_1!\, X_2!\, X_3!]\}\; p_1^{X_1}\, p_2^{X_2}\, p_3^{X_3} ,$$

and the probability distribution of all possible types of samples is

$$(p_1 + p_2 + p_3)^k \;=\; \sum_{X_1=0}^{k}\; \sum_{X_2=0}^{(k-X_1)} \{k!/[X_1!\, X_2!\, X_3!]\}\; p_1^{X_1}\, p_2^{X_2}\, p_3^{X_3} \;=\; 1.0 .$$

The distribution above is called *trinomial* because each simple event can happen in three ways. Summation does not have to be carried out with respect to the last variate X_3 because the latter is completely determined by the preceding variates: $X_3 = (k - X_1 - X_2)$. The binomial distribution is the particular case of the polynomial distribution where each simple event may happen in only two ways,

$$(p_1 + p_2)^k \;=\; \sum_{X_1=0}^{k} \{k!/[X_1!\, X_2!\,]\}\; p_1^{X_1}\, p_2^{X_2} \;=\; 1.0 ,$$

as may be seen by reverting to the notation of the earlier sections and carrying out the substitutions $p_1 = p$, $p_2 = q$, $X_1 = X$, and $X_2 = (k - X)$. Moreover, a polynomial distribution may be considered provisionally as a binomial distribution if the several ways in which the k simple events may happen are regrouped into two classes. It may be shown in this way that each variate X_i of a polynomial distribution has a mean equal to kp_i and a variance equal to $kp_i(1 - p_i)$, and the covariance of two polynomial variates X_i and X_j may be shown to be $-kp_i p_j$.

The polynomial distribution and related distributions have been used by Mosimann (1962, 1963) to analyze correlations (chapter 19) between the proportions of pollen grains coming from various plant genera in palynological surveys. While the polynomial distribution is not easy to apply in practice, it is theoretically important because it can be used to show that Pearson's criterion, utilized in the analysis of frequency tables (chapters 15 and 16), follows the χ^2 distribution approximately (section 29.9).

Chapter 18
The Poisson distribution

Section 18.1: introduction

The Poisson distribution is a discontinuous distribution which was described by the French mathematician Siméon Denis Poisson (1781-1840) in a book published in 1837. The Poisson distribution is a particular case of the binomial distribution (chapter 17) where the probability p becomes smaller and smaller ($p \rightarrow 0$) while the exponent k becomes larger and larger ($k \rightarrow \infty$) in such a way that their product, the mean $\mu = kp$, keeps a finite value ($0 < \mu = kp < \infty$). The Poisson distribution might thus be defined by the following equation:

$$\underset{\substack{k \rightarrow \infty,\, p \rightarrow 0 \\ 0 < \mu = kp < \infty}}{\text{Limit}} \{(p+q)^k\} = 1 .$$

As will be seen in the rest of this chapter, however, the probability p and the exponent k do not have to appear explicitly in the Poisson distribution because the latter may be expressed more simply with a single parameter, the mean μ. The Poisson distribution is appropriate in the analysis of the frequency of an event of which the probability is very small but which occurs nevertheless because there is a large number of occasions for it to happen. The Poisson distribution is used in many fields of applied as well as of pure science. In the biological sciences, the Poisson distribution is useful in the analysis of the occurrence of organisms within the space-time continuum, in ecology, in bacteriology and epidemiology, and in the study of accidental phenomena, such as genetic mutations.

Section 18.2: an artificial example

Assume that a finite biological population comprising $k = 100$ individual organisms occupies a finite area divisible into 100 equal and similar quadrats. Assume further that each organism has equal chances of being in any quadrat irrespective of whether other organisms are already present within it or not (this implies that there are neither positive nor negative interactions between individual organisms). The probability that one of the individual organisms is within a particular quadrat is then $p = 0.01$, and the number X of organisms per quadrat may theoretically take any integer value from 0 to 100. The probability that there are X organisms within a particular quadrat is given by the $(k+1) = 101$ successive terms of the expansion of the binomial

$$(p+q)^k = (0.01 + 0.99)^{100} = 1 .$$

The mean of this binomial distribution, which is markedly skewed, is

$$\mu = kp = 100(0.01) = 1 ,$$

which means that there is on the average 1 organism per quadrat.

Assume now that the area occupied by the population is a thousand times larger and is divisible into 100,000 equal and similar quadrats, in such a way that the probability that there is an individual organism within a particular quadrat is $p = 0.00001$. At the same time, the total number of organisms is also a thousand times larger, $k = 100{,}000$, so that the mean number of organisms per quadrat, the *population density* as it is called in ecology, is still 1 organism per quadrat:

$$kp = 100{,}000\,(0.00001) = 1 \,.$$

The probability that there is an individual organism within a particular quadrat is still assumed to be unaffected by the possible presence or absence of other organisms in the same quadrat, which implies that individual organisms are small enough and that 100,000 of them could be present within a single quadrat without interfering with each other. Under these conditions, the number X of organisms present in a particular quadrat may now vary from 0 to 100,000, and the probability distribution of X is

$$(p+q)^k = (0.00001 + 0.99999)^{100{,}000} = 1 \,.$$

This binomial distribution contains $(k+1) = 100{,}001$ terms and would be exceedingly difficult to use. Fortunately, Poisson found that, when $k \to \infty$ and $p \to 0$ but $0 < kp < \infty$, the binomial distribution becomes more and more similar to a simpler distribution of which, even though the number of terms is theoretically infinite, only a finite number of terms need to be considered in practice.

Section 18.3: terms of the Poisson distribution

In this section, an attempt will be made to see how the *general term* of the binomial distribution (section 17.2),

$$\frac{k!}{X!\,(k-X)!}\; p^X q^{(k-X)} \,,$$

may be simplified when $k \to \infty$ and $p \to 0$ but $0 < kp < \infty$, in order to obtain a simpler expression for the Poisson distribution. Dividing $k!$ by $(k-X)!$, multiplying each of the X factors in the numerator of the binomial coefficient by p, and replacing q by $(1-p)$ yields

$$\frac{kp(kp-p)\,\ldots\,(kp-pX+p)}{X!}\;\frac{(1-p)^k}{(1-p)^X} \,.$$

But, since $\mu = kp$ and $p = \mu/k$, the above becomes

$$\frac{\mu(\mu-p)\,\ldots\,(\mu-pX+p)}{X!}\;\frac{(1-\mu/k)^k}{(1-p)^X} \,.$$

Finally, since $\lim_{p \to 0} \{ [\mu(\mu - p) \ldots (\mu - pX + p)] \} = \mu^X,$

$$\lim_{p \to 0} \{ (1-p)^X \} = 1^X = 1,$$

and $\lim_{k \to \infty} \{ (1 - \mu/k)^k \} = e^{-\mu},$

the general term is

$$\frac{\mu^X}{X! \quad e^{\mu}},$$

and the Poisson distribution may be expressed by the equation

$$\underset{\substack{k \to \infty,\, p \to 0 \\ 0 < \mu = kp < \infty}}{\text{Limit}} \{ (p+q)^k \} = \sum_{X=0}^{\infty} \frac{\mu^X}{X! \; e^{\mu}} = 1.$$

Even though the number of terms of the Poisson distribution is theoretically infinite, the magnitude of successive terms usually becomes rapidly negligible, particularly when the mean μ is small, and computations can be ended when the sum of the terms already calculated is close enough to unity.

Section 18.4: geometrical aspect of the Poisson distribution

The Poisson distribution is strongly skewed when the mean μ is small (figure 18.4.1, top) but its degree of skewness decreases when the mean increases (figure 18.4.1, center and bottom). The evaluation of the general term (section 18.3) shows that the latter is equal to $1/e^{\mu}$ when $X=0$ and to μ/e^{μ} when $X=1$. The mode of the distribution is therefore the class $X=0$ when $\mu<1$ (figure 18.4.1, top) but is located more to the right when $\mu>1$. The probabilities of the classes $X=0$ and $X=1$ are equal when $\mu=1$. The Poisson distribution becomes less and less skewed when the mean increases: when $\mu=10$, the skewness is already barely visible (figure 18.4.1, bottom). Like the binomial distribution (sections 17.3 and 17.6), of which it is a particular case, the Poisson distribution is therefore often replaced in practice by the normal distribution, particularly when the mean μ is large. However, the lognormal distribution often provides an even better approximation because it is nonnegative and positively skewed (section 14.4).

Section 18.5: parametric mean and variance of the Poisson distribution

The parametric mean of the Poisson distribution is

$$\underset{\substack{k \to \infty,\, p \to 0 \\ 0 < \mu = kp < \infty}}{\text{Limit}} \{kp\} = \mu_X = \mu.$$

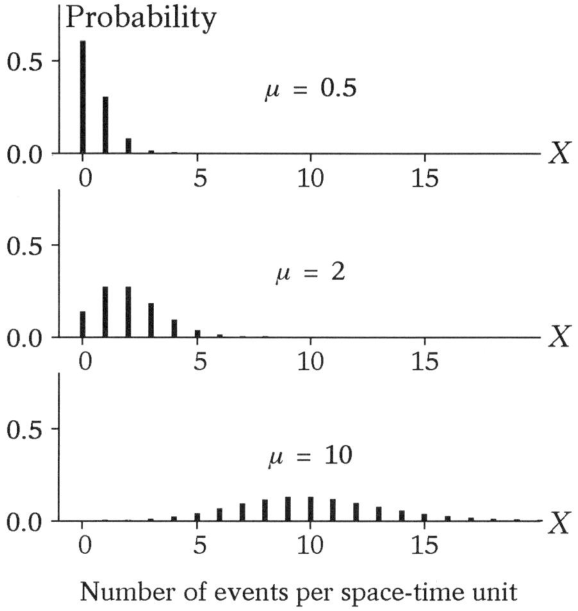

Figure 18.4.1
Geometrical aspect of the Poisson distribution according to the value of the mean μ

As for the variance, since the variance of the binomial distribution is

$$\sigma_X^2 = kpq = kp(1-p),$$

the variance of the Poisson distribution is

$$\sigma_X^2 = \underset{\substack{k\to\infty,\, p\to 0\\ 0<\mu=kp<\infty}}{\text{Limit}} \{kp(1-p)\} = \underset{\substack{k\to\infty,\, p\to 0\\ 0<\mu=kp<\infty}}{\text{Limit}} \{kp\} \times \underset{\substack{k\to\infty,\, p\to 0\\ 0<\mu=kp<\infty}}{\text{Limit}} \{(1-p)\}$$

$$= \mu \times 1 = \mu .$$

The variance and the mean of the Poisson distribution are therefore equal:

$$\sigma_X^2 = \mu .$$

In comparison with the Poisson distribution, the variance is smaller than the mean in the binomial distribution (section 17.4),

$$\sigma_X^2 = kpq = \mu(1-p) < \mu,$$

and unrelated to the mean in the normal distribution (section 5.3).

Section 18.6: estimating the mean and the variance

The mean and the variance of the Poisson distribution are estimated according to methods applicable to grouped data (section 4.7) in the same way as in the case of the binomial distribution (section 17.7). It should be noted that, when the data really follow a Poisson distribution, $\overline{X}$ and S_X^2 are estimates of the same parameter, $\mu = \sigma_X^2$, but these two estimates always differ somewhat from each other in practice because of sampling variation.

Section 18.7: tests of Poissonianity[1] (sporadicity)

When the estimates $\overline{X}$ and S_X^2 differ much from each other, this may suggest that the data do not really follow a Poisson distribution. The acceptability of the preliminary hypothesis $H_0: \sigma_X^2 = \mu$ may be checked by a *test of Poissonianity* through a method analogous to the one used for the binomiality test (section 17.9). Since the Poisson distribution resembles the normal distribution (particularly when the mean is large), the ratio of the centered sum of squares to the observed mean follows the χ^2 distribution approximately:

$$\nu S_X^2 / \overline{X} = \Sigma[(X-\overline{X})^2 F(X)]/\overline{X} \doteq \chi^2_{(\nu)}, \quad \text{where } \nu = \Sigma F(X) - 1 .$$

The ratio $S_X^2/\overline{X}$ (or more generally S_X^2/σ_X^2) is called the *dispersion index* by some writers. When the data really follow a Poisson distribution, the dispersion index is expected to be close to unity. The acceptance region of H_0 is

$$\nu S_X^2 / \overline{X} \leq \chi^2_{(1-\alpha;\, \nu)} \quad \text{if the alternative hypothesis is } H_1\colon\ \sigma_X^2 > \mu ,$$

$$\chi^2_{(\alpha;\, \nu)} \leq \nu S_X^2 / \overline{X} \quad \text{if the alternative hypothesis is } H_1\colon\ \sigma_X^2 < \mu ,$$

and

$$\chi^2_{(\alpha/2;\, \nu)} \leq \nu S_X^2 / \overline{X} \leq \chi^2_{(1-\alpha/2;\, \nu)} \quad \text{if } H_1\colon\ \sigma_X^2 \neq \mu .$$

When the preliminary hypothesis $H_0: \sigma_X^2 = \mu$ must be rejected, the data are concluded not to follow the Poisson distribution. This might be due to various reasons but, like the binomial distribution, the Poisson distribution involves the particularly important assumption that the simple events making up the compound event are independent from each other. If the preliminary hypothesis H_0 is unacceptable, this may suggest, among other things, that the presence of individual organisms within a particular quadrat may not be independent. Figure 18.7.1 illustrates three artificial examples of spatial arrangements of organisms in quadrats, for which the Poisson distribution is suitable (figure 18.7.1, center) or unsuitable (figure 18.7.1, top and bottom).

When $\sigma_X^2 < \mu$ (figure 18.7.1, top), there are fewer small and large observations than expected, and there are more observations close to the mean. This suggests that there

[1] The adjective Poissonian may be derived from the name of Siméon Denis Poisson in the same manner as the adjectives Gaussian and Hertzian from the names of Gauss and Hertz. It follows that the agreement of a frequency or probability distribution with the Poisson distribution may be called Poissonianity. Because the Poisson distribution applies to sporadic phenomena, the agreement with the Poisson distribution may also be called sporadicity.

may be a negative association between simple events: in the case of the spatial arrangement of organisms, the probability of having organisms in a quadrat would be decreased if other organisms are already present. From an ecological viewpoint, this could be due to *competition* for light, water or minerals in the case of plants and for food or space itself *(territoriality)* in the case of animals. A spatial arrangement in which $\sigma_X^2 < \mu$ is said to be *regular*. In the extreme case of a perfectly regular arrangement (figure 18.7.1, top), because of negative interactions between organisms, the number X of individual organisms per quadrat would be constant and equal to the mean and the variance would be null: $X = \mu_X$ and $\sigma_X^2 = 0$. Note that, in this special case, the number of organisms per quadrat would no longer be a random variable (variate). A good example of a regular arrangement may be found in a rookery of gannets, those large maritime birds of which the nests are more or less equally spaced because the adults of each breeding pair keep neighbors away as far as they can reach with their bill while sitting on their nest.

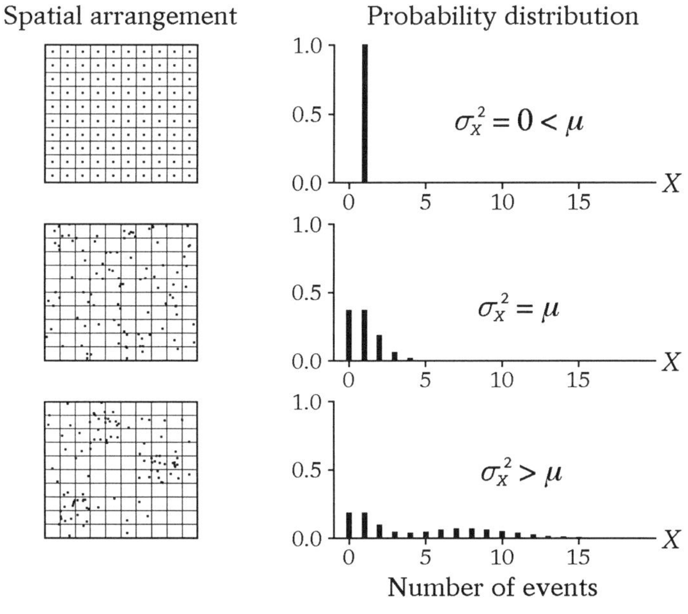

Figure 18.7.1
Examples of spatial arrangements corresponding to a regular distribution (top), to a sporadic or Poisson distribution (center), and to an aggregative distribution (bottom)

If $\sigma_X^2 > \mu$, there are more small and large observations than expected and fewer observations close to the mean. This suggests that there may be a positive association between simple events: in the case of the spatial arrangement of organisms, the probability of having organisms in a quadrat would be increased if other organisms are already present. In practice, this would indicate that organisms tend to be present in groups or bunches, and the distribution is said to be *aggregative, contagious* or *clumped*. In hematology, when red blood cells are counted, their tendency to agglutinate may result in an aggregative arrangement which increases the variance and makes the estimated mean number of cells per unit inaccurate: this may be prevented by stirring the solution carefully before examining it.

From an ecological viewpoint, the *aggregativity (contagiousness, clumpiness)* of a distribution may be due either to the heterogeneity of the environment or to the fact that organisms tend to be associated in reality. When an attempt is made to detect real associations between organisms, the possibility must first be checked that apparent associations might be due completely to environmental heterogeneity. If some parts of the environment provide more favorable habitats than others, organisms may be more abundant there even though they may not otherwise tend to associate. However, when the aggregativity of a distribution does not appear to be due to environmental heterogeneity, it may be interpreted as a result of real associative tendencies, such as gregariousness in animals or the quest of multilateral or unilateral benefits in plants.

When a space-time arrangement is strongly aggregative, this may lead to bimodality (figure 18.7.1, bottom) or even to the cleavage of the probability distribution into two separate distributions, one where the number X of organisms per space-time unit has low or null values and the other where X has high values (figure 18.7.2, next page). This kind of phenomenon occurs indeed in epidemiology in the case of the so-called contagious diseases: when one member of a family or of a group of individuals is affected by a contagious disease, the production of infectious germs by the pathogenic organism increases the probability that other members of the family or of the group will contract the disease. There is then a risk of the occurrence of a large number of cases of the disease within a short time in a particular locality, which is called an *epidemic* in human beings, an *epizootic* in animals and an *epiphytotic* in plants. On the contrary, when the pathogenic organism is absent, there are no cases of the disease.

As opposed to the epidemic phenomena just discussed, *sporadic* phenomena (from the Greek σποραδικος, *dispersed;* σπορα, *seed;* σπειρειν, to *sow*) are those which occur mostly as isolated and rare cases and which happen irregularly and without ever reaching epidemic frequencies. The Poisson distribution is the probability distribution of sporadic phenomena. However, it may happen that the space-time occurrence of some phenomena, such as the abundance of marine plankton, is sporadic at some levels but regular or aggregative at other levels.

Section 18.8: tests of goodness of fit

In addition to the Poissonianity test discussed in section 18.7, a test of goodness of fit (chapter 16) may also be used to check the hypothesis that data come from a Poisson distribution. The test of goodness of fit of a Poisson distribution is done in the same way as that of a binomial distribution (section 17.10) except that, in biology, a precise hypothetical value is practically never available for the parametric mean μ of a Poisson distribution. The estimated mean $\overline{X}$ is therefore practically always used to evaluate the probabilities of the successive classes, and the number of degrees of freedom of Pearson's criterion is $(R' - 2)$, where R' denotes the number of classes after pooling.

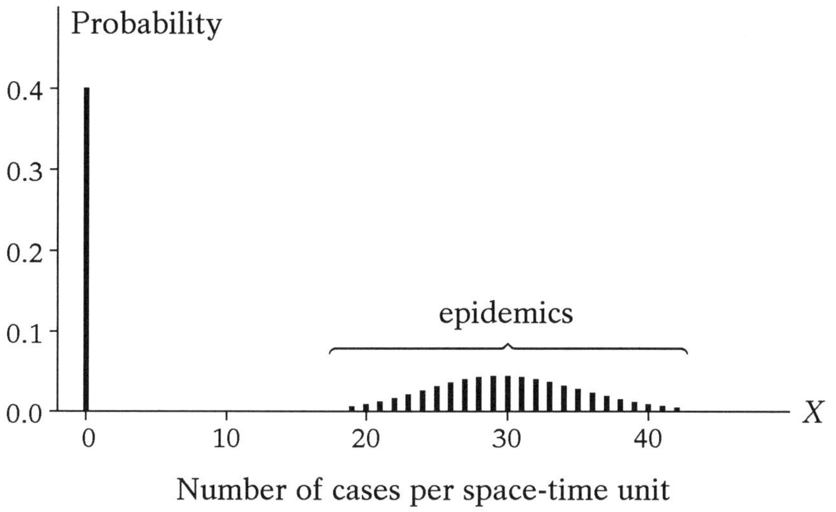

Figure 18.7.2
Cleavage of the strongly aggregative distribution of an epidemic phenomenon into two separate distributions

In general, the results of a test of Poissonianity and of a test of goodness of fit do not agree completely. Like the binomiality test (section 17.9), the Poissonianity test is sensitive mostly to the intensity of dispersion. As for the test of goodness of fit, it requires a large total frequency, and detecting weak or irregular associations may be difficult.

Section 18.9: an ecological example

Breedlove and Ehrlich (1968) have studied the predation of the larvae of the lepidopteran *Glaucopsyche lygdamus* on the leguminous plant *Lupinus amplus* in the Rocky Mountains of the state of Colorado. The number X of eggs laid was counted on a sample of inflorescences on which all flowers were still closed. Table 18.9.1 (next page) summarizes an attempt to fit the Poisson distribution to Breedlove and Ehrlich's data. The estimated mean number of eggs per inflorescence is

$$\bar{X} = \Sigma[XF(X)]/\Sigma F(X) = 1.112$$

and is used to estimate the probability of each X value according to the expression

$$\bar{X}^X/(X!\ e^{\bar{X}})\ ;$$

in order to prevent the occurrence of expected frequencies $NP(X)$ lower than unity, classes $X = 4, 5, 6, 7, 8$ are pooled before calculating Pearson's criterion, and the latter has $(R' - 2) = (5 - 2) = 3$ degrees of freedom $(\chi^2_{(3)} = 8.537 > 7.8147 = \chi^2_{(0.95;\ 3)})$.

Table 18.9.1
Poisson distribution fitted to the data of Breedlove and Ehrlich (1968)
(data reproduced with the permission of the *American Association for the Advancement of Science*)

X	$P(X)$	$F(X)$	$NP(X)$	$F(X) - NP(X)$	χ^2
0	0.328901	43	41.1126	+ 1.88744	0.086650
1	0.365737	53	45.7172	+ 7.28283	1.160168
2	0.203350	14	25.4187	− 11.4187	5.129591
3	0.075375	9	9.42188	− 0.421882	0.018891
4	0.020954	3	2.61928		
5	0.004660	1	0.582529		
6	0.000864	1	0.107962	2.6704	2.1417
7	0.000137	0	0.017151		
8	0.000019	1	0.002384		
Sums	1	125	125	0	8.537

According to the test of goodness of fit, the hypothesis that the number X of eggs per inflorescence follows a Poisson distribution can barely be rejected ($P < \alpha = 0.05$). However, as indicated in the preceding section (18.8), a test of goodness of fit is not very sensitive when several end classes have to be pooled like here. Nevertheless, the observed variance $S_X^2 = 1.6809$ is clearly greater than the mean $\overline{X} = 1.112$ and the Poissonianity test allows the hypothesis that the data have a Poisson distribution to be rejected very conclusively ($\chi^2_{(124)} = 187.439 > 182.445 = \chi^2_{(0.9995;\ 124)}$; $P < 0.0005$).

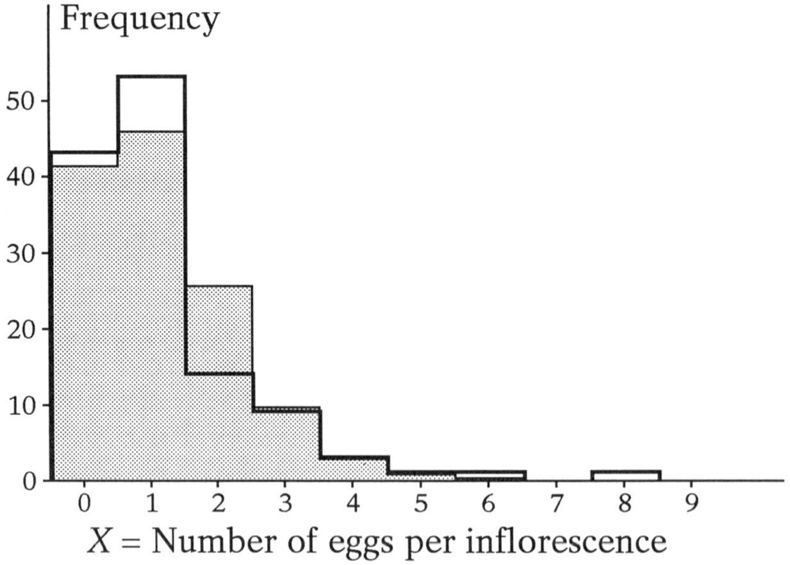

Figure 18.9.1
Histograms of observed frequencies (clear, heavy outline) and of frequencies expected according to the Poisson distribution (stippled, light outline) for the data of Breedlove and Ehrlich (1968)

The histograms of observed and expected frequencies are represented in figure 18.9.1. It should be noted that, if the number X of eggs per inflorescence really followed a Poisson distribution with a mean $\mu = 1.112$, the probability of having an observation at the value $X = 8$ would be very low ($P = 0.000019$). By itself, the presence of that observation thus suggests that the data are not distributed completely at random. The readers who wonder about the reasons for which some lepidopterans lay their eggs in clusters and others lay them singly may wish to consult the discussion by Stamp (1980). As for the species *Glaucopsyche lygdamus*, it does seem that all females do not lay their eggs in the same way.

Chapter 19
The bivariate normal distribution and the correlation coefficient, *r*

Section 19.1: introduction

Except in chapter 1 and in sections 3.6, 3.9, 3.10, and 15.3 to 15.5, the concepts and methods discussed up to now in this book concern the analysis of a single variate at a time. The simultaneous analysis of two variates will be taken up in this chapter. In order to provide as concrete an example as possible, figure 19.1.1 represents the scatter diagram of skull length X and skull width Y in 76 male (black dots) and 53 female (white dots) North American martens. The measurements were taken with vernier calipers and have an accuracy of 0.1 mm (Jolicoeur, 1963b, 1963c). The fact that black dots are located mostly higher and more to the right than white dots in the scatter diagram appears to indicate that males are larger than females.

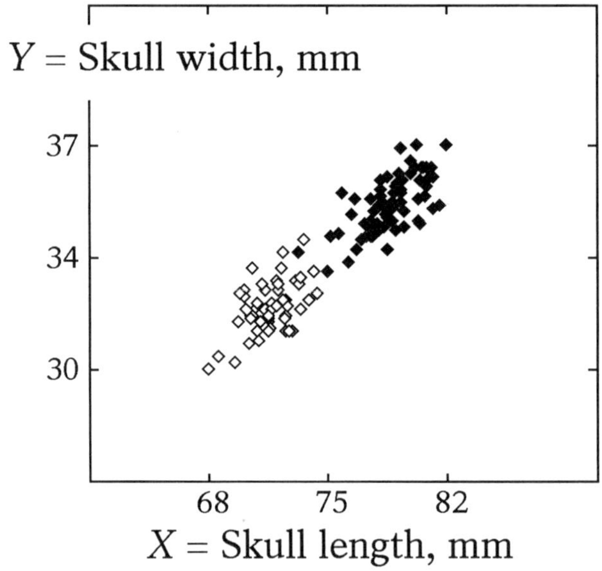

Figure 19.1.1
Scatter diagram of the skull length and width of 76 male (black dots) and 53 female (white dots) North American martens, according to data from Jolicoeur (1963b, 1963c)

Because individual dots tend to overlap in a scatter diagram when they are very numerous, it is sometimes preferable to group observations like the preceding into classes broader than the unit of measurement, which yields a two-way frequency table (sections 15.3 to 15.5), also called a *correlation table* in this context. When two variates

X and Y are analyzed simultaneously, their graphical representation already requires two dimensions, so that the illustration of the frequency or of the probability density calls for a third coordinate axis (figures 19.1.2 and 19.2.1). The graphical representation of a two-way frequency table is called a *frequency stereogram* (figure 19.1.2). The frequency stereogram is a three-dimensional version, a generalization, of the frequency histogram used in the analysis of a single variate (figures 3.3.1.b, 3.3.2 and 3.4.1).

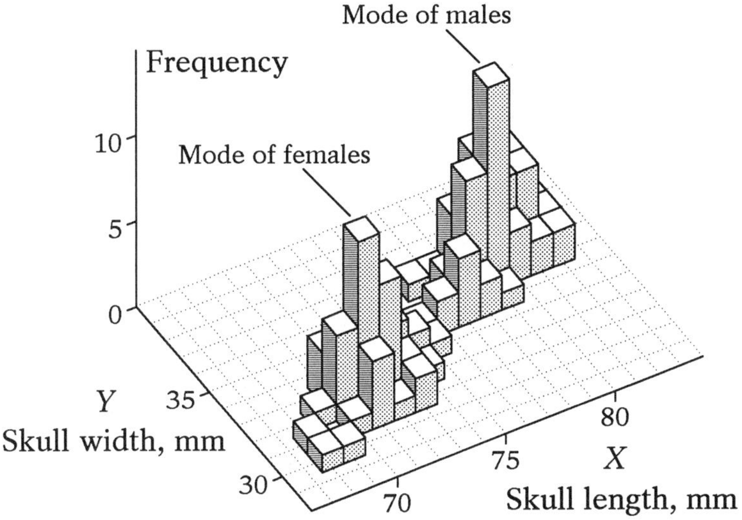

Figure 19.1.2
Stereogram of the bivariate frequency distribution of the skull length and width of 76 male and 53 female North American martens, after observations have been grouped into one-millimeter-wide classes (see also figures 3.4.1 and 19.1.1)

Like the frequency histogram of the skull length of North American martens (figure 3.4.1), the frequency stereogram of skull length and skull width clearly reveals the existence of two peaks, two modes corresponding to males and females respectively (figure 19.1.2). The scatter diagram of the skull dimensions of martens in figure 19.1.1 and the frequency stereogram in figure 19.1.2 are particularly interesting because they reflect the presence of the sexual dimorphism of body size in this species of carnivore. In a statistical analysis, however, as already noted in section 3.4, it is generally preferable to separate data into subsets which are as homogeneous as possible, by distinguishing males from females for instance.

A second example of bivariate data which do not give indications of heterogeneity will therefore be considered: the scatter diagram of calcium concentration X, in µg/ml, and phosphorus concentration Y, in hundredths of mg/ml, of specimens of morning urine from underweight adult men (figure 19.1.3, next page; data from Brown and Beerstecher, 1951; see also table 19.3.1).

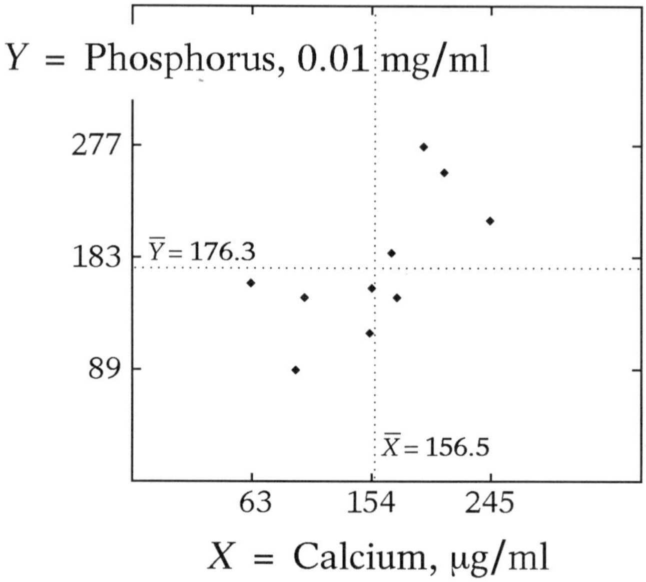

Figure 19.1.3
Scatter diagram of the concentrations of calcium (µg/ml) and phosphorus (0.01 mg/ml) in morning urine specimens from underweight adult men; the average concentrations of calcium ($\bar{X}$ = 156.5) and phosphorus ($\bar{Y}$ = 176.3) are represented by dotted lines (data from Brown and Beerstecher, 1951)

The mean concentrations of calcium $\bar{X}$ and phosphorus $\bar{Y}$ are represented by dotted lines (figure 19.1.3), and the examination of this scatter diagram shows that there are more individual dots in the lower left and upper right than in the upper left and lower right quadrants. This suggests that there might be an increasing statistical relationship between calcium and phosphorus concentrations in those human urine specimens.

Section 19.2: the bivariate normal probability density

While the univariate normal distribution can be represented by a curve having the shape of a bell which is flattened on a sheet of paper (figures 5.1.1 and 5.2.1), the bivariate normal distribution may be represented by a surface which has the shape of a genuine bell floating in three-dimensional space above the plane on which individual observations are dispersed (figure 19.2.1).

The probability density $f(X_0, Y_0)$, corresponding to a possible pair of measurements (X_0, Y_0), is given by the following equation:

$$f(X_0, Y_0) = \frac{1}{2\pi \sqrt{(\sigma_X^2 \sigma_Y^2 - \sigma_{XY}^2)}} \exp\left\{ \frac{-\sigma_X^2 \sigma_Y^2}{2(\sigma_X^2 \sigma_Y^2 - \sigma_{XY}^2)} \left[\frac{(X_0 - \mu_X)^2}{\sigma_X^2} - \frac{2\sigma_{XY}}{\sigma_X \sigma_Y} \frac{(X_0 - \mu_X)}{\sigma_X} \frac{(Y_0 - \mu_Y)}{\sigma_Y} + \frac{(Y_0 - \mu_Y)^2}{\sigma_Y^2} \right] \right\}.$$

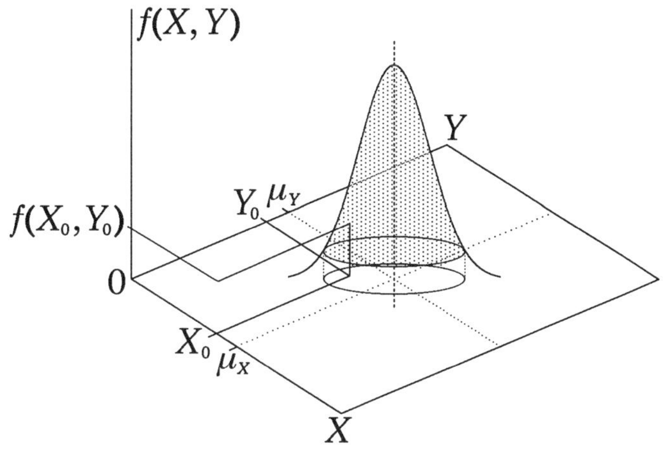

Figure 19.2.1
The probability density surface of the bivariate normal distribution has the shape of a genuine bell suspended in three-dimensional space; the projection on the plane of variates *X* and *Y* of a circular or elliptical contour line on which the probability density *f*(*X*, *Y*) is constant is called an *ellipse of equal probability density*

A careful examination of the probability density shows that, in addition to the means μ_X and μ_Y and to the variances σ_X^2 and σ_Y^2 of the two variates, it contains the covariance σ_{XY}, a quantity previously mentioned in section 3.10 and which appears when two or more variates are analyzed simultaneously. The mean μ_Y and the variance σ_Y^2 relative to the second variate Y are defined by equations similar to those used for variate X in section 5.3. As for the parametric covariance σ_{XY}, it is defined by an expression analogous to that for the variance except that it corresponds to the mean value of the product of the deviations of the two variates and that it requires a double instead of a simple integral, because integration must be carried out in two dimensions in this case:

$$\sigma_{XY} = \int_{-\infty}^{+\infty} \int_{-\infty}^{+\infty} (X-\mu_X)(Y-\mu_Y)\, f(X, Y)\, dX\, dY \ .$$

Section 19.3: bivariate statistical computations

In practice, the means, the variances and the standard deviations of the two variates are estimated from a sample of size N in the manner explained in sections 4.6 and 4.7; in the case of the second variate, Y is merely substituted for X in the formulae of tables 4.6.1 and 4.7.1. As for the estimate S_{XY} of the parametric covariance σ_{XY}, it is obtained by dividing the sum of the products of the deviations from the means of the two variates, $\Sigma(X-\overline{X})(Y-\overline{Y})$, by the number of degrees of freedom $(N-1)$:

$$S_{XY} = \sum_{h=1}^{N} (X_h-\overline{X})(Y_h-\overline{Y})/(N-1) \ .$$

The detailed steps of the calculations of the means and variances and of the covariance of the two variates are illustrated in table 19.3.1. While the sums of squared deviations, $\Sigma(X-\overline{X})^2$ and $\Sigma(Y-\overline{Y})^2$, and the variances, S_X^2 and S_Y^2, of the two variates are necessarily positive, the covariance S_{XY} and the sum of the products of the deviations, $\Sigma(X-\overline{X})(Y-\overline{Y})$, may be positive, null or negative depending on whether the terms of the latter are mostly positive, equally positive and negative, or mostly negative.

The sums of squares and of products of deviations, $\Sigma(X-\overline{X})^2$, $\Sigma(Y-\overline{Y})^2$, and $\Sigma(X-\overline{X})(Y-\overline{Y})$, are also called the *centered sums* and may also be calculated rapidly from the *raw sums* N, ΣX, ΣY, ΣX^2, ΣY^2, and ΣXY, through the following formulae:

$$\Sigma(X-\overline{X})^2 = \Sigma X^2 - (\Sigma X)^2/N\,; \quad \Sigma(Y-\overline{Y})^2 = \Sigma Y^2 - (\Sigma Y)^2/N\,;$$

$$\Sigma(X-\overline{X})(Y-\overline{Y}) = \Sigma XY - (\Sigma X)(\Sigma Y)/N\,.$$

However, as noted in sections 4.6, 12.3 and 13.6, deducing centered sums from raw sums is sometimes numerically inaccurate, and it is generally safer to calculate centered sums directly on the deviations from the means, a procedure which is not more difficult when computations are done on a computer.

Table 19.3.1
Detailed calculations of the means, variances and covariance of the concentrations X of calcium and Y of phosphorus in the morning urine of underweight adult men; data from Brown and Beerstecher (1951)
(data reproduced with the permission of Dr. D. R. Davis, Dept. Chem. and Biochem., University of Texas at Austin)

X	$(X-\overline{X})$	$(X-\overline{X})^2$	$(X-\overline{X})(Y-\overline{Y})$	$(Y-\overline{Y})^2$	$(Y-\overline{Y})$	Y
153	−3.5	12.25	+197.05	3169.69	−56.3	120
245	+88.5	7832.25	+3424.95	1497.69	+38.7	215
210	+53.5	2862.25	+4210.45	6193.69	+78.7	255
170	+13.5	182.25	+144.45	114.49	+10.7	187
194	+37.5	1406.25	+3776.25	10140.49	+100.7	277
104	−52.5	2756.25	+1380.75	691.69	−26.3	150
155	−1.5	2.25	+27.45	334.89	−18.3	158
63	−93.5	8742.25	+1337.05	204.49	−14.3	162
97	−59.5	3540.25	+5194.35	7621.29	−87.3	89
174	+17.5	306.25	−460.25	691.69	−26.3	150
Sums						
1565	0	27642.5	+19232.5	30660.1	0	1763

$\overline{X} = \Sigma X/N = 1565/10 = 156.5$

$\overline{Y} = \Sigma Y/N = 1763/10 = 176.3$

$S_X^2 = \Sigma(X-\overline{X})^2/(N-1) = 27642.5/9 = 3071.39$

$S_Y^2 = \Sigma(Y-\overline{Y})^2/(N-1) = 30660.1/9 = 3406.68$

$S_{XY} = \Sigma(X-\overline{X})(Y-\overline{Y})/(N-1) = +19232.5/9 = +2136.94$

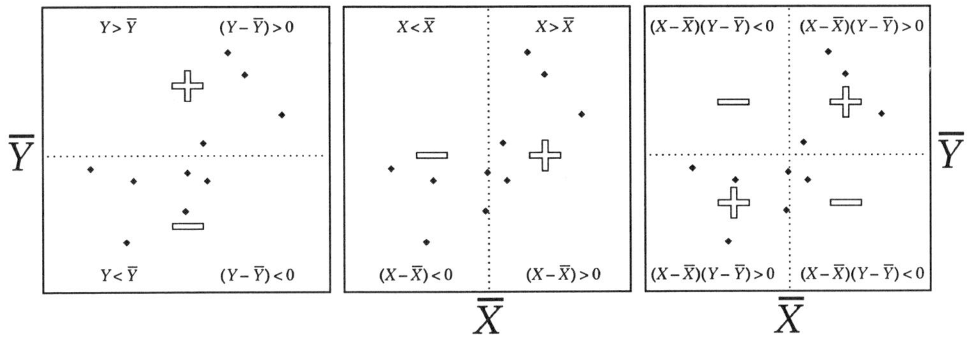

Figure 19.3.1
The product of deviations, $(X-\bar{X})(Y-\bar{Y})$, is negative in the upper left and lower right quadrants but positive in the lower left and upper right quadrants (data from figure 19.1.3 and table 19.3.1)

Figure 19.3.1 illustrates the circumstances in which the product $(X-\bar{X})(Y-\bar{Y})$ of the deviations corresponding to a particular pair of data (X, Y) is positive or negative. The deviation $(Y-\bar{Y})$ is negative when $Y<\bar{Y}$, that is when the dot representing a pair of data falls below the horizontal line passing through the mean $\bar{Y}$ in the scatter diagram (figure 19.3.1, left); as for the deviation $(X-\bar{X})$, it is negative when $X<\bar{X}$, that is when the dot (X, Y) falls to the left of the vertical line passing through the mean $\bar{X}$ in the scatter diagram (figure 19.3.1, center). The product of deviations, $(X-\bar{X})(Y-\bar{Y})$, is thus positive in the lower left and upper right quadrants, where the deviations of both variates have the same sign, but it is negative in the upper left and lower right quadrants where the deviations of the two variates have opposite signs (figure 19.3.1, right). The sum of the products of deviations, $\Sigma(X-\bar{X})(Y-\bar{Y})$, and the covariance, S_{XY}, are therefore negative when individual dots are more numerous and are located farther from the mean dot $(\bar{X}, \bar{Y})$ in the negative than in the positive quadrants; conversely, $\Sigma(X-\bar{X})(Y-\bar{Y})$ and S_{XY} are positive when individual dots are more numerous and located farther from the mean dot $(\bar{X}, \bar{Y})$ in the positive quadrants. Finally, $\Sigma(X-\bar{X})(Y-\bar{Y})$ and S_{XY} are approximately null when individual dots are as numerous and located as far from the mean dot $(\bar{X}, \bar{Y})$ in the positive as in the negative quadrants.

Section 19.4: the covariance of standardized variates or correlation coefficient

The above discussion indicates that the covariance S_{XY} reflects the tendency of individual dots to go down ($S_{XY}<0$), to remain at the same height ($S_{XY}=0$), or to go up ($S_{XY}>0$) toward the right in a scatter diagram. The numerical value of the covariance might therefore be used to detect or to describe the presence of a statistical relationship between two variates, but that utilization would not be fully satisfactory because the numerical value of the covariance is affected not only by the presence of a positive or negative relationship but also by the extent of the variation of the two variates.

Fortunately, the effects of the extent of the variation of the two variates may be eliminated by considering the covariance σ_{xy} of standardized variates $x=(X-\mu_X)/\sigma_X$ and $y=(Y-\mu_Y)/\sigma_Y$, usually called the *correlation coefficient* but also known as the *ordinary correlation coefficient* or as *Pearson's product-moment correlation coefficient*. The parametric (population) correlation coefficient is denoted by ρ (the lower case Greek letter *rho*, corresponding to the Roman r) and has the value $\rho=\sigma_{XY}/(\sigma_X\sigma_Y)$. The population correlation coefficient ρ is estimated by the sample correlation coefficient $r=S_{XY}/(S_X S_Y)$.

Figure 19.4.1 illustrates scatter diagrams obtained by getting random samples of size $N=200$ in bivariate normal statistical populations in which the parametric correlation ρ has the following values: -1.0, -0.8, 0.0, $+0.8$ and $+1.0$. The corresponding statistical relationships range from a perfect straight line with a negative slope ($\rho=-1.0$) to a perfect straight line with a positive slope ($\rho=+1.0$), through elliptical clusters of dots in which the major axis has a negative ($-1.0<\rho<0$) or a positive slope ($0<\rho<+1.0$). When the correlation coefficient ρ is null, the cluster of individual dots has the shape either of a circle (figure 19.4.1, center) or, if original (unstandardized) variates X and Y are considered, of a horizontal or vertical ellipse. Figure 19.4.2 illustrates the changing aspect of the probability density surface depending on whether the correlation coefficient is strongly negative ($\rho=-0.9$), null ($\rho=0.0$) or strongly positive ($\rho=+0.9$). The correlation coefficient provides a measure of the degree of intensity of a straight-line relationship.

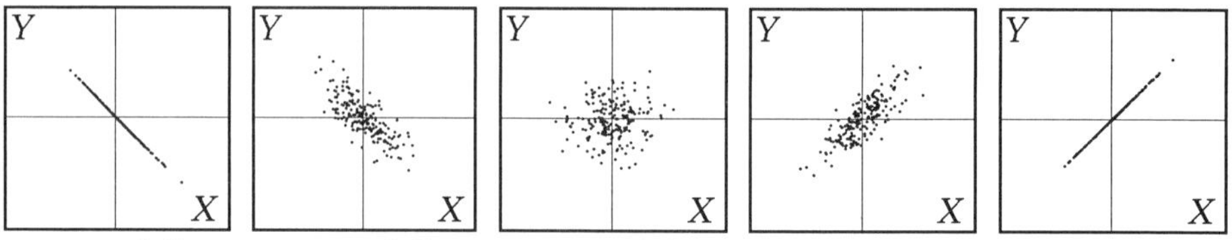

$\rho=-1.0$ $\rho=-0.8$ $\rho=+0.0$ $\rho=+0.8$ $\rho=+1.0$

Figure 19.4.1

Scatter diagrams of samples of size $N=200$ coming from bivariate normal populations having various values of the parametric correlation coefficient ρ

Section 19.5: testing hypotheses of null correlation

Of course, while it is the parametric correlation coefficient ρ about which knowledge is sought and concerning which hypotheses are made, it is the sample correlation coefficient r which is obtained from the data and which must be used to draw statistical conclusions. The preliminary hypothesis which is usually made at first concerning the parametric correlation coefficient is the null hypothesis H_0: $\rho=0$. This hypothesis may be considered as the simplest one from a logical viewpoint (section 9.2) because it implies that the conditional means of each of the two variates (Y for instance) with respect to the other one (X) are all equal to each other (section 20.1). Moreover, if the existence of a relationship must be confirmed on the basis of observed data, the presence of that relationship must naturally not be postulated *a priori*.

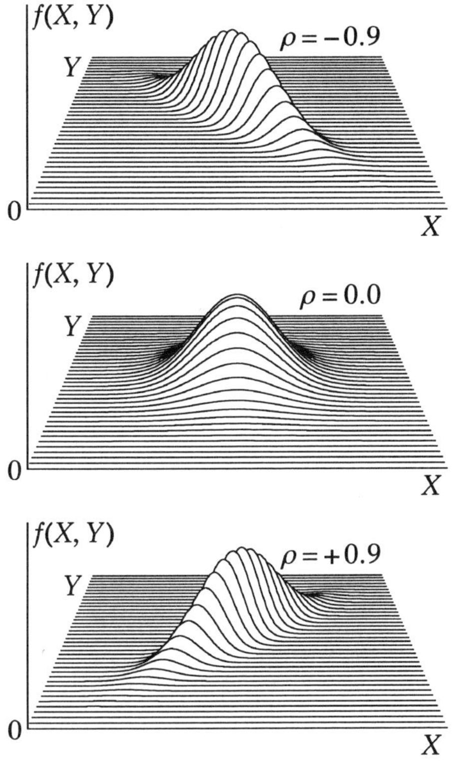

Figure 19.4.2
Various aspects of the probability density surface of the bivariate normal distribution according to the value of the parametric correlation coefficient ρ

As for the alternative hypothesis H_1, it should be one-sided in the cases where the existence of a positive relationship (H_1: $\rho > 0$) or of a negative relationship (H_1: $\rho < 0$) is suspected *a priori*, but it should be two-sided (H_1: $\rho \neq 0$) if the sign of the relationship is doubtful. In practice, even when the parametric correlation coefficient ρ is null, the sample correlation coefficient r is practically never exactly equal to 0 because of sampling variation, and its probability distribution resembles the normal distribution except that, instead of ranging from $-\infty$ to $+\infty$, it is limited to the interval $-1.0 < r < +1.0$ (figure 19.5.1, page 143). Observed nonzero values of the correlation coefficient r may therefore be considered as indications of the possible existence of relationships, but all of those indications must be sifted and only those that are too large to be attributed to random must be retained as worthy of attention.

When the preliminary hypothesis tested is the null hypothesis, H_0: $\rho = 0$, there is a superfluity of choices, for the distribution of r is then known exactly and is related to Student's t distribution as well as to the variance ratio (F) distribution. If a table of the distribution of r when $\rho = 0$ is used (at the end of this book), the limits of the acceptance and rejection regions of H_0 are determined according to the nature of the alternative hypothesis H_1, as illustrated in figure 19.5.1. The expression $r_{(P;\,\nu)}$ denotes a value of r of which the cumulative probability is P with ν degrees of freedom. In the case of the ordinary correlation coefficient r considered here, the number of degrees of freedom is $\nu=(N-2)$, where N is sample size. However, an observed value of the correlation coefficient r may also be transformed into either a Student's t,

$$t_{(\nu)} = r\sqrt{[\nu/(1-r^2)]}\,,$$

or a variance ratio,

$$F_{(1,\,\nu)} = r^2\,\nu/(1-r^2)\,.$$

While the table of the distribution of r when $\rho=0$ or Student's t distribution may be used to test either one-sided or two-sided alternative hypotheses, the variance ratio (F) distribution does not allow the sign of the correlation coefficient r to be taken into account because the latter is squared.

In the case of the data considered in table 19.3.1 and in figures 19.1.3 and 19.3.1, the correlation coefficient $r = S_{XY}/(S_X\,S_Y) = +2136.94/(55.4201\times 58.3668) = +0.660631$. Since there are $\nu=(N-2)=8$ degrees of freedom, if the alternative hypothesis is H_1: $\rho\neq 0$, the acceptance region of H_0 : $\rho=0$ extends from -0.63190 to $+0.63190$ at the significance level $\alpha=0.05$ and from -0.76459 à $+0.76459$ at $\alpha=0.01$. If P denotes the probability of observing a correlation coefficient differing as much (or more) from 0 as $r=+0.660631$ when a sample is picked at random in a population where $\rho=0$, then, $0.01<P<0.05$. This set of data thus provides a provisional indication ($P<0.05$) that there is a positive correlation between the calcium (X) and phosphorus (Y) concentrations in the morning urine of underweight adult men, but this cannot be considered as a well-established conclusion ($P>0.01$).

If a biological interpretation is sought for the preceding statistical conclusion, one should wonder whether the positive correlation between X and Y might not be due to trivial fluctuations of the density of morning urine. For instance, an individual drinking much water before going to bed would tend to produce the following morning a dilute urine specimen in which the concentrations of all substances, including calcium and phosphorus, would be low; on the contrary, an individual drinking little water would tend to produce a concentrated urine specimen. Such fluctuations could explain the presence of positive correlations between the concentrations of urinary constituents. If such a trivial explanation could be excluded, however, a more interesting interpretation could be considered: would the observed positive correlation between calcium and phosphorus concentrations reflect the existence of a biochemical or physiological association between those two substances? – For instance, individual variations in the intensity of the catabolism of bony tissue (possibly related to the action of calcitonin) could produce a joint increase or decrease of calcium and phosphorus concentrations.

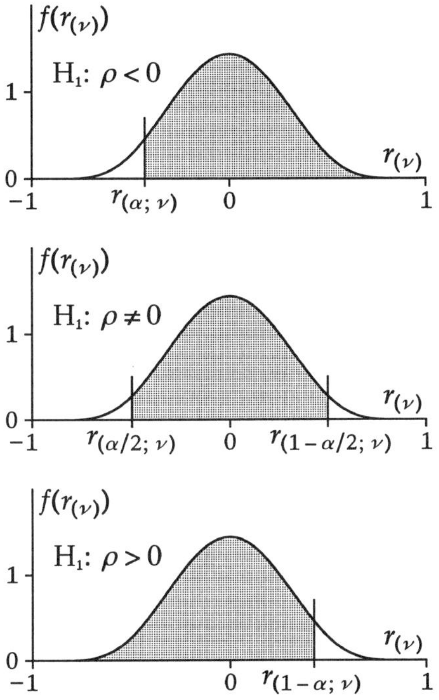

Figure 19.5.1
Acceptance (stippled) and rejection regions of preliminary hypothesis H_0: $\rho = 0$ according to the type of alternative hypothesis H_1

Section 19.6: nonnull (nonzero) preliminary hypotheses, confidence intervals

Unfortunately, even though the methods of the preceding section are excellent, they can be used only to test the null hypothesis H_0: $\rho = 0$. When the parametric correlation coefficient ρ differs from 0, the distribution of the sample correlation coefficient r loses its symmetry because it is no longer centered on 0 while being limited to the interval $-1.0 < r < +1.0$ (figure 19.6.1, top, page 145). But, in biology like in other fields, once the null hypothesis has been rejected, it may be interesting to test other (nonnull) preliminary hypotheses, such as H_0: $\rho = 0.5$ or H_0: $\rho = 0.25$. Confidence limits may also be desired for ρ, a context in which assuming that $\rho = 0$ would be contradictory. One should then use either the exact tables and nomograms of David (1938) or Fisher's transformation, $z = \tanh^{-1}(r) = (1/2)\log_e[(1+r)/(1-r)]$. The latter restores the symmetry of the distribution by stretching the extent of variation from $-\infty$ to $+\infty$ (figure 19.6.1, bottom) and yields an approximately normal variate of which the mean is $\zeta = \tanh^{-1}(\rho)$ and the standard deviation $\sigma_z = 1/\sqrt{(N-3)}$. The functions tanh and $\tanh^{-1}$ are the hyperbolic tangent and arc tangent, respectively. The Greek letter ζ (*zeta*) is the equivalent of the lower case Roman z.

If a nonnull preliminary hypothesis must be tested, according to which the parametric correlation coefficient ρ has some particular numerical value ρ_0, Fisher's transformation may then be applied to the sample correlation coefficient r on the one hand and to the hypothetical value ρ_0 on the other hand, which yields $z = \tanh^{-1}(r)$ and $\zeta = \tanh^{-1}(\rho_0)$, and the criterion,

$$t_{(\infty)} = \frac{z - \zeta_0}{1/\sqrt{(N-3)}} = (z - \zeta_0)\sqrt{(N-3)},$$

follows a standardized normal distribution $\mathcal{N}(0, 1)$ approximately. The limits of the acceptance and rejection regions of H_0: $\rho = \rho_0$, that is H_0: $\zeta = \zeta_0$, are determined according to the nature of the alternative hypothesis H_1, as in section 9.6, except that the normal distribution is used here instead of Student's t distribution (as indicated by the notation $t_{(\infty)}$ above). Fisher's transformation may also be used to compare two independent estimates, r_A and r_B, of the correlation coefficient: the criterion is

$$t_{(\infty)} = \frac{z_A - z_B}{\sqrt{\{[1/(N_A - 3)] + [1/(N_B - 3)]\}}},$$

where $z_A = \tanh^{-1}(r_A)$ and $z_B = \tanh^{-1}(r_B)$. Fisher's transformation may also be used to combine several independent estimates of the correlation coefficient, but the same result may be obtained more simply and more exactly by calculating a pooled correlation coefficient from the within-groups centered sums of squares and products of deviations (see the case of parallel straight lines in section 21.2).

When a confidence interval is needed for ρ, Fisher's transformation is applied at first to the sample correlation coefficient r, the confidence interval of ζ is determined next by using the normal distribution, and the inverse transformation is finally applied to the confidence limits of ζ in order to reexpress them with respect to ρ. In the case where, if a significance test were involved, the hypotheses would be H_0: $\rho = \rho_0$ and H_1: $\rho \neq \rho_0$, for instance, one has approximately

$$P[\, t_{(\alpha/2;\, \infty)} \leq (z - \zeta)\sqrt{(N-3)} \leq t_{(1-\alpha/2;\, \infty)} \,] = (1 - \alpha),$$

whence, after terms are rearranged,

$$P[\, z + t_{(\alpha/2;\, \infty)}/\sqrt{(N-3)} \leq \zeta \leq z + t_{(1-\alpha/2;\, \infty)}/\sqrt{(N-3)} \,] = (1 - \alpha).$$

The confidence interval of ζ is therefore

$$\zeta_1 \leq \zeta \leq \zeta_2,$$

where

$$\zeta_1 = z + t_{(\alpha/2;\, \infty)}/\sqrt{(N-3)}, \qquad \zeta_2 = z + t_{(1-\alpha/2;\, \infty)}/\sqrt{(N-3)},$$

and

$$z = \tanh^{-1}(r) = (1/2)\log_e[(1 + r)/(1 - r)].$$

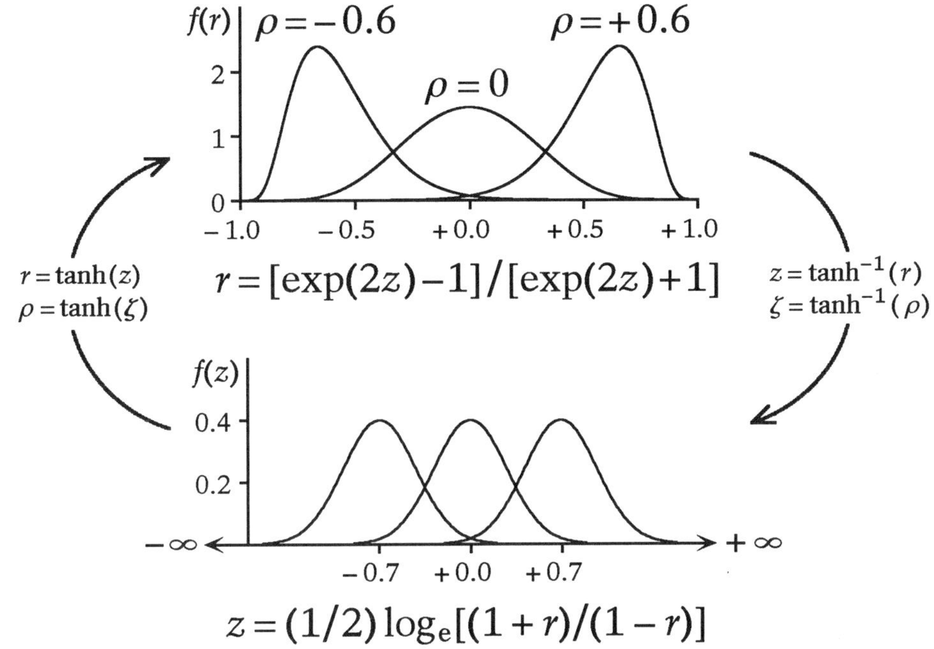

Figure 19.6.1
Probability distribution of the sample correlation coefficient r for various values of the parametric correlation coefficient ρ (top) and distribution of Fisher's transformed variate z (bottom)

The confidence limits ζ_1 and ζ_2 are finally reexpressed with respect to ρ by using the inverse function of Fisher's transformation, which yields $\rho_1 = \tanh(\zeta_1)$ and $\rho_2 = \tanh(\zeta_2)$. The confidence interval of ρ is then

$$\rho_1 \leq \rho \leq \rho_2 .$$

It should be noted that the hyperbolic tangent of a particular value ζ_i is equal to

$$\rho_i = \tanh(\zeta_i) = [\exp(2\zeta_i) - 1] / [\exp(2\zeta_i) + 1] .$$

In order to determine one-sided confidence limits for the correlation coefficient ρ, the reader should consult section 9.7 in addition to the present section.

Section 19.7: prediction ellipses

When two variates following a bivariate normal distribution are analyzed simultaneously, the two-sided confidence intervals and prediction intervals discussed in sections 9.7 and 9.11 are replaced by elliptical regions. The manner of drawing such ellipses will be described in chapter 30 (on Hotelling's T^2), but an example is provided in figure 19.7.1 (next page), which illustrates a prediction ellipse at the $\alpha = 0.05$ significance level for the body weight X in kg and the skull length Y in mm of 60 adult female wolves captured from 1954 to 1956 in the Northwest Territories of Canada (Jolicoeur, 1959).

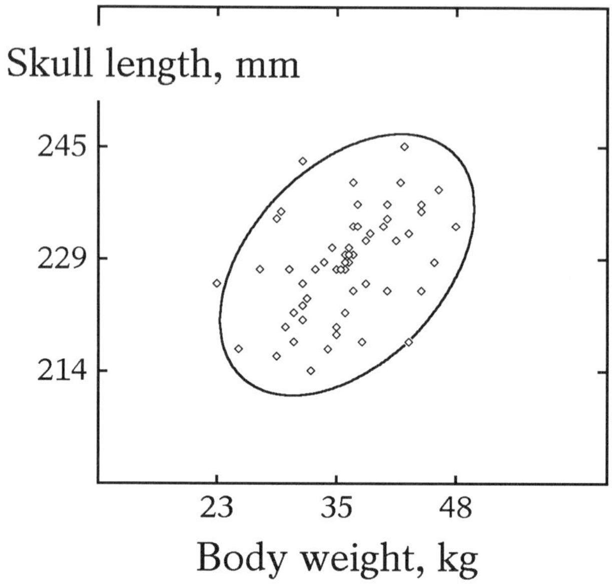

Figure 19.7.1
Prediction ellipse at significance level $\alpha = 0.05$ of body weight and skull length in 60 adult female wolves captured from 1954 to 1956 in the Northwest Territories of Canada (Jolicoeur, 1959)

This ellipse delimits the region within which a new individual observation $(X,\ Y)$ would be expected to fall if it came from the same statistical population as the sample in hand. The prediction ellipse is centered on the mean point $(\overline{X},\ \overline{Y})$ of the sample and is defined by the following inequality:

$$\frac{N}{(N+1)}\frac{S_X^2 S_Y^2}{(S_X^2 S_Y^2 - S_{XY}^2)}\left[\frac{(X-\overline{X})^2}{S_X^2} - 2r\frac{(X-\overline{X})(Y-\overline{Y})}{S_X \quad S_Y} + \frac{(Y-\overline{Y})^2}{S_Y^2}\right] \leq \frac{2(N-1)}{(N-2)} F_{(1-\alpha;\,2,\,N-2)} \, .$$

In the case of figure 19.7.1, three of the 60 individual dots, that is 5%, fall outside of the prediction ellipse, which agrees exactly here with the $\alpha = 0.05$ significance level used for drawing the ellipse. When the size of a sample is large enough, a prediction ellipse is often a good way to describe the estimated extent of individual variation in a bivariate scatter diagram, and it is often used for that purpose in morphology, in taxonomy, in animal and plant paleontology, in quantitative genetics, in ecology, in physiology, etc. . When sample size is small, however, the prediction ellipse is often too large in comparison with the extent of individual variation and should be replaced by a variation ellipse (section 30.6).

Section 19.8: the bivariate lognormal distribution

As already mentioned in the case of a single variate (chapter 14), the lognormal distribution is often useful in biology because many biological variates are nonnegative and have positively skewed and heteroscedastic probability distributions. When two variates are analyzed jointly, there is often an additional reason to use the lognormal distribution, for many biological variates are related to each other by power functions (such as $Y = aX^b$) of which the analysis is made simpler by the logarithmic transformation of data (section 20.7). In this section, original variates will be denoted by subscripted variables X_1 and X_2 instead of X and Y as previously done in order to keep the letter Y for logarithmically transformed data. The original variates X_1 and X_2 will thus be said to follow a bivariate lognormal distribution if their logarithms $Y_1 = \log_e(X_1)$ and $Y_2 = \log_e(X_2)$ follow a bivariate normal distribution (section 19.2).

The example considered here deals with a sample of measurements of head circumference in cm and brain weight in g of 78 male fetuses and babies (Cabana, Jolicoeur and Michaud, 1993). Figure 19.8.1 illustrates the prediction ellipses obtained, at significance level $\alpha = 0.05$, when the bivariate normal distribution is assumed to apply either to the original variates (figure 19.8.1, left) or to their logarithms (figure 19.8.1, center). The bivariate normal distribution obviously does not fit the original data satisfactorily (figure 19.8.1, left). Not only does the prediction ellipse disagree with the visible curvature of the cluster of individual dots, but also it excludes the largest observation as if it were an *outlier* and it extends below the abscissa, where brain weight would be negative, which is impossible in reality. However, the bivariate normal distribution fits logarithmically transformed data almost perfectly (figure 19.8.1, center): the curvature and the skewness present in original data have been completely eliminated by the logarithmic transformation and there does not appear to be any outlying observation. Finally, if the prediction ellipse of logarithms is reexpressed with respect to original variates, there is an excellent agreement between the shape of the cluster of observed dots and the antilogarithmic transform of the ellipse, which reflects remarkably well the nonnegativity, the positive skewness, the heteroscedasticity and the curvature of the data (figure 19.8.1, right).

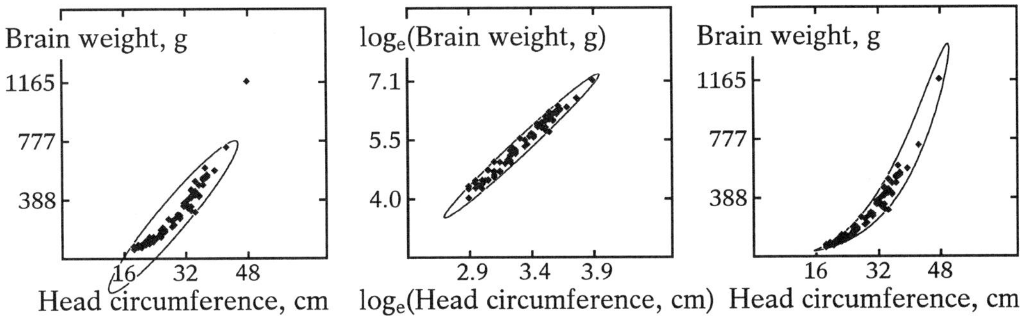

Figure 19.8.1

Bivariate normal prediction ellipse ($\alpha =$ 0.05) fitted to the original data (left) and to the logarithmically transformed data (center), and antilogarithmic transform of the prediction ellipse of logarithms (right); brain weight and head circumference of 78 male fetuses and babies (Cabana, Jolicoeur and Michaud, 1993)

Provided the logarithmic transformation or other transformations are used when needed, the bivariate normal distribution may be suitable for the analysis of many kinds of data for which it would be inappropriate otherwise (see also an ecological example in section 30.9).

Section 19.9: the need for caution when interpreting correlations

Biological interpretations of correlations must be thought out cautiously in order to avoid simplistic or erroneous explanations. When several variates are bodily dimensions of animal or plant organisms belonging to the same species, as in figures 1.2.1, 1.2.2, 19.1.1, 19.7.1, and 19.8.1, positive correlations are frequently observed. This often reflects a simple but fundamental biological fact: when an organism grows, all of its parts and all of its dimensions tend to increase approximately at equal rates, which makes shape more or less constant in spite of size differences due to age or individual variation.

In some fields, however, some measurements having closely similar definitions may be used jointly, such as fork length and standard length in ichthyology. The presence of a strong correlation between such variates is obviously meaningless since it is due to their redundancy rather than to biological reasons. Moreover, in quantitative morphology, indices, ratios or proportions may involve the same character in various ways, and it may not be justifiable to test a null hypothesis about the correlation between such variates. Several writers have discussed the so-called *spurious correlations* in this context, but it is careless interpretations rather than the correlations themselves which can be wrong.

Similarly, physiologists are sometimes interested in variates between which there appears to be an approximate inverse relationship: it should be checked whether that inverse relationship might not be due mostly to a mathematical artefact, one of the two variates being proportional to a third variate and the other being proportional to its inverse. In section 19.5, for instance, it was noted that uncontrolled fluctuations of the overall concentration of urine could create meaningless correlations between the concentrations of various substances, such as calcium and phosphorus. Fortunately, such misleading appearances can easily be avoided by multiplying the concentrations by the volume of the urine specimen and obtaining the total amounts of calcium and phosphorus excreted.

It may also happen, for instance in ecology, that several variates increase or decrease more or less in a parallel manner with respect to space or time. One must be careful not to attribute correlations between such variates too hastily to causal relationships. A classical but caricature-like example is the contention that, if the price of beer and the salaries of university professors have increased gradually during the last fifty years, it is because of direct causal relationships (supply and demand) between these two variates! When there is a general increasing or decreasing trend in several variates, the first step of a statistical analysis is usually the description of that trend in order to see whether, after its removal, residual deviations (section 20.3) are still correlated.

To sum up, if the discovery of interesting biological relationships is contemplated, one must choose variates which may be independent from each other and between which relationships due to trivial mathematical, spatial or chronological reasons are not expected.

Section 19.10: curved statistical relationships

While the correlation coefficient provides a measure of the intensity of straight-line statistical relationships (section 19.4), it is appropriate also when the relationship between variates is curved but can be straightened by the logarithmic transformation or other transformations (sections 19.8, 20.6, 20.7). However, some curved relationships cannot be straightened by simple transformations of the data. For instance, figure 19.10.1 represents the scatter diagram of a sample of 1000 pairs of observations coming from a simulated population in which there is a symmetrical relationship between the variates X and Y and the correlation coefficient is null ($\rho = 0$). This relationship is strong but it cannot be straightened through a simple variate transformation and its intensity cannot be measured by using the ordinary correlation coefficient. Other examples of relationships which cannot be straightened satisfactorily or easily are given by sigmoid or nonsigmoid growth curves, such as the logistic curve, the Gompertz curve and the curve of Pütter (also known as the curve of von Bertalanffy), and some models of human growth (Jolicoeur, Pontier, Pernin and Sempé, 1988; Jolicoeur, Baron and Cabana, 1988; Jolicoeur, Pontier and Abidi, 1992; Jolicoeur and Pontier, 1993; Pontier and Jolicoeur, 1996). Such relationships may be studied by so-called *nonlinear* methods which will be discussed in chapter 35.

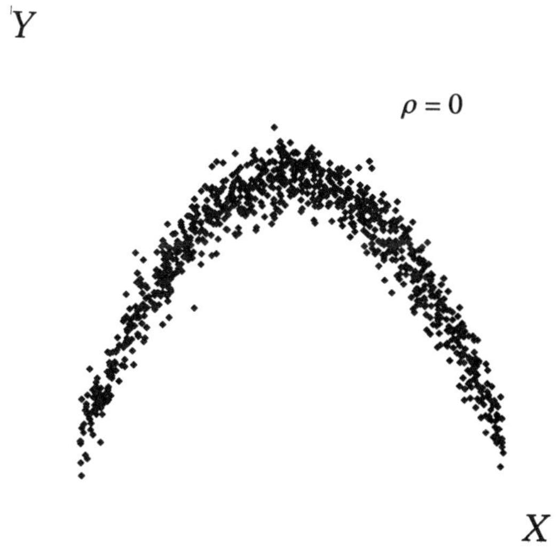

Figure 19.10.1
Some curved relationships cannot be straightened by simple variate transformations, and the linear correlation coefficient (ρ or r) should not be used to measure the intensity of such relationships (diagram based on random simulations)

Chapter 20
Estimation lines (the so-called "regression" lines)

Section 20.1: introduction

In the preceding chapter, the correlation coefficient has been treated as a tool for detecting the existence of a straight-line statistical relationship between two variates and for measuring the intensity of such a relationship. In this chapter, the presence of a straight-line relationship will be taken advantage of for estimating the value of one of the two variates, called the *predicted variate,* from that of the other, called the *predictor variate.* For instance, the relationship between the latitude X and the mean length Y in mm reached at the age of four years by the small-mouth black bass *Micropterus dolomieui* is illustrated in figure 20.1.1. The analysis was done originally by Gérard Pageau (1967), an ichthyologist who kindly made these interesting data available. Since the lower left and upper right corners of this scatter diagram are empty, the mean length reached at the age of four years by the various local populations appears to decrease when the latitude increases. This phenomenon seems easy enough to interpret since the increase of latitude entails a decrease of the mean water temperature and a shortening of the annual growth season for this *poikilothermal* fish (of which the body temperature follows that of the environment).

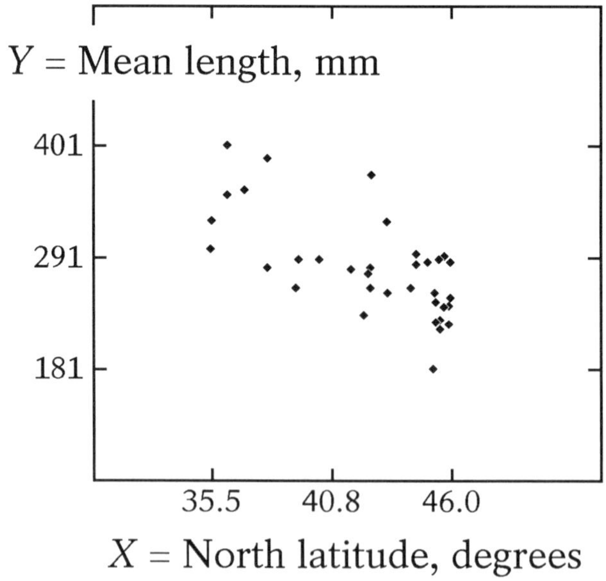

Figure 20.1.1
Scatter diagram of the mean length Y in mm reached at four years of age by the small-mouth black bass *Micropterus dolomieui* according to latitude X (data compiled and analyzed by Pageau, 1967)

The simplest way of confirming whether the mean length changes when the latitude increases is perhaps to classify the data into several subgroups according to latitude and to calculate the mean length for each subgroup (figure 20.1.2). This yields a sequence of *conditional means* of the length of the small-mouth black bass, the condition being that the latitude is restricted to a specified interval within each subgroup. Thus, in figure 20.1.2, the data are divided into three subgroups of approximately equal sizes (comprising 12, 11 and 12 pairs of observations) and the mean lengths of these three successive subgroups are 313.1, 279.7 and 250.6 mm. Therefore, the mean length reached at the age of four years by the small-mouth black bass seems to decrease when latitude increases.

However, the preceding approach has the drawback of suggesting that the change of mean length with latitude is discontinuous, something which the original data certainly do not indicate (figure 20.1.1). The decrease of the mean length of the small-mouth black bass could be described in a more gradual manner by dividing the data into a greater number of narrower subgroups but, since the whole sample contains only 35 pairs of observations, the sizes of narrower subgroups would be very small and their means would be strongly affected by sampling fluctuations (section 9.5). Subgroup means would then vary in an irregular manner and would poorly reflect the relationship between mean length and latitude. Fortunately, it may be shown theoretically that, when two variates follow a bivariate normal distribution (section 19.2), the range of the predictor variate X can be divided into a greater and greater number of narrower and narrower subgroups, and the parametric means of those subgroups all lie on a straight line of which the equation is

$$\hat{Y} = [\mu_Y - (\sigma_{XY}/\sigma_X^2)\mu_X] + (\sigma_{XY}/\sigma_X^2)X .$$

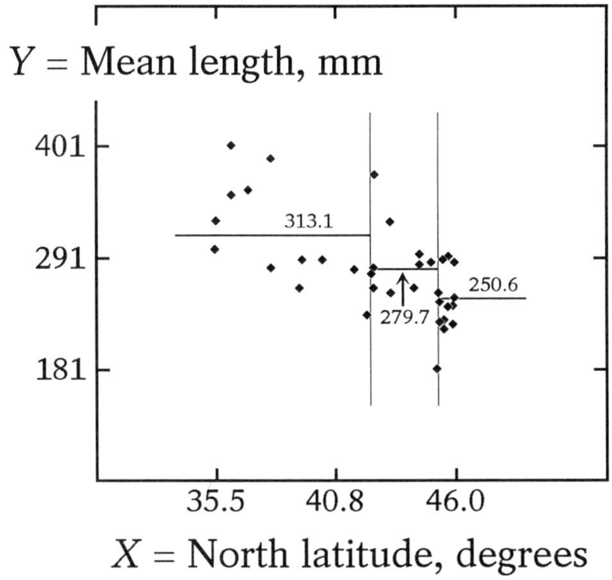

Figure 20.1.2
Mean length Y in mm of the small-mouth black bass at four years of age in three subgroups of local populations at increasing latitudes

Such a line is commonly called a *regression line,* but the name *estimation line* is preferable, since the expression *regression line* is used for historical reasons and goes back to a misunderstanding at the time of Francis Galton (1822-1911). The preceding equation is clearly theoretical since it is expressed with respect to the parameters μ_X, μ_Y, σ_X^2 , and σ_{XY}. In practice, those parameters are estimated on the basis of the total (undivided) sample by the quantities $\overline{X}$, $\overline{Y}$, S_X^2 and S_{XY}, and the equation

$$\hat{Y} = a + bX$$

is used where $a = \overline{Y} - b\overline{X}$ and $b = S_{XY}/S_X^2$. The straight line corresponding to this equation may be considered as a *changing conditional mean* of variate Y with respect to variate X, as opposed to the ordinary *(marginal)* mean $\overline{Y}$ which would be represented by a horizontal line. The circumflex accent above the letter Y indicates that the *predicted variate* is Y while the *predictor variate* is X.

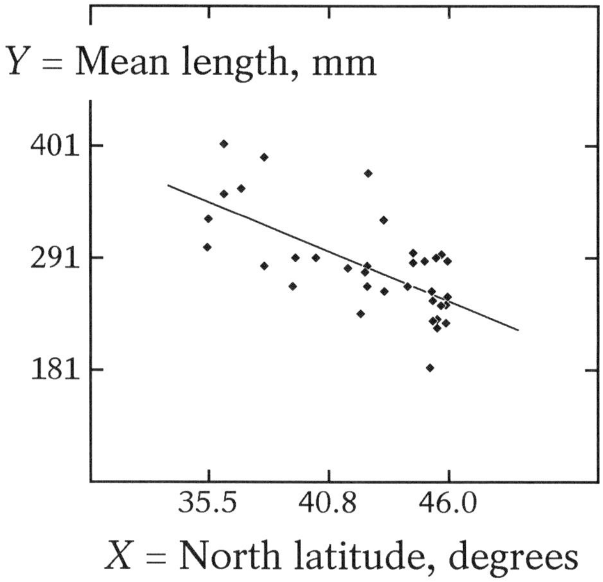

Figure 20.1.3
Least-squares estimation line of the mean length Y in mm of the small-mouth black bass at four years of age according to latitude X (Pageau, 1967)

In the past, Y has usually been called the *dependent variate* and X the *independent variate,* but those terms are ambiguous and should be avoided. The values given above to coefficients a and b satisfy the *least-squares principle,* which means that the sum of squared deviations, $\Sigma[Y-(a+bX)]^2$, between observed values Y and estimated values $\hat{Y}$ is minimized when $a = \overline{Y} - b\overline{X}$ and $b = S_{XY}/S_X^2$. The principle of least squares is satisfied also by the arithmetic mean in the case of a single variate (section 4.2). The coefficients of the estimation line of the mean length Y of the small-mouth black bass at four years of age predicted from latitude X have been calculated for the data plotted in figures 20.1.1 and 20.1.2 and the line is represented in figure 20.1.3. Its equation is

$$\hat{Y} = 676.29 - 9.2985\,X.$$

The slope $b = -9.2985$ indicates that, within the range covered by the data, the mean length of the small-mouth black bass at four years of age decreases by 9.2985 mm for each degree of latitude. As for the ordinate $a = 676.29$, one might be tempted to interpret it as the mean length in mm reached at four years of age by the small-mouth black bass in localities where the latitude is 0° ! – Such an interpretation would be unjustifiable, however, because the geographical range of the small-mouth black bass is far from reaching the equator. Moreover, the present data do not go further south than the 35th degree of north latitude, and the utilization of the estimation line at a value of X equal to 0 would be an unreasonably excessive extrapolation. This example brings out the fact that, when mathematical or statistical methods are applied to biological phenomena, the results should always be interpreted with caution, and an attempt should be made to discover the *biological meaning* of conclusions, as opposed to their mere *statistical significance* (section 9.2).

Section 20.2: the duality of estimation lines

In the preceding section, it was rather reasonable to treat X as the predictor variate and Y as the predicted variate because, while it is likely that the mean length of the small-mouth black bass may be influenced by latitude, the converse relationship is unlikely. However, there are other sets of data for which each of the two variates may, depending on the context, be reasonably considered either as the predicted or the predictor variate. For instance, figure 20.2.1 illustrates the scatter diagram of the head length in mm of the father X and of the oldest adult son Y in 73 families.

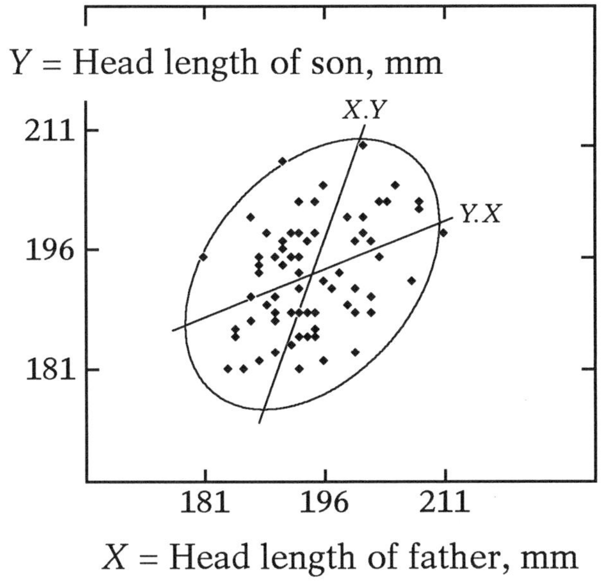

Figure 20.2.1
Scatter diagram of the head length in mm of the father X and of the oldest adult son Y in 73 families; the two possible estimation lines as well as a prediction ellipse at significance level $\alpha = 0.05$ are represented; data from Frets (1921)

In that diagram (preceding page), a prediction ellipse (at the significance level $\alpha = 0.05$) as well as the two possible estimation lines have been drawn: the estimation line of Y from X (also called the *Y on X regression line* and denoted by the abbreviation $Y.X$) and the estimation line of X from Y (denoted by the abbreviation $X.Y$ and also called the *X on Y regression line*). The estimation line of Y from X, of which the equation is $\hat{Y} = a + bX$ and which has already been discussed in section 20.1, should be used if the head length of the father is known and the head length of an adult son must be estimated. However, if the head length of the oldest adult son is known and the head length of the father must be estimated, one should rather use the estimation line of X from Y, of which the equation is $\hat{X} = c + dY$, where $c = \overline{X} - d\overline{Y}$ and $d = S_{XY}/S_Y^2$.

Except when the correlation coefficient is equal to ± 1, the two estimation lines have only one point in common, the mean point $(\overline{X}, \overline{Y})$, and differ considerably from each other: the $Y.X$ estimation line minimizes the sum of squares $\Sigma(Y - \hat{Y})^2$ of vertical deviations and intersects the prediction ellipse at its extreme left and extreme right, while the $X.Y$ estimation line minimizes the sum of squares $\Sigma(X - \hat{X})^2$ of horizontal deviations and intersects the prediction ellipse at its lowest point and at its highest point (figure 20.2.1). When the correlation is imperfect, like here ($r = +0.378617$), the $X.Y$ line has a much larger slope in absolute value than the $Y.X$ line. For the slopes of the two lines to be comparable, however, the equation of the $X.Y$ line ($\hat{X} = c + dY$) must be rewritten in the same way as the slope of the $Y.X$ line ($\hat{Y} = a + bX$), that is with Y as an explicit function of X: this yields

$$Y = -(c/d) + (1/d)\,\hat{X}\,.$$

In figure 20.2.1, $(1/d) = +2.80887$ while $b = +0.402655$, which reflects the marked divergence of the two lines. An estimation line must therefore be carefully chosen with due attention to the nature of the questions to be answered.

Historically, the unfortunate use of the name *regression line* goes back to an ambiguous interpretation of a scatter diagram analogous to figure 20.2.1 by Francis Galton (1822-1911). If the mean head length of sons whose fathers have very long heads is estimated, by using the line $\hat{Y} = a + bX$ for instance, those sons are found to have shorter heads on the average than their fathers. At first, Galton believed that this surprising result showed that there was a "regression", a return toward the mean, at each generation. Unfortunately, that erroneous interpretation conflicted with the obvious fact that, in human, animal and plant populations, quantitative or qualitative variation does not clearly decrease from one generation to the next one.

If the estimation line $\hat{X} = c + dY$ is used instead, it seems indeed that a conclusion completely opposite to that of Galton might be drawn, for the fathers whose sons have very long heads have shorter heads than their sons on the average! – The existence of two diverging estimation lines which could lead to contradictory conclusions is a serious problem in the study of imperfect statistical relationships, and erroneous interpretations should be carefully avoided. In some cases, when both variates are subject to random fluctuations and when the object of the analysis is to describe their joint relationship rather than to estimate one from the other, it may happen that neither one nor the other of the two ordinary estimation lines is appropriate, and it may be preferable to use the *orthogonal estimation line* or *major axis* (section 20.7 and chapter 22).

Section 20.3: residual deviations and variance

The utilization of an estimation line such as $\hat{Y} = a + bX$ in the analysis of a particular set of data may be justified by decomposing the total deviation $[Y - \overline{Y}]$ into two parts (figure 20.3.1) and checking whether the line $\hat{Y} = a + bX$ passes significantly closer to the data points (figure 20.3.2, next page, upper right) than a horizontal line representing the ordinary mean $\overline{Y}$ would do (figure 20.3.2, upper left). The total vertical deviation $[Y - \overline{Y}]$ between an individual observation (X, Y) (figure 20.3.1, black dot) and the horizontal (dotted) line passing through the mean $\overline{Y}$ can be separated into two parts: the so-called residual deviation $[Y - \hat{Y}] = [Y - (a + bX)]$ between the observed dot (X, Y) and the dot $(X, \hat{Y})$ predicted by the estimation line $\hat{Y} = a + bX$ (figure 20.3.1, white dot) and the so-called estimation deviation $[\hat{Y} - \overline{Y}] = [(a + bX) - \overline{Y}]$ between the predicted dot $(X, \hat{Y})$ and another dot $(X, \overline{Y})$ of which the abscissa is the same but the ordinate is the ordinary mean $\overline{Y}$ of the predicted variate. When there is a statistical relationship between two variates X and Y, the estimation deviation is the part of the total deviation which is being subtracted, "eliminated", "explained", by taking the presence of the relationship into account, that is by using the estimation line $\hat{Y} = a + bX$ instead of the ordinary mean $\overline{Y}$ as a measure of central tendency.

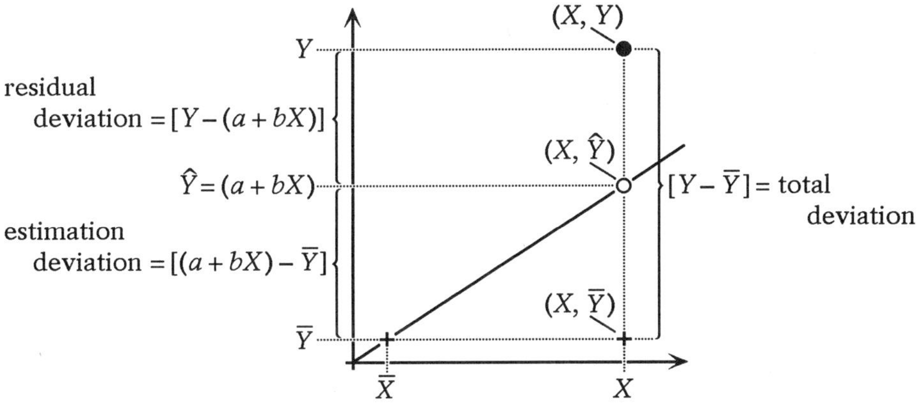

Figure 20.3.1
Decomposition of the total deviation $[Y - \overline{Y}]$ between an observation (X, Y) and the mean $\overline{Y}$ by the corresponding point $(X, \hat{Y})$ located on the estimation line $\hat{Y} = a + bX$

As for the *residual deviation,* it is the remainder of the total deviation which cannot be eliminated or explained by taking the relationship into account. As indicated in figure 20.3.2 (next page), the sum of squared deviations $\Sigma(Y - \overline{Y})^2$ of the total variation, as well as the corresponding number of degrees of freedom $(N - 1)$, decompose in the same way as the total deviation, since

$$\Sigma(Y - \overline{Y})^2 = \Sigma\{[Y - (a + bX)] + [(a + bX) - \overline{Y}]\}^2,$$

$$= \Sigma[Y - (a + bX)]^2 + 2\Sigma[Y - (a + bX)][(a + bX) - \overline{Y}] + \Sigma[(a + bX) - \overline{Y}]^2,$$

but $\Sigma[Y - (a + bX)][(a + bX) - \overline{Y}]$, the sum of products of residual and estimation deviations, cancels out when coefficients a and b are given the numerical values obtained following the least-squares principle (section 20.1, $a = \overline{Y} - b\overline{X}$ and $b = S_{XY}/S_X^2$).

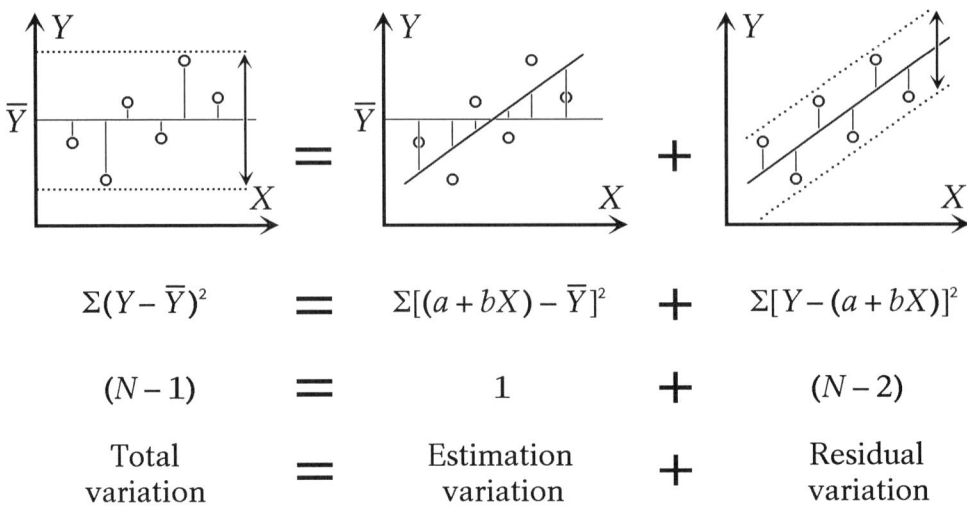

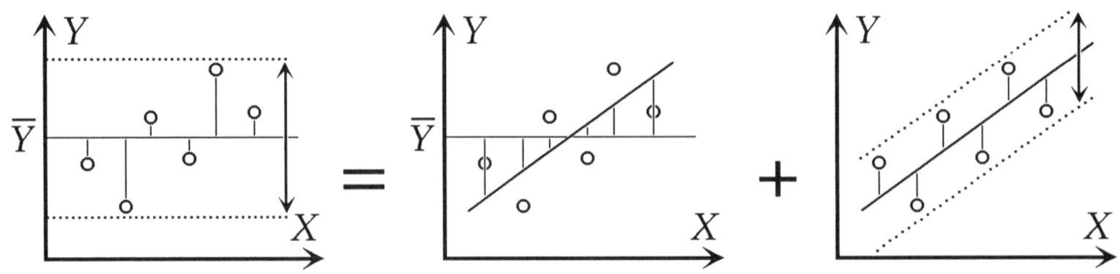

$\Sigma(Y-\overline{Y})^2$	=	$\Sigma[(a+bX)-\overline{Y}]^2$	+	$\Sigma[Y-(a+bX)]^2$
$(N-1)$	=	1	+	$(N-2)$
Total variation	=	Estimation variation	+	Residual variation

Figure 20.3.2
Decomposition of the total centered sum of squares $\Sigma[Y-\overline{Y}]^2$ into an estimation sum of squares $\Sigma[(a+bX)-\overline{Y}]^2$ and a residual sum of squares $\Sigma[Y-(a+bX)]^2$

The analysis of variance, of which the simplest kind has been discussed in chapter 12, may therefore be used to check whether the estimation line of a sample deserves to be used. The resulting analysis of variance table (table 20.3.1) leads to a variance ratio possessing (1, $N-2$) degrees of freedom:

$$F_{(1,\,N-2)} = S^2_{\text{estimation}} / S^2_{Y.X} = (N-2)\,r^2/(1-r^2)\,.$$

The preliminary hypothesis tested here is a null hypothesis according to which the slope of the parametric estimation line of Y from X would be null: $H_0: \sigma_{XY}/\sigma_X^2 = 0$. This analysis of variance is not discussed further here because it is strictly equivalent to the test of a null hypothesis about the slope of the estimation line with Student's t distribution (section 20.4), an approach which will be more profitable because it will allow considering one-sided alternative hypotheses about the slope.

Table 20.3.1
Analysis of variance corresponding to an estimation line

Variation	SS	D. of F.	Variance estimate
residual	$\Sigma[Y-(a+bX)]^2$	$(N-2)$	$S^2_{Y.X} = \Sigma[Y-(a+bX)]^2/(N-2)$
estimation	$\Sigma[(a+bX)-\overline{Y}]^2$	1	$S^2_{\text{estimation}} = \Sigma[(a+bX)-\overline{Y}]^2/1$
total	$\Sigma[Y-\overline{Y}]^2$	$(N-1)$	$S^2_Y = \Sigma[Y-\overline{Y}]^2/(N-1)$

However, the residual variance $S^2_{Y.X}$ (or its square root, the residual standard deviation $S_{Y.X}$) is particularly important because it reflects the dispersion of data dots about the estimation line and will be used to test hypotheses and to determine confidence limits for the slope (section 20.4) as well as for the predicted variate Y (section 20.5). When the residual variance is evaluated in practice, the *residual sum of squares*, which may be denoted by the abbreviations SS_{residual} or RSS and which could also be called the sum of squares of residual deviations, is calculated from centered sums by the formula

$$SS_{\text{residual}} = RSS = \Sigma(Y-\overline{Y})^2 - [\Sigma(X-\overline{X})(Y-\overline{Y})]^2/\Sigma(X-\overline{X})^2.$$

While the expression $\Sigma[Y-(a+bX)]^2$ is algebraically equivalent, it is not used for the present purpose because it would require heavier calculations. It may also be shown that $RSS = (1-r^2)\,\Sigma(Y-\overline{Y})^2$, where r denotes the sample correlation coefficient (section 19.4). When there is a strong straight-line relationship between two variates, the residual variance $S^2_{Y.X}$ is clearly smaller than the total variance S^2_Y, which indicates that the estimation line passes closer to the observations than the ordinary mean $\overline{Y}$ would do.

Section 20.4: the slope *b* : hypotheses and confidence intervals

The numerical value of the slope $b = S_{XY}/S^2_X$ of the estimation line of the sample, $\hat{Y} = a + bX$, is the estimate of the slope $\beta = \sigma_{XY}/\sigma_X^2$ of the parametric estimation line $\hat{Y} = [\mu_Y - (\sigma_{XY}/\sigma_X^2)\mu_X] + (\sigma_{XY}/\sigma_X^2)X$ (of the statistical population). The parametric slope is denoted here by the variant β of the Greek letter *beta* to avoid confusion with the more usual form β used to represent the probability of type II errors (section 9.4). Hypotheses are thus set up concerning the parametric slope $\beta = \sigma_{XY}/\sigma_X^2$, while it is the estimated slope $b = S_{XY}/S^2_X$ which must be evaluated to know whether a hypothesis can be accepted or should be rejected. Theoretically, it may be shown that the estimated slope b follows a normal distribution of which the mean is the parametric slope $\beta = \sigma_{XY}/\sigma_X^2$ and the standard deviation σ_b may be estimated by the expression

$$S_b = S_{Y.X}/\sqrt{\Sigma(X-\overline{X})^2},$$

where $S_{Y.X}$ is the estimated residual standard deviation and $\sqrt{\Sigma(X-\overline{X})^2}$ is the square root of the sum of squared deviations from the mean of the predictor variate. In practice, the parametric standard deviation σ_b is unknown and the estimate S_b must be used, so that the criterion

$$t_{(N-2)} = (b - \beta_0)/S_b\,,$$

where β_0 denotes the value of the parametric slope according to the preliminary hypothesis H_0: $\beta = \beta_0$, follows Student's t distribution with $(N-2)$ degrees of freedom.

In the case of the null hypothesis H_0: $\beta = \beta_0 = 0$, which is usually set up at first to check whether the estimation line deserves to be used, the criterion takes the simplified form $t_{(N-2)} = b/S_b$ and may be shown to be equal to the value $t_{(N-2)} = r\sqrt{[(N-2)/(1-r^2)]}$ found when testing hypothesis H_0 : $\rho = 0$ in section 19.5. Moreover, the square of the present criterion, b^2/S^2_b, is equal to the variance ratio $F_{(1,\,N-2)} = S^2_{\text{estimation}}/S^2_{Y.X}$ found in section 20.3. Whether Student's t or the variance ratio distribution is used, the tests of hypotheses H_0: $\beta = \beta_0 = 0$ and H_0: $\rho = 0$ are therefore equivalent.

The limits of the acceptance and rejection regions of the preliminary hypothesis are determined according to the nature of the alternative hypothesis (one-sided or two-sided) as in section 9.6, except that S_b is used here instead of $S_{\bar{Y}}$ and there are $(N-2)$ instead of $(N-1)$ degrees of freedom. In some fields, such as the study of allometry (section 20.7), it may happen that, once the null hypothesis has been rejected, other preliminary hypotheses must be considered, such as H_0: $\beta = \beta_0 = 2/3$, H_0: $\beta = \beta_0 = 1$, H_0: $\beta = \beta_0 = 2$, or H_0: $\beta = \beta_0 = 3$, and β_0 is then replaced by the pertinent numerical value during the calculation of the criterion.

When there is no particular hypothetical value to be tested for the parametric slope β, the sample is used to determine confidence limits like in section 9.7. In the case where, if a hypothesis were tested, the alternative hypothesis would be two-sided, the confidence interval of the slope of the estimation line is

$$b + t_{(\alpha/2;\, N-2)} S_b \leq \beta \leq b + t_{(1-\alpha/2;\, N-2)} S_b \,.$$

The reader may refer again to section 9.7 concerning one-sided confidence intervals.

When the estimation line $\hat{X} = c + dY$ of X from Y is used, one merely has to substitute X for Y and vice versa everywhere in the formulae. Thus, the standard deviation of the slope d is estimated by the expression

$$S_d = S_{X.Y} / \sqrt{\Sigma(Y - \bar{Y})^2},$$

where $S_{X.Y}$ is the residual standard deviation of X predicted from Y and $\sqrt{\Sigma(Y-\bar{Y})^2}$ is the square root of the sum of squared deviations from the mean of Y, which is the predictor variate in this case.

Before the confidence interval of the slope of the $X.Y$ estimation line can be compared with that of the $Y.X$ estimation line, however, the equation of the $X.Y$ line must necessarily be rewritten as in section 20.2, with Y as an explicit function of X. Thus, while the two-sided confidence interval of the parametric slope δ (*delta*) of the $X.Y$ line is

$$d_1 = d + t_{(\alpha/2;\, N-2)} S_d \leq \delta \leq d + t_{(1-\alpha/2;\, N-2)} S_d = d_2 \,,$$

the confidence interval of the inverse $1/\delta$ of the slope is

$$1/d_2 \leq 1/\delta \leq 1/d_1 \,.$$

It should finally be noted that, while the confidence interval of the slope of an estimation line is in one piece even when it covers the point 0 (because the relationship is not significant), the confidence interval of the inverse of that slope is then *disjoint* and made up of two parts which straddle the direction of the ordinate ($\pm\infty$), a kind of phenomenon which occurs also in the case of the slope of the major axis (figure 22.3.1, bottom, center):

$$-\infty \leq 1/\delta \leq 1/d_1$$

and

$$1/d_2 \leq 1/\delta \leq +\infty \,.$$

Section 20.5: confidence or prediction intervals of the estimated variate

When the methods of the preceding section indicate that the estimation line deserves to be used, it may also be desirable to provide confidence or prediction intervals for the variate predicted through that line. The study of Working and Hotelling (1929) has shown that these intervals become broader when the value X_0 of the predictor variate at which predictions are made deviates from the mean $\overline{X}$ toward the left or the right. The limits obtained are thus represented by hyperbolic curves (figure 20.5.1).

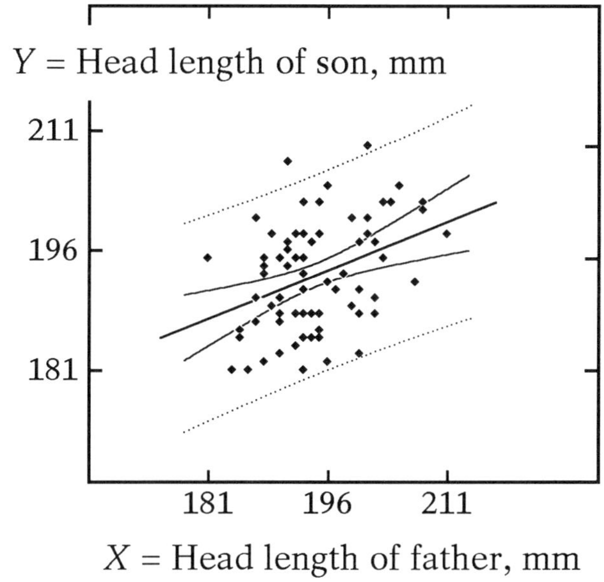

Figure 20.5.1
Confidence limits (continuous lines) and prediction limits (dotted lines) at level $\alpha = 0.05$ above and below the estimation line $\hat{Y} = a + bX$; head length in mm of the father X and of the oldest adult son Y in 73 families; data from Frets (1921)

When the estimation line is interpreted as a *changing mean*, an *adjusted mean* of the predicted variate Y, the confidence interval of the mean Y corresponding to a particular value X_0 of the predictor variate X is delimited (figure 20.5.1, continuous lines) by the pair of curves

$$Y = a + bX_0 + t_{(\alpha/2;\, N-2)}\, S_{Y.X} \sqrt{[(1/N) + (X_0 - \overline{X})^2/\Sigma(X-\overline{X})^2]}$$

and

$$Y = a + bX_0 + t_{(1-\alpha/2;\, N-2)}\, S_{Y.X} \sqrt{[(1/N) + (X_0 - \overline{X})^2/\Sigma(X-\overline{X})^2]}\ .$$

However, the prediction interval, within which a new individual Y value located at $X = X_0$ is expected to fall if it comes from the same population as the sample, is much broader (figure 20.5.1, dotted lines) and is delimited by the following pair of curves:

$$Y = a + bX_0 + t_{(\alpha/2;\, N-2)}\, S_{Y.X} \sqrt{[1 + (1/N) + (X_0 - \overline{X})^2/\Sigma(X-\overline{X})^2]}$$

and

$$Y = a + bX_0 + t_{(1-\alpha/2;\, N-2)}\, S_{Y.X} \sqrt{[1 + (1/N) + (X_0 - \overline{X})^2/\Sigma(X-\overline{X})^2]}\ .$$

If, for instance, an estimation line is used to predict basal metabolism Y from body weight X in a sample of mammals, the confidence interval shows what range of values is likely for the mean basal metabolism corresponding to a given body weight X_0, while the prediction interval indicates outside of which limits a new individual observation should be considered as too extreme to belong to the same population as the sample.

The confidence and prediction belts discussed in this section are appropriate only within the contexts where it is really justifiable to estimate one variate from the other. In other cases, where joint confidence or prediction regions are desired for two variates believed to follow a bivariate normal distribution, the elliptical regions discussed in sections 19.7, 19.8, 20.2, 22.1, 23.1 and 23.2 and in chapter 30 are often preferable.

Section 20.6: straightening exponential relationships

As briefly indicated in sections 19.8 and 19.10, some curved relationships can be transformed into straight lines through simple variate transformations. This is the case for exponential relationships, such as

$$\hat{Y} = a \exp(bX) ,$$

for which taking the logarithm of the predicted variate Y gives

$$\log_e(\hat{Y}) = \log_e(a) + bX .$$

In practice, the predictor variate X is time (t) in many exponential relationships,which are thus written

$$Y = a e^{bt},$$

where the constant e = 2.71828182846... is the base of natural logarithms. Coefficient b is positive in exponential growth and negative in exponential decrease.

Figure 20.6.1 illustrates the accelerated growth of a (biological) population of duck weed (*Lemna minor*) in experimental conditions where the increase in the number of individual plants goes faster and faster and is not slowed by any shortage of space, light or food. Exponential growth is very important in theoretical ecology because it is the way in which most populations of animals, plants, protists, bacteria and viruses would increase if they were placed in optimal conditions and did not face any environmental constraints. The number of individuals present in a population would then increase in geometric progression, like a capital invested at compound interest, and would rapidly become too large for planet Earth. From a mathematical viewpoint, since the derivative (the rate of increase) of the exponential function $f(t) = e^{bt}$ with respect to time is proportional to that function, $de^{bt}/dt = be^{bt}$, the exponential function is suitable to describe or to simulate the ability of living matter to self-reproduce.

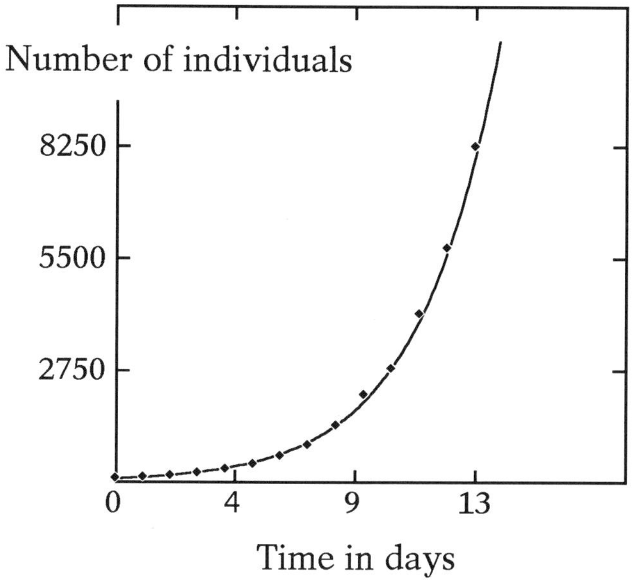

Figure 20.6.1
Exponential growth of an experimental population of duck weed (*Lemna minor*), according to data from Ashby and Oxley (1935)

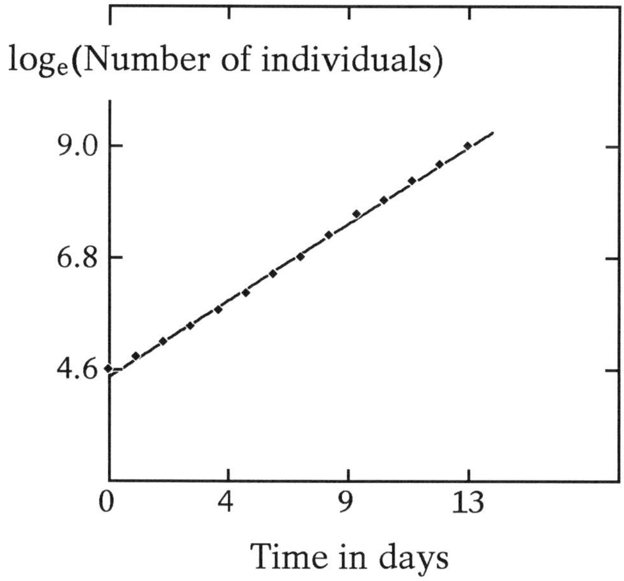

Figure 20.6.2
Exponential growth of an experimental population of duck weed (*Lemna minor*), represented in a semi-logarithmic scatter diagram

This type of increase is sometimes called *Malthusian growth* because the English political economist Thomas Robert Malthus (1766-1834) was the first writer to discuss the problems of uncontrolled population growth. Nowadays, the *zero-growth* objective is still the subject of much talk, paradoxically in the most prosperous countries. It is evident that, even if most biological populations begin growing exponentially when they invade a new environment, they rapidly face nutritional or spatial constraints which slow down and eventually stop their increase.

At first sight, the exponential curve drawn in figure 20.6.1 seems completely different from a straight line but, if the scatter diagram is redone with a logarithmically scaled ordinate or, what amounts to the same thing, by inscribing the logarithm of the predicted variate Y on the ordinate, the observed dots line up almost perfectly along a straight line which has been fitted here by least squares (figure 20.6.2). In other words, once the predicted variate Y has been replaced by its logarithm $\log_e(Y)$, the exponential relationship may be described simply by the straight estimation line of the transformed predicted variate $\log_e(Y)$ from the untransformed predictor variate X or t. In the case of the experimental population of *Lemna minor,* the equation of that line is

$$\log_e(\hat{Y}) = 4.45545 + 0.348643\, t\,,$$

which corresponds to the exponential curve

$$\hat{Y} = 86.0952 \exp(\, 0.348643\, t)$$

when the relationship is reexpressed with respect to the original ordinate.

Since the logarithmic transformation is applied here to only one of the two variates, as opposed to the double logarithmic transformation done in the case of the *power function* (sections 19.8 and 20.7), the scatter diagram obtained by straightening an exponential function is said to be *semi-logarithmic*. For the logarithmic transformation to be truly satisfactory, it is the transformed variate $\log_e(Y)$, rather than the original variate Y, of which the probability distribution should resemble the normal distribution, but that assumption is generally reasonable enough since most biological variates are nonnegative and have positively skewed distributions (chapters 13 and 14, and section 19.8). Moreover, in the case of exponential growth or decrease, an ordinary estimation line (as opposed to the orthogonal estimation line treated in chapter 22) is usually appropriate because the predictor variate X or t generally does not vary at random (and should thus be called a variable rather than a variate) and is known without errors.

Instead of growing, a biological population may have high death rates and low birth rates and decrease when it faces adverse environmental conditions. Figure 20.6.3 illustrates the exponential decrease of the number of surviving bacteria of the species *B. typhosus*[1] after treatment with 0.2% phenol, according to data compiled by Lee and Gilbert (1918).

[1] This is undoubtedly the species now known as *Salmonella typhi,* which causes typhoid fever.

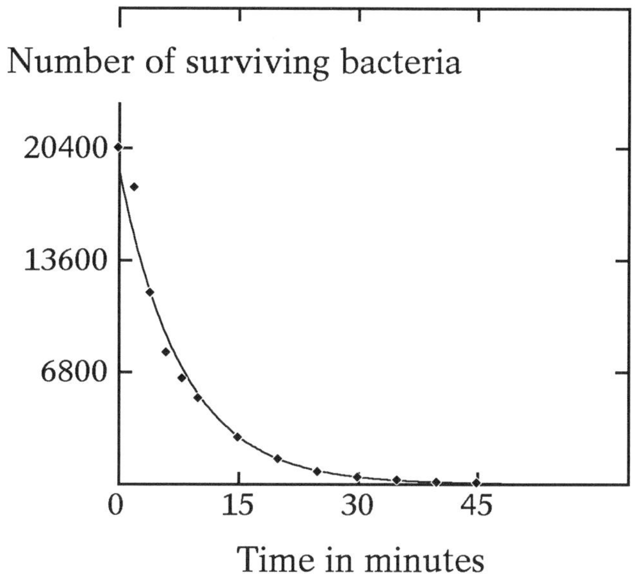

Figure 20.6.3
Exponential decrease of a culture of the bacterium *B. typhosus* after treatment with 0.2% phenol (data compiled by Lee and Gilbert, 1918)

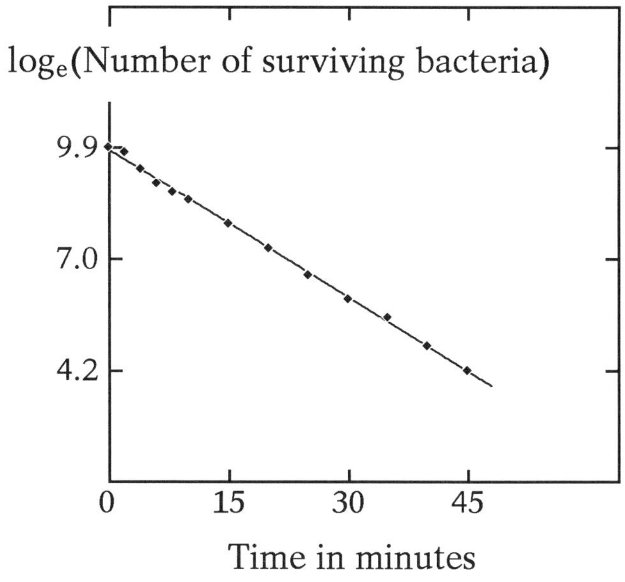

Figure 20.6.4
Exponential decrease of a culture of the bacterium *B. typhosus* after treatment with 0.2% phenol represented in a semi-logarithmic scatter diagram

The decreasing exponential curve drawn in figure 20.6.3 has also been obtained by least squares after replacing the predicted variate Y by its logarithm $\log_e(Y)$ and has the equation

$$\hat{Y} = 19262.5 \exp(-0.127246\, t)\,.$$

The straightened scatter diagram is shown in figure 20.6.4. Decreasing exponential curves are also found in several other fields of biology, for instance when an experimentally injected substance is filtered out or metabolically degraded by an organism. The concentration of that substance then decreases exponentially with time. If the substance has been labelled with a radioactive isotope, the decrease of radioactivity may be due not only to falling concentrations but also to the decomposition of the isotope, which is also exponential (although generally slower). Figure 20.6.5 represents the decreasing radioactivity of glycerophosphates in the liver of rats injected with Na_2HPO_4 labelled with radioactive ^{32}P phosphorus. The equation of this decreasing exponential curve is

$$\hat{Y} = 2.62378 \exp(-0.000826738\, t)\,.$$

While the coefficient b, which is multiplied by the predictor variate t in the exponent, is positive in the case of the exponential growth of the duck weed population (figures 20.6.1 and 20.6.2), it is negative in the cases of the exponential decrease of the bacteria *B. typhosus* (figures 20.6.3 and 20.6.4) and of glycerophosphates (figure 20.6.5). This coefficient (b) reflects the speed of an exponential increase or decrease and is descriptively important.

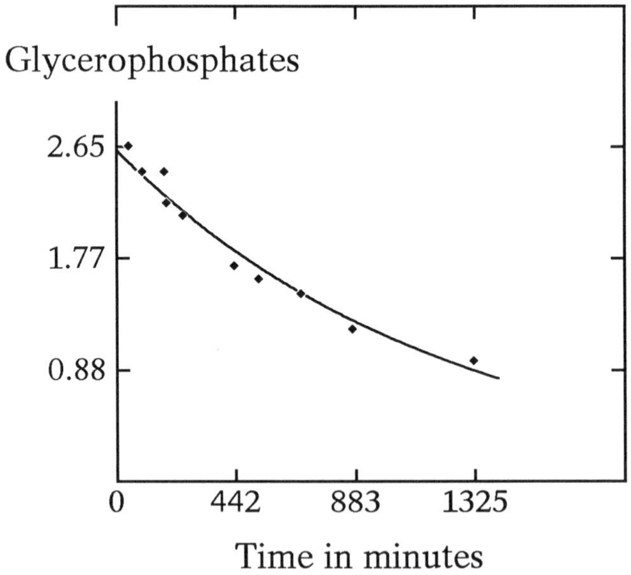

Figure 20.6.5
Exponential decrease of the radioactivity of glycerophosphates in the liver of rats injected with Na_2HPO_4 labelled with radioactive ^{32}P phosphorus (data from Popjak and Muir, 1950)

Thus, the expression $\log_e(2)/|b|$ is called *doubling time* in the case of exponential growth and *half-life* in the case of exponential decrease, because it is equal to the period of time necessary for the size of a population or for the concentration of a substance to be multiplied or divided by 2. Because doubling time and half-life are proportional to the inverse of the slope of an estimation line, it is easy to test hypotheses or to determine confidence intervals about them (section 20.4). Thus, the point estimate of the doubling time of the duck weed culture illustrated in figures 20.6.1 and 20.6.2 is $\log_e(2)/|0.348643| = 1.98813 \doteq 2$ days and, provided the relationship is statistically significant, the two-sided confidence interval could be determined from

$$\log_e(2)/[\,|b| + t_{(1-\alpha/2;\,N-2)}S_b\,] \le \log_e(2)/|\beta| \le \log_e(2)/[\,|b| + t_{(\alpha/2;\,N-2)}S_b\,]\,.$$

Section 20.7: straightening allometry relationships

In biology, the word *allometry* is used to indicate that two variates X and Y are related by a *power function* such as

$$\hat{Y} = aX^b\,.$$

While the predictor variate, X or t, appears in the exponent of the exponential function $\hat{Y} = ae^{bX}$ (section 20.6), it is affected in the power function by an exponent b called the *allometry exponent*. Taking the logarithm of both members of the allometry equation yields

$$\log_e(\hat{Y}) = \log_e(a) + b\log_e(X)\,,$$

which shows that the logarithms of the two variates X and Y are related by the equation of a straight line. Unlike the straightening of exponential relationships, which requires the logarithmic transformation to be applied only to the predicted variate Y, the straightening of an allometry relationship thus requires the logarithmic transformation of both variates, so that the straightened scatter diagram is said to be *bilogarithmic* (figure 20.7.3).

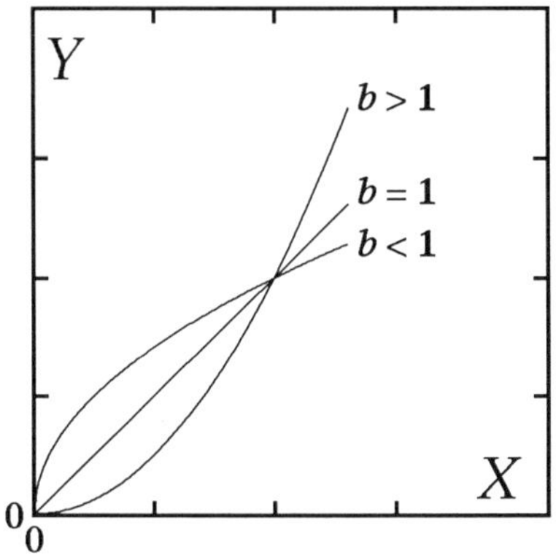

Figure 20.7.1
The allometry curve passes through the origin (0, 0) and is concave upwards when the allometry exponent $b > 1$, straight when $b = 1$, and concave downwards when $b < 1$

The allometry curve passes through the origin (0, 0) and takes the shape of a curve which is concave upwards when $b > 1$, of a straight line when $b = 1$, and of a curve which is concave downwards when $b < 1$ (figure 20.7.1). Following the current terminology, it is said that there is

positive allometry of Y with respect to X when the allometry exponent $b > 1$,

isometry when $b = 1$,

and *negative allometry* when $b < 1$.

However, that terminology is not truly satisfactory because the expression *negative allometry* should be reserved for the cases where $b < 0$, as when a bodily dimension decreases during growth, such as the tail length of a frog tadpole during metamorphosis. When $b < 1$, this indicates that variate Y increases relatively less rapidly than variate X, so that the ratio (Y/X) is smaller in large than in small specimens. Isometry corresponds to the case where the ratio (Y/X) remains constant ($Y = aX$, whence $Y/X = a$) and may be considered as a particular case of allometry.

For instance, figure 20.7.2 illustrates the relationship between the body weight in kg and the basal metabolism in kilojoules per day of a sample of adult placental mammals covering a broad size range, from the house mouse to the elephant. The allometry curve, represented by a continuous line, differs clearly from the straight line corresponding to the isometry hypothesis, represented by a dotted line. It has been well known for many years that a unit of biomass has a much higher metabolic rate in bird or mammal species of small than of large body size. This phenomenon is reflected not only by metabolic measurements but also by daily food consumption and by the rates of various processes, such as heart contraction and respiration, which are faster in small than in large organisms.

While the principal explanation given by physiologists may not take into account all aspects of the phenomenon, it undoubtedly contains much truth: if metabolic rates did not vary, the amount of heat produced by a *homeothermal* organism (of which the temperature remains constant) would be proportional to its mass, while heat losses would be proportional to body area. But, if organisms of similar shapes and densities are compared, their area is proportional to the square while their volume and their mass are proportional to the cube of their linear dimensions (length, diameter, etc.). Therefore, the area of the body is proportional to power (2/3) of its mass and is relatively more important in small than in large organisms. Consequently, in habitats of which the mean temperature is low, small organisms must produce more heat than large ones in order to keep their body temperature constant and higher than that of the environment. Conversely, in habitats of which the mean temperature is high, large organisms must produce less heat and find ways of cooling themselves in order not to overheat. While the fact that the metabolic rate of a unit of biomass is higher in small than in large organisms undoubtedly contributes to balance their thermal exchanges, it is likely to be also a way to balance other exchanges with the environment (respiratory, nutritional, etc.), which also tend to be approximately proportional to areas.

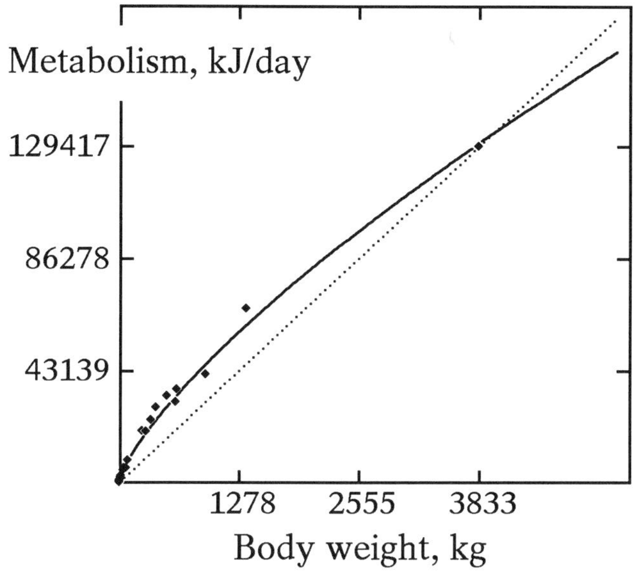

Figure 20.7.2
Allometry relationship (continuous line) of the basal metabolism in kilojoules per day and of the body weight in kg of adult placental mammals; the dotted line represents a hypothetical isometry line passing through the largest specimen; data compiled by Brody (1945)

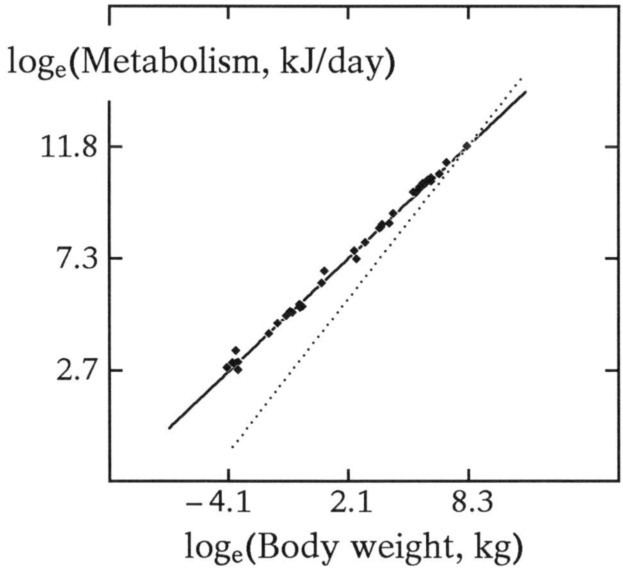

Figure 20.7.3
Bilogarithmic scatter diagram of the basal metabolism in kilojoules per day and of the body weight in kg of adult placental mammals; the continuous line corresponds to the allometry curve and the dotted line to the hypothetical isometry line passing through the largest specimen

In theory, basal metabolism Y would thus be expected to be proportional to body area rather than to mass, that is to power 2/3 of body weight X, and to correspond to the following equation:

$$Y = aX^{2/3}.$$

The parameters of the allometry relationship have been estimated here by fitting a straight line to the logarithms of the two variates,

$$\log_e(\hat{Y}) = 5.6989 + 0.737189 \log_e(X) ,$$

which corresponds to the curve

$$\hat{Y} = 298.539\, X^{0.737189} .$$

The observed value of the allometry exponent, 0.737189, thus does not seem extremely different from the theoretical value, $2/3 \doteq 0.666667$, and a comparison could be made by a significance test or by a confidence interval following the methods of section 20.4. From a graphical viewpoint, the estimation line of $\log_e(Y)$ from $\log_e(X)$, represented by a continuous line in the bilogarithmic scatter diagram (figure 20.7.3), seems clearly different from the line of slope 1 corresponding to the isometry hypothesis, represented by a dotted line.

It must be emphasized, however, that the utilization of an ordinary estimation line is more difficult to justify in the case of an allometry relationship than in the case of an exponential relationship, because both variates of an allometry relationship are usually affected by random fluctuations. In the study of allometry, the *orthogonal estimation line* or *major axis* (section 20.2 and chapter 22) is therefore often preferable to ordinary estimation lines. As for the double logarithmic transformation, it is satisfactory provided the probability distribution of original variates is more similar to the lognormal than to the normal distribution (chapter 14, sections 19.8 and 30.9), but that assumption is often justifiable in biology. When the double logarithmic transformation fails to completely eliminate the curvature of the cluster of observed dots, however, the relationship is said to be a case of *complex allometry* (section 35.14), as opposed to the *simple allometry* discussed in the present section. Allometry may also be studied with respect to more than two variates jointly (sections 31.6 to 31.9, 35.12, 35.14 and 35.18).

Many studies of allometry have been done also on plants (for instance Niklas, 1994). The first detailed discussion of allometric growth was published by Huxley (1932). The utilization of the bivariate and multivariate lognormal distributions in growth studies has been reviewed by Mosimann and Campbell (1988).

Section 20.8: causal interpretations

In section 19.9, the reader has already been urged to be cautious concerning the interpretation of correlation coefficients. A similar warning must be given in the case of estimation lines. This highly controversial question may be summarized by noting that, in general, the presence of a statistical relationship between two variates does not provide by itself sufficient grounds to conclude that there is also a causal relationship.

When the existence of a causal relationship is already known independently in advance, however, a correlation coefficient or an estimation line can provide an approximate quantitative description of that relationship. For instance, figure 20.8.1 illustrates the increasing statistical relationship observed during an agricultural experiment between the quantity X of fertilizer applied in pounds (1 pound = 454 grams) and the amount Y of straw harvested per lot in grams.

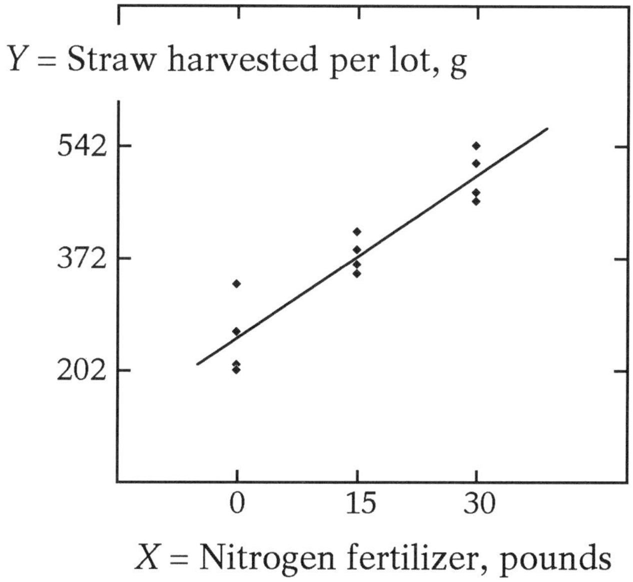

Figure 20.8.1
Quantitative description by an estimation line of the causal relationship between the quantity of fertilizer applied (1 pound = 454 grams) and the amount of straw harvested in g during an agricultural experiment; according to data published by Anderson (1946)

But it has already been shown many times and it is now well known that, during agricultural experiments, the production of plant matter can be increased by applying natural or chemical fertilizers. The existence of that causal relationship being already established, the estimation line of Y from X may describe the effect of fertilization provided the relationship is truly linear. In this kind of context, the predictor variate X, which is interpreted as a cause, is usually controlled by the experimenter and known accurately, so that the ordinary $Y.X$ estimation line is generally satisfactory. If extrapolations were carried out at very large values of the predictor variate X, however, it is almost certain that the relationship would no longer be linear and that the estimation line would yield incorrect predictions, because the addition of excessive amounts of fertilizers could have toxic effects and straw production might stop increasing or even decrease (see section 35.9).

Chapter 21
The analysis of covariance or "ANCOVA": comparing estimation lines

Section 21.1: introduction

When there is a straight-line statistical relationship between a predictor variate X and a predicted variate Y, the estimation line $\hat{Y} = a + bX$ is equivalent to a *conditional mean of Y given X*, that is to a measure of central tendency of Y which changes according to the value of X (section 20.1). The analysis of variance, which has been used in chapter 12 to compare the mean values of several groups of data, can therefore be adapted to the comparison of several estimation lines, since the latter are essentially modified means. The version of the analysis of variance which applies to estimation lines is currently known as the *analysis of covariance* (acronym: ANCOVA), a name of which the meaning is regrettably liable to be misinterpreted. While the ordinary analysis of variance (chapter 12) corresponds basically to the decomposition of the sum of squared deviations between individual observations Y and the mean $\overline{Y}$,

$$SS = \Sigma(Y - \overline{Y})^2 = \Sigma Y^2 - (\Sigma Y)^2/N,$$

the analysis of covariance corresponds to the decomposition of the sum of squares of residual deviations ($SS_{residual}$ or RSS, section 20.3) between individual observations Y and the estimation line $\hat{Y} = a + bX$,

$$RSS = \Sigma(Y - \hat{Y})^2 = \Sigma[Y - (a + bX)]^2 = \Sigma(Y - \overline{Y})^2 - [\Sigma(X - \overline{X})(Y - \overline{Y})]^2/\Sigma(X - \overline{X})^2.$$

When there is a marked straight-line statistical relationship between variates X and Y, the residual variances are perceptibly smaller that the total variances, and the analysis of covariance is more sensitive than the analysis of variance.

Section 21.2: decomposing the total residual sum of squares

Like the analysis of variance (chapter 12), the analysis of covariance can be used to do an overall comparison of several groups. To keep things simple, let us consider first the case where there are only two groups. Figure 21.2.1 and table 21.2.1 illustrate three possible ways to calculate residual sums of squares.

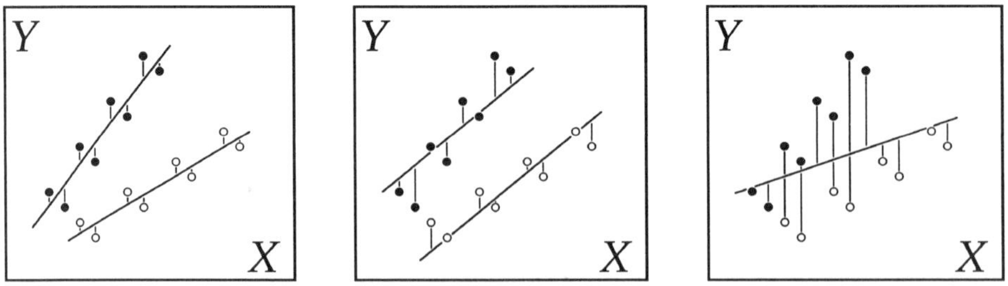

Figure 21.2.1
Residual sums of squares with two different straight lines (left), two parallel straight lines (center), and two identical straight lines, that is a single straight line (right)

Table 21.2.1

Three ways of calculating residual sums of squares (RSS)

(1) Groups pooled at the stage of residual sums: different straight lines

$$\left.\begin{array}{l}\begin{bmatrix} N & \Sigma X & \Sigma Y \\ & \Sigma X^2 & \Sigma XY \\ & & \Sigma Y^2 \end{bmatrix}_1 \rightarrow \begin{bmatrix} \Sigma(X-\bar{X})^2 & \Sigma(X-\bar{X})(Y-\bar{Y}) \\ & \Sigma(Y-\bar{Y})^2 \end{bmatrix}_1 \rightarrow \Sigma[Y-(a+bX)]_1^2 \\ \begin{bmatrix} N & \Sigma X & \Sigma Y \\ & \Sigma X^2 & \Sigma XY \\ & & \Sigma Y^2 \end{bmatrix}_2 \rightarrow \begin{bmatrix} \Sigma(X-\bar{X})^2 & \Sigma(X-\bar{X})(Y-\bar{Y}) \\ & \Sigma(Y-\bar{Y})^2 \end{bmatrix}_2 \rightarrow \Sigma[Y-(a+bX)]_2^2 \end{array}\right\} \rightarrow \sum_{g=1}^{C} \Sigma[Y-(a+bX)]_g^2 = \mathrm{RSS}_{\text{within-groups}}$$

(2) Groups pooled at the stage of centered sums: parallel straight lines

$$\left.\begin{array}{l}\begin{bmatrix} N & \Sigma X & \Sigma Y \\ & \Sigma X^2 & \Sigma XY \\ & & \Sigma Y^2 \end{bmatrix}_1 \rightarrow \begin{bmatrix} \Sigma(X-\bar{X})^2 & \Sigma(X-\bar{X})(Y-\bar{Y}) \\ & \Sigma(Y-\bar{Y})^2 \end{bmatrix}_1 \\ \begin{bmatrix} N & \Sigma X & \Sigma Y \\ & \Sigma X^2 & \Sigma XY \\ & & \Sigma Y^2 \end{bmatrix}_2 \rightarrow \begin{bmatrix} \Sigma(X-\bar{X})^2 & \Sigma(X-\bar{X})(Y-\bar{Y}) \\ & \Sigma(Y-\bar{Y})^2 \end{bmatrix}_2 \end{array}\right\} \rightarrow \begin{bmatrix} \sum_{g=1}^{C} \Sigma(X-\bar{X})^2 & \sum_{g=1}^{C} \Sigma(X-\bar{X})(Y-\bar{Y}) \\ & \sum_{g=1}^{C} \Sigma(Y-\bar{Y})^2 \end{bmatrix}_{//} \rightarrow \Sigma[Y-(a+bX)]_{//}^2 = \mathrm{RSS}_{//}$$

(3) Groups pooled at the stage of raw sums: identical straight lines (= one single straight line)

$$\left.\begin{array}{l}\begin{bmatrix} N & \Sigma X & \Sigma Y \\ & \Sigma X^2 & \Sigma XY \\ & & \Sigma Y^2 \end{bmatrix}_1 \\ \begin{bmatrix} N & \Sigma X & \Sigma Y \\ & \Sigma X^2 & \Sigma XY \\ & & \Sigma Y^2 \end{bmatrix}_2 \end{array}\right\} \rightarrow \begin{bmatrix} N & \Sigma X & \Sigma Y \\ & \Sigma X^2 & \Sigma XY \\ & & \Sigma Y^2 \end{bmatrix}_{.} \rightarrow \begin{bmatrix} \Sigma(X-\bar{X})^2 & \Sigma(X-\bar{X})(Y-\bar{Y}) \\ & \Sigma(Y-\bar{Y})^2 \end{bmatrix}_{.} \rightarrow \Sigma[Y-(a+bX)]_{.}^2 = \mathrm{RSS}_{\text{total}}$$

If the samples were known in advance to come from different statistical populations, on the one hand, the most logical procedure would be to calculate residual sums of squares (RSS) separately for each group [figure 21.2.1, left; table 21.2.1, (1)]. If the samples were known to come from identical populations (or equivalently from a single population), on the other hand, groups should be pooled right at the beginning of calculations, that is at the stage of raw sums [figure 21.2.1, right; table 21.2.1, (3)]. Figure 21.2.1 shows how residual sums of squares are smaller when they are calculated for each group separately before being added up (figure 21.2.1, left) than when data are pooled into a single group before calculations are carried out (figure 21.2.1, right).

As in the analysis of variance (chapter 12), the results of an analysis of covariance are usually presented in a table, which has been set up here for the general case of any number, C, of groups (table 21.2.2). However, this table can easily be adapted to the case where there are only two groups by letting $C=2$. Adding up the residual sums of squares calculated for each group separately yields the *within-groups residual sum of squares,* also called the *pooled residual sum of squares* and recorded on the row labelled "C different lines" in table 21.2.2,

$$\text{RSS}_{\text{within-groups}} = \sum_{g=1}^{C} \Sigma[Y-(a+bX)]_g^2 .$$

Subtracting that within-groups RSS from the total RSS, $\Sigma[Y-(a+bX)]^2_{\cdot}$, would yield a *between-groups* RSS corresponding to $(2C-2)=2(C-1)$ degrees of freedom, that is 2 degrees of freedom when $C=2$. Between the two extreme ways of carrying out calculations [left and right sides of figure 21.2.1; (1) and (3) in table 21.2.1; rows labelled "C different lines" and "one single line" in table 21.2.2], however, there is an intermediate way in which groups are pooled at the stage of centered sums, and the RSS obtained in this way,

$$\text{RSS}_{//} = \Sigma[Y-(a+bX)]^2_{//} ,$$

can be shown to correspond to the fitting of parallel estimation lines, which may have different positions but have equal slopes. Since the slope of estimation line $\hat{Y}=a+bX$ of any group is estimated by expression

$$b = S_{XY}/S_X^2 = \Sigma(X-\overline{X})(Y-\overline{Y})/\Sigma(X-\overline{X})^2 ,$$

the same slope is being used implicitly indeed for all groups when centered sums are pooled before residual sums of squares are calculated. The sum of residual sums of squares with "C parallel lines" is obtained then from the following expression:

$$\Sigma[Y-(a+bX)]^2_{//} = \sum_{g=1}^{C} \Sigma(Y-\overline{Y})_g^2 - [\sum_{g=1}^{C} \Sigma(X-\overline{X})(Y-\overline{Y})_g]^2 / \sum_{g=1}^{C} \Sigma(X-\overline{X})_g^2 .$$

Table 21.2.2
Analysis of covariance

Variation	Centered sums of squares and products			D. of F.	Residual SS (RSS)	D. of F.	Variance
group 1	$\Sigma(X-\bar{X})^2_1$	$\Sigma(X-\bar{X})(Y-\bar{Y})_1$	$\Sigma(Y-\bar{Y})^2_1$	N_1-1	$\Sigma[Y-(a+bX)]^2_1$	N_1-2	$S^2_{Y.X_1}$
. . .	. . .	. . .	. . .	. . .	. . .	. . .	. . .
group C	$\Sigma(X-\bar{X})^2_C$	$\Sigma(X-\bar{X})(Y-\bar{Y})_C$	$\Sigma(Y-\bar{Y})^2_C$	N_C-1	$\Sigma[Y-(a+bX)]^2_C$	N_C-2	$S^2_{Y.X_C}$
C different lines					$RSS_{within} = \sum_{g=1}^{C} \Sigma[Y-(a+bX)]^2_g$	$N.-2C$	$S^2_{Y.X\,within\text{-}groups}$
slope differences					$RSS_{//} - RSS_{within\text{-}groups}$	$C-1$	S^2_{slopes}
C parallel lines	$\sum_{g=1}^{C} \Sigma(X-\bar{X})^2_g$	$\sum_{g=1}^{C} \Sigma(X-\bar{X})(Y-\bar{Y})_g$	$\sum_{g=1}^{C} \Sigma(Y-\bar{Y})^2_g$	$N.-C$	$RSS_{//} = \Sigma[Y-(a+bX)]^2_{//}$	$N.-C-1$	$S^2_{Y.X//}$
position differences					$RSS_{total} - RSS_{//}$	$C-1$	$S^2_{positions}$
one single line	$\Sigma(X-\bar{X})^2_.$	$\Sigma(X-\bar{X})(Y-\bar{Y})_.$	$\Sigma(Y-\bar{Y})^2_.$	$N.-1$	$RSS_{total} = \Sigma[Y-(a+bX)]^2_.$	$N.-2$	$S^2_{Y.X\,total}$

The comparison of $RSS_{//}$ with the *within-groups* RSS on the one hand and with the *total* RSS on the other hand thus decomposes the *between-groups* RSS into two parts,

$$RSS_{\text{slopes}} \text{ and } RSS_{\text{positions}} ,$$

which have $(C-1)$ degrees of freedom each and which reflect respectively slope differences and position differences between the estimation lines of the groups compared.

The variance estimate is obtained in the same way on each row of the analysis of covariance table (table 21.2.2): by dividing the quantity entered in the RSS column by the corresponding number of degrees of freedom entered in the next column. In the case of group 1, for instance,

$$S^2_{Y.X_1} = \Sigma[Y-(a+bX)]^2_1/(N_1-2) .$$

Section 21.3: testing hypotheses

The preliminary hypothesis that the estimation lines of the groups compared have equal parametric slopes may be tested with the variance ratio

$$S^2_{\text{slopes}}/S^2_{Y.X\text{ within-groups}} ,$$

the acceptance region being

$$S^2_{\text{slopes}}/S^2_{Y.X\text{ within-groups}} \leq F_{(1-\alpha;\ C-1,\ N.-2C)} .$$

When the preceding hypothesis (of equal parametric slopes) is accepted, the hypothesis that estimation lines also have identical positions is tested next with the variance ratio

$$S^2_{\text{positions}}/S^2_{Y.X//} ,$$

the acceptance region being

$$S^2_{\text{positions}}/S^2_{Y.X//} \leq F_{(1-\alpha;\ C-1,\ N.-C-1)} .$$

The residual variance with parallel lines, $S^2_{Y.X//}$ is used as a denominator to test positions because there would be no reasons to separate the RSS corresponding to slope differences from within-groups variation when the hypothesis of equal parametric slopes is accepted. When the hypothesis of equal parametric slopes is rejected, however, the hypothesis of identical positions is usually not tested, because two converging lines necessarily have one point in common and their positions could not be compared without ambiguity.

Section 21.4: a limnological example

Water *p*H, X, and the percentage of organic matter of sediments, Y, have been determined in several Wisconsin lakes (Juday, Birge and Meloche, 1941) and in lakes of the Atlantic coast of Canada (Hayes and Anthony, 1958). The data and the estimation lines of Y from X are illustrated in figure 21.4.1, and the estimation lines of the two groups are compared by an analysis of covariance in table 21.4.1.

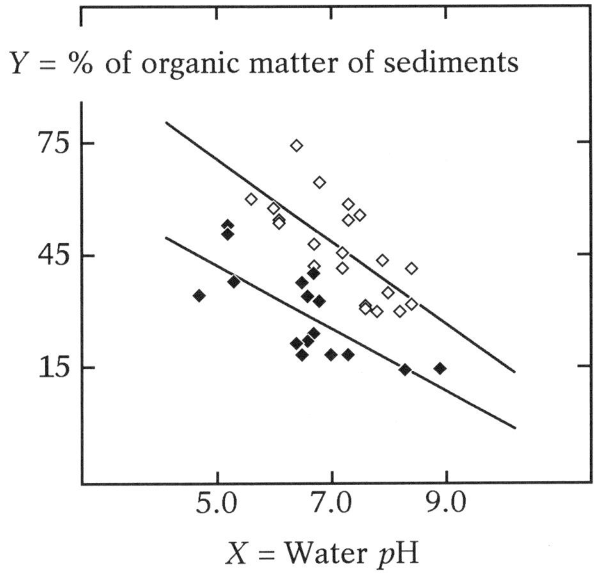

Figure 21.4.1
Relationship between the percentage of organic matter of sediments *Y* and water *p*H *X* of Wisconsin lakes (white dots) and lakes of the Atlantic coast of Canada (black dots); data from Juday, Birge and Meloche (1941) and Hayes and Anthony (1958)

Table 21.4.1
Analysis of covariance: comparison of the estimation lines of the percentage of organic matter of sediments from the water *p*H of Wisconsin lakes and lakes of the Atlantic coast of Canada

Nature of variation	$\Sigma(X-\bar{X})^2$	$\Sigma(X-\bar{X})(Y-\bar{Y})$	$\Sigma(Y-\bar{Y})^2$	Degrees of freedom	RSS $\Sigma[Y-(a+bX)]^2$	Degrees of freedom	Variance
Atlantic coast	18.1194	−152.533	2301.6	15	1017.54	14	72.6818
Wisconsin	13.5124	−149.015	3283.66	20	1640.33	19	86.3331
within-groups	–	–	–	–	2657.87	33	80.5416
slope differences	–	–	–	–	52.7176	1	52.7176
parallel lines	31.6318	−301.548	5585.27	35	2710.59	34	79.7233
position differ.	–	–	–	–	4481.66	1	4481.66
total variation	35.3189	−200.864	8334.6	36	7192.26	35	205.493

The hypothesis that the parametric slopes of the two lines are equal can be accepted ($F_{(1, 33)} = 0.654539$, $P > 0.05$), but the hypothesis that their positions are identical must be rejected ($F_{(1, 34)} = 56.2153$, $P < 0.001$). It may therefore be concluded that the percentage of organic matter of sediments decreases in the same way in the lakes of the two regions but that, when water *p*H is the same, that percentage is higher in Wisconsin than on the Atlantic coast of Canada. The higher percentage of organic matter in the lake sediments of Wisconsin might be explained by a higher productivity of Wisconsin lakes or by a greater abundance of inorganic sediments on the Atlantic coast of Canada.

As for the increase in the percentage of organic matter of sediments when water *p*H is low, it might be interpreted as a result of the inhibition of bacterial activities (including the decay of plant and animal debris) by environmental acidity. An analogous explanation has indeed been proposed for the extraordinary preservation of corpses of Pleistocene men discovered in peat bogs in Denmark.

The estimation lines of the two samples, established through least squares, have the following equations:

Atlantic coast: $\hat{Y} = 84.2056 - 8.41823\,X$;

Wisconsin: $\hat{Y} = 125.71 - 11.0280\,X$.

The numerical estimate of the Y-intercept of the estimation line of Wisconsin (125.71) calls for a comment: it would obviously be nonsensical to state that the percentage of organic matter in sediments Y would reach 125.71% when water *p*H would be $X = 0$. Since variate Y is a percentage, its value must clearly be neither lower than 0 nor higher than 100. Here, as in sections 20.1 and 20.8, the estimation line must therefore not be used to extrapolate far outside of the range covered by the data. For a mathematical model allowing the relationship between the percentage of organic matter of lake sediments and water *p*H to be described correctly over a broader range of numerical values, a nonlinear asymptotic function should be used, such as a decreasing logistic function

$$\hat{Y} = a/\{1 + \exp[(X - b)/c]\}$$

(for the increasing logistic function, see section 35.10). However, such a nonlinear function would be more difficult to fit, and the comparison of two such functions through an analysis of covariance would be approximate. The above estimation lines provide a satisfactory analysis provided they are used with caution.

Section 21.5: possibilities and limitations of the analysis of covariance

When there is no doubt that one of the two variates is to be used as a predictor (X) and that the other is to be predicted (Y), the analysis of covariance is not only appropriate but also often very powerful. The analysis of covariance may also be used in the case of *multiple regression* (chapters 23 and 25), where there are several predictor variates. However, when both variates are subject to random fluctuations and when one seeks a description of their relationship rather than the prediction of one of the two variates from the other, the analysis of covariance is of doubtful usefulness: two different analyses of covariance are indeed then possible, depending on whether the predicted variate is Y or X (section 20.2), and the conclusions based on those analyses are likely to be contradictory. In the case of data for the analysis of which ordinary estimation lines are not suitable, it may be appropriate to compare two or several groups of observations with respect to two or several variates by using multivariate statistical techniques such as Hotelling's T^2 distribution (chapter 30) or discriminant functions (chapters 32 and 33). A special number of the periodical *Biometrics*, beginning with an introductory paper by Cochran (1957), was devoted to the analysis of covariance.

Chapter 22
The orthogonal estimation line or *major axis*

Section 22.1: introduction

As already mentioned earlier (sections 20.2, 20.7, and 21.5), ordinary estimation lines minimize the sums of squares of residual deviations with respect to only one of two variates: in a direction parallel to the ordinate in the case of the $Y.X$ estimation line of Y from X (figure 22.1.1, left) but parallel to the abscissa in the case of the $X.Y$ estimation line of X from Y (figure 22.1.1, right). Ordinary estimation lines are therefore appropriate either (1) when random fluctuations affect only one or mostly one of the two variates (sections 20.6 and 20.8) or (2) when random fluctuations affect both variates but it is clearly the estimation of one of the two variates from the other that is required. But it occurs often in biology (see section 20.7, for instance), as well as in other fields, that variates X and Y are both affected by random fluctuations and that what is needed is a symmetrical description of their relationship rather than a one-way estimation of one variate from the other. Ordinary estimation lines are then unsatisfactory because there are no more reasons to choose the $Y.X$ line than the $X.Y$ line and they can yield contradictory answers (section 20.2). When the random fluctuations of both variates are thought to have approximately the same magnitude, the *orthogonal estimation line* (also known as the *orthogonal regression line*) may be used. Because that line minimizes the sum of squares of residual deviations perpendicularly to itself (figure 22.1.1, center), it is taking into account the fluctuations of both variates jointly. The orthogonal estimation line is also called the *major axis* because its parametric (population) version coincides with the major axis of the equal probability density ellipses of a bivariate normal distribution (section 19.2; see also sections 30.8 and 31.1).

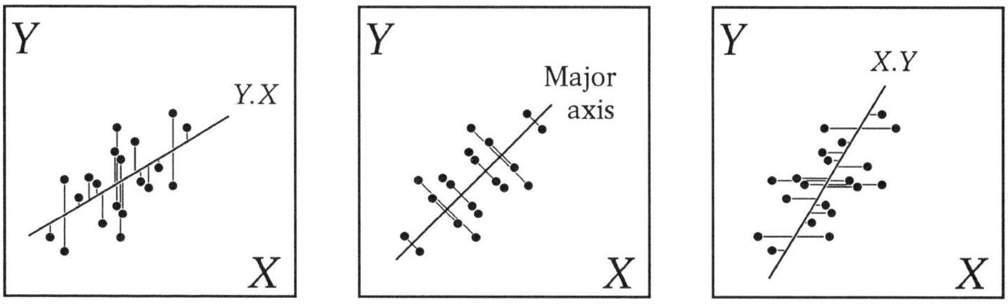

Figure 22.1.1
Three ways of fitting a straight estimation line according to the least-squares principle: minimizing the sum of squares of vertical deviations with the $Y.X$ line (left), the sum of squares of orthogonal deviations with the major axis (center), and the sum of squares of horizontal deviations with the $X.Y$ line (right)

Figure 22.1.2 illustrates the major axis of the set of data already considered in section 20.2 and figure 20.2.1. The slope of the major axis here is $k = +1.17582$ and is thus intermediate between the slope ($b = +0.402655$) of the $Y.X$ estimation line and the inverse ($1/d = +2.80887$) of the slope of the $X.Y$ estimation line. In the case of these data, on the relationship between the head lengths of father and son, using the major axis instead of ordinary estimation lines would help to solve Francis Galton's paradox (section 20.2; see also section 22.3).

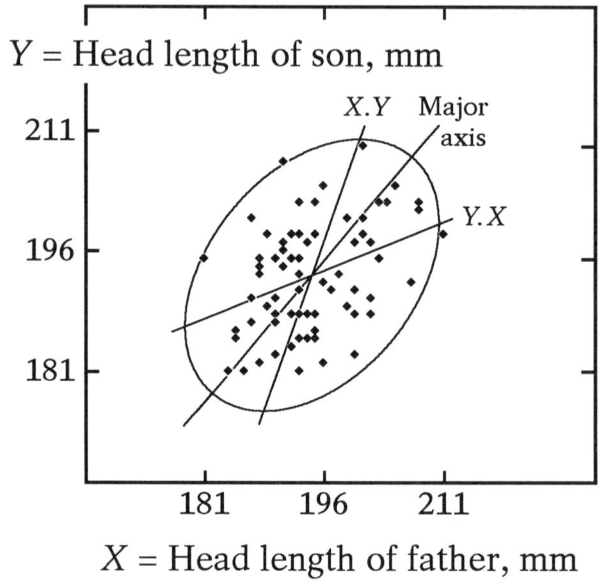

Figure 22.1.2
Prediction ellipse, at level $\alpha = 0.05$, of the head length of the father X and of the oldest adult son Y, in mm, in 73 families; the slope of the orthogonal estimation line or *major axis* is intermediate between the slope of the $Y.X$ estimation line and the inverse of the slope of the $X.Y$ estimation line; data from Frets (1921)

Section 22.2: orthogonal least squares

Figure 22.2.1 illustrates the particular manner in which the least-squares principle is applied in the case of the major axis. The predicted point $(\hat{X}, \hat{Y})$ is the intersection between the orthogonal estimation line of which the slope $k = \tan(f)$ must be estimated and the perpendicular falling on that estimation line from the observed point (X, Y). The deviation between the observed and the predicted points is thus the hypotenuse of a right triangle of which the other two sides are the deviations in the direction of the abscissa and of the ordinate, $(X - \hat{X})$ and $(Y - \hat{Y})$. Therefore, the absolute value of the orthogonal deviation is

$$\sqrt{[(X - \hat{X})^2 + (Y - \hat{Y})^2]}$$

but, since the slope of the perpendicular to the major axis is $-1/k$, the coordinates of the predicted point $(\hat{X}, \hat{Y})$ can be expressed entirely with respect to those of the observed point (X, Y) and of the slope k to be estimated.

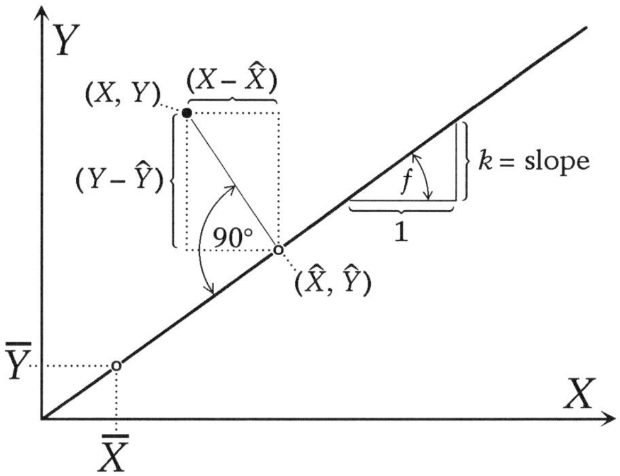

Figure 22.2.1
Decomposition into two components $(X-\hat{X})$ and $(Y-\hat{Y})$ of the orthogonal deviation between an observation (X, Y) and the corresponding point $(\hat{X}, \hat{Y})$ on the major axis

Consequently, the sum of squares of orthogonal deviations may finally be shown to be

$$\sum_{h=1}^{N} [(X-\hat{X})^2 + (Y-\hat{Y})^2] = \frac{(N-1)[S_X^2 k^2 - 2S_{XY}k + S_Y^2]}{(k^2+1)}$$

and, if the derivative of that sum with respect to the slope k is set equal to 0, a second degree equation is obtained,

$$S_{XY}k^2 + (S_X^2 - S_Y^2)k - S_{XY} = 0 \, ,$$

of which the two roots are

$$k_{\mathrm{I}} = \{(S_Y^2 - S_X^2) + \sqrt{[(S_Y^2 - S_X^2)^2 + 4\,S_{XY}^2]}\}/(2\,S_{XY})$$

and

$$k_{\mathrm{II}} = \{(S_Y^2 - S_X^2) - \sqrt{[(S_Y^2 - S_X^2)^2 + 4\,S_{XY}^2]}\}/(2\,S_{XY}) \, .$$

A scrutiny of those two roots shows that k_{I} has the same sign as the sample covariance S_{XY} and minimizes the sum of squares of orthogonal deviations, while $k_{\mathrm{II}} = -1/k_{\mathrm{I}}$ has the opposite sign and maximizes the sum of squares. k_{I} is thus the slope of the sample major axis, which must be estimated, while k_{II} is the slope of the minor axis, which is perpendicular to the major axis. The major and the minor axis have one point in common (but only one) with ordinary estimation lines: all of those lines pass through the mean point $(\overline{X}, \overline{Y})$. The slope of the major axis is often denoted by k without subscript, the subscript I being implied. If the least-squares principle is applied at the level of the statistical population, the slope $\varkappa$ (*kappa*, the Greek letter corresponding to the Roman k) of the parametric major axis is obtained from an equation analogous to that for k_{I}:

$$\varkappa = \{(\sigma_Y^2 - \sigma_X^2) + \sqrt{[(\sigma_Y^2 - \sigma_X^2)^2 + 4\sigma_{XY}^2]}\}/(2\,\sigma_{XY}) \, .$$

When the parametric correlation ρ is low, the slope k of the sample major axis is very variable, for its numerical value can range from strongly negative to strongly positive. Moreover, the angle f made by the major axis with the abscissa is not then an unbiased estimator of the corresponding angle φ *(phi)* of the statistical population, because f has an axial distribution of which the origin is arbitrary. However, Jolicoeur and Ducharme (1992) have noted that the usual concepts of (scalar) mean and of bias are not appropriate in the case of a directional variate like angle f. The same authors have also shown that the direction of the sample major axis is a satisfactory estimator of the direction of the population major axis.

Section 22.3: confidence intervals of the slope

Following earlier works by Creasy (1956), Mallows (1961) and Rempe (1962), Jolicoeur (1968) and Jolicoeur and Mosimann (1968) have shown that an approximate confidence interval may be obtained for the slope $\varkappa$ of the parametric major axis of a bivariate normal distribution by postulating that, if the directions of the parametric major axis and conjugate minor axis are known, the parametric correlation between the coordinates of observations on those two axes is null. The confidence interval $k_1 \leq \varkappa \leq k_2$ thus contains all hypothetical values of $\varkappa$ such that the observed correlation between the corresponding major and minor axes does not reach statistical significance. The limits k_1 and k_2 have the following values:

$$k_1 = \{(S_Y^2 - S_X^2) + \sqrt{[(S_Y^2 - S_X^2)^2 + 4S_{XY}^2 - 4Q]}\}/(2\,S_{XY} + 2\sqrt{Q})$$

and

$$k_2 = \{(S_Y^2 - S_X^2) + \sqrt{[(S_Y^2 - S_X^2)^2 + 4S_{XY}^2 - 4Q]}\}/(2\,S_{XY} - 2\sqrt{Q})\,.$$

where $Q = F_{(1-\alpha;\,1,\,N-2)}(S_X^2 S_Y^2 - S_{XY}^2)/(N-2)$, S_X^2, S_Y^2 and S_{XY} are the sample variances and covariance, and $F_{(1-\alpha;\,1,\,N-2)}$ is the value of the variance ratio which has cumulative probability $(1-\alpha)$ with $(1, N-2)$ degrees of freedom.

Jolicoeur (1973, 1990) has shown through random simulations that, even though interval $[k_1, k_2]$ is approximate, the significance level obtained in practice is closely similar to the nominal level (α) even when the parametric correlation is as low as $\rho = 0.4$ and sample size is as small as $N = 10$. However, the sampling behavior of interval $[k_1, k_2]$ is more complicated than that of the slope of an ordinary estimation line (figure 22.3.1), for the expression $[(S_Y^2 - S_X^2)^2 + 4S_{XY}^2 - 4Q]$ may be negative and limits k_1 and k_2 may often be imaginary when the correlation coefficient of variates X and Y is not statistically significant. Moreover, interval $[k_1, k_2]$ is sometimes disjoint (figure 22.3.1, bottom, center), in the same way as the confidence interval of the inverse of the slope of the $X.Y$ estimation line (section 20.4). Even when the confidence interval of the slope of the major axis is disjoint, though, its two parts (positive and negative) correspond to contiguous directions on both sides of the vertical direction of the ordinate. Jolicoeur and Mosimann have also emphasized that even imaginary values of limits k_1 and k_2 may receive a reasonable geometric interpretation, if the corresponding confidence interval $[k_1, k_2]$ is considered as infinite (figure 22.3.1, bottom, right), following the suggestion of Fieller (1954).

Finally, limits k_1 and k_2 are always real and always have the same sign as the sample correlation coefficient r when the latter is statistically significant (figure 22.3.1, top), so that the utilization of the confidence interval $[k_1, k_2]$ of the slope of the major axis should not give rise to difficulties in practice.

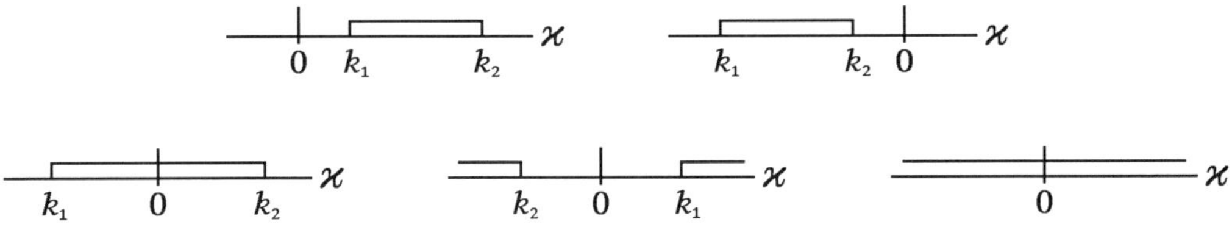

Figure 22.3.1
Representative types of confidence intervals of the slope of the major axis when the correlation is statistically significant (top) and when it is not (bottom)

In the case of the data of figures 20.2.1 and 22.1.2, for instance, the limits of the confidence interval $[k_1, k_2]$ of the slope of the major axis at significance level $\alpha = 0.05$ are [0.630002, 2.35915], while those of the slope of the $Y.X$ estimation line are [0.169729, 0.635581] and those of the inverse of the slope of the $X.Y$ estimation line (section 20.4) are [1.77949, 6.6636]. The hypothesis that the slope is 1.0 can therefore be accepted in the case of the major axis while it should be rejected in the case of ordinary estimation lines. The utilization of the major axis instead of ordinary estimation lines would thus contribute to solve Galton's paradox (sections 20.2 and 22.1), since the hypothesis that an increase ΔX mm of the father's head length goes along with an equal increase ΔY of the son's head length and vice versa would be acceptable. However, the latter hypothesis is overly simple and ordinary estimation lines should not completely be banned since only half of the son's genotype comes from his father, so that an increase ΔX of the father's head length should theoretically entail an increase $\Delta Y = \Delta X/2$ of the son's head length (section 25.9; Fisher, 1970, page 131).

Section 22.4: pros and cons of the major axis

Unlike ordinary estimation lines, the orthogonal estimation line has the advantage of being unique and of involving both variates in the residual sum of squares. These two qualities make the major axis preferable to ordinary estimation lines when both variates are subject to random fluctuations. Unfortunately, the major axis has a defect from which ordinary estimation lines are free: the slope of the major axis is not invariant when variates X and Y are submitted to unequal linear changes of scale, such as the replacement of X by $b_X X$ and of Y by $b_Y Y$ where $b_X \neq b_Y$. While the slope of ordinary estimation lines is then simply multiplied by the ratio b_Y / b_X of scaling coefficients, the slope k of the major axis is affected in a complicated manner and must be completely recalculated. That situation is obviously troublesome, for the results of an analysis based on the major axis are modified when a simple change of units of measurements is done: when a straight line is fitted to the basal metabolism (in kilojoules per day) and body weight of mammals for instance (figure 20.7.2), the slope of the major axis changes in an irreconcilable manner if body weight is expressed in grams instead of kilograms.

A second problem, related to the preceding one, also mars the slope of the major axis: the direction of the major axis tends to approach that of the coordinate axis of the variate having the largest variance and thus, if variates are expressed on the arithmetic scale, of the variate having the largest mean and the greatest order of magnitude (section 4.9 and chapter 14). If a relationship is studied between a very large and a very small element of the human skeleton, for instance, such as the length of the femur and that of the terminal phalanx of the fifth toe, the direction of the major axis would tend to be excessively close to that of the coordinate axis of the femur. Fortunately, two procedures (of unequal worth) are available for correcting the above defects of the major axis: the logarithmic transformation of both variates (section 22.5) or their standardization (section 22.6).

Section 22.5: the major axis of logarithmically transformed data

The advantages of the logarithmic transformation of biological data have already been emphasized several times (sections 4.3, 4.9, chapter 14, sections 19.8, 20.6 and 20.7). The logarithmic transformation of two variates X and Y has the additional advantage of making the slope k of the major axis invariant with respect to unequal linear changes of scale of the original (untransformed) variates. To make the discussion clearer, the original variates X and Y will be represented by the subscripted variables X_1 and X_2 in the rest of this section, while their logarithms will be denoted by $Y_1 = \log_e(X_1)$ and $Y_2 = \log_e(X_2)$. It can easily be shown that the variances and the covariance of logarithms Y_1 and Y_2 are not affected when the original variates X_1 and X_2 are multiplied by unequal scaling coefficients b_1 and b_2. In the case of the covariance of logarithms, for instance, if the geometric means of original variates X_1 and X_2 are denoted by G_1 and G_2 respectively,

$$s_{Y_1Y_2} = \Sigma(Y_1 - \overline{Y}_1)(Y_2 - \overline{Y}_2)/(N-1) = \Sigma[\log_e(X_1/G_1)\log_e(X_2/G_2)]/(N-1)\,,$$

and this is unchanged when the original variates (X_1, X_2) are replaced by (b_1X_1, b_2X_2) because the geometric means of these new variates are b_1G_1 and b_2G_2 and the scaling coefficients b_1 and b_2 cancel out:

$$b_1X_1/b_1G_1 = X_1/G_1 \qquad \text{and} \qquad b_2X_2/b_2G_2 = X_2/G_2\,.$$

Similarly, when the equation of the major axis of logarithms Y_1 and Y_2,

$$(Y_2 - \overline{Y}_2) = k(Y_1 - \overline{Y}_1)\,,$$

is rewritten with respect to the original variates X_1 and X_2,

$$(X_2/G_2) = (X_1/G_1)^k\,,$$

each original variate is divided by its geometric mean, so that each of the two members of the equation is a *pure (dimensionless) number* which is unaffected by any linear change of scale. Moreover, since the geometric mean of each variate reflects its order of magnitude, the major axis of logarithms tends not to be affected by the latter. It should be noted that k here denotes the slope of the major axis of logarithms Y_1 and Y_2, not of the original variates X_1 and X_2. In conclusion, the logarithmic transformation of the two original variates corrects both defects of the major axis mentioned in section 22.4.

Since the logarithmic transformation also transforms the equation of a power function into that of a straight line, and since the variates of an allometry relationship usually have positively skewed distributions (chapters 13 and 14) and are generally both affected by random fluctuations, the major axis of logarithmically transformed data is often suitable in allometry studies. In the case of the data on the metabolism of mammals (section 20.7), for instance, the point estimate and the confidence interval of the slope of the major axis of logarithms at level $\alpha = 0.01$ are

$$k = +0.738129 \qquad \text{and} \qquad +0.717211 \leq \varkappa \leq +0.759474\,.$$

According to the present analysis of this particular set of data, the hypothesis that the allometry exponent of basal metabolism with respect to body weight is $\varkappa = 2/3$ thus does not seem acceptable. However, the hypothesis that $\varkappa = 3/4$, proposed by some physiologists, could be retained. The question of the allometry of basal metabolism with respect to the body weight of animals is still debated by physiologists (Feldman, 1995; Heusner, 1991).

Section 22.6: the major axis of standardized variates

In addition to the logarithmic transformation discussed in the preceding section, there is a second way of making the slope of the major axis invariant to unequal linear changes of scale and to differences in the order of magnitude of the two variates (section 22.4): replacing observed variates X and Y by standardized variates (section 5.4),

$$x = (X - \mu_X)/\sigma_X \qquad \text{and} \qquad y = (Y - \mu_Y)/\sigma_Y\,.$$

It can be shown that the slope of the parametric major axis of standardized variates, also called the *standardized major axis,* is equal to $+1.0$ or -1.0 depending on whether the parametric correlation coefficient ρ is positive or negative. The slope of the standardized major axis is undetermined when the correlation ρ is null. The equation of the parametric major axis of standardized variates is thus $y = \text{sign}(\rho)x$ but this is usually reexpressed with respect to observed variates X and Y, which yields

$$(Y - \mu_Y)/\sigma_Y = \text{sign}(\rho)(X - \mu_X)/\sigma_X\,.$$

In practice, the parameters μ_X, μ_Y, σ_X, σ_Y and ρ, which are usually unknown, are replaced by their sample estimates $\overline{X}$, $\overline{Y}$, S_X, S_Y and r, and the equation of the sample major axis is used:

$$(Y - \overline{Y})/S_Y = \text{sign}(r)(X - \overline{X})/S_X\,.$$

Since the deviation from the mean of each variate is divided by its standard deviation, both members of the equation of the standardized major axis are *pure numbers* which are unaffected by any linear change of scale.

Moreover, because the coefficient of variation of biological data is approximately constant (sections 4.9 and 14.3), the standard deviation is usually more or less proportional to the mean, so that dividing each member of the equation by the corresponding standard deviation tends to make the standardized major axis approximately independent from the order of magnitude of variates. When the equation of the standardized major axis of the sample is expressed with Y as an explicit function,

$$Y = \overline{Y} + \operatorname{sign}(r)(S_Y/S_X)(X - \overline{X}) \, ,$$

its slope $\operatorname{sign}(r)(S_Y/S_X)$ is equal to the product of the ratio of standard deviations by the sign of the correlation coefficient.

Advocated initially in 1948 by the French biometrician Georges Teissier (1900-1972), the standardized major axis has become popular and is presently known under various names (*standardized major axis, geometric mean functional relationship, geometric mean regression, line of organic correlation, reduced major axis* etc.). However, Jolicoeur (1975, 1990) and Kimura (1992) have emphasized that the statistical performance of the standardized major axis is less satisfactory than that of the major axis of observed variates (called the *ordinary major axis* to prevent confusion), perhaps because the ratio (σ_Y/σ_X) of standard deviations is a quotient of two measures of dispersion rather than a measure of relationship. When the parametric correlation ρ is low and sample size N is small, the actual significance level achieved by the confidence interval of the slope becomes very inaccurate in the case of the standardized major axis while it remains approximately correct in the case of the ordinary major axis. Moreover, unlike the confidence intervals of the slopes of the ordinary major axis and of the ordinary estimation lines, the confidence interval of the slope of the standardized major axis reaches neither 0 nor ∞ when the sample correlation r is not significant and thus does not warn the user to refrain from linear estimation. For the preceding reasons, Jolicoeur (1990) advocates the utilization of the ordinary major axis of logarithmically transformed data (section 22.5) whenever observed variates may be shown or assumed to have a bivariate lognormal distribution (section 19.8), which is often the case in biology. The standardized major axis should be used only when sample size is large enough ($N \geq 20$), when the correlation is thought to be large enough in absolute value ($|\rho| \geq 0.6$) and when the data follow a bivariate normal rather than a bivariate lognormal distribution. In a suitable context, the utilization of the standardized major axis may be justified by the work of Kruskal (1953), who showed that, when two variates follow a bivariate normal distribution and a single line is needed to estimate X_1 from X_2 as often as X_2 from X_1, the standardized major axis maximizes the probability of correct predictions. Formulae for the confidence limits of the slope of the standardized major axis were developed by Jolicoeur and Mosimann (1968) and discussed by Jolicoeur (1990).The approximate confidence interval of the slope of the standardized major axis is $[b_1, b_2]$,

where $\quad b_1 = \operatorname{sign}(r)(S_Y/S_X)[\sqrt{(B+1)} - \sqrt{B}] \, , \quad b_2 = \operatorname{sign}(r)(S_Y/S_X)[\sqrt{(B+1)} + \sqrt{B}] \, ,$

$$B = F_{(1-\alpha;\, 1,\, N-2)}(1 - r^2)/(N - 2) \, ,$$

and $\quad r =$ the sample correlation coefficient.

Section 22.7: a more general model, the structural relationship

For a linear relationship between two variates to be correctly described by the major axis, as noted in section 22.1, the random fluctuations of the two variates must be equally important. If the random fluctuations of one of the two variates are known or assumed to be more important than those of the other, the major axis should be replaced by another estimation line known as the *structural relationship* (or, in full, *bivariate linear structural relationship*). The structural relationship corresponds to a more general model of which the major axis and ordinary estimation lines may be considered as particular cases. According to this model, the numerical values taken by observed variates X_1 and X_2 arise as follows:

(1) The value of observed variate X_i in the hth observation of the sample is the sum of a so-called structural variate ξ_i (ξ is the lower case Greek letter *xi*, which corresponds to the Roman x) and of a normal centered and independent random deviation e_{hi}; therefore

$$X_{h1} = \xi_{h1} + e_{h1} , \quad X_{h2} = \xi_{h2} + e_{h2} ,$$

where

$$e_{h1} \leftarrow \mathcal{N}(0, \omega_1^2) , \quad e_{h2} \leftarrow \mathcal{N}(0, \omega_2^2) ,$$

and

$$\mathcal{E}(e_{h1}\, e_{h2}) = 0 .$$

ω is the lower case Greek letter *omega*.

(2) The first structural variate ξ_1 follows a normal distribution of which the mean is $\overline{\xi}_1$ and the variance σ^2:

$$\xi_1 \leftarrow \mathcal{N}(\overline{\xi}_1, \sigma^2) .$$

(3) The two structural variates ξ_1 and ξ_2 are linked by an exact relationship,

$$(\xi_2 - \overline{\xi}_2) = \varkappa(\xi_1 - \overline{\xi}_1),$$

of which the slope $\varkappa$ is to be estimated from the data.

According to this model, the only values which are directly observed are those of X_1 and X_2 : all other symbols denote theoretical quantities which are assumed to exist in order to obtain a logical and coherent way of analyzing the data. In particular, the independent deviations e_{h1} and e_{h2} stand for the random fluctuations of observed variates X_1 and X_2, of which the importance is reflected by the *residual variances* ω_1^2 and ω_2^2. Those residual variances are called *error variances* in several fields where they result from observational or experimental inaccuracies, but that name is not very suitable in biology, where residual variances often reflect objective variations due to real genetical, physiological or environmental differences. Inasmuch as the structural relationship model corresponds to reality, the parametric variances $\sigma_{X_1}^2$ and $\sigma_{X_2}^2$ and covariance $\sigma_{X_1X_2}$ of observed variates are linked to model parameters by the following equations:

$$\sigma_{X_1}^2 = \sigma^2 + \omega_1^2 ; \quad \sigma_{X_2}^2 = \varkappa^2\sigma^2 + \omega_2^2 ; \quad \sigma_{X_1X_2} = \varkappa\sigma^2 .$$

In practice, model parameters, which appear in the right members of the preceding equations, are estimated by replacing the parametric variances and covariance of observed variates, which appear in the left members, by their sample estimates $S^2_{X_1}$, $S^2_{X_2}$ and $S_{X_1X_2}$. However, the problem can be solved only if additional information is available, since there are only three equations but there are four parameters to be estimated: σ^2, ω_1^2, ω_2^2, and $\varkappa$. This difficulty is usually bypassed by assuming that the ratio $\lambda = \omega_2^2/\omega_1^2$ of residual variances is known. The slope $\varkappa$ of the structural relationship may then be estimated through the expression

$$k = \{(S^2_{X_2} - \lambda S^2_{X_1}) + \sqrt{[(S^2_{X_2} - \lambda S^2_{X_1})^2 + 4\lambda S^2_{X_1X_2}]}\}/(2\,S_{X_1X_2})\,;$$

or, when the ratio λ is very large, through

$$k = 2\,S_{X_1X_2}/\{(S^2_{X_1} - S^2_{X_2}/\lambda) + \sqrt{[(S^2_{X_1} - S^2_{X_2}/\lambda)^2 + 4S^2_{X_1X_2}/\lambda]}\}\,.$$

The lower case Greek letter λ (*lambda*) corresponds to the lower case L of the Roman alphabet. The preceding expressions show that the structural relationship is a generalization not only of the major axis (sections 22.1 and 22.2) but also of ordinary estimation lines (section 20.2), for the slope k of the structural relationship becomes equal to the slope of the major axis when $\lambda = 1$, to the slope of the $X_2.X_1$ estimation line when $\lambda = \infty$, and to the inverse of the slope of the $X_1.X_2$ estimation line when $\lambda = 0$. The slope of the structural relationship thus increases in absolute value when the ratio λ of residual variances decreases, and its position can range gradually from that of the $X_2.X_1$ estimation line to that of the $X_1.X_2$ estimation line depending on whether random fluctuations affect only the second variate ($\omega_1^2 = 0$ and $\omega_2^2 > 0$, whence $\lambda = \infty$), both variates ($0 < \lambda < \infty$), or only the first variate ($\lambda = 0$). If the residual variance ω_2^2 of the second variate X_2 is known or assumed to be λ times more important (or $1/\lambda$ times less important) than the residual variance ω_1^2 of the first variate X_1, the structural relationship should therefore be used to describe their relationship.

Following the early work of Creasy (1956), formulae for the confidence limits of the slope of the structural relationship were developed by Jolicoeur and Heusner (1971) and discussed by Jolicoeur (1990). The limits of the approximate confidence interval $[k_1, k_2]$ of the slope of the structural relationship may be determined by the following expressions:

$$k_1 = \{(S^2_{X_2} - \lambda S^2_{X_1}) + \sqrt{[(S^2_{X_2} - \lambda S^2_{X_1})^2 + 4\lambda S^2_{X_1X_2} - 4\lambda Q]}\}/(2S_{X_1X_2} + 2\sqrt{Q})$$

and

$$k_2 = \{(S^2_{X_2} - \lambda S^2_{X_1}) + \sqrt{[(S^2_{X_2} - \lambda S^2_{X_1})^2 + 4\lambda S^2_{X_1X_2} - 4\lambda Q]}\}/(2S_{X_1X_2} - 2\sqrt{Q})\,,$$

where

$$Q = F_{(1-\alpha;\,1,\,N-2)}\,(S^2_{X_1}S^2_{X_2} - S^2_{X_1X_2})/(N-2)\,.$$

However, when the value of ratio λ is very large, the following formulae are more accurate:

$$k_1 = (2S_{X_1X_2} - 2\sqrt{Q}) / \{(S^2_{X_1} - S^2_{X_2}/\lambda) + \sqrt{[(S^2_{X_1} - S^2_{X_2}/\lambda)^2 + 4S^2_{X_1X_2}/\lambda - 4Q/\lambda]}\}$$

and

$$k_2 = (2S_{X_1X_2} + 2\sqrt{Q}) / \{(S^2_{X_1} - S^2_{X_2}/\lambda) + \sqrt{[(S^2_{X_1} - S^2_{X_2}/\lambda)^2 + 4S^2_{X_1X_2}/\lambda - 4Q/\lambda]}\}.$$

Depending on the value of the ratio λ, the confidence interval $[k_1, k_2]$ coincides with those of ordinary estimation lines ($\lambda = 0$ or $\lambda = \infty$) or of the major axis ($\lambda = 1$).

A more elaborate version of the structural relationship, the *structural relationship with replicates,* is mentioned in section 33.11.

Chapter 23
The trivariate normal distribution: partial and multiple correlations and regressions

Section 23.1: introduction

In this chapter, the joint analysis of three variates is introduced in the simplest possible manner (without recourse to matrix and vector algebra, which will be briefly reviewed in chapter 24). The new notions with which the reader will become acquainted here will recur in later chapters on the so-called *multivariate analysis* of any number of variates (chapters 25 and 29 to 34). When the joint probability distribution of three variates is considered, the three dimensions of classical Euclidean geometry are already all occupied by the variates (see figures 23.2.1 and 23.3.1, for instance). Consequently, there is no dimension left for the coordinate axis of the frequency (compare with figure 19.1.2) or of the probability density (figure 19.2.1). However, the limitations of three-dimensional geometry do not impede the statistical analysis.

In the case of the trivariate normal distribution, the swarm of dots representing the data has the approximate shape of a spindle or a cigar and may be represented by an ellipsoid floating in three-dimensional space (figure 23.2.1). An ellipsoid is a three-dimensional generalization of an ellipse. In the special case where its three *principal axes* (the major, the intermediate and the minor axis) have the same length, an ellipsoid degenerates into a sphere, which is a three-dimensional generalization of a circle. A biological example is provided by data obtained by Haldane and Priestley (1905), who pioneered the study of human respiration. These data were obtained in a physiological experiment during which one of the two experimenters (J. G. P.) inspired and expired air in a closed box. The CO_2 concentration of the air therefore increased gradually from the beginning to the end of the experiment, which caused the subject to breathe more deeply and more rapidly.

The three variates analysed here are X = the percentage of CO_2 in inspired air, Y = a combined measurement of lung ventilation, and Z = the percentage of CO_2 in expired air. In order to make relationships linear and facilitate the analysis of these interesting data, the Y variate was defined as the cubic root of the increase in lung ventilation, the latter being expressed in percentage of its initial value (measured prior to the beginning of the experiment and to the rise of CO_2 concentration). The conclusions reached using true trivariate methods will be found to differ completely from those obtained through the ordinary bivariate methods presented in earlier chapters. The three possible bivariate scatter diagrams of variates X, Y, and Z are illustrated in figure 23.1.1.

Section 23.2: ordinary or partial correlations?

A cursory examination of the scatter diagrams in figure 23.1.1 suggests that the increase X of CO_2 concentration in inspired air strongly stimulates lung ventilation Y, but that these two variates are practically unrelated to the concentration Z of CO_2 in expired air. The numerical values of ordinary product-moment correlation coefficients (discussed in chapter 19) confirm these graphical indications, since we have

$$r_{XY} = +0.96876, \quad r_{XZ} = +0.02019 \quad \text{and} \quad r_{YZ} = -0.19148 .$$

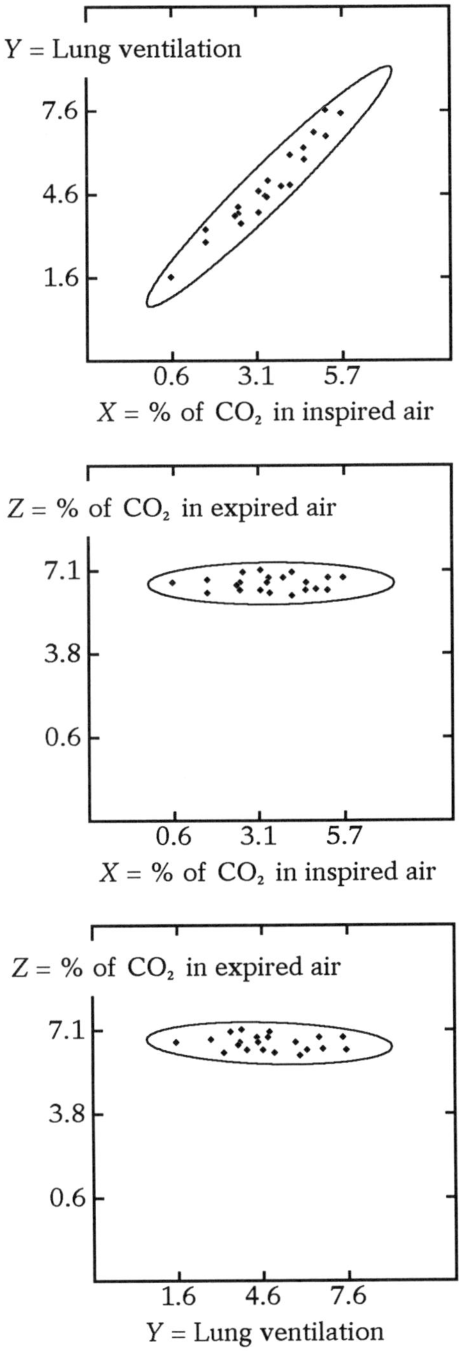

Figure 23.1.1
Scatter diagrams, with prediction ellipses at level $\alpha = 0.05$, of the three variates of the example on human respiration (data from Haldane and Priestley, 1905)

Since sample size is $N = 21$, there are $(N-2) = 19$ degrees of freedom and, if the null hypothesis H_0: $\rho = 0$ is tested against a two-sided alternative hypothesis H_1: $\rho \neq 0$, the acceptance region of H_0 extends from -0.43286 to $+0.43286$ at significance level $\alpha = 0.05$ and from -0.66521 to $+0.66521$ at level $\alpha = 0.001$. While the r_{XY} correlation is highly significant statistically ($P < 0.001$), the other two correlations are not significant at all ($P > 0.05$). Moreover, even if a one-sided alternative hypothesis were made concerning ρ_{YZ} (H_1: $\rho_{YZ} < 0$), the acceptance region of H_0 would go from $r_{(0.05;\,19)} = -0.36874$ to $+1.0$, and r_{YZ} would still not differ significantly from 0.

While the strong positive correlation between X and Y confirms the known effect of CO_2 on respiration (a ventilation increase), the r_{XZ} and r_{YZ} correlations do not seem to agree with physiological knowledge. One would indeed expect an increase of the concentration of CO_2 in inspired air to be followed by a measurable increase of the concentration of CO_2 in expired air, and one would also expect an increase of lung ventilation to promote CO_2 expulsion and to decrease the concentration of CO_2 in expired air (which reflects the composition of internal gases within the lung).

However, detailed examination of the data seems to indicate that ordinary bivariate correlations and scatter diagrams do not extract all available information. It could indeed be shown by numbering observations that, when a dot representing a data triplet (X, Y, Z) falls relatively higher than its neighbors in the scatter diagram of X and Y (figure 23.1.1, top), it generally falls lower in the scatter diagram of X and Z (figure 23.1.1, center). This suggests that a better analytical technique might be obtained by carrying out statistical adjustments on variates Y and Z and compensating for the inconstancy of the X variate. Following this principle, a new kind of correlation coefficient may be defined, the so-called *partial correlation coefficient*.

Geometrically speaking, the partial correlation coefficient between Y and Z corresponds to the scatter diagram obtained by projecting all observations on a plane on which X is kept constant, the projection being oblique with respect to that plane but parallel to the $Y.X$ and $Z.X$ estimation (regression) lines. The prediction ellipse corresponding to this partial correlation is illustrated in figure 23.2.1. Each of the 21 trivariate observations (X, Y, Z) is represented by a dot possessing a vertical stem, the length of the stem indicating the height Z of the dot above the plane of variates X and Y.

A careful scrutiny of figures 23.2.1 and 23.3.1 shows that stem length and dot height are generally greater for observations which are farther from the viewer (and thus correspond to lower values of the Y variate). Moreover, the major axis of the partial correlation ellipse appears to have a strongly negative slope, unlike the major axis of the ordinary correlation ellipse which was practically horizontal (figure 23.1.1, bottom). This seems to confirm that there is between Y and Z a relationship which could not be detected with the ordinary correlation coefficient r_{YZ}.

Unlike the ordinary correlation r_{YZ}, which may be called a *total correlation* in order to prevent any ambiguity, the partial correlation between Y and Z when the X variate is held constant is denoted by the symbol $r_{YZ.X}$ and may be calculated from ordinary correlations by the following formula:

$$r_{YZ.X} = (r_{YZ} - r_{XY} r_{XZ}) / \sqrt{[(1 - r_{XY}^2)(1 - r_{XZ}^2)]} \, .$$

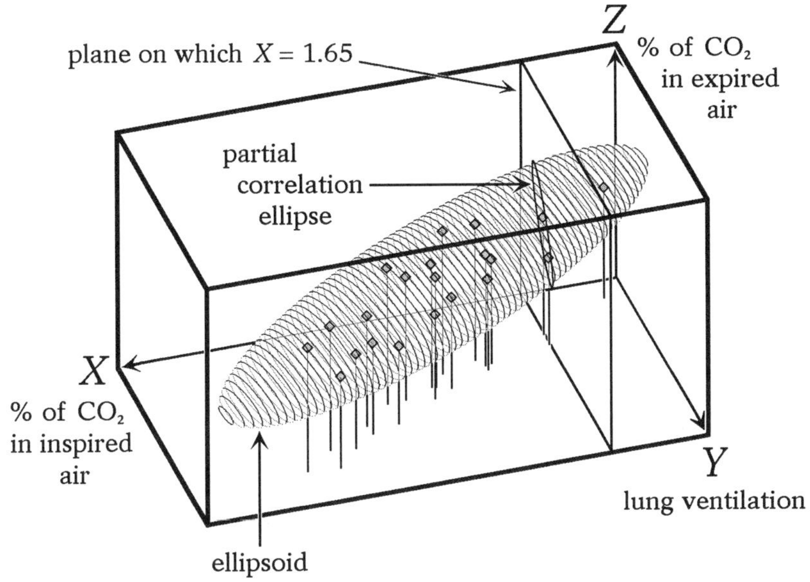

Figure 23.2.1
Trivariate scatter diagram of human respiratory data; each observation in the sample is represented by a dot possessing a vertical stem; the partial correlation ellipse of variates *Y* and *Z* is an oblique projection of the three-dimensional ellipsoid on a plane on which the *X* variate is constant ($X = 1.65$)

The probability distribution of the partial correlation coefficient is similar to that of the total correlation coefficient except that the number of degrees of freedom is decreased by 1 for each variate held constant. In the simplest case, the only one considered in this chapter, a single variate is held constant and the number of degrees of freedom of a partial correlation coefficient is $[(N-2)-1] = (N-3)$. The partial correlation coefficient between lung ventilation and the percentage of CO_2 in expired air is $r_{YZ.X} = -0.85110$ with 18 degrees of freedom. If the null hypothesis H_0: $\rho_{YZ.X} = 0$ is tested against a two-sided alternative hypothesis, the acceptance region of H_0 extends from -0.67878 to $+0.67878$ at level $\alpha = 0.001$; therefore the partial correlation is strongly significant statistically ($P < 0.001$). When the concentration of CO_2 in inspired air is held constant, an increase of lung ventilation thus tends to strongly decrease the concentration of CO_2 in expired air.

This conclusion is the opposite of the one based on the ordinary total correlation and agrees better with current knowledge in respiratory physiology. The total correlation r_{YZ} did not detect a relationship between *Y* and *Z* because it reflected the general position of the ellipsoid, which is practically horizontal in three-dimensional space (figure 23.1.1, center and bottom). Contrariwise, the partial correlation $r_{YZ.X}$ confirms the existence of a strong relationship because it shows that the ellipsoid is obliquely inclined and relatively thin, somewhat like a surfboard (figure 23.2.1).

One may also calculate other partial correlations, such as $r_{XZ.Y}$ or $r_{XY.Z}$, by interchanging variates within the formula of $r_{YZ.X}$. In the example of human respiratory physiology, for instance, the partial correlation

$$r_{XZ.Y} = (r_{XZ} - r_{XY}r_{YZ})/\sqrt{[(1 - r_{XY}^2)(1 - r_{YZ}^2)]}$$

reflects the increase in the concentration of CO_2 in expired air which would occur if that of inspired air increased but lung ventilation were held constant. In the case of the data of Haldane and Priestley (1905), $r_{XZ.Y} = +0.84499$ with $(N-3) = 18$ degrees of freedom, which is strongly significant ($P < 0.001$). Therefore, this partial correlation too brings out a relationship which could not be detected with the ordinary total correlation r_{XZ}, in addition to being difficult to observe in practice.

Section 23.3: the estimation ("multiple regression") plane

One possible three-dimensional generalization of the $\hat{Y} = a + bX$ estimation line (chapter 20) is an *estimation plane* (usually called *multiple regression*) of which the equation $\hat{Z} = a + bX + cY$ is used to estimate one variate, Z, on the basis of other two variates, X and Y. Multiple regression is appropriate when the nature of the variates or of the questions to be solved is such that one of the variates, Z, must be estimated or may be interpreted as an effect while the other variates, X and Y, may be used as predictors or interpreted as causes. In the case of human respiration, for instance, the concentration Z of CO_2 in expired air (which reflects the chemical composition of internal pulmonary gases) may naturally be considered as affected on the one hand by the concentration X of CO_2 in inspired air, an external condition, and on the other hand by the intensity Y of lung ventilation, a physiological response.

The fact that, in everyday reality, the percentage Z of CO_2 within the lungs tends to be kept constant, because the increase in the percentage X of CO_2 in inspired air tends to be compensated for by an increase of lung ventilation Y, is an excellent example of *physiological regulation*. The coefficients a, b and c of the multiple regression can be estimated according to the *least-squares principle* by minimizing the *residual sum of squares* (or RSS)

$$\sum_{h=1}^{N} (Z - \hat{Z})^2 = \sum_{h=1}^{N} [Z - (a + bX + cY)]^2 .$$

If the partial derivatives of this RSS with respect to the three coefficients a, b, and c are set equal to 0, a system of three equations known as the *Gauss equations* is obtained :

$$aN + b\Sigma X + c\Sigma Y = \Sigma Z ,$$

$$a\Sigma X + b\Sigma X^2 + c\Sigma XY = \Sigma XZ ,$$

$$a\Sigma Y + b\Sigma XY + c\Sigma Y^2 = \Sigma YZ .$$

Solving this system shows that the coefficients a, b and c must have the following values:

$$b = (S_{XZ} S_Y^2 - S_{YZ} S_{XY}) / (S_X^2 S_Y^2 - S_{XY}^2) \,,$$

$$c = (S_{YZ} S_X^2 - S_{XZ} S_{XY}) / (S_X^2 S_Y^2 - S_{XY}^2) \,,$$

and

$$a = \overline{Z} - b\overline{X} - c\overline{Y} \,.$$

The concentration Z of CO_2 in expired air can then be estimated from the concentration X of CO_2 in inspired air and lung ventilation Y through the following equation:

$$\hat{Z} = 7.09583 + 0.747978X - 0.647592Y \,.$$

The least-squares condition ensures that the estimation plane passes as close to the data dots as possible, inasmuch as the sum of squared vertical deviations (parallel to the Z coordinate axis) between the dots and the plane is minimum. The average squared distance between the observations and the plane is estimated by the residual variance,

$$S_{Z.XY}^2 = [(N - 1)/(N - 3)] (S_Z^2 - b S_{XZ} - c S_{YZ}) \,,$$

which possesses $(N-3)$ degrees of freedom, since it presupposes that three coefficients, a, b and c, are estimated from the data.

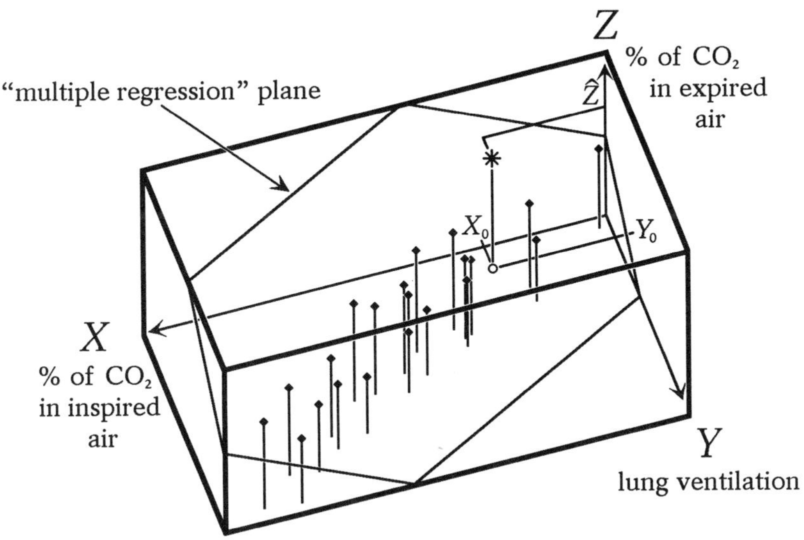

Figure 23.3.1
The so-called "multiple regression" plane corresponds to equation $\hat{Z} = a + bX + cY$ and is inclined approximately the same way as the thin ellipsoid illustrated in figure 23.2.1; the estimated value $\hat{Z}$ of the Z variate is the height of the intersection (*) between the regression plane and the vertical straight line raised from the dot (o) of which the coordinates are the values X_0 and Y_0 of the predictor variates

Section 23.4: hypotheses concerning the b and c coefficients

In general, the estimation plane $\hat{Z}=a+bX+cY$ deserves to be used only when the preliminary hypotheses that the parametric values b_0 and c_0 of the coefficients b and c are null must be rejected. Moreover, when those null hypotheses are rejected, one may wish to test the hypotheses that the parameters b_0 and c_0 have other nonnull theoretical values. These two kinds of questions can be solved using the criteria

$$t = (b - b_0)/S_b \qquad \text{and} \qquad t = (c - c_0)/S_c ,$$

which follow Student's t distribution with $(N-3)$ degrees of freedom. In the case of the null hypotheses, the above criteria take the simpler form

$$t = b/S_b \qquad \text{and} \qquad t = c/S_c ,$$

since $b_0=0$ and $c_0=0$. The estimated standard deviations (also called "standard errors") S_b and S_c of coefficients b and c are obtained by taking the square roots of the corresponding variances,

$$S_b^2 = S_{Z.XY}^2 S_Y^2/[(N - 1)(S_X^2 S_Y^2 - S_{XY}^2)] ,$$

and

$$S_c^2 = S_{Z.XY}^2 S_X^2/[(N - 1)(S_X^2 S_Y^2 - S_{XY}^2)] .$$

In order to determine the limits of acceptance and rejection regions, one must take into account the nature of the alternative hypothesis H_1 (one-sided or two-sided), as in sections 9.6, 19.5 and 20.4. Student's t distribution may be used also to obtain confidence limits. For instance, in the case where, if one were testing a hypothesis, one would have a two-sided alternative, the confidence interval of the parametric coefficient b_0 would be

$$b + t_{(\alpha/2;\, N-3)}\, S_b \leq b_0 \leq b + t_{(1-\alpha/2;\, N-3)}\, S_b .$$

It may be shown that the null hypotheses H_0: $b_0=0$ and H_0: $c_0=0$ are strictly equivalent to hypotheses that the partial correlation coefficients $\rho_{XZ.Y}$ and $\rho_{YZ.X}$ are null (discussed in section 23.2). In fact, the b and c coefficients of a *multiple regression* may also be called *partial regression coefficients,* since b reflects the influence of X on Z when Y is held constant while c reflects the influence of Y on Z with X held constant. Joint tests of the null hypotheses of both regression coefficients b_0 and c_0 are equivalent to testing the hypothesis of nullity of the multiple correlation coefficient, discussed in the next section.

Section 23.5: the multiple correlation coefficient

In the case of *partial correlation* (section 23.2) as well as in that of *multiple regression* coefficients (sections 23.3 and 23.4), one is describing the influence on the estimated variate Z of one of the predictor variates (X or Y) while keeping the other predictor variate constant. However, one sometimes wishes to measure the overall relationship of the estimated variate Z on the one hand with all predictor variates jointly on the other hand. This measure of overall relationship is denoted by the symbol $R_{Z.XY}$ and is called the *multiple correlation coefficient* of Z with X and Y. Since the upper case Greek letter *rho* (ρ) is similar to our upper case P, the parametric multiple correlation coefficient is often denoted by $\overline{R}_{Z.XY}$ to avoid ambiguity but will be represented here by $\mathfrak{R}_{Z.XY}$, $\mathfrak{R}$ being the gothic (German) R. The sample multiple correlation coefficient is the positive square root of

$$R^2_{Z.XY} = (bS_{XZ} + cS_{YZ})/S^2_Z.$$

From a technical viewpoint, the multiple correlation coefficient can be defined as the ordinary correlation coefficient between the observed value Z and the estimated value $\hat{Z} = (a + bX + cY)$. In spite of this definition and unlike the partial correlation coefficient, the multiple correlation coefficient has a probability distribution which differs radically from that of the ordinary correlation coefficient: the multiple correlation is never negative ($R_{Z.XY} \geqq 0$), because the least-squares condition satisfied by coefficients a, b and c compels the estimated value $\hat{Z}$ to resemble the corresponding observed values of Z as much as possible. The probability distribution of the multiple correlation coefficient is nevertheless related to the variance ratio (F) distribution, and an observed value $R_{Z.XY}$ can be transformed into a variance ratio:

$$F_{(2,\,N-3)} = (N - 3)R^2_{Z.XY}/[2(1 - R^2_{Z.XY})].$$

The acceptance region of the null hypothesis H_0: $\mathfrak{R}_{Z.XY} = 0$ is

$$F_{(2,\,N-3)} \leqq F_{(1-\alpha;\,2,\,N-3)}.$$

In the example of human respiratory data, $R_{Z.XY} = +0.85117$, which is only slightly larger in absolute value than $r_{YZ.X} = -0.85110$. The variance ratio corresponding to this multiple correlation coefficient is 23.6667 and has $(2, N-3) = (2, 18)$ degrees of freedom. The null hypothesis H_0: $\mathfrak{R}_{Z.XY} = 0$ can thus be rejected without hesitation ($P < 0.001$), which means that the percentage Z of CO_2 in expired air is related jointly to the percentage X of CO_2 in inspired air and to lung ventilation Y. This relationship had not been detected with the ordinary correlation coefficients r_{XZ} and r_{YZ} (section 23.2).

Section 23.6: possibilities and limitations of the methods in this chapter

Even if, in this chapter, we considered only the simplest case where the number of variates is limited to three, partial and multiple correlations and regressions may be used also when more than three variates are included in an analysis. When there are more than three variates, however, computations can no longer be carried out through simple formulas as in the previous sections, and require numerical methods based on vector and matrix algebra (see chapters 24 and 25). While the example considered here is primarily physiological, partial and multiple correlations and regressions can be useful also in other fields, such as quantitative genetics and ecology. Nevertheless, partial and multiple correlations and regressions are not appropriate for all kinds of problems occurring in multivariate analysis. In general, partial and multiple correlations and regressions are adequate when some variates must be estimated or may be considered as effects while the others can be used as predictors or may be interpreted as causes. When all variates are thought to be subject to random fluctuations and when the aim of a study is to describe relationships or to compare groups of observations rather than to predict some of the variates from the others, other multivariate methods should be considered, such as Hotelling's T^2 distribution (chapter 30), principal components (chapter 31), discriminant analysis (chapters 32 and 33) or canonical correlations (chapter 34). For instance, the estimation ("multiple regression") plane discussed in section 23.3 is but one way to generalize the estimation lines used in bivariate analysis (chapters 20, 21 and 22). The equation of a three-dimensional straight line passing through the mean dot $[\bar{X}, \bar{Y}, \bar{Z}]$,

$$(X - \bar{X})/a = (Y - \bar{Y})/b = (Z - \bar{Z})/c,$$

may be preferable when a description of a joint linear relationship between three variates is required. In some cases, the *direction cosines* $[a, b, c]$ of such a line (sections 24.1 and 24.6) may be estimated using the *trivariate major axis* or *first principal component* (chapter 31), a generalization of the bivariate orthogonal estimation line (chapter 22). Multivariate statistical methods are discussed from chapter 29 to chapter 34.

Chapter 24
Elementary linear calculations (vectors and matrices)

Section 24.1: introduction

The direct computing methods used in the preceding chapter for the trivariate normal distribution (sections 23.2 to 23.5) are not practical when there are more than two predictor variates. The most efficient manner of carrying out calculations is then based on vector and matrix algebra. Most variates considered in earlier chapters involved a single quantity, known as a *scalar.* A scalar quantity is generally denoted by an italic letter, such as

$$X = \text{the stature (height) of a human adult} = 170\text{ cm}\,.$$

Unlike scalars, vectors and matrices, which will be briefly described in this chapter, are ordered *sets* containing several *elements* which can be identified by their respective position. The elements are separated from each other either by commas or more simply by spaces. For instance, the following is a *vector* comprising three elements:

$$[X,\ Y,\ Z] = [170\text{ cm},\ 50\text{ cm},\ 65\text{ kg}]\ ,$$

where $X=$ the stature, $Y=$ the shoulder width and $Z=$ the body weight of a human adult. For the sake of conciseness and coherence, the vector as a whole is often denoted by a bold type letter and its successive elements by the same letter in italic type bearing inferior *indices* (order numbers, which are also called *subscripts*). The above vector may thus be written

$$\mathbf{X} = [X_1\ \ X_2\ \ X_3] = [170\text{ cm}\ \ 50\text{ cm}\ \ 65\text{ kg}]\,,$$

where X_1, X_2 and X_3 stand for the variates X, Y and Z respectively. The elements of vectors and matrices are often scalars but may occasionally themselves be sets (see the case of subdivided matrices in section 24.14). Vector and matrix algebra is particularly efficient when several variates must be analyzed jointly because it allows these variates to be handled in an integrated manner. Moreover, *algorithms* (computational procedures) developed for vectors and matrices are extensible, which means that they may be used with any number of variates (provided that number is specified).

The reader may already know the geometrical meaning of the word vector: a straight-line segment which has a given length, a direction and a sense and which may be represented by an arrow. In algebra, a vector is an ordered set of two or several elements arranged in a row or in a column. Even though the algebraical and the geometrical meanings of the word vector are clearly different, both may often be considered as two aspects of a single entity. In this way, relationships may be perceived between mathematical entities which would otherwise appear to be unrelated.

In two-dimensional analytic geometry, for instance, an algebraic vector $[a, b]$ may be interpreted (figure 24.1.1) as the set of coordinates of a point, or else as the components of a vector going from the origin [0, 0] to the point [a, b], or also as the *direction numbers* (numbers proportional to the *direction cosines,* see section 24.6) of a straight line passing through the origin [0, 0] as well as through point [a, b]. Several of the words used in linear algebra have geometrical connotations which facilitate understanding.

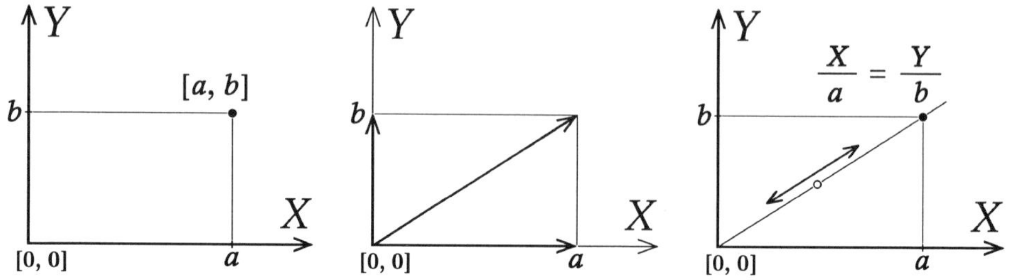

Figure 24.1.1
Three possible geometrical interpretations of an algebraic vector [a, b]: the coordinates of a dot (left), the components of a vector (center), and the direction numbers of a straight line (right)

Section 24.2: row vectors and column vectors, transposition

In linear algebra, a vector is called a *row vector* if it is written horizontally but a *column vector* if it is written vertically, for these two kinds of vectors behave differently in some operations. However, a row vector may be transformed into a column vector or vice versa through an operation known as *transposition,* which may be indicated by placing a tilde (~) above the letter denoting the vector. In this textbook, both kinds of vectors may readily be recognized, for row vectors are written without a tilde and column vectors with a tilde (many authors place a transposition sign on row vectors rather than column vectors, but it may be more natural to treat row vectors as untransposed). Thus, transposing the row vector

$$\mathbf{X} = [X_1 \;\; X_2 \;\; X_3] = [170 \text{ cm} \;\; 50 \text{ cm} \;\; 65 \text{ kg}]$$

yields the column vector

$$\tilde{\mathbf{X}} = \begin{bmatrix} X_1 \\ X_2 \\ X_3 \end{bmatrix} = \begin{bmatrix} 170 \text{ cm} \\ 50 \text{ cm} \\ 65 \text{ kg} \end{bmatrix} .$$

Section 24.3: vector equality, addition and subtraction

The equality, the addition or the subtraction of two vectors implies that these two vectors are of the same kind (both are row vectors, or both are column vectors), and that they possess the same number of elements. Two vectors are equal if all corresponding elements are equal. For instance, if

$$\mathbf{X} = [X_1 \; X_2 \; X_3], \quad \mathbf{Y} = [Y_1 \; Y_2 \; Y_3] \quad \text{and} \quad \mathbf{X} = \mathbf{Y},$$

this means that $X_1 = Y_1$, $X_2 = Y_2$ and $X_3 = Y_3$.

Besides, if a vector **Z** is equal to the sum or to the difference of two vectors **X** and **Y**, each element of vector **Z** is equal to the sum or to the difference of the corresponding elements of vectors **X** and **Y** :

thus, if $\mathbf{X}+\mathbf{Y}=\mathbf{Z}$,

$$\mathbf{Z} = [Z_1 \;\; Z_2 \;\; Z_3] = [X_1+Y_1 \quad X_2+Y_2 \quad X_3+Y_3]\,,$$

and, if $\mathbf{X}-\mathbf{Y}=\mathbf{Z}$,

$$\mathbf{Z} = [Z_1 \;\; Z_2 \;\; Z_3] = [X_1-Y_1 \quad X_2-Y_2 \quad X_3-Y_3]\,.$$

Subtracting a vector from itself yields a null vector **0** of which all elements are null:

$$\mathbf{X} - \mathbf{X} = \mathbf{0} = [0,\; 0,\; 0]\,.$$

Section 24.4: product of a vector by a scalar

The product of a vector **X** by a scalar a is a vector of which each element is the product of the corresponding element of vector **X** by the scalar a :

$$a\,\mathbf{X} = a\,[X_1 \;\; X_2 \;\; X_3] = [aX_1 \;\; aX_2 \;\; aX_3]\,.$$

Section 24.5: product of a row vector by a column vector

The product (also called the *inner product*) of a row vector **X** by a column vector $\tilde{\mathbf{Y}}$ is the sum of the products of the corresponding (scalar) elements. Unlike the product of a vector by a scalar (section 24.4), the inner product of a row vector by a column vector is therefore a scalar, not a vector:

$$\mathbf{X}\tilde{\mathbf{Y}} = [X_1 \;\; X_2 \;\; X_3]\begin{bmatrix} Y_1 \\ Y_2 \\ Y_3 \end{bmatrix} = X_1Y_1 + X_2Y_2 + X_3Y_3 = c\,.$$

A numerical example:

$$[1,\; 2]\begin{bmatrix} 3 \\ 4 \end{bmatrix} = 1\times 3 + 2\times 4 = 3 + 8 = 11\,.$$

For the inner product to be possible, the two vectors must comprise the same number of elements. Moreover, in the inner product of a row vector by a column vector, the row vector conventionally comes first. A product would also be obtained if the row vector were premultiplied by the column vector, but that product would be a matrix instead of a scalar, as will be seen later (section 24.11). However, the value of the inner product of two vectors remains the same if the two vectors are interchanged, provided the first vector is written as a row vector and the second one as a column vector:

$$\mathbf{X}\tilde{\mathbf{Y}} = \mathbf{Y}\tilde{\mathbf{X}} = X_1Y_1 + X_2Y_2 + X_3Y_3 = c\,.$$

It is interesting to note that an equation like

$$a_1 X_1 + a_2 X_2 + a_3 X_3 = 0$$

may be expressed by setting the inner product of a vector $\mathbf{a} = [a_1 \ \ a_2 \ \ a_3]$ of coefficients by a vector $\mathbf{X} = [X_1 \ \ X_2 \ \ X_3]$ of variables equal to zero:

$$\mathbf{a}\tilde{\mathbf{X}} = [a_1 \ \ a_2 \ \ a_3] \begin{bmatrix} X_1 \\ X_2 \\ X_3 \end{bmatrix} = a_1 X_1 + a_2 X_2 + a_3 X_3 = 0 \, .$$

When two nonnull vectors like $\mathbf{X} = [+1, \ +1, \ +1]$ and $\mathbf{Y} = [+2, \ -1, \ -1]$ have a null inner product,

$$\mathbf{X}\tilde{\mathbf{Y}} = \mathbf{Y}\tilde{\mathbf{X}} = (+1)\times(+2) + (+1)\times(-1) + (+1)\times(-1) = +2-1-1 = 0 \, ,$$

these two vectors are said to be *orthogonal* to each other.

Section 24.6: length, angles and direction cosines of a vector

In a two-dimensional Euclidean space, the *length* (also called the *norm*) $||\mathbf{X}||$ of a vector $\mathbf{X} = [a, b] = [X_1 \ \ X_2]$ is equal to the distance between the origin $\mathbf{0} = [0, \ 0]$, and the point $\mathbf{X}$, that is to the hypotenuse of a right triangle of which the sides have the lengths $(a-0) = (X_1 - 0) = X_1$ and $(b-0) = (X_2 - 0) = X_2$ respectively (see figure 24.1.1); thus

$$||\mathbf{X}|| = +\sqrt{(X_1^2 + X_2^2)} = +\sqrt{\mathbf{X}\tilde{\mathbf{X}}} \, ;$$

the length of a vector $\mathbf{X}$ is therefore equal to the positive square root of the inner product $\mathbf{X}\tilde{\mathbf{X}}$ (section 24.5) of that vector by its own transpose. Similarly, in a three-dimensional Euclidean space, the length of vector $\mathbf{X}$ is $||\mathbf{X}|| = +\sqrt{(X_1^2 + X_2^2 + X_3^2)}$, and, in a q-dimensional space, $||\mathbf{X}|| = +\sqrt{(X_1^2 + X_2^2 + \ldots + X_q^2)}$.

From a vector $\mathbf{X}$ of any length, may be obtained a vector of unit length,

$$\mathbf{x} = [x_1 \ \ x_2 \ \ \ldots \ \ x_q] \, ,$$

by dividing each element of vector $\mathbf{X}$ by the length of that vector $||\mathbf{X}||$:

$$\mathbf{x} = \left(1/||\mathbf{X}||\right) \mathbf{X} = \left(1/\sqrt{\mathbf{X}\tilde{\mathbf{X}}}\right) \mathbf{X} = \left(1/\sqrt{(X_1^2 + X_2^2 + \ldots + X_q^2)}\right) [X_1 \ \ X_2 \ \ \ldots \ \ X_q] \, .$$

Therefore, $\mathbf{x}\tilde{\mathbf{x}} = 1$. Each element (x_i) of the vector of unit length is equal to the cosine of the angle f_i between the vector $\mathbf{X}$ and the coordinate axis OX_i and is called a direction cosine. The set of all direction cosines,

$$\mathbf{x} = [x_1 \ \ x_2 \ \ldots \ x_i \ \ldots \ x_q] = [\cos(f_1), \ \cos(f_2), \ \ldots, \ \cos(f_i), \ \ldots, \ \cos(f_q)],$$

provides a description of the direction (and also of the sense) of vector $\mathbf{X}$ (and also of vector $\mathbf{x}$) in space with respect to the coordinate axes OX_1, OX_2, ..., OX_i, ..., OX_q (which are mutually perpendicular). The elements of the vector $\mathbf{X} = [X_1 \ \ X_2 \ \ldots \ X_q]$, of which the length differs from unity, are proportional to the direction cosines and may be called *direction numbers*.

The cosine of the angle f_{XY} between two vectors **X** and **Y** is equal to the inner product of the corresponding unit length vectors **x** and **y** :

$$\cos(f_{XY}) = \mathbf{x}\tilde{\mathbf{y}} = \mathbf{X}\tilde{\mathbf{Y}} \,/ \big(\|\mathbf{X}\| \times \|\mathbf{Y}\|\big) .$$

When two algebraic vectors **X** and **Y** are orthogonal (section 24.5), $\cos(f_{XY}) = 0$, $f_{XY} = 90°$, and the corresponding geometrical vectors are perpendicular to each other.

Section 24.7: several rows and several columns: matrices

Unlike vectors, of which the elements are placed in a single row or in a single column, matrices generally contain several rows and several columns: matrix **A**, for instance, possesses three rows and three columns, and its elements bear double subscripts:

$$\mathbf{A} = \begin{bmatrix} a_{11} & a_{12} & a_{13} \\ a_{21} & a_{22} & a_{23} \\ a_{31} & a_{32} & a_{33} \end{bmatrix} .$$

However, matrix **A** could also be considered either as a row vector of which the elements are three column vectors,

$$\mathbf{A} = \begin{bmatrix} \begin{bmatrix} a_{11} \\ a_{21} \\ a_{31} \end{bmatrix} \begin{bmatrix} a_{12} \\ a_{22} \\ a_{32} \end{bmatrix} \begin{bmatrix} a_{13} \\ a_{23} \\ a_{22} \end{bmatrix} \end{bmatrix} ,$$

or as a column vector of which the elements are three row vectors:

$$\mathbf{A} = \begin{bmatrix} \begin{bmatrix} a_{11} & a_{12} & a_{13} \end{bmatrix} \\ \begin{bmatrix} a_{21} & a_{22} & a_{23} \end{bmatrix} \\ \begin{bmatrix} a_{31} & a_{32} & a_{33} \end{bmatrix} \end{bmatrix} .$$

The first and second subscripts are called row subscript and column subscript respectively because they indicate the row and the column to which each element belongs. Thus, the upper left element of matrix **A** above should be read as "*a* one one" rather than "*a* eleven", while the lower left element should be read as "*a* three one" rather than"*a* thirty-one". A matrix is sometimes represented by its *general element* a_{ij}, i being the row subscript and j the column subscript:

$$\mathbf{A} = [a_{ij}] .$$

A matrix possessing m rows and n columns is said to be of order $(m \times n)$. Thus matrix **A** above is of order (3×3), and one may write

$$\mathbf{A}_{(3\times 3)} = \begin{bmatrix} a_{11} & a_{12} & a_{13} \\ a_{21} & a_{22} & a_{23} \\ a_{31} & a_{32} & a_{33} \end{bmatrix} .$$

The notation generally used for a matrix of order $(m \times n)$ is

$$\mathbf{A}_{(m\times n)} = \begin{bmatrix} a_{11} & a_{12} & \dots & a_{1j} & \dots & a_{1n} \\ a_{21} & a_{22} & \dots & a_{2j} & \dots & a_{2n} \\ \dots & \dots & \dots & \dots & \dots & \dots \\ a_{i1} & a_{i2} & \dots & a_{ij} & \dots & a_{in} \\ \dots & \dots & \dots & \dots & \dots & \dots \\ a_{m1} & a_{m2} & \dots & a_{mj} & \dots & a_{mn} \end{bmatrix} .$$

When $m = n$, the number of rows of matrix **A** is equal to its number of columns, and matrix **A** is said to be a *square matrix*.

As particular limiting cases, a matrix $\mathbf{X}_{(1\times n)} = [X_{11} \;\; X_{12} \; \dots \; X_{1n}]$ possessing a single row is a row vector, a matrix $\tilde{\mathbf{Y}}_{(m\times 1)}$ possessing a single column is a column vector, and a matrix $\mathbf{A}_{(1\times 1)} = [a_{11}]$ possessing a single row and a single column is a scalar.

Section 24.8: matrix transposition

When a matrix **A** is transposed, the columns of the transposed matrix $\tilde{\mathbf{A}}$ are equal to the rows of the original matrix **A** and vice versa:

thus, if $\mathbf{A} = \begin{bmatrix} a_{11} & a_{12} & a_{13} \\ a_{21} & a_{22} & a_{23} \end{bmatrix}$, $\tilde{\mathbf{A}} = \begin{bmatrix} a_{11} & a_{21} \\ a_{12} & a_{22} \\ a_{13} & a_{23} \end{bmatrix}$.

Section 24.9: matrix equality, addition and subtraction

In the same manner as vectors (section 24.3), two matrices **A** and **B** are equal if each element of the first matrix is equal to the corresponding element of the second matrix :

thus, if $\mathbf{A} = \begin{bmatrix} a_{11} & a_{12} \\ a_{21} & a_{22} \end{bmatrix} = \mathbf{B} = \begin{bmatrix} b_{11} & b_{12} \\ b_{21} & b_{22} \end{bmatrix}$,

$$a_{11} = b_{11} , \quad a_{12} = b_{12} , \quad a_{21} = b_{21} , \quad \text{and} \quad a_{22} = b_{22} .$$

Besides, if the sum (or the difference) of two matrices **A** and **B** is equal to a third matrix **C**, each element of this third matrix is equal to the sum (or the difference) of the corresponding elements of the first two matrices:

thus, if $\mathbf{A} + \mathbf{B} = \mathbf{C}$, $\mathbf{C} = \begin{bmatrix} a_{11} + b_{11} & a_{12} + b_{12} \\ a_{21} + b_{21} & a_{22} + b_{22} \end{bmatrix}$

while, if $\mathbf{A} - \mathbf{B} = \mathbf{C}$, $\mathbf{C} = \begin{bmatrix} a_{11} - b_{11} & a_{12} - b_{12} \\ a_{21} - b_{21} & a_{22} - b_{22} \end{bmatrix}$.

In order for the equality, the addition or the subtraction of two matrices to be possible, these two matrices must have the same order (section 24.7). Subtracting a matrix from itself yields a *null matrix*, of which all elements are null:

$$\mathbf{A} - \mathbf{A} = \mathbf{0} \,,$$

that is

$$\begin{bmatrix} a_{11} & a_{12} \\ a_{21} & a_{22} \end{bmatrix} - \begin{bmatrix} a_{11} & a_{12} \\ a_{21} & a_{22} \end{bmatrix} = \begin{bmatrix} 0 & 0 \\ 0 & 0 \end{bmatrix} .$$

Section 24.10: product of a matrix by a scalar

In the same manner as for vectors (section 24.4), the product of a matrix **B** by a scalar a is a matrix of which each element is the product of the corresponding element of matrix **B** by the scalar a:

$$a\mathbf{B} = \begin{bmatrix} ab_{11} & ab_{12} \\ ab_{21} & ab_{22} \end{bmatrix} .$$

Section 24.11: product of a matrix by another matrix

The product of a first matrix **A** by a second matrix **B** is a third matrix **C** of which each element is the *inner product* (section 24.5) of the corresponding row vector of the first matrix by the corresponding column vector of the second matrix:

$$\mathbf{AB} = \mathbf{C} \,,$$

that is

$$\begin{bmatrix} a_{11} & a_{12} \\ a_{21} & a_{22} \end{bmatrix} \begin{bmatrix} b_{11} & b_{12} \\ b_{21} & b_{22} \end{bmatrix} = \begin{bmatrix} c_{11} & c_{12} \\ c_{21} & c_{22} \end{bmatrix} = \begin{bmatrix} [a_{11} \; a_{12}] \\ [a_{21} \; a_{22}] \end{bmatrix} \begin{bmatrix} \begin{bmatrix} b_{11} \\ b_{21} \end{bmatrix} \begin{bmatrix} b_{12} \\ b_{22} \end{bmatrix} \end{bmatrix} = \begin{bmatrix} a_{11}b_{11} + a_{12}b_{21} & a_{11}b_{12} + a_{12}b_{22} \\ a_{21}b_{11} + a_{22}b_{21} & a_{21}b_{12} + a_{22}b_{22} \end{bmatrix} .$$

A numerical example:

$$\begin{bmatrix} +1 & +4 \\ +4 & +9 \end{bmatrix} \begin{bmatrix} +1 & +1 \\ -1 & +1 \end{bmatrix} = \begin{bmatrix} -3 & +5 \\ -5 & +13 \end{bmatrix} .$$

For matrix multiplication to be possible, the number of columns of the first matrix must be equal to the number of rows of the second matrix:

$$\mathbf{A}_{(m \times n)} \, \mathbf{B}_{(n \times o)} = \mathbf{C}_{(m \times o)} \, .$$

For instance,

$$\mathbf{A}_{(4 \times 3)} \, \mathbf{B}_{(3 \times 2)} = \mathbf{C}_{(4 \times 2)}$$

that is

$$\begin{bmatrix} a_{11} & a_{12} & a_{13} \\ a_{21} & a_{22} & a_{23} \\ a_{31} & a_{32} & a_{33} \\ a_{41} & a_{42} & a_{43} \end{bmatrix} \begin{bmatrix} b_{11} & b_{12} \\ b_{21} & b_{22} \\ b_{31} & b_{32} \end{bmatrix} = \begin{bmatrix} c_{11} & c_{12} \\ c_{21} & c_{22} \\ c_{31} & c_{32} \\ c_{41} & c_{42} \end{bmatrix} = \begin{bmatrix} a_{11}b_{11} + a_{12}b_{21} + a_{13}b_{31} & a_{11}b_{12} + a_{12}b_{22} + a_{13}b_{32} \\ a_{21}b_{11} + a_{22}b_{21} + a_{23}b_{31} & a_{21}b_{12} + a_{22}b_{22} + a_{23}b_{32} \\ a_{31}b_{11} + a_{32}b_{21} + a_{33}b_{31} & a_{31}b_{12} + a_{32}b_{22} + a_{33}b_{32} \\ a_{41}b_{11} + a_{42}b_{21} + a_{43}b_{31} & a_{41}b_{12} + a_{42}b_{22} + a_{43}b_{32} \end{bmatrix} .$$

Unlike the inner product of a row vector by a column vector (section 24.5), which is a scalar, the product of a column vector $\tilde{\mathbf{Y}}_{(m \times 1)}$ by a row vector $\mathbf{X}_{(1 \times n)}$ is a rectangular matrix, $\tilde{\mathbf{Y}}_{(m \times 1)} \mathbf{X}_{(1 \times n)} = \mathbf{C}_{(m \times n)}$:

$$\begin{bmatrix} Y_1 \\ Y_2 \\ \dots \\ Y_m \end{bmatrix} \begin{bmatrix} X_1 \, X_2 \, \dots \, X_n \end{bmatrix} = \begin{bmatrix} X_1 Y_1 & X_2 Y_1 & \dots & X_n Y_1 \\ X_1 Y_2 & X_2 Y_2 & \dots & X_n Y_2 \\ \dots & \dots & \dots & \dots \\ X_1 Y_m & X_2 Y_m & \dots & X_n Y_m \end{bmatrix} .$$

In the case of the product of three matrices,

$$\mathbf{A}_{(m \times n)} \, \mathbf{B}_{(n \times o)} \, \mathbf{C}_{(o \times p)} = \mathbf{D}_{(m \times p)} ,$$

the number of rows m of the resulting matrix **D** is equal to the number of rows of the first matrix **A** while the number of columns p is equal to the number of columns of the last matrix **C** of the triple product.

Section 24.12: premultiplication and postmultiplication of matrices

Unlike scalar multiplication, matrix multiplication is generally not commutative, which means that the product of two matrices **A** and **B** usually differs according to the order in which the two matrices enter into the product. Matrix **A** is said to be premultiplied by **B** in **BA** but postmultiplied by **B** in **AB** and, except in a few special cases like that of diagonal matrices (section 24.17), $\mathbf{AB} \neq \mathbf{BA}$.

A numerical example:

If $\mathbf{A} = \begin{bmatrix} +1 & +4 \\ +4 & +9 \end{bmatrix}$ and $\mathbf{B} = \begin{bmatrix} +1 & +1 \\ -1 & +1 \end{bmatrix}$, $\mathbf{AB} = \begin{bmatrix} -3 & +5 \\ -5 & +13 \end{bmatrix}$, while $\mathbf{BA} = \begin{bmatrix} +5 & +13 \\ +3 & +5 \end{bmatrix}$.

There are even cases, like $\mathbf{A} = \mathbf{A}_{(3 \times 2)}$ and $\mathbf{B} = \mathbf{B}_{(2 \times 2)}$, where only one of the two products is possible.

Section 24.13: transposition of a matrix product

The matrix obtained by transposing the product of two or several matrices (the "*transpose*" of the product) is equal to the product of the transposes in reverse order:

$$\widetilde{(\mathbf{AB})} = \tilde{\mathbf{B}} \, \tilde{\mathbf{A}} ,$$

and

$$\widetilde{(\mathbf{ABC})} = \tilde{\mathbf{C}} \, \tilde{\mathbf{B}} \, \tilde{\mathbf{A}} .$$

Section 24.14: subdivided matrices

In some cases, there may be an advantage in subdividing a matrix, for instance when some of its elements correspond to variates which differ in nature. Thus, if data are analyzed concerning four variates of which the first two are ecological while the last two are physiological, it may be appropriate to subdivide a matrix **A** of order (4×4) into four submatrices of order (2×2) in the following manner:

$$\mathbf{A} = \begin{bmatrix} a_{11} & a_{12} & a_{13} & a_{14} \\ a_{21} & a_{22} & a_{23} & a_{24} \\ a_{31} & a_{32} & a_{33} & a_{34} \\ a_{41} & a_{42} & a_{43} & a_{44} \end{bmatrix} = \left[\begin{array}{cc:cc} a_{11} & a_{12} & a_{13} & a_{14} \\ a_{21} & a_{22} & a_{23} & a_{24} \\ \hdashline a_{31} & a_{32} & a_{33} & a_{34} \\ a_{41} & a_{42} & a_{43} & a_{44} \end{array}\right] = \left[\begin{array}{c:c} \mathbf{A}_{11} & \mathbf{A}_{12} \\ \hdashline \mathbf{A}_{21} & \mathbf{A}_{22} \end{array}\right] .$$

Assume now that a second matrix, **B**, also of order (4×4), must be added to **A**, or subtracted from it, or else multiplied by it. If matrix **B** is subdivided in a manner which is compatible with the subdivision of **A**,

$$\mathbf{B} = \begin{bmatrix} b_{11} & b_{12} & b_{13} & b_{14} \\ b_{21} & b_{22} & b_{23} & b_{24} \\ b_{31} & b_{32} & b_{33} & b_{34} \\ b_{41} & b_{42} & b_{43} & b_{44} \end{bmatrix} = \left[\begin{array}{cc:cc} b_{11} & b_{12} & b_{13} & b_{14} \\ b_{21} & b_{22} & b_{23} & b_{24} \\ \hdashline b_{31} & b_{32} & b_{33} & b_{34} \\ b_{41} & b_{42} & b_{43} & b_{44} \end{array}\right] = \left[\begin{array}{c:c} \mathbf{B}_{11} & \mathbf{B}_{12} \\ \hdashline \mathbf{B}_{21} & \mathbf{B}_{22} \end{array}\right] ,$$

the desired operation can be carried out on the subdivided matrices and the final result is the same as if matrices **A** and **B** had been divided in another conformable manner or had not been subdivided at all. For instance,

$$\mathbf{AB} = \left[\begin{array}{c:c} \mathbf{A}_{11} & \mathbf{A}_{12} \\ \hdashline \mathbf{A}_{21} & \mathbf{A}_{22} \end{array}\right] \left[\begin{array}{c:c} \mathbf{B}_{11} & \mathbf{B}_{12} \\ \hdashline \mathbf{B}_{21} & \mathbf{B}_{22} \end{array}\right] = \begin{bmatrix} \mathbf{A}_{11}\mathbf{B}_{11} + \mathbf{A}_{12}\mathbf{B}_{21} & \mathbf{A}_{11}\mathbf{B}_{12} + \mathbf{A}_{12}\mathbf{B}_{22} \\ \mathbf{A}_{21}\mathbf{B}_{11} + \mathbf{A}_{22}\mathbf{B}_{21} & \mathbf{A}_{21}\mathbf{B}_{12} + \mathbf{A}_{22}\mathbf{B}_{22} \end{bmatrix} = \begin{bmatrix} a_{11} & a_{12} & a_{13} & a_{14} \\ a_{21} & a_{22} & a_{23} & a_{24} \\ a_{31} & a_{32} & a_{33} & a_{34} \\ a_{41} & a_{42} & a_{43} & a_{44} \end{bmatrix} \begin{bmatrix} b_{11} & b_{12} & b_{13} & b_{14} \\ b_{21} & b_{22} & b_{23} & b_{24} \\ b_{31} & b_{32} & b_{33} & b_{34} \\ b_{41} & b_{42} & b_{43} & b_{44} \end{bmatrix}$$

$$= \begin{bmatrix} a_{11}b_{11} + a_{12}b_{21} + a_{13}b_{31} + a_{14}b_{41}, & a_{11}b_{12} + a_{12}b_{22} + a_{13}b_{32} + a_{14}b_{42}, & a_{11}b_{13} + a_{12}b_{23} + a_{13}b_{33} + a_{14}b_{43}, & a_{11}b_{14} + a_{12}b_{24} + a_{13}b_{34} + a_{14}b_{44} \\ a_{21}b_{11} + a_{22}b_{21} + a_{23}b_{31} + a_{24}b_{41}, & a_{21}b_{12} + a_{22}b_{22} + a_{23}b_{32} + a_{24}b_{42}, & a_{21}b_{13} + a_{22}b_{23} + a_{23}b_{33} + a_{24}b_{43}, & a_{21}b_{14} + a_{22}b_{24} + a_{23}b_{34} + a_{24}b_{44} \\ a_{31}b_{11} + a_{32}b_{21} + a_{33}b_{31} + a_{34}b_{41}, & a_{31}b_{12} + a_{32}b_{22} + a_{33}b_{32} + a_{34}b_{42}, & a_{31}b_{13} + a_{32}b_{23} + a_{33}b_{33} + a_{34}b_{43}, & a_{31}b_{14} + a_{32}b_{24} + a_{33}b_{34} + a_{34}b_{44} \\ a_{41}b_{11} + a_{42}b_{21} + a_{43}b_{31} + a_{44}b_{41}, & a_{41}b_{12} + a_{42}b_{22} + a_{43}b_{32} + a_{44}b_{42}, & a_{41}b_{13} + a_{42}b_{23} + a_{43}b_{33} + a_{44}b_{43}, & a_{41}b_{14} + a_{42}b_{24} + a_{43}b_{34} + a_{44}b_{44} \end{bmatrix} .$$

Section 24.15: square matrices, diagonal elements and trace

A matrix **A** of order $(m \times n)$ possessing equal numbers of rows and columns $(m = n)$ is said to be a *"square matrix of order n"*, and its elements $a_{ij} = a_{ii}$ having equal row and column subscripts i and j are called its *diagonal elements*. The set of the diagonal elements of a square matrix is known as its *main diagonal*,

$$\mathbf{A} = \begin{bmatrix} a_{11} & \cdots & a_{1i} & a_{1j} & \cdots & a_{1n} \\ \cdots & \cdots & \cdots & \cdots & \cdots & \cdots \\ a_{i1} & \cdots & a_{ii} & a_{ij} & \cdots & a_{in} \\ a_{j1} & \cdots & a_{ji} & a_{jj} & \cdots & a_{jn} \\ \cdots & \cdots & \cdots & \cdots & \cdots & \cdots \\ a_{n1} & \cdots & a_{ni} & a_{nj} & \cdots & a_{nn} \end{bmatrix} ,$$

and the sum of these diagonal elements is a scalar and an important characteristic called the *trace*:

$$\text{Trace}(\mathbf{A}) = \sum_{i=1}^{n} a_{ii} .$$

Section 24.16: symmetric matrices

A square matrix **A** is said to be *symmetric* if the elements which are similarly placed with respect to the main diagonal are equal:

$$a_{ij} = a_{ji} \,;$$

this implies that matrix **A** is equal to its transpose $\tilde{\mathbf{A}}$:

$$\mathbf{A} = \tilde{\mathbf{A}} \,.$$

When a matrix is symmetric, this is sometimes indicated by writing first the smallest of the two subscripts, rather than the row subscript as usually done. This implies that symmetrically placed elements are equal, since they are represented by the same symbol. Sometimes, the elements below (or above) the main diagonal of a symmetric matrix are also simply omitted:

$$\mathbf{A} = \begin{bmatrix} a_{11} & a_{12} & a_{13} \\ a_{21} & a_{22} & a_{23} \\ a_{31} & a_{32} & a_{33} \end{bmatrix} = \begin{bmatrix} a_{11} & a_{12} & a_{13} \\ a_{12} & a_{22} & a_{23} \\ a_{13} & a_{23} & a_{33} \end{bmatrix} = \begin{bmatrix} a_{11} & a_{12} & a_{13} \\ & a_{22} & a_{23} \\ & & a_{33} \end{bmatrix} .$$

However, the reader should not confuse a symmetric matrix **A**, of which the elements placed below or above the main diagonal are omitted, with a so-called *triangular matrix*, of which the elements placed below or above the main diagonal would be null.

Square symmetric matrices often occur in statistics: when three variates $[X, Y, Z] = [X_1, X_2, X_3]$ are analyzed, for instance, the sample variances and covariances of those three variates make up a square symmetric matrix **S** called the *covariance matrix:*

$$\mathbf{S} = [S_{ij}] = \begin{bmatrix} S_X^2 & S_{XY} & S_{XZ} \\ & S_Y^2 & S_{YZ} \\ & & S_Z^2 \end{bmatrix} = \begin{bmatrix} S_{11} & S_{12} & S_{13} \\ S_{12} & S_{22} & S_{23} \\ S_{13} & S_{23} & S_{33} \end{bmatrix} = \begin{bmatrix} S_{11} & S_{12} & S_{13} \\ & S_{22} & S_{23} \\ & & S_{33} \end{bmatrix} .$$

The elements of the main diagonal are the variances, and the others are the covariances. As for the covariance matrix of standardized variates

$$[x, y, z] = [(X-\bar{X})/S_X \,,\ \ (Y-\bar{Y})/S_Y \,,\ \ (Z-\bar{Z})/S_Z \,] \,,$$

it is known as the *correlation matrix* $[r_{ij}]$:

$$[r_{ij}] = \begin{bmatrix} 1 & r_{XY} & r_{XZ} \\ & 1 & r_{YZ} \\ & & 1 \end{bmatrix} = \begin{bmatrix} 1 & r_{12} & r_{13} \\ r_{12} & 1 & r_{23} \\ r_{13} & r_{23} & 1 \end{bmatrix} = \begin{bmatrix} 1 & r_{12} & r_{13} \\ & 1 & r_{23} \\ & & 1 \end{bmatrix} .$$

Section 24.17: diagonal, scalar and unit (identity) matrices

A square symmetric matrix **D** of which all elements placed outside of the main diagonal are null is said to be a *diagonal matrix.* Because there is no ambiguity in this case, the elements of the main diagonal of a diagonal matrix are generally written with a single subscript instead of double subscripts:

$$\mathbf{D} = \begin{bmatrix} d_{11} & 0 & 0 \\ 0 & d_{22} & 0 \\ 0 & 0 & d_{33} \end{bmatrix} = \begin{bmatrix} d_1 & 0 & 0 \\ 0 & d_2 & 0 \\ 0 & 0 & d_3 \end{bmatrix} .$$

A diagonal matrix **C** of which the elements of the main diagonal all have the same value c is said to be a *scalar matrix*:

$$\mathbf{C} = \begin{bmatrix} c & 0 & 0 \\ 0 & c & 0 \\ 0 & 0 & c \end{bmatrix}.$$

Finally, a scalar matrix of which all diagonal elements are equal to unity is called a *unit matrix* (also known as an *identity matrix*) and is usually denoted by an upper-case bold **I**:

$$\mathbf{I} = \begin{bmatrix} 1 & 0 & 0 \\ 0 & 1 & 0 \\ 0 & 0 & 1 \end{bmatrix}.$$

It may easily be verified that multiplying any other matrix **A** by a unit matrix **I** of suitable order leaves that other matrix unchanged:

$$\mathbf{A}_{(2\times3)}\,\mathbf{I}_{(3\times3)} = \mathbf{A}_{(2\times3)} \qquad \text{and} \qquad \mathbf{I}_{(2\times2)}\,\mathbf{A}_{(2\times3)} = \mathbf{A}_{(2\times3)}\,,$$

that is

$$\begin{bmatrix} a_{11} & a_{12} & a_{13} \\ a_{21} & a_{22} & a_{23} \end{bmatrix} \begin{bmatrix} 1 & 0 & 0 \\ 0 & 1 & 0 \\ 0 & 0 & 1 \end{bmatrix} = \begin{bmatrix} a_{11} & a_{12} & a_{13} \\ a_{21} & a_{22} & a_{23} \end{bmatrix} \qquad \text{and} \qquad \begin{bmatrix} 1 & 0 \\ 0 & 1 \end{bmatrix} \begin{bmatrix} a_{11} & a_{12} & a_{13} \\ a_{21} & a_{22} & a_{23} \end{bmatrix} = \begin{bmatrix} a_{11} & a_{12} & a_{13} \\ a_{21} & a_{22} & a_{23} \end{bmatrix}.$$

Therefore, provided a unit matrix of suitable order is chosen, $\mathbf{AI} = \mathbf{A}$ and $\mathbf{IA} = \mathbf{A}$. Because multiplication by a unit matrix does not change another matrix, the unit matrix plays in matrix algebra a role analogous to that of number "1" in scalar algebra.

Section 24.18: determinant and rank of a square regular or singular matrix

The determinant $|\mathbf{A}|$ of a square matrix **A** is a scalar which reflects the values of all elements of that matrix. Like the *trace* (section 24.15), the determinant is an important characteristic of a matrix. The evaluation of determinants is a vast topic of which a detailed treatment would require several chapters (or even a complete book). Therefore, a simple and direct computing method is given here only for square matrices of order 2 and 3:

if matrix $\mathbf{A} = \mathbf{A}_{(2\times2)} = \begin{bmatrix} a_{11} & a_{12} \\ a_{21} & a_{22} \end{bmatrix}$, the determinant $|\mathbf{A}| = \begin{vmatrix} a_{11} & a_{12} \\ a_{21} & a_{22} \end{vmatrix} = (a_{11}a_{22} - a_{21}a_{12})$,

and,

if $\mathbf{A} = \mathbf{A}_{(3\times3)} = \begin{bmatrix} a_{11} & a_{12} & a_{13} \\ a_{21} & a_{22} & a_{23} \\ a_{31} & a_{32} & a_{33} \end{bmatrix}$, the determinant $|\mathbf{A}| = a_{11}\begin{vmatrix} a_{22} & a_{23} \\ a_{32} & a_{33} \end{vmatrix} - a_{21}\begin{vmatrix} a_{12} & a_{13} \\ a_{32} & a_{33} \end{vmatrix} + a_{31}\begin{vmatrix} a_{12} & a_{13} \\ a_{22} & a_{23} \end{vmatrix}$.

For square matrices of order larger than 3, the evaluation of a determinant is tedious and should be done by using a computer algorithm. A matrix **A** is said to be *regular* (or *nonsingular*) if its determinant $|\mathbf{A}| \neq 0$ but *singular* if its determinant $|\mathbf{A}| = 0$. A square matrix can have a true *inverse* (section 24.19) only if it is regular. The *rank* of a square matrix is the order of its largest regular square submatrix (section 24.14). The rank of a square matrix is equal to its order if that matrix is regular (it is then said to have *full rank*) but smaller than its order if the matrix is singular (see also section 24.21).

Section 24.19: inverse of a square regular (nonsingular) matrix

The division of a matrix **B** by another matrix **A** does not exist as such, but it is replaced by an equivalent operation: instead of being divided by **A**, matrix **B** is multiplied by the inverse of **A**, yielding the product $\mathbf{BA}^{-1}$. The inverse $\mathbf{A}^{-1}$ of a square matrix **A** is another matrix such that the product $\mathbf{AA}^{-1}$ is a unit matrix: $\mathbf{AA}^{-1} = \mathbf{I}$. It may easily be shown that, if $\mathbf{A}^{-1}$ is the right inverse of **A**, it is also its left inverse. If the left inverse of **A** were indeed another matrix **E**, the equation $\mathbf{EA} = \mathbf{I}$ would be satisfied; but the postmultiplication of both members of that equation by $\mathbf{A}^{-1}$ shows that $\mathbf{EAA}^{-1} = \mathbf{EI} = \mathbf{E} = \mathbf{IA}^{-1} = \mathbf{A}^{-1}$; hence the left inverse is $\mathbf{E} = \mathbf{A}^{-1}$. Only a square matrix may have a true inverse, but even a square matrix does not necessarily have an inverse. Like the evaluation of determinants (section 24.18), the inversion of matrices is tedious and is outlined here only for matrices of order (2 × 2) and (3 × 3):

if matrix $\mathbf{A} = \mathbf{A}_{(2\times 2)} = \begin{bmatrix} a_{11} & a_{12} \\ a_{21} & a_{22} \end{bmatrix}$, the inverse matrix $\mathbf{A}^{-1} = \dfrac{1}{|\mathbf{A}|}\begin{bmatrix} a_{22} & -a_{12} \\ -a_{21} & a_{11} \end{bmatrix}$,

and,

if $\mathbf{A} = \mathbf{A}_{(3\times 3)} = \begin{bmatrix} a_{11} & a_{12} & a_{13} \\ a_{21} & a_{22} & a_{23} \\ a_{31} & a_{32} & a_{33} \end{bmatrix}$, the inverse $\mathbf{A}^{-1} = \dfrac{1}{|\mathbf{A}|}\begin{bmatrix} \begin{vmatrix} a_{22} & a_{23} \\ a_{32} & a_{33} \end{vmatrix} & -\begin{vmatrix} a_{12} & a_{13} \\ a_{32} & a_{33} \end{vmatrix} & \begin{vmatrix} a_{12} & a_{13} \\ a_{22} & a_{23} \end{vmatrix} \\ -\begin{vmatrix} a_{21} & a_{23} \\ a_{31} & a_{33} \end{vmatrix} & \begin{vmatrix} a_{11} & a_{13} \\ a_{31} & a_{33} \end{vmatrix} & -\begin{vmatrix} a_{11} & a_{13} \\ a_{21} & a_{23} \end{vmatrix} \\ \begin{vmatrix} a_{21} & a_{22} \\ a_{31} & a_{32} \end{vmatrix} & -\begin{vmatrix} a_{11} & a_{12} \\ a_{31} & a_{32} \end{vmatrix} & \begin{vmatrix} a_{11} & a_{12} \\ a_{21} & a_{22} \end{vmatrix} \end{bmatrix}$.

It is noteworthy that the above expressions require the value of the determinant, which may be obtained as indicated in section 24.18. The presence of the inverse of the determinant in the expression of the inverse matrix explains why a singular matrix does not have a true inverse, for the inverse of a null determinant is infinite. Even when the determinant $|\mathbf{A}|$ is not exactly null, the evaluation of the inverse matrix may be difficult or even impossible in practice if the determinant is very small: matrix **A** is then said to be *nearly singular*.

For matrices of order larger than (3 × 3), matrix inversion is usually difficult and should be done by using computer algorithms, which are generally satisfactory except in the case of nearly singular matrices. As for matrices which do not possess true inverses, the specialists of linear algebra have developed *pseudoinverses* or *generalized inverses*.

Matrix inversion is particularly useful to solve a system of linear (first degree) equations:

$$a_{11}X_1 + a_{12}X_2 = c_1 ,$$
$$a_{21}X_1 + a_{22}X_2 = c_2 ;$$

that is
$$\mathbf{A}\tilde{\mathbf{X}} = \tilde{\mathbf{c}},$$

where $\mathbf{A} = \begin{bmatrix} a_{11} & a_{12} \\ a_{21} & a_{22} \end{bmatrix}$ is the matrix of coefficients, $\mathbf{X} = [X_1 \;\; X_2]$ is the vector of unknowns, and $\mathbf{c} = [c_1 \;\; c_2]$ is the vector of known quantities. The unknowns may be evaluated by premultiplying both members of equation $\mathbf{A}\tilde{\mathbf{X}} = \tilde{\mathbf{c}}$ by the inverse matrix $\mathbf{A}^{-1}$, which yields:

$$\mathbf{A}^{-1}\mathbf{A}\tilde{\mathbf{X}} = \mathbf{I}\tilde{\mathbf{X}} = \tilde{\mathbf{X}} = \mathbf{A}^{-1}\tilde{\mathbf{c}}.$$

Once the inverse matrix $\mathbf{A}^{-1}$ has been determined, the hardest part of computations has been done, and the unknowns $\mathbf{X}=[X_1\ X_2]$ may easily and rapidly be reevaluated if new known quantities $\mathbf{c}=[c_1\ c_2]$ are available. Matrix inversion may be used in many statistical methods, such as the estimation of the coefficients of a multiple regression by solving the *Gauss equations* (sections 23.3 and 25.3), the evaluation of Hotelling's T^2 criterion (chapter 30), or the comparison of the mean vectors of two samples through Fisher's linear discriminant function (chapter 32).

Section 24.20: orthogonal matrices

A square (usually non symmetric) matrix **A** is said to be *orthogonal* if its product by its transpose $\tilde{\mathbf{A}}$ is a unit matrix **I** :

$$\mathbf{A}\tilde{\mathbf{A}} = \mathbf{I}\,.$$

Therefore the inverse $\mathbf{A}^{-1}$ of an orthogonal matrix **A** is simply its transpose $\tilde{\mathbf{A}}$, and the vectors of an orthogonal matrix have unit length and are mutually orthogonal (section 24.6). Geometrically speaking, the elements of an orthogonal matrix can be interpreted as the direction cosines of a new set of mutually perpendicular coordinate axes with respect to previous coordinate axes. Orthogonal matrices are used in the diagonalization of a square symmetric matrix (section 24.21), as well as in principal component analysis (chapter 31) and in the multiple discriminant analysis of several groups of observations (chapter 33).

Section 24.21: diagonalization of a square symmetric matrix

A square symmetric matrix **A** may be expressed as a triple matrix product, $\mathbf{A}=\tilde{\mathbf{U}}\mathbf{D}\mathbf{U}=\mathbf{U}^{-1}\mathbf{D}\mathbf{U}$, where **D** is a diagonal matrix and **U** is an orthogonal matrix (section 24.20), of which the inverse $\mathbf{U}^{-1}$ is thus simply the transpose $\tilde{\mathbf{U}}$. The row vectors of matrix **U** are known as the *latent vectors,* the *characteristic vectors,* or the *eigen-vectors* (a word coming in part from the German), while the diagonal elements of matrix **D** are called the *latent roots,* the *characteristic roots,* or the *eigen-values* of matrix **A**. Each latent vector $\mathbf{U}_i$ satisfies the equation $\mathbf{U}_i\mathbf{A} = d_i\mathbf{U}_i$, which shows that the product of a matrix **A** by one of its latent vectors, $\mathbf{U}_i$, is equal to the product of that latent vector by the corresponding latent root d_i. The number of nonzero latent roots is equal to the rank of matrix **A** (section 24.18) and shows how many dimensions are really occupied by that matrix in multidimensional space. Since the inverse $\mathbf{U}^{-1}$ of matrix **U** is simply its transpose $\tilde{\mathbf{U}}$, premultiplying both members of equation $\mathbf{A} = \tilde{\mathbf{U}}\mathbf{D}\mathbf{U}$ by **U** and postmultiplying them by $\tilde{\mathbf{U}}$ shows that matrix **D** can be expressed as a function of matrix **A** by the matrix product

$$\mathbf{U}\mathbf{A}\tilde{\mathbf{U}} = \mathbf{U}\tilde{\mathbf{U}}\mathbf{D}\mathbf{U}\tilde{\mathbf{U}} = \mathbf{I}\mathbf{D}\mathbf{I} = \mathbf{D}\,.$$

The so-called *diagonalization* of matrix **A** is the series of operations through which the values of the latent vectors $\mathbf{U}_i$ and of the latent roots d_i, $i=1, 2, \ldots q$, are obtained from matrix **A**. Two important characteristics are *invariant,* which means that their values are unchanged, when matrix **A** is replaced by its diagonal form **D**, the trace (section 24.15) and the determinant (section 24.18): one then has $\mathrm{trace}(\mathbf{D}) = \mathrm{trace}(\mathbf{A})$ and $|\mathbf{D}| = |\mathbf{A}|$.

The computation of latent vectors and latent roots must be done in many fields of applied mathematics and has long been considered a difficult problem, but efficient algorithms are now available for most computers (see section 31.3 and Appendix A). In statistics, latent roots and latent vectors are used in the construction of confidence or prediction ellipses or ellipsoids, in the study of the principal components or principal axes of a multivariate normal distribution (chapter 31), as well as in multiple discriminant analysis (chapter 33) and canonical correlation analysis (chapter 34).

Section 24.22: linear combinations

A *linear combination* of variates $X_1, X_2, \ldots, X_q$ is a scalar equal to the inner product (section 24.5) of a vector of coefficients $\mathbf{a} = [a_1 \ a_2 \ \ldots \ a_q]$ by the transpose $\tilde{\mathbf{X}}$ of the vector of variables:

$$\mathbf{a}\tilde{\mathbf{X}} = \mathbf{X}\tilde{\mathbf{a}} = a_1X_1 + a_2X_2 + \ldots + a_qX_q = [a_1 \ a_2 \ \ldots \ a_q] \begin{bmatrix} X_1 \\ X_2 \\ \ldots \\ X_q \end{bmatrix}.$$

Linear combinations are used in the case of multiple regression (chapters 23 and 25), of the analysis of variance with contrasts (chapter 26), of principal component analysis (chapter 31), of discriminant functions (chapters 32 and 33) and of canonical correlations (chapter 34).

Section 24.23: bilinear forms

A *bilinear form* is a scalar obtained by premultiplying a square or rectangular matrix $\mathbf{A}$ by a row vector $\mathbf{X}$ and by postmultiplying it by a column vector $\tilde{\mathbf{Y}}$:

$$\mathbf{XA\tilde{Y}} = a_{11}X_1Y_1 + \ldots + a_{1q}X_1Y_q + a_{q1}X_qY_1 + \ldots + a_{qq}X_qY_q = [X_1 \ldots X_q] \begin{bmatrix} a_{11} & \ldots & a_{1q} \\ \ldots & \ldots & \ldots \\ a_{q1} & \ldots & a_{qq} \end{bmatrix} \begin{bmatrix} Y_1 \\ \ldots \\ Y_q \end{bmatrix}.$$

A quadratic form (next section) is a particular kind of bilinear form which is widely used in statistics.

Section 24.24: quadratic forms

A *quadratic form* is a scalar obtained by premultiplying a square and symmetric matrix $\mathbf{A}$ by a row vector $\mathbf{X}$ and by postmultiplying it by the transpose $\tilde{\mathbf{X}}$ of the same vector:

$$\mathbf{XA\tilde{X}} = a_{11}X_1^2 + \ldots + 2a_{1q}X_1X_q + \ldots + a_{qq}X_q^2 = [X_1 \ldots X_q] \begin{bmatrix} a_{11} & \ldots & a_{1q} \\ \ldots & \ldots & \ldots \\ a_{1q} & \ldots & a_{qq} \end{bmatrix} \begin{bmatrix} X_1 \\ \ldots \\ X_q \end{bmatrix}.$$

When the matrix $\mathbf{A}$ of a quadratic form is diagonal,

$$\mathbf{A} = \mathbf{D} = \begin{bmatrix} d_1 & \ldots & 0 \\ \ldots & \ldots & \ldots \\ 0 & \ldots & d_q \end{bmatrix},$$

the product terms (like $2a_{1q}X_1X_q$) vanish:

$$\mathbf{XD\tilde{X}} = d_1X_1^2 + d_2X_2^2 + \ldots + d_qX_q^2 .$$

Moreover, when the matrix of a quadratic form is a unit matrix $\mathbf{I}$, the quadratic form is equal to the sum of the squares of the variables X_1 , X_2 , ... , X_q , that is to the square of the length $||\mathbf{X}||$ of vector $\mathbf{X}$ (section 24.6):

$$\mathbf{X}\mathbf{I}\tilde{\mathbf{X}} = \mathbf{X}\tilde{\mathbf{X}} = X_1^2 + X_2^2 + \ldots + X_q^2 = ||\mathbf{X}||^2 .$$

If an equation is set according to which the quadratic form $\mathbf{X}\mathbf{I}\tilde{\mathbf{X}} = \mathbf{X}\tilde{\mathbf{X}}$ is equal to a constant c^2,

$$\mathbf{X}\mathbf{I}\tilde{\mathbf{X}} = \mathbf{X}\tilde{\mathbf{X}} = X_1^2 + X_2^2 + \ldots + X_q^2 = c^2,$$

this equation therefore defines the locus of the points lying at the same distance (c) from the origin $\mathbf{0} = [0, 0, \ldots 0]$. This equation thus determines two points if there is a single dimension, a circle centered on the origin if there are two dimensions, a sphere if there are three dimensions, and a hypersphere if there are more than three dimensions. As for the equation

$$\mathbf{X}\mathbf{D}\tilde{\mathbf{X}} = d_1 X_1^2 + d_2 X_2^2 + \ldots + d_q X_q^2 = c^2,$$

it corresponds in two dimensions to an ellipse which is centered on the origin and of which the principal axes coincide with the coordinate axes, to an ellipsoid in three dimensions, and to a hyperellipsoid in more than three dimensions. Finally, the equation

$$\mathbf{X}\mathbf{A}\tilde{\mathbf{X}} = c^2$$

corresponds to ellipses (figures 19.7.1, 19.8.1, 22.1.2) or ellipsoids (figure 23.2.1) of which the principal axes are generally not lined up with the coordinate axes. The preceding examples illustrate the advantages of matrix notation, thanks to which a single concise equation like $\mathbf{X}\mathbf{I}\tilde{\mathbf{X}} = \mathbf{X}\tilde{\mathbf{X}} = c^2$, $\mathbf{X}\mathbf{D}\tilde{\mathbf{X}} = c^2$ or $\mathbf{X}\mathbf{A}\tilde{\mathbf{X}} = c^2$, remains geometrically meaningful regardless of the number of variables considered.

In statistics, quadratic forms are involved in the probability density of the normal distribution (sections 5.2, 19.2, 23.1, 23.2, 29.2) and are used to determine the value of Hotelling's T^2 criterion or to delimit circular or elliptic confidence or prediction regions (chapter 30).

Section 24.25: spectral decomposition of a square symmetric matrix

A square symmetric matrix $\mathbf{A}$ may be expressed with respect to its latent roots and latent vectors if matrices $\mathbf{U}$ and $\tilde{\mathbf{U}}$ are subdivided according to the latter in the equation $\mathbf{A} = \tilde{\mathbf{U}}\mathbf{D}\mathbf{U}$ discussed at the beginning of section 24.21: the matrix product $\tilde{\mathbf{U}}\mathbf{D}\mathbf{U}$ is then analogous to a diagonalized quadratic form in which the row vector is the matrix $\tilde{\mathbf{U}}$ and has the column vectors $\tilde{\mathbf{U}}_1$, $\tilde{\mathbf{U}}_2$, ..., $\tilde{\mathbf{U}}_q$ (instead of scalars) as elements:

$$\mathbf{A} = \tilde{\mathbf{U}}\mathbf{D}\mathbf{U} = d_1 \tilde{\mathbf{U}}_1 \mathbf{U}_1 + d_2 \tilde{\mathbf{U}}_2 \mathbf{U}_2 + \ \ldots \ + d_i \tilde{\mathbf{U}}_i \mathbf{U}_i + \ \ldots \ + d_q \tilde{\mathbf{U}}_q \mathbf{U}_q \ .$$

The preceding equation indicates the contribution of each latent root d_i and of the corresponding latent vector $\mathbf{U}_i$ to matrix $\mathbf{A}$, and discloses the so-called *spectral decomposition* of matrix $\mathbf{A}$ with respect to its latent vectors and latent roots.

Section 24.26: further readings, computing tools

The reader who wishes to improve his knowledge of linear algebra should consult the works of Aitken (1956), Beaumont (1965), Graybill (1969), Murdoch (1957), Schreier and Sperner (1951), Searle (1966) and Wade (1951). Numerical applications are considerably facilitated by using mathematical software like *Mathematica®* (Wolfram Research, Inc., 100 Trade Center Drive, Champaign, Illinois 61820-7237, U.S.A.) or a high-level programming language possessing matrix instructions like *True BASIC™* (True BASIC Inc., 12 Commerce Avenue, West Lebanon, New Hampshire 03784-1669, U.S.A.).

The following box illustrates the simplicity and conciseness of a matrix inversion program written in "*True BASIC*". The first statement, the title of the program, is a comment, as shown by the initial exclamation mark.

```
! Matrix inversion
DIM A(4,4),B(4,4)
MAT READ A
MAT B = INV(A)
MAT PRINT B
DATA 2,1,1,1
DATA 1,2,1,1
DATA 1,1,2,1
DATA 1,1,1,2
PAUSE 180
END
```

The second statement specifies the order of the matrix to be inverted,

$$\mathbf{A}_{(4\times 4)} = \begin{bmatrix} 2 & 1 & 1 & 1 \\ 1 & 2 & 1 & 1 \\ 1 & 1 & 2 & 1 \\ 1 & 1 & 1 & 2 \end{bmatrix},$$

as well as of its inverse $\mathbf{A}^{-1} = \mathbf{B}$. The third statement, `MAT READ A`, triggers the transfer into matrix **A** of the numerical values of its elements, given in the four `DATA` statements. The fourth statement, `MAT B = INV(A)`, causes the inverse of matrix **A** to be calculated, and the fifth statement, `MAT PRINT B`, controls the display of the results,

$$\mathbf{A}^{-1} = \mathbf{B} = \begin{bmatrix} +0.8 & -0.2 & -0.2 & -0.2 \\ -0.2 & +0.8 & -0.2 & -0.2 \\ -0.2 & -0.2 & +0.8 & -0.2 \\ -0.2 & -0.2 & -0.2 & +0.8 \end{bmatrix}.$$

The penultimate statement, `PAUSE 180`, ensures that, in the new 5.0 version of *True BASIC*, the user has enough time to write down results (here approximately 180 seconds), and the last statement, `END`, signals the end of the program. Other examples of *True BASIC* programs are given in sections 29.3 and 30.8 and in appendices A and B.

Chapter 25
Partial and multiple correlations and regressions: matrix calculations

Section 25.1: introduction

Thanks to vectors and matrices (chapter 24), the partial and multiple correlations and regressions introduced in chapter 23 can still easily be used when there are more than two predictor variates. In order to broaden the discussion, let us subdivide each observed vector $\mathbf{X}$ into two subvectors $\mathbf{X}_1$ and $\mathbf{X}_2$ of which the elements are the predictor and the predicted variates respectively:

$$\mathbf{X} = [\mathbf{X}_1 | \mathbf{X}_2] = [X_1 \dots X_i, X_j \dots X_k | X_{(k+1)} \dots X_u, X_v \dots X_q] .$$

Subvector $\mathbf{X}_1$ then contains the *predictor variates,* which are kept constant or which are used to estimate or adjust other variates, while subvector $\mathbf{X}_2$ contains the *predicted variates,* which are estimated or adjusted on the basis of the others. Predictor and predicted variates are sometimes called *independent variates* and *dependent variates* respectively, but that terminology is misleading and should be avoided because a variate which is statistically independent from another one cannot be used to estimate, predict or adjust it.

Section 25.2: partial correlation coefficients

According to the notation introduced in the preceding section, the partial correlation between two predicted variates X_u and X_v when a predictor variate X_k is kept constant (section 23.2) can be expressed in the following manner:

$$r_{uv.k} = (r_{uv} - r_{ku} r_{kv}) / \sqrt{[(1 - r_{ku}^2)(1 - r_{kv}^2)]} ,$$

with $\nu = [(N-2) - 1] = (N-3)$ degrees of freedom. However, the concept of a partial correlation may be extended to the case where several predictor variates are kept constant. If all predictor variates in subvector $\mathbf{X}_1$ are kept constant, for instance, the partial correlation coefficient $r_{uv.1 \dots k}$ is equal to

$$r_{uv.1 \dots k} = (r_{uv.1 \dots (k-1)} - r_{ku.1 \dots (k-1)} r_{kv.1 \dots (k-1)}) / \sqrt{[(1 - r_{ku.1 \dots (k-1)}^2)(1 - r_{kv.1 \dots (k-1)}^2)]}$$

and has $\nu = [(N-2) - k]$ degrees of freedom. The above formula is said to be *recursive* inasmuch as a partial correlation coefficient with k variates held constant is defined using other partial correlation coefficients with $(k-1)$ variates held constant, which are themselves defined using partial correlation coefficients with $(k-2)$ variates held constant, and so on until total correlation coefficients r_{uv}, r_{ku} and r_{kv} (with 0 variate held constant) are reached. Such calculations would obviously be tedious and subject to rounding errors. Fortunately, the longest and hardest part of these computations corresponds to the inversion of the correlation or covariance matrix of predictor variates (sections 24.19, 24.26, 25.3 and 25.5), and the latter can be done easily, rapidly and accurately with a computer program.

Section 25.3: matrix calculations with two predictor variates

To keep things initially as simple as possible, the application of vector and matrix methods to multiple regression (section 23.3) is illustrated first in the case where there are only three variates, including two predictor variates. The equation of multiple regression may be expressed using subscripted variates,

$$\hat{X}_3 = a + b_1 X_1 + b_2 X_2 \,,$$

and the Gauss equations, obtained from the least-squares principle, may be written using matrices:

$$\begin{bmatrix} N & \Sigma X_1 & \Sigma X_2 \\ \Sigma X_1 & \Sigma X_1^2 & \Sigma X_1 X_2 \\ \Sigma X_2 & \Sigma X_1 X_2 & \Sigma X_2^2 \end{bmatrix} \begin{bmatrix} a \\ b_1 \\ b_2 \end{bmatrix} = \begin{bmatrix} \Sigma X_3 \\ \Sigma X_1 X_3 \\ \Sigma X_2 X_3 \end{bmatrix} .$$

Once the matrix of raw sums has been inverted,

$$\begin{bmatrix} N & \Sigma X_1 & \Sigma X_2 \\ \Sigma X_1 & \Sigma X_1^2 & \Sigma X_1 X_2 \\ \Sigma X_2 & \Sigma X_1 X_2 & \Sigma X_2^2 \end{bmatrix}^{-1} ,$$

the vector of coefficients $[a \;\; b_1 \;\; b_2]$ may thus be obtained through a simple matrix multiplication:

$$\begin{bmatrix} a \\ b_1 \\ b_2 \end{bmatrix} = \begin{bmatrix} N & \Sigma X_1 & \Sigma X_2 \\ \Sigma X_1 & \Sigma X_1^2 & \Sigma X_1 X_2 \\ \Sigma X_2 & \Sigma X_1 X_2 & \Sigma X_2^2 \end{bmatrix}^{-1} \begin{bmatrix} \Sigma X_3 \\ \Sigma X_1 X_3 \\ \Sigma X_2 X_3 \end{bmatrix} .$$

However, if the first Gauss equation is solved separately at the beginning,

$$a = \bar{X}_3 - b_1 \bar{X}_1 - b_2 \bar{X}_2 \,,$$

the other equations,

$$\begin{bmatrix} \Sigma(X_1 - \bar{X}_1)^2 & \Sigma(X_1 - \bar{X}_1)(X_2 - \bar{X}_2) \\ \Sigma(X_1 - \bar{X}_1)(X_2 - \bar{X}_2) & \Sigma(X_2 - \bar{X}_2)^2 \end{bmatrix} \begin{bmatrix} b_1 \\ b_2 \end{bmatrix} = \begin{bmatrix} \Sigma(X_1 - \bar{X}_1)(X_3 - \bar{X}_3) \\ \Sigma(X_2 - \bar{X}_2)(X_3 - \bar{X}_3) \end{bmatrix} ,$$

require inverting only the (smaller) matrix of centered sums of predictor variates:

$$\begin{bmatrix} b_1 \\ b_2 \end{bmatrix} = \begin{bmatrix} \Sigma(X_1 - \bar{X}_1)^2 & \Sigma(X_1 - \bar{X}_1)(X_2 - \bar{X}_2) \\ \Sigma(X_1 - \bar{X}_1)(X_2 - \bar{X}_2) & \Sigma(X_2 - \bar{X}_2)^2 \end{bmatrix}^{-1} \begin{bmatrix} \Sigma(X_1 - \bar{X}_1)(X_3 - \bar{X}_3) \\ \Sigma(X_2 - \bar{X}_2)(X_3 - \bar{X}_3) \end{bmatrix} .$$

Finally, the value of the vector of coefficients, $\mathbf{b} = [b_1 \;\; b_2]$, may be expressed most concisely with respect to variances and covariances:

$$\begin{bmatrix} b_1 \\ b_2 \end{bmatrix} = \begin{bmatrix} S_{11} & S_{12} \\ S_{12} & S_{22} \end{bmatrix}^{-1} \begin{bmatrix} S_{13} \\ S_{23} \end{bmatrix} .$$

The least-squares principle ensures that, geometrically speaking, the plane of multiple regression passes close to observed dots and that the sum of squares of residual deviations,

$$\text{RSS} = \sum_{h=1}^{N} (X_3 - \hat{X}_3)^2 = \sum_{h=1}^{N} [X_3 - (a + b_1 X_1 + b_2 X_2)]^2 ,$$

is as small as possible (section 23.3). If the coefficients $\mathbf{b} = [b_1 \;\; b_2]$ are given the values found previously through the least-squares principle, the minimum value of the residual sum of squares (RSS) may be found in a single operation, without having to consider each individual deviation:

$$\text{RSS} = \Sigma(X_3 - \hat{X}_3)^2 = \Sigma(X_3 - \bar{X}_3)^2 - \Sigma(\hat{X}_3 - \bar{X}_3)^2$$

that is

$$\text{RSS} = \Sigma(X_3 - \hat{X}_3)^2 = \Sigma(X_3 - \bar{X}_3)^2 - [b_1 \;\; b_2]\begin{bmatrix} \Sigma(X_1 - \bar{X}_1)(X_3 - \bar{X}_3) \\ \Sigma(X_2 - \bar{X}_2)(X_3 - \bar{X}_3) \end{bmatrix} .$$

This residual sum of squares possesses $\nu = [(N-1) - 2] = (N-3)$ degrees of freedom, and the estimate of the residual variance of X_3 with respect to the plane of multiple regression in the direction of the OX_3 coordinate axis is

$$S^2_{3.12} = \Sigma(X_3 - \hat{X}_3)^2 / [(N-1) - 2] .$$

The estimate $[S_{b_i b_j}]$ of the covariance matrix of the coefficients $\mathbf{b} = [b_1 \;\; b_2]$, which also has $\nu = [(N-1) - 2] = (N-3)$ degrees of freedom, is obtained next by multiplying the inverse of the matrix of centered sums of predictor variates by the residual variance $S^2_{3.12}$, an instance of the product of a matrix by a scalar (section 24.10):

$$[S_{b_i b_j}] = S^2_{3.12}\begin{bmatrix} \Sigma(X_1 - \bar{X}_1)^2 & \Sigma(X_1 - \bar{X}_1)(X_2 - \bar{X}_2) \\ \Sigma(X_1 - \bar{X}_1)(X_2 - \bar{X}_2) & \Sigma(X_2 - \bar{X}_2)^2 \end{bmatrix}^{-1} .$$

The reader must note that the above matrix of centered sums of predictor variates must be inverted before being multiplied by the residual variance $S^2_{3.12}$. It may be verified that the diagonal elements of matrix $[S_{b_i b_j}]$ are indeed equal to the variances of coefficients b and c given in section 23.4. As for the multiple correlation coefficient (section 23.5), it may be evaluated using vectors and matrices through one of the following expressions:

$$R_{3.12} = +\sqrt{\frac{[b_1 \; b_2]\begin{bmatrix} \Sigma(X_1 - \bar{X}_1)(X_3 - \bar{X}_3) \\ \Sigma(X_2 - \bar{X}_2)(X_3 - \bar{X}_3) \end{bmatrix}}{\Sigma(X_3 - \bar{X}_3)^2}} = +\sqrt{\frac{[b_1 \; b_2]\begin{bmatrix} S_{13} \\ S_{23} \end{bmatrix}}{S_{33}}} ,$$

or

$$R_{3.12} = +\sqrt{[r_{13} \; r_{23}]\begin{bmatrix} 1 & r_{12} \\ r_{12} & 1 \end{bmatrix}^{-1}\begin{bmatrix} r_{13} \\ r_{23} \end{bmatrix}} .$$

Section 25.4: a more complete and explicit notation

In the preceding section, the multiple regression equation was given in a simplified form,

$$\hat{X}_3 = a + b_1X_1 + b_2X_2 ,$$

and now is the time to use a more complete and more explicit notation:

$$\hat{X}_{3.12} = a_{3.12} + b_{31.2}X_1 + b_{32.1}X_2 .$$

This notation is interpreted as follows:

$\hat{X}_{3.12}$ is the value of X_3 such as estimated using X_1 and X_2 ;

$a_{3.12}$ is the intercept (the X_3-intercept) of the estimation equation of X_3 from variates X_1 and X_2 ;

$b_{31.2}$ is the estimation coefficient of X_3 from X_1, the X_2 variate being kept constant;

$b_{32.1}$ is the estimation coefficient of X_3 from X_2, the X_1 variate being kept constant.

The coefficients $b_{31.2}$ and $b_{32.1}$ of the multiple regression equation are related to the partial correlation coefficients $r_{31.2}$ and $r_{32.1}$ and are in fact partial regression coefficients.

Section 25.5: the general case of several *(k)* predictor variates

When there are more than two predictor variates, the procedures described in section 25.3 can still be used except that the equations involve more numerous variates and the vectors and the matrices contain greater numbers of elements. The predictor variates are the k components of vector $\mathbf{X}_1 = [X_1 \; ... \; X_i \; ... \; X_k]$ while the predicted variate is $X_{(k+1)}$. The multiple regression equation may then be written:

$$\hat{X}_{(k+1)} = a + b_1X_1 + ... + b_kX_k ,$$

that is, in full notation (as in section 25.4),

$$\hat{X}_{(k+1).1...k} = a_{(k+1).1...k} + b_{(k+1)1.2...k}X_1 + ... + b_{(k+1)k.12...(k-1)}X_k ,$$

where $\hat{X}_{(k+1).1...k}$ is the value of $X_{(k+1)}$ such as estimated from variates $[X_1 \; ... \; X_i \; ... \; X_k]$;

$a_{(k+1).1...k}$ is the intercept (the $X_{(k+1)}$-intercept) of the estimation equation of $X_{(k+1)}$ from variates $[X_1 \; ... \; X_i \; ... \; X_k]$;

$b_{(k+1)1.2...k}$ is the estimation coefficient of $X_{(k+1)}$ from X_1, all other predictor variates $[X_2 \; ... \; X_i \; ... \; X_k]$ being kept constant;

$b_{(k+1)k.12...(k-1)}$ is the estimation coefficient of $X_{(k+1)}$ from X_k, all other predictor variates $[X_1 \; ... \; X_i \; ... \; X_{(k-1)}]$ being kept constant.

The coefficients of the multiple regression equation are then estimated through simple extensions of the procedures used in section 25.3:

$$a_{(k+1).1...k} = \overline{X}_{(k+1)} - b_{(k+1)1.2...k}\overline{X}_1 - ... - b_{(k+1)k.12...(k-1)}\overline{X}_k$$

and

$$\begin{bmatrix} b_{(k+1)\,1.2\ldots k} \\ \cdots \\ b_{(k+1)\,k.12\ldots(k-1)} \end{bmatrix} = \begin{bmatrix} \Sigma(X_1-\bar{X}_1)^2 & \cdots & \Sigma(X_1-\bar{X}_1)(X_k-\bar{X}_k) \\ \cdots & \cdots & \cdots \\ \Sigma(X_1-\bar{X}_1)(X_k-\bar{X}_k) & \cdots & \Sigma(X_k-\bar{X}_k)^2 \end{bmatrix}^{-1} \begin{bmatrix} \Sigma(X_1-\bar{X}_1)(X_{(k+1)}-\bar{X}_{(k+1)}) \\ \cdots \\ \Sigma(X_k-\bar{X}_k)(X_{(k+1)}-\bar{X}_{(k+1)}) \end{bmatrix}.$$

The residual sum of squares (RSS) is, in simplified notation,

$$\Sigma(X_{(k+1)}-\hat{X}_{(k+1)})^2 = \Sigma(X_{(k+1)}-\bar{X}_{(k+1)})^2 - [b_1 \;\ldots\; b_k] \begin{bmatrix} \Sigma(X_1-\bar{X}_1)(X_{(k+1)}-\bar{X}_{(k+1)}) \\ \cdots \\ \Sigma(X_k-\bar{X}_k)(X_{(k+1)}-\bar{X}_{(k+1)}) \end{bmatrix}$$

and possesses $\nu = [(N-1)-k]$ degrees of freedom.

The residual variance with respect to the multiple regression hyperplane is

$$S^2_{(k+1).1\ldots k} = \Sigma(X_{(k+1)}-\hat{X}_{(k+1)})^2 \,/\, [(N-1)-k] \;.$$

The estimated covariance matrix of coefficients $\mathbf{b} = [b_1 \;\ldots\; b_k]$ is

$$[S_{b_i b_j}] = S^2_{(k+1).1\ldots k} \begin{bmatrix} \Sigma(X_1-\bar{X}_1)^2 & \cdots & \Sigma(X_1-\bar{X}_1)(X_k-\bar{X}_k) \\ \cdots & \cdots & \cdots \\ \Sigma(X_1-\bar{X}_1)(X_k-\bar{X}_k) & \cdots & \Sigma(X_k-\bar{X}_k)^2 \end{bmatrix}^{-1}.$$

The multiple correlation coefficient between the predicted variate $X_{(k+1)}$ and the k predictor variates $[X_1 \ldots X_i \ldots X_k]$ may be evaluated using one of the following expressions:

$$R_{(k+1).1\ldots k} = +\sqrt{\frac{[b_1 \ldots b_k] \begin{bmatrix} \Sigma(X_1-\bar{X}_1)(X_{(k+1)}-\bar{X}_{(k+1)}) \\ \cdots \\ \Sigma(X_k-\bar{X}_k)(X_{(k+1)}-\bar{X}_{(k+1)}) \end{bmatrix}}{\Sigma(X_{(k+1)}-\bar{X}_{(k+1)})^2}}\,,$$

$$R_{(k+1).1\ldots k} = +\sqrt{\frac{[b_1 \ldots b_k] \begin{bmatrix} S_{1(k+1)} \\ \cdots \\ S_{k(k+1)} \end{bmatrix}}{S_{(k+1)(k+1)}}}\,,$$

or

$$R_{(k+1).1\ldots k} = +\sqrt{[r_{1(k+1)} \;\cdots\; r_{k(k+1)}] \begin{bmatrix} 1 & \cdots & r_{1k} \\ \cdots & \cdots & \cdots \\ r_{1k} & \cdots & 1 \end{bmatrix}^{-1} \begin{bmatrix} r_{1(k+1)} \\ \cdots \\ r_{k(k+1)} \end{bmatrix}}\,,$$

and the corresponding variance ratio is

$$F_{(k,\,N-1-k)} = \frac{(N-1-k)\,R^2_{(k+1).1\ldots k}}{k\,[1-R^2_{(k+1).1\ldots k}]}\;.$$

The acceptance region of the preliminary hypothesis H_0: $\mathfrak{R}_{(k+1).1\ldots k} = 0$, a null hypothesis in this case, is

$$F_{(k,\, N-1-k)} \leq F_{(1-\alpha;\, k,\, N-1-k)} \;.$$

Section 25.6: analysis of variance corresponding to a multiple regression

As in the case of the bivariate estimation line $Y = a + bX$ (section 20.3 and table 20.3.1), the decomposition of the total sum of squares of the predicted variate corresponding to a multiple regression may be described by an analysis of variance (table 25.6.1).

Table 25.6.1
Analysis of variance corresponding to a multiple regression

Variation	RSS	D. of F.	Variance estimate
residual	$\Sigma(X_{(k+1)} - \hat{X}_{(k+1)})^2$	$N-1-k$	$S^2_{(k+1).1\ldots k} = \Sigma(X_{(k+1)} - \hat{X}_{(k+1)})^2/(N-1-k)$
regression	$\Sigma(\hat{X}_{(k+1)} - \bar{X}_{(k+1)})^2$	k	$\Sigma(\hat{X}_{(k+1)} - \bar{X}_{(k+1)})^2/k$
total	$\Sigma(X_{(k+1)} - \bar{X}_{(k+1)})^2$	$N-1$	$S^2_{(k+1)} = \Sigma(X_{(k+1)} - \bar{X}_{(k+1)})^2/(N-1)$

In such an analysis, the first preliminary hypothesis to be tested usually specifies the nullity of the vector of parametric regression coefficients $\boldsymbol{\beta} = [\beta_1 \;\ldots\; \beta_k]$ estimated by $\mathbf{b} = [b_1 \;\ldots\; b_k]$: $\quad H_0: \boldsymbol{\beta} = [\beta_1 \ldots \beta_k] = \mathbf{0} = [0 \;\ldots\; 0]$; $H_1: \boldsymbol{\beta} \neq \mathbf{0}$.

The variance ratio obtained in this analysis of variance has exactly the same numerical value and the same numbers of degrees of freedom as the variance ratio corresponding to the multiple correlation coefficient (section 25.5). As in the bivariate case (section 20.4), the hypothesis of nullity of the vector $\boldsymbol{\beta} = [\beta_1 \;\ldots\; \beta_k]$ is thus strictly equivalent to the hypothesis of nullity of the parametric multiple correlation coefficient $\mathfrak{R}_{(k+1).1\ldots k}$. Moreover, provided the number k of degrees of freedom corresponding to the multiple regression is taken into account, the analysis of covariance, previously applied to estimation lines (chapter 21), can also be used to compare two or several multiple regressions.

Section 25.7: hypotheses concerning some coefficients in particular

The hypothesis tested in section 25.6 was that all coefficients $\boldsymbol{\beta} = [\beta_1 \;\ldots\; \beta_k]$ were null jointly. In order to test hypotheses concerning some coefficients in particular, the methods discussed in section 23.4 must be generalized. The variance of each estimated coefficient b_i is itself estimated with $\nu = [(N-1) - k]$ degrees of freedom by multiplying the inverse of the matrix of centered sums of predictor variates by the residual variance $S^2_{(k+1).1\ldots k}$ and taking the corresponding diagonal element $S^2_{b_i}$ (sections 25.3 and 25.5). The preliminary hypothesis $H_0: \beta_i = \beta_{i_0}$ that the parametric coefficient β_i has a particular numerical value β_{i_0} (0 in the case of a null hypothesis, or another theoretical value) may be tested using the criterion

$$t_{(N-1-k)} = (b_i - \beta_{i_0}) / S_{b_i} \;,$$

which follows Student's t distribution with $(N-1-k)$ degrees of freedom.

When theoretical considerations do not provide any precise hypothetical value β_{i_0}, the significance test may be replaced by a confidence interval such as, for instance, the two-sided version:

$$b_i + t_{(\alpha/2;\, N-1-k)}\, S_{b_i} \;\leq\; \beta_i \;\leq\; b_i + t_{(1-\alpha/2;\, N-1-k)}\, S_{b_i} \;.$$

Section 25.8: confidence or prediction intervals of the estimated variate

As in the case of the bivariate estimation line $Y = a + bX$ (section 20.5), the values $\hat{X}_{(k+1).1\ldots k}$ predicted through the multiple regression are affected by the uncertainty of the coefficients $a_{(k+1).1\ldots k}$ and $\mathbf{b} = [b_{(k+1)1.2\ldots k} \;\; \ldots \;\; b_{(k+1)k.12\ldots(k-1)}]$, since the latter are estimated from a sample of finite size. In the case where there are two predictor variates, the uncertainty region of predicted values $\hat{X}_{(k+1).1\ldots k}$ is limited by two curved surfaces which are convex upwards below the multiple regression plane and convex downwards above it. In the general case of k predictor variates, the confidence or prediction limits are:

$$\hat{X}_{(k+1).1\ldots k} \pm t_{(1-\alpha/2;\, N-1-k)}\, S_{(k+1).1\ldots k}\, H \,,$$

where H is a *broadening factor* (that is a factor making the interval broader). When the predicted value $\hat{X}_{(k+1).1\ldots k}$ is interpreted as a *changing mean* or an *adjusted mean*, the confidence limits of that mean are obtained by making the broadening factor H equal to

$$H = \sqrt{(1/N) + \left[X_{1_0} - \bar{X}_1 \,,\; \ldots \; X_{k_0} - \bar{X}_k\right] \begin{bmatrix} \Sigma(X_1-\bar{X}_1)^2 & \ldots & \Sigma(X_1-\bar{X}_1)(X_k-\bar{X}_k) \\ \ldots & \ldots & \ldots \\ \Sigma(X_1-\bar{X}_1)(X_k-\bar{X}_k) & \ldots & \Sigma(X_k-\bar{X}_k)^2 \end{bmatrix}^{-1} \begin{bmatrix} X_{1_0} - \bar{X}_1 \\ \ldots \\ X_{k_0} - \bar{X}_k \end{bmatrix}} \;.$$

The vector $\mathbf{X}_0 = [X_{1_0} \;\ldots\; X_{k_0}]$ is the set of values of the predictor variates put into the multiple regression equation in order to estimate the predicted variate. When prediction limits are needed between which a new individual observation is likely to fall if it comes from the same population as the sample, however, the prediction interval is much broader than the above confidence interval and is obtained by making the broadening factor H equal to:

$$H = \sqrt{1 + (1/N) + \left[X_{1_0} - \bar{X}_1 \,,\; \ldots \; X_{k_0} - \bar{X}_k\right] \begin{bmatrix} \Sigma(X_1-\bar{X}_1)^2 & \ldots & \Sigma(X_1-\bar{X}_1)(X_k-\bar{X}_k) \\ \ldots & \ldots & \ldots \\ \Sigma(X_1-\bar{X}_1)(X_k-\bar{X}_k) & \ldots & \Sigma(X_k-\bar{X}_k)^2 \end{bmatrix}^{-1} \begin{bmatrix} X_{1_0} - \bar{X}_1 \\ \ldots \\ X_{k_0} - \bar{X}_k \end{bmatrix}} \;.$$

The matrix appearing in the broadening factor is the inverse of the matrix of centered sums of predictor variates, already mentioned in sections 25.3 and 25.5.

Section 25.9: an example from quantitative human genetics

Since the utilization of partial and multiple correlations and regressions has been illustrated on physiological data in chapter 23, an example will be drawn here from the field of quantitative human genetics. The head length in mm of the father (X_1), of the mother (X_2) and of the first adult son (X_3) in 73 Dutch families has been extracted from the monograph of Frets (1921), precautions being taken in the manner discussed in section 34.5. The mean vector $\overline{\mathbf{X}}$, the matrix of centered sums and the correlation matrix $[r_{ij}]$ are given for all variates in table 25.9.1.

Table 25.9.1
Mean vector $\overline{\mathbf{X}}$, matrix of centered sums and correlation matrix $[r_{ij}]$ of the head length in mm of the father (X_1), of the mother (X_2) and of the first adult son (X_3) in 73 Dutch families according to data extracted from the study of Frets (1921)

Mean vector $\overline{\mathbf{X}}$

$$[194.507 \quad 184.096 \quad 192.890]$$

Matrix of centered sums

$$\begin{bmatrix} 2846.247 & 396.452 & 1146.055 \\ 396.452 & 1730.329 & 956.767 \\ 1146.055 & 956.767 & 3219.123 \end{bmatrix}$$

Correlation matrix $[r_{ij}]$

$$\begin{bmatrix} 1.000000 & 0.178645 & 0.378617 \\ 0.178645 & 1.000000 & 0.405390 \\ 0.378617 & 0.405390 & 1.000000 \end{bmatrix}$$

The correlation between the head length of the father and of the mother does not reach significance (H_1: $\rho \neq 0$; $r = +0.178645$, $P > 0.05$), but the head length of the first adult son is significantly correlated with the head length of his father ($r = +0.378617$, $P < 0.001$) and mother ($r = +0.405390$, $P < 0.001$). Moreover, the multiple correlation of the head length of the first adult son (X_3) with those of his father and mother jointly,

$$R_{3.12} = +0.511066,$$

is even higher and also strongly significant ($F_{(2,\,70)} = 12.3734$, $P < 0.001$).

The multiple regression equation is

$$\hat{X}_{3.12} = 39.8582 + 0.336371\, X_1 + 0.475870\, X_2 \; .$$

The estimated covariance matrix $[S_{b_i b_j}]$ of the coefficients $[b_1 \; b_2]$ of the multiple regression is obtained by multiplying the inverse of the matrix of centered sums of predictor variates by the residual variance $S^2_{3.12} = 33.9761$, which yields

$$[S_{b_i b_j}] = \begin{bmatrix} 0.0123307 & -0.00282520 \\ -0.00282520 & 0.0202829 \end{bmatrix}.$$

The standard deviations (also called standard errors) of the multiple regression coefficients are thus $S_{b_1} = 0.111044$ and $S_{b_2} = 0.142418$, and the Student's t values obtained by testing null hypotheses are $t_1 = 3.02918$ and $t_2 = 3.34136$ ($\nu = 70$, $P < 0.01$). Both coefficients b_1 and b_2 thus differ significantly from 0, and the use of multiple regression is warranted.

At the $\alpha = 0.05$ significance level, the corresponding parametric regression coefficients have the following two-sided confidence intervals

$$+0.114902 \leq \mathscr{b}_1 \leq +0.557841 \quad \text{and} \quad +0.191826 \leq \mathscr{b}_2 \leq +0.759914.$$

Even at this least rigorous of usual significance levels, the hypotheses that $\mathscr{b}_1 = 1/2$ and that $\mathscr{b}_2 = 1/2$, mentioned at the end of section 22.3, are therefore acceptable. The present results thus agree with the theoretical view that, allowance being made for sexual dimorphism, a quantitative character such as the head length of an adult son is determined half by paternal and half by maternal inheritance. At the more demanding $\alpha = 0.01$ level, the two-sided confidence intervals are

$$+0.0423383 \leq \mathscr{b}_1 \leq +0.630404 \quad \text{and} \quad +0.0987606 \leq \mathscr{b}_2 \leq +0.852980,$$

and still do not reach 0, confirming the statistical significance of both coefficients established above by testing null hypotheses.

Section 25.10: possibilities and limitations of the methods in this chapter

Partial and multiple correlations and regressions may be useful in many different fields, including physiology, quantitative genetics and ecology. The possibilities of those methods are obviously broadened by the application of vectors and matrices, which facilitate calculations when there are many variates. However, the user must beware of including many predictor variates of which some would be highly correlated or redundant (a case of *collinearity*), because the matrix of centered sums of such variates might then be singular or nearly singular (sections 24.18 and 24.19). This could not only hinder computations but also make interpretations difficult, because coefficients having opposite signs might be obtained for predictor variates known a priori to influence the predicted variate in the same way.

In ichthyology, for instance, *total length* and *fork length* should generally not be included simultaneously among predictor variates, for these two variates differ very little from each other and, if one of them is already used, adding the second one would bring little additional information. Similarly, in ecology, when many environmental conditions have been measured, it is usually preferable to make a reasoned choice and to include initially in the analysis just a few predictor variates which are not expected to be too closely related (see again section 19.9). Tomassone, Audrain, Lesquoy-de Turckheim and Millier (1992) have devoted several chapters to recent aspects of multiple regression.

Chapter 26
One-way type I analysis of variance with contrasts

Section 26.1: obtaining more information from an analysis of variance

The importance of comparing the confidence intervals of group means at the end of an analysis of variance has been emphasized in section 12.8. The overall significance test done in an analysis of variance (section 12.6) indicates whether there are conclusive differences between group means but reveals neither the nature nor the extent of those differences. Therefore, the inspection of a series of confidence intervals, such as illustrated in figure 12.8.1, is very useful from a descriptive point of view.

From a statistical viewpoint, however, the graphical comparison of confidence intervals may give ambiguous indications. For instance, two confidence intervals which overlap only slightly are illustrated in figure 26.1.1. Intuitively, the reader would be inclined to accept the hypothesis that the means μ_1 and μ_2 of the populations from which the two samples have come are equal. The probability that the upper end of the first confidence interval covers the mean μ_1 and, similarly, the probability that the lower end of the second confidence interval covers the mean μ_2 are indeed greater than the significance level (let us say approximately $P = 0.06 > \alpha = 0.05$). At first sight, it would thus seem that a common value $\mu = \mu_1 = \mu_2$ could be the mean of both populations.

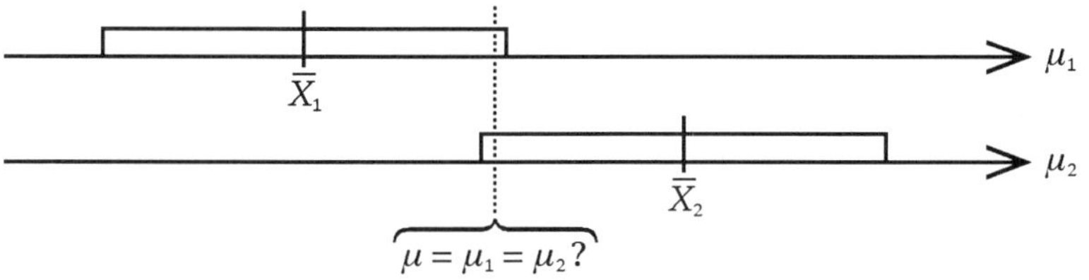

Figure 26.1.1
Does a slight overlap of two confidence intervals indicate that the parametric means may be considered as equal?

However, the above reasoning would not take into account the fact that, if each of the two samples has been drawn at random from its own population, it is very unlikely that one of the confidence intervals would have fallen much to the left of its parametric mean while the other confidence interval would have fallen simultaneously much to the right of the mean of its own population. In fact, even if the probability of each of these two independent events is not excessively small ($P = 0.06$), the probability that both events occur jointly is equal to the product of the probabilities that they occur separately (section 3.9), that is to $P = 0.06 \times 0.06 = 0.0036$, which is smaller even than the $\alpha = 0.01$ significance level. By itself, therefore, the graphical comparison of two confidence intervals does not indicate explicitly whether the means $\overline{X}_1$ and $\overline{X}_2$ of the corresponding samples differ in a statistically significant manner or not.

Even though, as just shown, the visual comparison of confidence intervals is not a sensitive manner of detecting differences between population means, that procedure would nevertheless increase the risk of type I errors (section 9.4). For instance, figure 26.1.2 shows that the simultaneous utilization of the confidence intervals of two means would correspond geometrically to a square which would have a probability of only $P = (1-\alpha)^2 \doteq 1 - 2\alpha$ of covering the point $[\mu_1, \mu_2]$ representing true means. More generally, the simultaneous utilization of C confidence intervals would correspond to a parallelepiped or a hyperparallelepiped having a probability of only $P = (1-\alpha)^C \doteq 1 - C\alpha$ of covering the set $[\mu_1, \mu_2, \dots, \mu_C]$ of true means. The risk of erroneously rejecting at least one true parametric mean is thus multiplied approximately by the number C of confidence intervals used jointly. If twenty confidence intervals based on the $\alpha = 0.05$ significance level are used simultaneously, for instance, there is an approximate probability of $P \doteq 20 \times 0.05 = 1.0$ that at least one true parametric mean will not be covered!

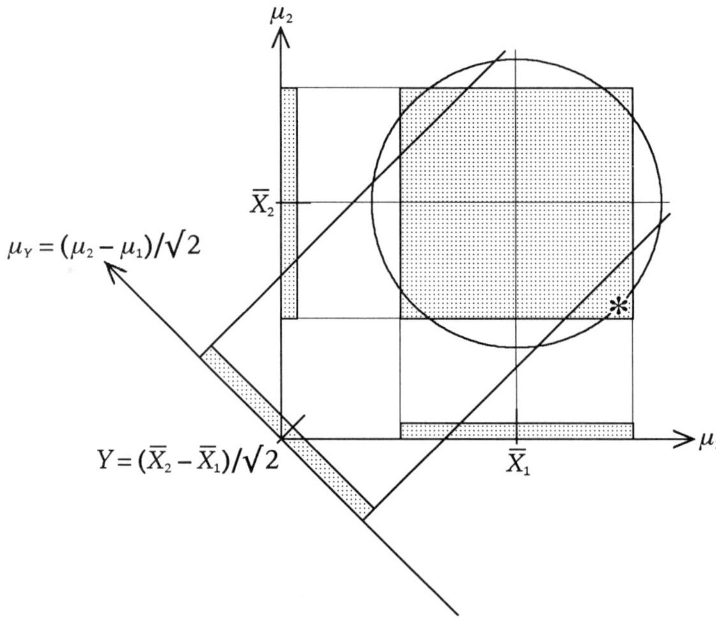

Figure 26.1.2
Confidence intervals corresponding to a probability $P = (1-\alpha) = 0.95$ for μ_1 (stippled, horizontal), for μ_2 (stippled, vertical), for $\mu_Y = (\mu_2 - \mu_1)/\sqrt{2}$ (stippled, inclined), and for μ_1 and μ_2 jointly (circular). The square corresponds to the joint utilization of two intersecting intervals and to a probability of only $P = (1-\alpha)^2 \doteq 1 - 2\alpha$, that is approximately 0.90

The most efficient statistical approach for comparing two sample means $\bar{X}_1$ and $\bar{X}_2$ is based on their difference $(\bar{X}_2 - \bar{X}_1)$, or else on the variate $Y = (\bar{X}_2 - \bar{X}_1)/\sqrt{2}$ which corresponds to a simple axis rotation without a change of scale. This is equivalent to using a single confidence interval (stippled and inclined in figure 26.1.2) which keeps the risk of type I errors equal to α but is oriented in such a way as to maximize the chances of detecting a possible difference between population means. This is basically what is being done in practice when Student's t distribution is used to test a null hypothesis or to determine a confidence interval (section 9.8) concerning the difference $(\mu_2 - \mu_1)$. Thus a pair of hypothetical mean values $[\mu_1^*, \mu_2^*]$, represented by an asterisk in the lower right corner of the rectangle (figure 26.1.2), would erroneously be considered as acceptable if the confidence intervals of μ_1 and μ_2 were used jointly but is discovered to be unacceptable with the more sensitive confidence interval of the difference $\mu_Y = (\mu_2 - \mu_1)/\sqrt{2}$.

To sum up, while the simultaneous utilization of the confidence intervals of several group means is a descriptively useful complement at the end of an analysis of variance, it is not a statistically efficient manner of checking which means differ significantly from each other.

Section 26.2: the so-called Bonferroni method of adjusting significance levels

When confidence intervals are used simultaneously for several means, there is a way to make sure that the overall probability of erroneously rejecting at least one true parametric mean does not exceed the desired significance level α: the significance level used for each interval may merely be divided by the total number C of intervals. This procedure is associated with the name of Bonferroni. The reader particularly interested in multiple comparisons should consult the work of Miller (1966). If population variances are assumed to be equal (section 12.6) and if the variance of each sample mean is estimated on the basis of the *pooled within-groups* variance,

$$S^2_{\bar{X}_g} = S^2_{\text{within-groups}} / N_g ,$$

each confidence interval may then be expressed as follows:

$$\bar{X}_g + t_{(\alpha/(2C);\, \nu)} S_{\bar{X}_g} \leq \mu_g \leq \bar{X}_g + t_{(1-\alpha/(2C);\, \nu)} S_{\bar{X}_g} ,$$

where $\nu = (N. - C)$ is the number of degrees of freedom of the within-groups sum of squares (section 12.4). The significance level is thus (α/C) for each interval but α for the whole set of C confidence intervals. The cumulative probabilities $(1 - \alpha/(2C))$ necessary for this method are often much closer to unity than values available in tables. The corresponding unusual values of Student's t were formerly obtained through approximations but may now be computed rapidly and accurately with specialized statistical software.

A similar procedure may be used when pairwise comparisons of several means have been planned a priori before examination of the data (see section 26.4). The significance level α then applies to each comparison if the confidence interval

$$(\bar{X}_j - \bar{X}_i) + t_{(\alpha/2;\, \nu)} S_X \sqrt{[(1/N_i) + (1/N_j)]} \leq (\mu_j - \mu_i) \leq (\bar{X}_j - \bar{X}_i) + t_{(1-\alpha/2;\, \nu)} S_X \sqrt{[(1/N_i) + (1/N_j)]}$$

is used, where S_X is the within-groups standard deviation. However, the level α can also be made to apply to a whole set of k comparisons if a level (α/k) is used for each one of them.

It must be emphasized that, while the adjustments discussed in this section have the advantage that they make the significance level (α) valid over a whole set of several confidence intervals or comparisons, they also have the drawback that they make each particular confidence interval or comparison less sensitive.

Section 26.3: contrasts: generalized comparisons

A more efficient manner to bring out the nature and the extent of differences between the means of several groups will now be discussed: the concept of a *contrast,* which constitutes a generalization of the difference between two means, found satisfactory in the case of two groups (section 26.1). Technically speaking, a *contrast* (also called a *"comparison"*) is an expression such as

$$Y = a_1\overline{X}_1 + a_2\overline{X}_2 + \; ... \; + a_g\overline{X}_g + \; ... \; + a_C\overline{X}_C = \sum_{g=1}^{C} a_g\overline{X}_g \; .$$

In the terminology of linear algebra, a contrast is thus a *linear combination* (section 24.22) of means obtained by taking the *inner product* (section 24.5) of the vector of means $\overline{\mathbf{X}} = [\overline{X}_1, \overline{X}_2, \ldots, \overline{X}_g, \ldots, \overline{X}_C]$ by a vector of coefficients $\mathbf{a} = [a_1, a_2, \ldots, a_g, \ldots, a_C]$ (or vice versa):

$$Y = \mathbf{a}\tilde{\overline{\mathbf{X}}} = \overline{\mathbf{X}}\tilde{\mathbf{a}} \, .$$

However, for a linear combination Y of means to be a contrast, the following condition must be fulfilled:

$$a_1 + a_2 + \; ... \; + a_g + \; ... \; + a_C = \sum_{g=1}^{C} a_g = 0 \, .$$

It may be noted that, when the vector of coefficients is $\mathbf{a} = [-1, +1, 0, \ldots, 0]$, the contrast is simply the difference $Y = (\overline{X}_2 - \overline{X}_1)$ between two means discussed in section 26.1.

Because of the properties of *expected values* (section 3.10), the parametric mean of a contrast Y of group means is equal to the contrast of the corresponding parametric means,

$$\mu_Y = a_1\mu_1 + a_2\mu_2 + \; ... \; + a_g\mu_g + \; ... \; + a_C\mu_C \, ,$$

and, if the preliminary hypothesis made in the analysis of variance is true,

$$H_0\colon \mu_1 = \mu_2 = \ldots = \mu_g = \ldots = \mu_C \, ,$$

the parametric variance of the contrast is

$$\sigma_Y^2 = a_1^2\sigma_{\overline{X}_1}^2 + \; ... \; + a_C^2\sigma_{\overline{X}_C}^2 = a_1^2\sigma_{X_1}^2/N_1 + \; ... \; + a_C^2\sigma_{X_C}^2/N_C = \sum_{g=1}^{C} a_g^2\sigma_{X_g}^2/N_g \, .$$

Moreover, if, as assumed in the analysis of variance (sections 12.6 and 12.7), the variances of the C populations are equal,

$$\sigma_{X_1}^2 = \; ... \; = \sigma_{X_g}^2 = \; ... \; = \sigma_{X_C}^2 = \sigma_X^2 = \sigma^2 ,$$

the parametric variance of the contrast may be expressed more simply:

$$\sigma_Y^2 = \sigma^2 \sum_{g=1}^{C} a_g^2/N_g .$$

Finally, if sample sizes also are equal, $N_1 = \ldots = N_g = \ldots = N_C = N$, the variance of the contrast may be expressed even more simply:

$$\sigma_Y^2 = \sigma^2 (1/N) \Sigma a_g^2 .$$

In practice, the variance of the contrast is estimated by replacing σ^2 by its within-groups estimate:

$$\hat{\sigma}_Y^2 = S_Y^2 = S^2_{\text{within-groups}} \sum_{g=1}^{C} a_g^2/N_g \quad \text{or} \quad \hat{\sigma}_Y^2 = S_Y^2 = S^2_{\text{within-groups}} (1/N) \sum_{g=1}^{C} a_g^2 .$$

As indicated in the following sections, the values of Y and S_Y^2 computed in an analysis of variance can easily be used to test hypotheses or to determine confidence intervals about the parametric means of contrasts.

Section 26.4: *a priori* and *a posteriori* contrasts

When, in an analysis of variance, the preliminary hypothesis $H_0: \mu_1 = \mu_2 = \mu_3 = \mu_4$ that means are equal is tested, the alternative hypothesis is that some means are unequal, but this includes many different possibilities, such as for instance $H_1: \mu_1 < \mu_2 < \mu_3 < \mu_4$, or $H_1: \mu_1 > \mu_2 > \mu_3 > \mu_4$, or $H_1: \mu_1 = \mu_2 < \mu_3 = \mu_4$, etc., etc. Contrasts may be used to find out which alternative hypothesis is most likely when the preliminary hypothesis has to be rejected.

Assume for instance that growth in body weight is studied until the age of four years in four (biological) populations of the same fish species inhabiting two lakes crossed by the St. Lawrence River, lake St. Louis (population 1) and lake St. Peter (population 2), and two other lakes situated further north in the Laurentian hills (populations 3 and 4). Even though, according to the preliminary hypothesis, mean body weight at four years of age should be equal in the four populations, it is suspected to be larger in populations 1 and 2 which are located further south and enjoy a longer growing season, a higher average temperature and more abundant food than populations 3 and 4. In this context, if the preliminary hypothesis $H_0: \mu_1 = \mu_2 = \mu_3 = \mu_4$ turns out to be unacceptable, the alternative hypothesis $H_1: \mu_1 = \mu_2 > \mu_3 = \mu_4$ corresponds to what is suspected and what one would like to detect. But, when the preceding alternative hypothesis is true, it may be detected most efficiently if $\mathbf{a} = [+1, +1, -1, -1]$ is taken as the vector of coefficients of the contrast $Y = a_1\overline{X}_1 + a_2\overline{X}_2 + a_3\overline{X}_3 + a_4\overline{X}_4$, since the latter's parametric mean, $\mu_Y = \mu_1 + \mu_2 - \mu_3 - \mu_4$, is null according to the preliminary hypothesis H_0 but positive according to the alternative hypothesis H_1. One indeed has $\mu_1 - \mu_3 = 0$ and $\mu_2 - \mu_4 = 0$, whence $\mu_Y = \mu_1 + \mu_2 - \mu_3 - \mu_4 = 0$, if the preliminary hypothesis is true but $\mu_1 - \mu_3 > 0$ and $\mu_2 - \mu_4 > 0$, whence $\mu_Y = \mu_1 + \mu_2 - \mu_3 - \mu_4 > 0$, if the alternative hypothesis is true.

Because the choice of the coefficients is based on earlier knowledge or suspicions, Y is said to be an *a priori* contrast and the preliminary hypothesis $H_0: \mu_Y = 0$ may be tested using Student's t distribution, the alternative hypothesis being $H_1: \mu_Y > 0$ and the acceptance region of H_0 being:

$$t_{(N.-C)} = (Y - \mu_Y)/S_Y = (Y - 0)/S_Y = Y/S_Y \leq t_{(1-\alpha;\, N.-C)} \ .$$

If the whole spectrum of likely values of the parametric mean μ_Y of the contrast is desired, a confidence interval (bilateral in this case, for instance) may be preferred:

$$Y + t_{(\alpha/2;\, N.-C)}\, S_Y \leq \mu_Y \leq Y + t_{(1-\alpha/2;\, N.-C)}\, S_Y \ .$$

In other contexts, however, there is no *a priori* basis for the choice of an alternative hypothesis or of the coefficients of a contrast. Nevertheless, inspection of the data may reveal unforeseen but marked differences which could be brought out using a so-called *a posteriori* contrast. Most current statistical methods are not appropriate to test *a posteriori* hypotheses because observations are usually assumed to be random and independent (section 9.6). But the independence assumption is violated when the decision to make a comparison is based merely on the fact that observations have been found to differ. Thus, among a group of 10 sample means, the smallest and the largest ones obviously differ more than two means which would have been picked completely at random.

In order to make valid comparisons of extreme values, to test hypotheses suggested by the inspection of data, or to draw conclusions concerning *a posteriori* contrasts, one must therefore use methods conceived specially for those purposes. Some of those methods call for tables of the range, that is of the difference between the smallest and the largest values within a group of observations (section 4.8). When such tables are adapted to small samples, they are said to concern the *studentized range*. In the case of observed means, range tables can only be used provided all samples have the same size. Otherwise, the intensity of dispersion would differ from one mean to another. The method designed by Scheffé (1959), an American statistician, is more general because it can be used to determine the confidence interval of an *a posteriori* contrast even when sample sizes are unequal:

$$Y - S_Y \sqrt{[(C-1)\, F_{(1-\alpha;\, C-1,\, \nu)}]} \leq \mu_Y \leq Y + S_Y \sqrt{[(C-1)\, F_{(1-\alpha;\, C-1,\, \nu)}]} \, ,$$

where $\nu = (N. - C)$ when the variance of the contrast is estimated using the within-groups sum of squares in a one-way analysis of variance. The reader may verify that the confidence interval yielded by Scheffé's method for the mean of an *a posteriori* contrast is broader than the confidence interval obtained earlier with Student's t for the mean of an *a priori* contrast. This greater breadth compensates for the fact that a comparison suggested by the inspection of data is not completely random.

Section 26.5: decomposing the between-group SS using orthogonal contrasts

Contrasts may sometimes be used in a particularly worthwile manner: several independent contrasts are chosen *a priori* in such a way that each contrast corresponds to one of the $(C-1)$ degrees of freedom of the between-groups sum of squares (SS) in a one-way analysis of variance. For this to work, the chosen contrasts must be mutually *orthogonal* and their covariances must be null. Let us then consider two contrasts:

$$Y_1 = a_{11}\bar{X}_1 + \ldots + a_{1C}\bar{X}_C \qquad \text{and} \qquad Y_2 = a_{21}\bar{X}_1 + \ldots + a_{2C}\bar{X}_C \ .$$

It may be shown that, if the variances of the C populations are equal as assumed in the analysis of variance, the covariance of the two contrasts is

$$\mathrm{covar}(Y_1, Y_2) = \sigma^2 \sum_{g=1}^{C} a_{1g}a_{2g} / N_g \ .$$

When sample sizes too are equal, the covariance of the contrasts may be expressed more simply as

$$\mathrm{covar}(Y_1, Y_2) = (\sigma^2/N) \sum_{g=1}^{C} a_{1g}a_{2g} = (\sigma^2/N)\, \mathbf{a}_1\tilde{\mathbf{a}}_2 \ .$$

For the two contrasts Y_1 and Y_2 to be orthogonal, their coefficients must therefore be such that their covariance vanishes in one of the above equations. The situation is simple enough when sample sizes are equal, since the contrasts are then orthogonal provided the vectors of coefficients $\mathbf{a}_1$ and $\mathbf{a}_2$ are themselves orthogonal (sections 24.5 and 24.6). Finding orthogonal contrasts is more difficult when sample sizes are unequal, because the latter are then involved in the relationships. More generally, in the analysis of variance, having equal sample sizes is very desirable because it prevents many technical difficulties.

When the preliminary hypothesis of equal means is true, the square of the observed value of a contrast, $(Y-\mu_Y)^2 = (Y-0)^2 = Y^2$, is an estimate of the variance of that contrast possessing one degree of freedom:

$$Y^2 = \hat{\sigma}_Y^2 = \hat{\sigma}^2 (1/N) \Sigma\, a_g^2 \qquad \text{or} \qquad Y^2 = \hat{\sigma}_Y^2 = \hat{\sigma}^2 \Sigma\, a_g^2/N_g \ .$$

Therefore, if the square of the observed value of a contrast, Y^2, is divided by the factor multiplying $\hat{\sigma}^2$ in the above expressions, an estimate of the variance σ^2 is obtained which possesses one degree of freedom and which is equal to the part of the between-groups sum of squares corresponding to that contrast:

$$\hat{\sigma}^2 = Y^2/[(1/N)\Sigma\, a_g^2] \qquad \text{or else} \qquad \hat{\sigma}^2 = Y^2/[\Sigma\, a_g^2/N_g] \ .$$

Carrying out the preceding operation for each of a series of $(C-1)$ orthogonal contrasts in a one-way analysis of variance is equivalent to breaking down the between-groups sum of squares into $(C-1)$ parts of which each one corresponds to one of the contrasts and may be used to test the preliminary hypothesis concerning it.

Section 26.6: a taxonomical and ecological example

An American entomologist (Brower, 1959) noted that the butterfly species *Papilio glaucus* and *Papilio rutulus* are so similar, particularly at the adult stage, that some authors hesitate to recognize them as distinct species. In the present analysis, the length of the right forewing in mm is compared in samples of 10 adult male specimens coming from four regions: (1) South Dakota, (2) Illinois, (3) Colorado, and (4) Utah. Groups (1) and (2) are believed to belong to species *Papilio glaucus* and groups (3) and (4) to species *Papilio rutulus*.

It seems natural to use the following three orthogonal contrasts:

$$\mathbf{a}_1 = [+1, +1, -1, -1],$$

$$\mathbf{a}_2 = [+1, -1, \ 0, \ 0],$$

and

$$\mathbf{a}_3 = [\ 0, \ 0, +1, -1],$$

since the first contrast should bring out the differences between the (biological) populations attributed to *Papilio glaucus* and those attributed to *Papilio rutulus,* the second contrast should bring out the difference between the two populations attributed to *Papilio glaucus,* and the third contrast should bring out the difference between the two populations attributed to *Papilio rutulus.*

The observed means are $[\overline{X}_1 \quad \overline{X}_2 \quad \overline{X}_3 \quad \overline{X}_4] = [44.4 \quad 48.7 \quad 46.9 \quad 46.6]$ mm, and the analysis of variance may be summarized in the following table.

Table 26.6.1
One-way type I analysis of variance with contrasts: right forewing length in mm of adult male butterfly specimens belonging to species *Papilio glaucus* and *P. rutulus* (Brower, 1959)

Variation	Sums of squares (SS)	Degrees of freedom	Variance estimate	F	P
Papilio glaucus, South Dakota	32.4	9	3.6		
Papilio glaucus, Illinois	52.1	9	5.78889		
Papilio rutulus, Colorado	50.9	9	5.65556		
Papilio rutulus, Utah	40.4	9	4.48889		
within-groups	175.8	36	4.88333		
between-groups	93.3	3	31.1	6.3686	$0.001 < P < 0.01$
contrast 1	0.4	1	0.4	0.08191	$P > 0.05$
contrast 2	92.45	1	92.45	18.9317	$P < 0.001$
contrast 3	0.45	1	0.45	0.09215	$P > 0.05$
total	269.1	39	6.9		

Surprisingly, the (biological) populations attributed to *Papilio glaucus* do not differ conclusively from those attributed to *Papilio rutulus* (contrast No. 1). The only contrast reaching statistical significance is the second one, which opposes the *Papilio glaucus* from South Dakota to those from Illinois. Finally, the *Papilio rutulus* from Colorado and Utah do not differ significantly (contrast No. 3).

The *Papilio glaucus* from South Dakota perhaps have short wings because the *"badlands"* of that state may be less favorable to the growth of caterpillars than the good agricultural lands of Illinois. But the lack of statistically significant differences in wing length between *Papilio glaucus* and *Papilio rutulus* does not necessarily mean that the two species are not valid: wing length is possibly affected by environmental conditions and may not be a very good taxonomical character. The study should include other characters as well as a comparison of larval stages, which may be less similar than adults.

From a methodological point of view, the decomposition of the between-groups sum of squares using orthogonal contrasts appears to be a good way of extracting more information from data in an analysis of variance. In the case of the present example, this procedure shows that differences between the means of the four butterfly populations are not structured as was expected.

Chapter 27

One-way type II analysis of variance with variance components

Section 27.1: introduction

In type I analysis of variance, discussed until now (chapters 12 and 26), the statistical populations from which samples are drawn are said to be *fixed,* because they are chosen in a systematic manner. Consequently, the means of those populations and the differences between them are of major interest. For instance, if an entomologist wishes to compare wing length in several butterfly species, such as *Papilio glaucus, P. eurymedon, P. multicaudatus, P. rutulus,* etc., he may get a sample of specimens from each population in order to estimate the wing length of each species, and the same populations would be sampled again if another study was done later. Similarly, a research worker wishing to check the effectiveness of several experimental or medical treatments would obtain a sample of individuals submitted to each treatment in addition to a sample of *control* individuals not submitted to any real treatment (except possibly a *placebo*) and used as a basis for comparison. If another research worker wishes to confirm earlier results in a later study, he will get new samples submitted to the same experimental treatments and representing the same statistical populations. This first kind of situation, in which the populations under study are chosen in a systematic (fixed) manner, is represented in figure 27.1.1.

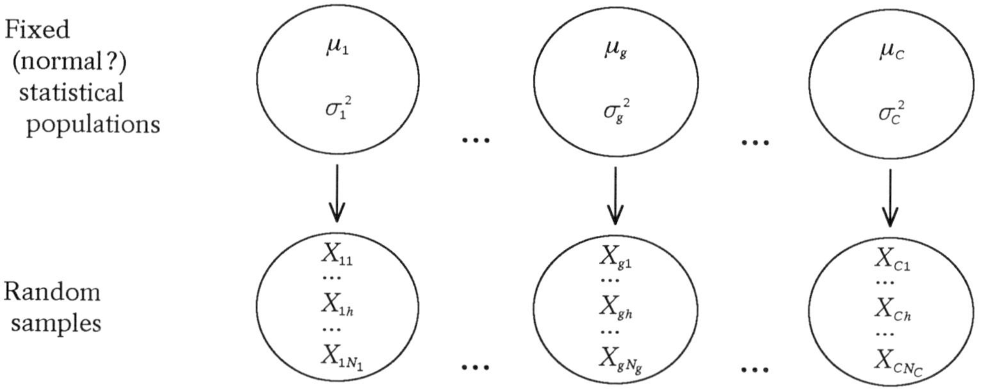

Figure 27.1.1
Putative genesis of observations in a type I analysis of variance

Type I analysis of variance may also be described by a *model equation:*

$$X_{gh} = \mu_g + e_{gh}, \quad \text{where} \quad e_{gh} \leftarrow \mathcal{N}(0, \sigma^2),$$

or, equivalently,

$$X_{gh} = \bar{\mu} + (\mu_g - \bar{\mu}) + e_{gh} = \bar{\mu} + \delta_g + e_{gh},$$

if $\bar{\mu} = (1/C) \sum_{g=1}^{C} \mu_g$ is the mean of population means and $\delta_g = (\mu_g - \bar{\mu})$ a deviation from it.

The major characteristic of type I analysis of variance is that the population means μ_g (or the differences δ_g) have fixed values which depend on the populations chosen for study: the μ_g's and the δ_g's are thus pure parameters.

In type II analysis of variance, on the contrary,the populations and their means μ_g are not chosen systematically: they are themselves drawn at random from what could be called a *"superpopulation"* of populations. This second kind of situation is outlined in figure 27.1.2.

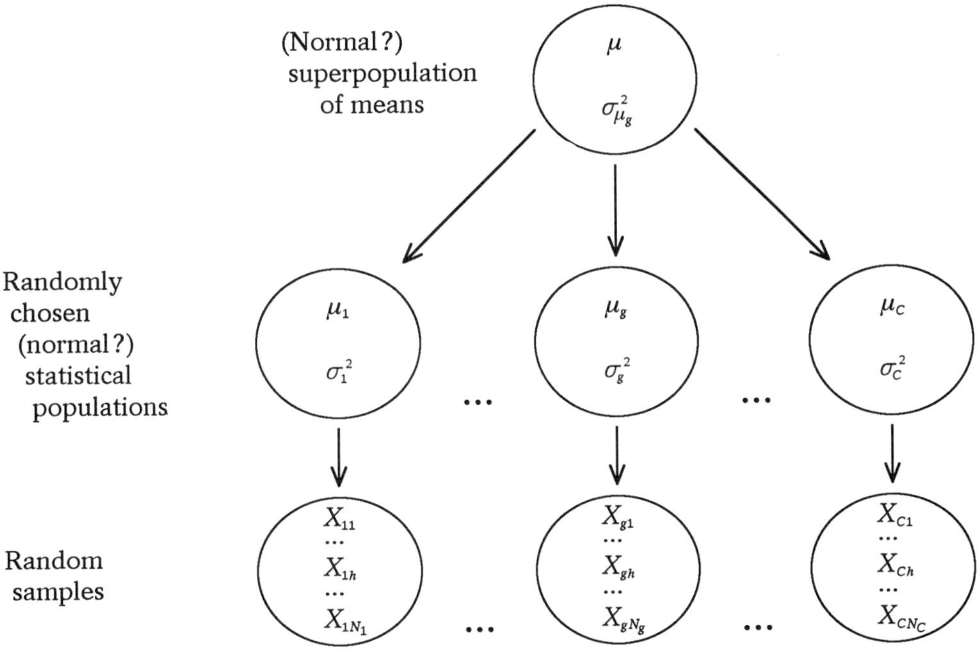

Figure 27.1.2
Putative genesis of observations in a type II analysis of variance

Type II analysis of variance may also be represented algebraically by the following model equation:

$$X_{gh} = \mu_g + e_{gh} ,$$

$$\text{where } \mu_g \leftarrow \mathcal{N}(\mu, \sigma^2_{\mu_g}) \text{ and } e_{gh} \leftarrow \mathcal{N}(0, \sigma^2) ,$$

that is

$$X_{gh} = \mu + (\mu_g - \mu) + e_{gh} = \mu + d_g + e_{gh} ,$$

$$\text{where } d_g \leftarrow \mathcal{N}(0, \sigma_d^2) \text{ and } e_{gh} \leftarrow \mathcal{N}(0, \sigma^2) ,$$

the variance of the deviation $d_g = (\mu_g - \mu)$ being $\sigma_d^2 = \sigma^2_{\mu_g}$. The deviations d_g and e_{gh} are statistically independent. It is preferable to denote the deviation $(\mu_g - \mu)$ by d_g rather than δ_g here in order to emphasize that, within the context of a type II analysis of variance, the means μ_g and the deviations $d_g = (\mu_g - \mu)$ are no longer pure parameters. While the mean μ_g of each population is a parameter with respect to the individual observations X_{gh} drawn from that population, it is indeed also a random variable with respect to the superpopulation from which it is randomly drawn. As for the variances σ^2 and σ_d^2, they are called *variance components*.

The reader should note that both types (I and II) of analysis of variance include the statement

$$e_{gh} \leftarrow \mathcal{N}(0, \sigma^2) \text{ rather than } e_{gh} \leftarrow \mathcal{N}(0, \sigma_g^2),$$

which implies that population variances are assumed to be equal (sections 12.6 and 12.7):

$$\sigma_1^2 = \dots = \sigma_g^2 = \dots = \sigma_C^2 = \sigma^2 .$$

The distinguishing characteristic of type II in comparison with type I of the analysis of variance is that the populations are themselves picked at random. This has practical consequences and sets clearly apart the cases where type II anova is more suitable than type I: in type II analysis of variance, it is neither possible nor even desirable to sample the very same populations again in subsequent studies or experiments. Therefore, the means μ_g and the differences $d_g = (\mu_g - \mu)$ have little interest in themselves: it is their variance $\sigma_d^2 = \sigma_{\mu_g}^2$ which is most important.

Section 27.2: some biological examples of type II analysis of variance

Our first example is based on data extracted from a study of Dutch families by Frets (1921). The head lengths in mm of the first three adult sons (adults, to make sure that growth was finished) in eighteen families are included in this analysis. If adult head length is unrelated to the order of birth (an assumption which could be checked independently), this is an instance where a type II analysis of variance is suitable. The variance of the head length of the sons belonging to the same family about the mean μ_g of their family is the component σ^2 while the variance of family means about the general mean μ of the superpopulation is the component σ_d^2. The within-families variance σ^2 must be partly hereditary for, apart from monozygotic twins, brothers do not receive identical halves of the paternal and maternal sets of genes (in part because of the recombination phenomenon). However, the within-families variance σ^2 may also have an environmental part, for children from a same family may be submitted to slightly different environmental conditions. As for the between-families component σ_d^2, it may reflect environmental differences between families but it should also have an important genetical part (except perhaps in highly inbred populations).

A second example may be found in the work of Brown and Beerstecher (1951), who determined the calcium concentration in μg/ml of the morning urine of 10 underweight men. If the order in which urine specimens were obtained is unrelated to calcium concentration, this is also a situation where a type II analysis of variance is appropriate. The within-men component σ^2 corresponds to the differences between urine specimens provided by the same person, and may reflect daily metabolic fluctuations, while the between-men component σ_d^2 reflects the fact that some persons excrete more calcium than others in their urine. The analysis of this data will be slightly more involved because Brown and Beerstecher (1951) failed to obtain three urine specimens from every individual: two persons provided only two urine specimens each, and one person provided only one specimen.

Type II analysis of variance may also be useful in the analysis of some ecological data. If the abundance of a marine or freshwater invertebrate species is studied, for instance, a series of stations may be chosen which are not expected to differ systematically because the environment is known or believed to be homogeneous, and a few quadrats are selected at random at each station. Since stations are expected to differ at random (because of the presumed homogeneity of the environment), type II analysis of variance is applicable. The component σ^2 measures abundance differences between the quadrats located at the same station, while the component σ_d^2 reflects abundance variations between stations.

Section 27.3: expected values of within-groups and between-groups variances

The first steps of a one-way type II analysis of variance are similar to those of a type I anova (chapter 12), except that the expected value of the between-groups variance estimate is different (table 27.3.1; see also sections 12.4 and 12.5).

Table 27.3.1
Expected values of within-groups and between-groups variances in one-way type I and type II analyses of variance

Type of analysis	$S^2_{\text{within-groups}}$	$S^2_{\text{between-groups}}$
I	σ^2	$\sigma^2 + [1/(C-1)]\{\sum_{g=1}^{C} N_g\delta_g^2 - (\sum_{g=1}^{C} N_g\delta_g)^2/N.\}$
II	σ^2	$\sigma^2 + [1/(C-1)]\{\sum_{g=1}^{C} N_g - (\sum_{g=1}^{C} N_g^2)/N.\}\,\sigma_d^2$

When sample sizes are unequal in a one-way type II analysis of variance, however, the expected value of the between-groups variance may be expressed more simply if a kind of a weighted mean sample size N_0 is defined:

$$N_0 = [1/(C-1)]\{\sum_{g=1}^{C} N_g - (\sum_{g=1}^{C} N_g^2)/N.\} .$$

It follows that
$$\mathscr{E}(S^2_{\text{between-groups}}) = \sigma^2 + N_0\,\sigma_d^2 .$$

In the particular case where all samples sizes are equal to a common value N, the weighted average N_0 is simply equal to that common value.

Section 27.4: preliminary (null) hypothesis and alternative hypothesis

Unlike the preliminary hypothesis of type I analysis of variance, in which the differences $\delta_g = (\mu_g - \bar{\mu})$ between means were asserted to be null (sections 12.2, 12.4 and 12.6), the preliminary hypothesis of type II analysis of variance consists in the statement that the variance component σ_d^2 is null:

$$H_0:\ \sigma_d^2 = 0\ ,\ H_1:\ \sigma_d^2 > 0\ .$$

The acceptance regions of the preliminary hypotheses are nevertheless similar in both types of analyses:

$$S^2_{\text{between-groups}} / S^2_{\text{within-groups}} \leq F_{(1-\alpha;\, C-1,\, N.-C)}\ .$$

Section 27.5: point estimates of variance components σ^2 and σ_d^2

The point estimation of component σ^2 is easy enough. Since $\mathscr{E}(S^2_{\text{within-groups}}) = \sigma^2$, the unbiased estimator $\hat{\sigma}^2 = S^2_{\text{within-groups}}$ is used. The estimation of component σ_d^2 requires more care. Since $\mathscr{E}(S^2_{\text{between-groups}}) = \sigma^2 + N_0\,\sigma_d^2$, it is possible to write

$$S^2_{\text{between-groups}} = \hat{\sigma}^2 + N_0\,\hat{\sigma}_d^2 = S^2_{\text{within-groups}} + N_0\,\hat{\sigma}_d^2\ ,$$

whence $$\hat{\sigma}_d^2 = (S^2_{\text{between-groups}} - S^2_{\text{within-groups}})/N_0\ .$$

However, the numerator of $\hat{\sigma}_d^2$ is a difference between two independent random variance estimators, and this observed difference may occasionally be negative even though the parametric variance σ_d^2 cannot be negative. When a negative point estimate $\hat{\sigma}_d^2 < 0$ is obtained, the preliminary hypothesis H_0, according to which $\sigma_d^2 = 0$, is therefore accepted.

Section 27.6: breakdown of the individual variance with respect to σ^2 and σ_d^2

If an individual observation X_{gh} is picked at random in a statistical population which is itself drawn at random from a superpopulation of which the mean is μ (section 27.1), how is the variance σ_X^2 of observations such as X_{gh} related to the components σ^2 and σ_d^2 ? Since the model equation is

$$X_{gh} = \mu + d_g + e_{gh}\ ,$$

$$(X_{gh} - \mu) = d_g + e_{gh}$$

whence $$(X_{gh} - \mu)^2 = d_g^2 + 2\,d_g\,e_{gh} + e_{gh}^2$$

and $\mathscr{E}[(X_{gh} - \mu)^2] = \sigma_X^2 = \mathscr{E}(d_g^2) + \mathscr{E}(2\,d_g\,e_{gh}) + \mathscr{E}(e_{gh}^2) = \sigma_d^2 + 0 + \sigma^2 = \sigma_d^2 + \sigma^2$.

The variance σ_X^2 of individual observations is thus equal to the sum of the two variance components σ^2 and σ_d^2 .

Section 27.7: relative importance of σ_d^2 and intraclass correlation

While this seems today to be mostly a historical curiosity, the problems now solved using type II analysis of variance were approached at the beginning of the twentieth century with the so-called *intraclass correlation coefficient.* Unlike the ordinary correlation coefficient (chapter 19), also called *interclass correlation,* the intraclass correlation coefficient is denoted by the symbol r_i (the subscript "i" being the initial letter of the adjective intraclass) and was formerly obtained by pairing all observations in all possible manners within a sample. Nowadays, it is better to follow Fisher (1970, pages 223-227) and define the population intraclass correlation coefficient as

$$\rho_i = \sigma_d^2/(\sigma^2 + \sigma_d^2) = 100\,\sigma_d^2/(\sigma^2 + \sigma_d^2)\,\% .$$

This ratio varies from 0 to 1 (or from 0% to 100%) and measures the relative importance of the component σ_d^2 in the individual variance $\sigma_x^2 = \sigma^2 + \sigma_d^2$. As for the point estimate of the parameter ρ_i, the following expression is now used:

$$r_i = \hat{\sigma}_d^2/(\hat{\sigma}^2 + \hat{\sigma}_d^2) = (S^2_{\text{between-groups}} - S^2_{\text{within-groups}})/[S^2_{\text{between-groups}} + (N_0 - 1)S^2_{\text{within-groups}}] .$$

Section 27.8: confidence interval of the ratio $\rho_i = \sigma_d^2/(\sigma^2 + \sigma_d^2)$

The bilateral confidence interval of a variance ratio (section 11.4),

$$(S_1^2/S_2^2)/F_{(1-\alpha/2;\ \nu_1,\ \nu_2)} \le (\sigma_1^2/\sigma_2^2) \le (S_1^2/S_2^2)/F_{(\alpha/2;\ \nu_1,\ \nu_2)} ,$$

may be used for the ratio $(\sigma^2 + N_0\,\sigma_d^2)/\sigma^2$, whence

$$(S^2_{\text{between}}/S^2_{\text{within}})/F_{(1-\alpha/2;\ C-1,\ N.-C)} \le (\sigma^2 + N_0\,\sigma_d^2)/\sigma^2 \le (S^2_{\text{between}}/S^2_{\text{within}})/F_{(\alpha/2;\ C-1,\ N.-C)} .$$

Subtracting the number 1 from the three members of the above inequality and dividing them next by N_0 yields

$$(1/N_0)\{[S^2_{\text{between}} - S^2_{\text{within}}\,F_{(1-\alpha/2;\ C-1,\ N.-C)}]/S^2_{\text{within}}\,F_{(1-\alpha/2;\ C-1,\ N.-C)}\} \le (\sigma_d^2/\sigma^2) \le \ldots$$

$$\ldots (1/N_0)\{[S^2_{\text{between}} - S^2_{\text{within}}\,F_{(\alpha/2;\ C-1,\ N.-C)}]/S^2_{\text{within}}\,F_{(\alpha/2;\ C-1,\ N.-C)}\} .$$

If the three terms of this new inequality are now inverted, rearranged in order of increasing magnitude, added to the number 1 and reinverted, the confidence interval of ratio ρ_i is finally obtained:

$$\rho_{i_1} \le \sigma_d^2/(\sigma^2 + \sigma_d^2) \le \rho_{i_2} ,$$

where

$$\rho_{i_1} = \{[S^2_{\text{between}} - S^2_{\text{within}}\,F_{(1-\alpha/2;\ C-1,\ N.-C)}]/[S^2_{\text{between}} + (N_0 - 1)S^2_{\text{within}}\,F_{(1-\alpha/2;\ C-1,\ N.-C)}]\}$$

and

$$\rho_{i_2} = \{[S^2_{\text{between}} - S^2_{\text{within}}\,F_{(\alpha/2;\ C-1,\ N.-C)}]/[S^2_{\text{between}} + (N_0 - 1)S^2_{\text{within}}\,F_{(\alpha/2;\ C-1,\ N.-C)}]\} .$$

Section 27.9: variance of the general mean estimate $\overline{X}.$

The estimate $\overline{X}.$ of the general mean μ summarizes all information about the latter contained by the data. In order to know how accurate is the information provided by $\overline{X}.$, it is therefore desirable to know the variance of $\overline{X}.$. According to the model equation of type II analysis of variance, $X_{gh} = \mu + d_g + e_{gh}$,

$$\overline{X}. = (\Sigma X)./N. = (1/N.) \sum_{g=1}^{C} \sum_{h=1}^{N_g} (\mu + d_g + e_{gh}) = \mu + (1/N.) \sum_{f=1}^{C} N_f\, d_f + (1/N.) \sum_{g=1}^{C} \sum_{h=1}^{N_g} e_{gh}$$

whence $(\overline{X}. - \mu)^2 = (1/N.^2)[(\Sigma N_f\, d_f)^2 + 2(\Sigma N_f\, d_f)(\Sigma\Sigma\, e_{gh}) + (\Sigma\Sigma\, e_{gh})^2]$.

When $\mathscr{E}[(\overline{X}. - \mu)^2]$ is evaluated, all product terms, such as $d_f d_g$ when $f \neq g$, $e_{gh} e_{gi}$ when $h \neq i$, and $d_f e_{gh}$ have null expectations because of the independence of deviations d_f, d_g, e_{gh} and e_{gi}. Following the ensuing simplifications,

$$\mathscr{E}[(\overline{X}. - \mu)^2] = \sigma^2_{\overline{X}.} = (1/N.^2)[\sum_{g=1}^{C} N_g^2\, \sigma_d^2 + N.\, \sigma^2].$$

When the sample sizes are all equal to a common value N, the total sample size $N. = CN$ and the variance of the general mean takes an even simpler form,

$$\mathscr{E}[(\overline{X}. - \mu)^2] = \sigma^2_{\overline{X}.} = (\sigma^2 + N\sigma_d^2)/(CN).$$

Section 27.10: optimal resource allocation (numbers of sampling units)

The simplest way of knowing the general mean μ with greater accuracy would obviously be to increase the number C of populations as well as the size N of samples, but this could increase the total cost of a study above any preassigned limit. It will be shown in this section that, by taking into account the relative magnitude of the two variance components, σ^2 and σ_d^2, it is possible to chose sample size N in such a way as to minimize the variance of $\overline{X}.$ without exceeding the available budget. However, let us first introduce a new notation (table 27.10.1) inspired by sampling theory (Snedecor and Cochran, 1967, pages 531-534).

Table 27.10.1
Equivalent notations in one-way type II analysis of variance

nature of variation	between-groups	within-groups
sample sizes (numbers of sampling units)	$C \rightarrow n_1$ (primary units)	$N \rightarrow n_2$ (secondary units)
variance components	$\sigma_d^2 \rightarrow \sigma_1^2$	$\sigma^2 \rightarrow \sigma_2^2$
unit cost	c_1	c_2
parametric variances	$\sigma^2 + N\sigma_d^2 \rightarrow \sigma_2^2 + n_2\,\sigma_1^2$	$\sigma^2 \rightarrow \sigma_2^2$
physiological example	$C \rightarrow n_1$ individuals	$N \rightarrow n_2$ specimens per individual
ecological example	$C \rightarrow n_1$ stations	$N \rightarrow n_2$ quadrats per station

The populations are considered as primary units and the samples as secondary units because, in human physiology, the participation of an individual must be arranged before urine specimens can be obtained. Similarly, in ecology, the location of a sampling station must be determined before that of quadrats. The costs c_1 and c_2 of primary and secondary units may consist either in monetary values or in working hours (or in a combination of both). This new notation (n_1 primary units, n_2 secondary units, variance components σ_1^2 and σ_2^2, unit costs c_1 and c_2) has the advantage that it may be more easily generalized when random sampling is carried out at more than two levels, such as in the so-called *nested anova.* Some important quantities are expressed using both the former and the new notations in table 27.10.2.

Table 27.10.2
Important quantities concerning resource allocation

	(C, N) notation	(n_1, n_2) notation
total number of secondary units	CN	$n_1 n_2$
variance of general mean	$\sigma_d^2/C + \sigma^2/(CN)$	$\sigma_1^2/n_1 + \sigma_2^2/(n_1 n_2)$
total cost of study	$Cc_1 + CNc_2$	$n_1 c_1 + n_1 n_2 c_2$

In practice, it is desirable to minimize both the variance of the estimate $\overline{X}.$ of the general mean and the total cost of the study. This may be done by minimizing the product of both quantities:

$$\text{variance} \times \text{cost} = [\sigma_1^2/n_1 + \sigma_2^2/(n_1 n_2)] \times [n_1 c_1 + n_1 n_2 c_2] ,$$

that is

$$\text{variance} \times \text{cost} = c_1\sigma_1^2 + c_1\sigma_2^2/n_2 + c_2 n_2 \sigma_1^2 + c_2\sigma_2^2 .$$

It should be noted that the number n_1 of primary units has vanished from this product, which will allow only the number of secondary units n_2 to be estimated. However, the number of primary units will be estimated afterward by taking into account the total budget available. Setting the derivative of the product (variance × cost) with respect to n_2 equal to zero yields:

$$n_2 = \sqrt{[(c_1\sigma_2^2)/(c_2\,\sigma_1^2)]} .$$

The same result may be obtained also by conditional derivation using an undetermined multiplier of Lagrange. Since the number of secondary units must necessarily be an integer, its estimate $\hat{n}_2$ is the integer which is closest to the value of $\sqrt{[(c_1\,\hat{\sigma}_2^2)/(c_2\,\hat{\sigma}_1^2)]}$. As for the number n_1 of primary units, it is determined so as to use up the available budget as completely as possible without exceeding it:

$$\hat{n}_1 = \text{the integer part of the ratio } [(\text{available budget})/(c_1 + \hat{n}_2 c_2)] .$$

It must be noted that the present approach requires variance components and unit costs to have been estimated earlier in a pilot study.

Section 27.11: an example from anthropometry

Type II analysis of variance may be applied to the data of Frets (1921) mentioned in section 27.2 in order to decompose the total variance of the head length of adult sons of Dutch families (table 27.11.1).

Table 27.11.1
Type II analysis of variance of the head length in mm of adult sons in Dutch families according to data from Frets (1921)

Variation	Sums of squares	Degrees of freedom	Observed variance	Expected variance
within-families	898.667	36	24.963	σ^2
between-families	1583.48	17	93.146	$\sigma^2+3\,\sigma_d^2$
total	2482.15	53	46.833	–

The preliminary hypothesis H_0 according to which the between-families variance component σ_d^2 would be null can be rejected very conclusively for

$$F_{(0.999;\,17,\,36)} < S^2_{\text{between-families}} / S^2_{\text{within-families}} = 3.73137 .$$

The point estimate of the between-families variance component is

$$\hat{\sigma}_d^2 = (93.146 - 24.963)/3 = 22.7277 ,$$

a value very close to the estimate $\hat{\sigma}^2 = 24.963$ of the within-families variance component. The estimated part of the total variance due to the between-families component, also known as the intraclass correlation, is

$$22.7277/(24.963 + 22.7277) = 0.476565 = 47.6565\,\% ,$$

and the confidence interval of the latter at significance level $\alpha = 0.01$ is

$$0.103709 \leq \sigma_d^2/(\sigma^2+3\,\sigma_d^2) \leq 0.792837,$$

that is from 10.3709 % to 79.2837 %.

Section 27.12: an example from human physiology

In the case of the data of Brown and Beerstecher (1951) on the calcium concentration of human urine (section 27.2), besides allowing the relative importance of the between-men variance component σ_d^2 to be estimated, type II analysis of variance could be used to optimize resource allocation in a later study. The analysis is summarized in table 27.12.1.

Table 27.12.1
Type II analysis of variance of the calcium concentration of human urine according to data from Brown and Beerstecher (1951)

Variation	Sums of squares	Degrees of freedom	Observed variance	Expected variance
within-men	72214.3	16	4513.4	σ^2
between-men	127265	9	14140.5	$\sigma^2+N_0\,\sigma_d^2$
total	199479	25	7979.16	–

The preliminary hypothesis $H_0 : \sigma_d^2 = 0$ can be rejected but less conclusively than in the preceding example, for

$$F_{(0.95;\,9,\,16)} < S^2_{\text{between-men}} / S^2_{\text{within-men}} = 3.13301 < F_{(0.99;\,9,\,16)} \,.$$

The coefficient N_0, determined from the unequal sample sizes (sections 27.2 and 27.3) is equal to 2.5812 and the point estimate of the between-men variance component σ_d^2 is

$$\hat{\sigma}_d^2 = (14140.5 - 4513.4) / 2.5812 = 3729.72 \,,$$

a numerical value which, in this second example too, is not very far from the within-men variance component $\hat{\sigma}^2 = 4513.4$. The estimate of the relative importance of the between-men variance components is

$$3729.72 / (4513.4 + 3729.72) = 0.452465 = 45.2465\,\% \,.$$

Since the preliminary hypothesis $H_0 : \sigma_d^2 = 0$ can be rejected at the $\alpha = 0.05$ level but not at $\alpha = 0.01$, the level $\alpha = 0.05$ is used for the confidence interval of the intraclass correlation coefficient:

$$0.0105939 \leq \rho_i \leq 0.806093 \,.$$

Assume finally that a new and larger study is planned on the same question and that a total budget of 2600.00 monetary units will be available. Each individual is offered 100.00 monetary units to participate in the study whatever its duration ($c_1 = 100$) and each urine analysis costs 10.00 monetary units ($c_2 = 10$). On the basis of the estimates obtained for variance components in the preceding analysis, what would be the optimal resource allocation for the new study?

The estimated number of secondary units (urine analyses per individual), $\hat{n}_2$, is the integer closest to

$$\sqrt{[(100 \times 4513.4) / (10 \times 3729.72)]} = 3.4787$$

that is $\hat{n}_2 = 3$.

The estimated number of primary units (individuals) is

$$\hat{n}_1 = \text{the integer part of the ratio } [2600 / (100 + 3 \times 10)] = 20 \,.$$

If 20 individuals participate in the study and if 3 urine samples are obtained and analyzed for each individual, it should thus be possible to estimate the average calcium concentration of human urine with maximum accuracy without exceeding the budget available for that research. It must be noted that the numbers of primary and secondary units $\hat{n}_1 = 20$ and $\hat{n}_2 = 3$ are estimated merely on the basis of a small-scale pilot study and should be revised once more data are available.

Chapter 28
Two-way type I analysis of variance with interaction

Section 28.1: introduction

The topic of this chapter will once again be type I analysis of variance, in which the populations under comparison are chosen systematically, and where the population means and the differences between them are of major interest. In the case considered here, however, observations can be cross-classified with respect to two classification criteria instead of only one criterion as in the one-way analyses of variance treated in chapters 12, 26 and 27. In order to make the situation as concrete as possible, biological examples will now be given without further delay.

In his detailed study of butterflies of the *Papilio glaucus* group, Brower (1959) measured the length of the eighth tergite of male specimens belonging to the two species *Papilio eurymedon* and *Papilio rutulus* and coming from two regions: the areas around Boulder, Colorado, and Salt Lake City, Utah. The measurements are expressed in mm but are eight times larger than reality because they were taken on photographs made on a microscope with an 8x magnification factor. Unlike the data considered in chapters 12 and 26, the measurements of Brower (1959) discussed here can be cross-classified in two ways: with respect to taxonomy as well as geographical origin (table 28.1.1). Each sample thus occupies a cell located at the intersection of one of the rows and one of the columns of a two-way table, rather than just one of the columns of a one-way table as in one-way analyses of variance (chapters 12 and 26).

Table 28.1.1
Two-way classification of the data of Brower (1959) on male butterflies of the species *Papilio eurymedon* and *Papilio rutulus* from Colorado and Utah

		Geographical origin	
		Colorado	Utah
Species	*Papilio eurymedon*	10 specimens	10 specimens
	Papilio rutulus	10 specimens	10 specimens

A second example may be found in the results of an agricultural experiment (Millikan, 1963) on the effects of phosphorus and zinc on the growth in weight of clover (*Trifolium subterraneum* L.). The data were kindly provided by Dr. C. R. Millikan, then Chief Biologist of the Department of Agriculture of the State of Victoria, in Australia. The results included in the present analysis were obtained on the *Edenhope* variety at three levels of phosphorus ($P_1 = 3$ p.p.m., $P_2 = 7.5$ p.p.m., and $P_3 = 15$ p.p.m.) and zinc ($Zn_1 = 0.005$ p.p.m., $Zn_2 = 0.05$ p.p.m., and $Zn_3 = 0.5$ p.p.m.) concentrations. Millikan obtained six measurements (also known technically as *"replicates"*) at each combination of phosphorus and zinc concentrations, and each measurement is the sum of the dry weights of the roots and upper parts of two plants 66 days after the beginning of the experiment (table 28.1.2).

Table 28.1.2
Two-way classification of the data of Millikan (1963) on the growth of clover *Trifolium subterraneum* L. following fertilization with phosphorus and zinc

		Zinc concentration		
		$Zn_1 = 0.005$ p.p.m.	$Zn_2 = 0.05$ p.p.m.	$Zn_3 = 0.5$ p.p.m.
Phosphorus concentration	$P_1 = 3$ p.p.m.	6 weights	6 weights	6 weights
	$P_2 = 7.5$ p.p.m.	6 weights	6 weights	6 weights
	$P_3 = 15$ p.p.m.	6 weights	6 weights	6 weights

The data of Millikan (1963) can thus also be cross-classified with respect to two criteria, phosphorus and zinc, but the nine samples contain six weight measurements each and occupy a total number of $3 \times 3 = 9$ cells located at the intersections of 3 rows and 3 columns.

Table 28.1.3
Two-way classification of the data obtained in the simultaneous trial of two medical treatments

		Treatment "B"	
		not administered	administered
Treatment "A"	not administered	N patients	N patients
	administered	N patients	N patients

Two-way analysis of variance may also be useful in medicine, when the simultaneous effects of two treatments are evaluated (table 28.1.3), and in ecology, when an attempt is made to discover the combinations of environmental conditions which are most favorable to individual or population growth (table 28.1.4).

Table 28.1.4
Two-way classification of data on individual or population growth according to two ecological conditions

		Humidity, precipitations or salinity		
		low	average	high
Environmental temperature	0°C	N data	N data	N data
	5°C	N data	N data	N data
	10°C	N data	N data	N data

More generally, in a two-way analysis of variance, the number of classes of the first and second classification criteria may be denoted by the letters R (the first letter of the word *row*) and C (the first letter of the word *column*) respectively, and the procedure may be referred to as a $(R \times C)$ anova. Thus the analysis of the data of table 28.1.1 is a (2×2) anova while that of table 28.1.2 is a (3×3) anova. However, an additional column and an additional row are usually placed at the right and at the bottom of the table of data (table 28.1.5). The extra column is used to record the sums of the quantities appearing on each row, while the extra row is used to record the sums of the quantities appearing in each column.

Table 28.1.5
Notation used for raw sums in a two-way analysis of variance

		Criterion 2		Raw sums within each row
		column 1	column 2	
Criterion 1	row 1	$N_{11}, (\Sigma X)_{11}, (\Sigma X^2)_{11}$	$N_{12}, (\Sigma X)_{12}, (\Sigma X^2)_{12}$	$N_{1\cdot}, (\Sigma X)_{1\cdot}, (\Sigma X^2)_{1\cdot}$
	row 2	$N_{21}, (\Sigma X)_{21}, (\Sigma X^2)_{21}$	$N_{22}, (\Sigma X)_{22}, (\Sigma X^2)_{22}$	$N_{2\cdot}, (\Sigma X)_{2\cdot}, (\Sigma X^2)_{2\cdot}$
Raw sums within each column		$N_{\cdot 1}, (\Sigma X)_{\cdot 1}, (\Sigma X^2)_{\cdot 1}$	$N_{\cdot 2}, (\Sigma X)_{\cdot 2}, (\Sigma X^2)_{\cdot 2}$	$N_{\cdot\cdot}, (\Sigma X)_{\cdot\cdot}, (\Sigma X^2)_{\cdot\cdot}$

As in one-way analysis of variance (section 12.2 of chapter 12), the quantities pertaining to each group are denoted by a subscripted variable, and a subscript with respect to which summation has been carried out is replaced by a dot (period). In two-way analysis of variance, however, the first (row) subscript f and the second (column) subscript g are followed by a third subscript h indicating the order number of each measurement (replicate) within the sample contained in a cell. Thus, the sum of squares of the measurements of the sample contained in the cell located at the intersection of the second row ($f=2$) and the first column ($g=1$) is

$$(\Sigma X^2)_{21} = \sum_{h=1}^{N_{21}} X_{21h}^2 \ ,$$

while the sum of the measurements contained within the second row is

$$(\Sigma X)_{2.} = \sum_{g=1}^{C} (\Sigma X)_{2g} = \sum_{g=1}^{C} \sum_{h=1}^{N_{2g}} X_{2gh} \ ,$$

the general (total) sample size is

$$N_{..} = \sum_{f=1}^{R} N_{f.} = \sum_{g=1}^{C} N_{.g} = \sum_{f=1}^{R} \sum_{g=1}^{C} N_{fg} \ ,$$

and the sum of all data is

$$(\Sigma X)_{..} = \sum_{f=1}^{R} \sum_{g=1}^{C} \sum_{h=1}^{N_{fg}} X_{fgh} \ .$$

The above system of subscripts is used also for means, centered sums and variances. The case where several observations (replicates) have been obtained in each cell and where sample sizes are equal in all cells ($N_{11} = \ldots = N_{fg} = \ldots = N_{RC} = N$) is the only one discussed in this book because it is statistically most tractable.

Section 28.2: parameters and model equations

For the sake of simplicity, the various possible systems of parameters and of model equations will be discussed first in the case of the (2 × 2) analysis of variance, where each classification criterion has only two classes (tables 28.1.1 and 28.1.3). The concepts will then be easily extended to the more general case of a $(R \times C)$ anova. The simplest set of parameters is obtained by postulating a mean for each cell of the data table (table 28.2.1).

Table 28.2.1
Cell means in a type I (2 × 2) analysis of variance

		Criterion 2	
		column 1	column 2
Criterion 1	row 1	μ_{11}	μ_{12}
	row 2	μ_{21}	μ_{22}

The corresponding *model equation* is

$$X_{fgh} = \mu_{fg} + e_{fgh}\,, \quad \text{where } e_{fgh} \leftarrow \mathcal{N}(0, \sigma^2)\,.$$

In the above scheme, there are basically four parameters of central tendency, μ_{11}, μ_{12}, μ_{21}, and μ_{22}, in addition to a fifth parameter concerning dispersion, σ^2. However, some aspects of the analysis can be made more explicit by defining additional means for each row and for each column as well as a general mean (table 28.2.2),

Table 28.2.2
Means of cells, rows and columns, and general mean in a type I (2 × 2) analysis of variance

		Criterion 2		
		column 1	column 2	
Criterion 1	row 1	μ_{11}	μ_{12}	$\mu_{1\cdot} = (\mu_{11} + \mu_{12})/2$
	row 2	μ_{21}	μ_{22}	$\mu_{2\cdot} = (\mu_{21} + \mu_{22})/2$
		$\mu_{\cdot 1} = (\mu_{11} + \mu_{21})/2$	$\mu_{\cdot 2} = (\mu_{12} + \mu_{22})/2$	$\mu_{\cdot\cdot}$

and by defining a set of deviations between the general mean $\mu_{\cdot\cdot}$ and the means of each row, each column and each cell (table 28.2.3).

Table 28.2.3
Deviations between cell, row and column means and the general mean in a type I (2 × 2) analysis of variance

		Criterion 2		
		column 1	column 2	
Criterion 1	row 1	$\delta_{11} = \mu_{11} - (\mu_{\cdot\cdot} + \delta_{1\cdot} + \delta_{\cdot 1})$	$\delta_{12} = \mu_{12} - (\mu_{\cdot\cdot} + \delta_{1\cdot} + \delta_{\cdot 2})$	$\delta_{1\cdot} = \mu_{1\cdot} - \mu_{\cdot\cdot}$
	row 2	$\delta_{21} = \mu_{21} - (\mu_{\cdot\cdot} + \delta_{2\cdot} + \delta_{\cdot 1})$	$\delta_{22} = \mu_{22} - (\mu_{\cdot\cdot} + \delta_{2\cdot} + \delta_{\cdot 2})$	$\delta_{2\cdot} = \mu_{2\cdot} - \mu_{\cdot\cdot}$
		$\delta_{\cdot 1} = \mu_{\cdot 1} - \mu_{\cdot\cdot}$	$\delta_{\cdot 2} = \mu_{\cdot 2} - \mu_{\cdot\cdot}$	$\mu_{\cdot\cdot}$

The means of cells, rows and columns can then be reexpressed in terms of the general mean and of row, column and cell deviations, and the *model equation* can be rewritten as follows:

$$X_{fgh} = \mu_{\cdot\cdot} + \delta_{f\cdot} + \delta_{\cdot g} + \delta_{fg} + e_{fgh} ,$$

where X_{fgh} = the hth observation (or *"replicate"*) of the cell located at the intersection of the fth row and of the gth column,
$\mu_{\cdot\cdot}$ = the general mean,
$\delta_{f\cdot}$ = the row deviation, also called *"row effect"*,
$\delta_{\cdot g}$ = the column deviation, also called *"column effect"*,
δ_{fg} = the cell deviation (or *interaction term*),
and e_{fgh} = the random deviation accounting for dispersion.

In opposition to the interaction term δ_{fg}, the row deviation $\delta_{f\cdot}$ and the column deviation $\delta_{\cdot g}$ are jointly called *"main effects"*. The expressions *"row effect"*, *"column effect"*, *"main effects"*, and *"interaction"* have causal connotations which seem more natural in the case of experimental (tables 28.1.2 and 28.1.3) than of descriptive data (tables 28.1.1 and 28.1.4). The utilization of that terminology goes back to the fact that two-way analysis of variance was developed originally in the context of agricultural experiments.

However, the cell, row and column deviations defined in table 28.2.3 are clearly not independent from each other, for they comply with numerous relationships. Thus, the sum of row deviations is null, $\delta_{1\cdot} + \delta_{2\cdot} = 0$, and so are the sum of column deviations, $\delta_{\cdot 1} + \delta_{\cdot 2} = 0$, and the sums of the deviations of cells located in the same row or in the same column: $\delta_{11} + \delta_{12} = 0$, $\delta_{11} + \delta_{21} = 0$, $\delta_{21} + \delta_{22} = 0$, and $\delta_{12} + \delta_{22} = 0$. Consequently, in a (2×2) type I analysis of variance, the following substitutions can be made: $\delta_{1\cdot} = -\delta_{2\cdot}$, $\delta_{\cdot 1} = -\delta_{\cdot 2}$, $\delta_{11} = \delta_{22}$, $\delta_{12} = -\delta_{22}$, and $\delta_{21} = -\delta_{22}$. Cell, row and column means may therefore be reexpressed more simply as indicated in table 28.2.4.

Table 28.2.4

Cell, row and column means reexpressed with respect to the general mean and cell, row, and column deviations in a type I (2 × 2) analysis of variance

		Criterion 2		
		column 1	column 2	
Criterion 1	row 1	$\mu_{11} = \mu_{\cdot\cdot} - \delta_{2\cdot} - \delta_{\cdot 2} + \delta_{22}$	$\mu_{12} = \mu_{\cdot\cdot} - \delta_{2\cdot} + \delta_{\cdot 2} - \delta_{22}$	$\mu_{1\cdot} = \mu_{\cdot\cdot} - \delta_{2\cdot}$
	row 2	$\mu_{21} = \mu_{\cdot\cdot} + \delta_{2\cdot} - \delta_{\cdot 2} - \delta_{22}$	$\mu_{22} = \mu_{\cdot\cdot} + \delta_{2\cdot} + \delta_{\cdot 2} + \delta_{22}$	$\mu_{2\cdot} = \mu_{\cdot\cdot} + \delta_{2\cdot}$
		$\mu_{\cdot 1} = \mu_{\cdot\cdot} - \delta_{\cdot 2}$	$\mu_{\cdot 2} = \mu_{\cdot\cdot} + \delta_{\cdot 2}$	$\mu_{\cdot\cdot}$

Replacing the four cell means in a (2×2) analysis of variance by the general mean $\mu_{\cdot\cdot}$, the row deviation $\delta_{2\cdot}$, the column deviation $\delta_{\cdot 2}$, and the cell deviation δ_{22} does not reduce by itself the total number (4) of independent central tendency parameters. While the new system is thus not more economical than the earlier one in general, it does become so in the particular case where the interaction term δ_{22} is null. That condition is then said to be a case of *additivity,* since cell means can then be reconstituted entirely by adding the main effects to the general mean.

The operation through which a set of parameters is replaced by another one may be called a *"reparameterization"*. In the general case of a $(R \times C)$ analysis of variance, there are $(R-1)$ independent row deviations, $(C-1)$ independent column deviations, and $(R-1)(C-1)$ independent interaction deviations. The main advantage of the reparameterization discussed in this section is that it brings out explicitly the trends of variation with respect to which questions may most naturally be asked: the main effects and the interaction.

Section 28.3: main effects, interaction, and orthogonal contrasts

In a (2×2) type I analysis of variance, there is an interesting relationship between the main effects and the interaction on the one hand and orthogonal contrasts (section 26.5) on the other hand. If the cell means of each row of the data table are placed consecutively in a row vector (sections 24.2, 24.5, 24.22) containing $(2 \times 2) = 4$ elements,

$$\boldsymbol{\mu} = [\mu_{11} \quad \mu_{12} \quad \mu_{21} \quad \mu_{22}] ,$$

it is possible to define three orthogonal contrasts, $Y_1 = \mathbf{a}_1 \tilde{\boldsymbol{\mu}}$, $Y_2 = \mathbf{a}_2 \tilde{\boldsymbol{\mu}}$, and $Y_3 = \mathbf{a}_3 \tilde{\boldsymbol{\mu}}$, which respectively reflect the differences between rows, the differences between columns, and the interaction. Thus, if the vectors of coefficients of these three contrasts are

$$\begin{aligned} \mathbf{a}_1 &= [-1 \quad -1 \quad +1 \quad +1] , \\ \mathbf{a}_2 &= [-1 \quad +1 \quad -1 \quad +1] , \\ \text{and} \quad \mathbf{a}_3 &= [+1 \quad -1 \quad -1 \quad +1] , \end{aligned}$$

it may be shown that their means are

$$\begin{aligned} \mu_{Y_1} &= 4\,\delta_{2\cdot} , \\ \mu_{Y_2} &= 4\,\delta_{\cdot 2} , \\ \text{and} \quad \mu_{Y_3} &= 4\,\delta_{22} . \end{aligned}$$

The between-groups sum of squares may therefore be broken down into three independent parts corresponding to the differences between rows, the differences between columns, and the interaction. However, the number of orthogonal contrasts is larger in the case of a $(R \times C)$ anova, where the differences between rows, between columns, and the interaction correspond to $(R-1)$, $(C-1)$, and $(R-1)(C-1)$ contrasts respectively.

Section 28.4: computing and breaking down sums of squares

The computation of the various sums of squares should be done in a methodical manner. While the utilization of *centering terms* may occasionally entail losses of numerical accuracy (sections 4.6 and 12.3), it does have one serious advantage in two-way analysis of variance, in that it makes the geometrical meaning of calculations easier to understand. However, the effects of rounding errors should be minimized by doing computations in high precision. A table analogous to the data table (table 28.1.5) should then be prepared and a centering term should be computed for each of the RC cells, of the R rows and of the C columns, as well as for the complete set of all data (table 28.4.1).

Table 28.4.1
Centering terms required in the calculations of a two-way type I analysis of variance

		Criterion 2		For each row
		column 1	column 2	
Criterion 1	row 1	$(\Sigma X)^2_{11}/N_{11}$	$(\Sigma X)^2_{12}/N_{12}$	$(\Sigma X)^2_{1\cdot}/N_{1\cdot}$
	row 2	$(\Sigma X)^2_{21}/N_{21}$	$(\Sigma X)^2_{22}/N_{22}$	$(\Sigma X)^2_{2\cdot}/N_{2\cdot}$
For each column		$(\Sigma X)^2_{\cdot 1}/N_{\cdot 1}$	$(\Sigma X)^2_{\cdot 2}/N_{\cdot 2}$	$(\Sigma X)^2_{\cdot\cdot}/N_{\cdot\cdot}$

The various sums of squares needed in a two-way type I analysis of variance may next be calculated in the following manner, where the abbreviation "c. t." stands for "centering term":

$$SS_{\text{total}} = (\Sigma X^2)_{\cdot\cdot} - (\Sigma X)^2_{\cdot\cdot}/N_{\cdot\cdot}\,, \text{ where } (\Sigma X)^2_{\cdot\cdot}/N_{\cdot\cdot} = \text{the general c. t.,}$$

$$SS_{\text{within-groups}} = (\Sigma X^2)_{\cdot\cdot} - \text{the sum of the c. t.'s of the } RC \text{ cells,}$$

$$SS_{\text{between-groups}} = SS_{\text{total}} - SS_{\text{within-groups}}\,,$$

$$SS_{\text{rows}} = \text{the sum of the c. t.'s of the } R \text{ rows} - \text{the general c. t.,}$$

$$SS_{\text{columns}} = \text{the sum of the c. t.'s of the } C \text{ columns} - \text{the general c. t.,}$$

$$SS_{\text{interaction}} = SS_{\text{between-groups}} - SS_{\text{rows}} - SS_{\text{columns}}\,.$$

Some of the above formulae are also expressed below using a more detailed notation:

$$SS_{\text{total}} = \sum_f \sum_g \sum_h X^2 - \left(\sum_f \sum_g \sum_h X\right)^2 / N_{\cdot\cdot}\,,$$

$$SS_{\text{within-groups}} = \sum_f \sum_g \sum_h X^2 - \sum_f \sum_g \left[\left(\sum_h X\right)^2 / N_{fg}\right],$$

$$SS_{\text{rows}} = \sum_f \left[\left(\sum_g \sum_h X\right)^2 / N_{f\cdot}\right] - \left(\sum_f \sum_g \sum_h X\right)^2 / N_{\cdot\cdot}\,,$$

$$SS_{\text{columns}} = \sum_g \left[\left(\sum_f \sum_h X\right)^2 / N_{\cdot g}\right] - \left(\sum_f \sum_g \sum_h X\right)^2 / N_{\cdot\cdot}\,.$$

Section 28.5: the analysis of variance of a $(R \times C)$ table

In the table of a $(R \times C)$ analysis of variance, the between-groups sum of squares is broken down into three parts corresponding to row differences, column differences and the interaction, and possessing $(R-1)$, $(C-1)$ and $(R-1)(C-1)$ degrees of freedom respectively (table 28.5.1). When the sample sizes of all cells are equal to a common value,

$$N_{11} = \ldots = N_{fg} = \ldots = N_{LC} = N ,$$

as assumed in the present chapter, the total sample size $N..$ is equal to the product of the common sample size N by the number of cells:

$$N.. = RCN .$$

Table 28.5.1
Two-way type I analysis of variance

Variation	Sum of squares (SS)	Degrees of freedom	Observed variance	Expected variance
within-groups	$SS_{\text{within-groups}}$	$(N..-RC)$	$SS_{\text{within-groups}}/(N..-RC)$	σ^2
rows	SS_{rows}	$(R-1)$	$SS_{\text{rows}}/(R-1)$	$\sigma^2 + \Delta^2_{\text{rows}}$
columns	SS_{columns}	$(C-1)$	$SS_{\text{columns}}/(C-1)$	$\sigma^2 + \Delta^2_{\text{columns}}$
interaction	$SS_{\text{interaction}}$	$(R-1)(C-1)$	$SS_{\text{interaction}}/[(R-1)(C-1)]$	$\sigma^2 + \Delta^2_{\text{interaction}}$
total	SS_{total}	$(N..-1)$	$SS_{\text{total}}/(N..-1)$	(generally not used)

The parametric quantities Δ^2_{rows}, $\Delta^2_{\text{columns}}$ and $\Delta^2_{\text{interaction}}$ appearing in the "Expected variance" column can be determined by procedures similar to those of section 12.4. They have the following values:

$$\Delta^2_{\text{rows}} = [CN/(R-1)] \sum_{f=1}^{R} \delta^2_{f\cdot} ,$$

$$\Delta^2_{\text{columns}} = [RN/(C-1)] \sum_{g=1}^{C} \delta^2_{\cdot g} ,$$

and

$$\Delta^2_{\text{interaction}} = \{N/[(R-1)(C-1)]\} \sum_{f=1}^{R} \sum_{g=1}^{C} \delta^2_{fg} .$$

When the preliminary hypothesis is true, there are neither main effects nor an interaction, the quantities Δ^2_{rows}, $\Delta^2_{\text{columns}}$ and $\Delta^2_{\text{interaction}}$ are all null, and the observed variances S^2_{rows}, S^2_{columns} and $S^2_{\text{interaction}}$ are all estimates of the same parameter, σ^2. On the contrary, as soon as one of the quantities Δ^2_{rows}, $\Delta^2_{\text{columns}}$ and $\Delta^2_{\text{interaction}}$ differs from zero, the corresponding observed variance tends to be larger than σ^2.

The three preliminary sub-hypotheses that there are no row effects, no column effects, and no interaction, may thus be tested by using the variance ratio distribution (table 28.5.2):

Table 28.5.2
Tests of preliminary sub-hypotheses in a two-way type I analysis of variance

Hypothesis	Acceptance region
$\delta_{f\cdot}^2 = 0$	$S^2_{\text{rows}} / S^2_{\text{within-groups}} \leq F_{(1-\alpha;\, R-1,\, N..-RC)}$
$\delta_{\cdot g}^2 = 0$	$S^2_{\text{columns}} / S^2_{\text{within-groups}} \leq F_{(1-\alpha;\, C-1,\, N..-RC)}$
$\delta_{fg}^2 = 0$	$S^2_{\text{interaction}} / S^2_{\text{within-groups}} \leq F_{(1-\alpha;\, (R-1)(C-1),\, N..-RC)}$

In practice, the hypothesis that there is no interaction is usually tested first and, when that hypothesis is rejected, hypotheses concerning row and column effects are ignored because, in the presence of an interaction, differences between cell means cannot be explained completely in terms of main effects and a more involved analytical approach would be necessary (section 28.7).

Section 28.6: a taxonomical and ecological example

A first example is provided by the data of Brower (1959) already mentioned in section 28.1 (table 28.1.1). The means of the four samples are given in table 28.6.1.

Table 28.6.1
Mean lengths of the eighth tergite × 8 in mm of 10 male butterflies of two species in two localities (Brower, 1959)

		Geographical origin	
		Colorado	Utah
Species	*Papilio eurymedon*	$\overline{X}_{11} = 17.75$	$\overline{X}_{12} = 19.65$
	Papilio rutulus	$\overline{X}_{21} = 19.95$	$\overline{X}_{22} = 20.25$

The analysis of variance is summarized in table 28.6.2.

Table 28.6.2
Analysis of variance of the length of the eighth tergite × 8 in mm of male butterflies of the species *P. eurymedon* and *P. rutulus* in Colorado and Utah (Brower, 1959)

Variation	SS	D. of F.	Variance	F	P
within-groups	56	36	1.55556		
species	19.6	1	19.6	12.6	$0.001 < P < 0.01$
localities	12.1	1	12.1	7.77857	$0.001 < P < 0.01$
interaction	6.4	1	6.4	4.11429	$0.01 < P \leq 0.05$
total	94.1	39	2.41282		

The interaction is barely significant from a statistical point of view, the observed value of the variance ratio (4.11429) being only very slightly larger than the critical value $F_{(0,95;\ 1,\ 36)} = 4.11317$. However, the differences between the two rows (species) and between the two columns (regions) are more conclusive ($P < 0.01$). The comparison of means (table 28.6.1) suggests that the eighth tergite is longer in *Papilio rutulus* than in *Papilio eurymedon*, and longer in Utah than in Colorado.

Section 28.7: an agronomical example

A second example may be found in the results of Millikan (1963) on the growth of clover *Trifolium subterraneum* L. (table 28.1.2). Because a preliminary analysis of raw data showed that the variance was higher in cells having large means, the logarithmic transformation $Y = \log_{10}(X)$ was carried out, where X is the sum of the dry weights in g of the roots and upper parts of two plants of the Edenhope variety 66 days after the beginning of the experiment. The means of the nine samples are given in table 28.7.1.

Table 28.7.1
Mean decimal logarithms of dry weight in g of two 66-day-old plants of clover (*Trifolium subterraneum* L.) of the Edenhope variety

		Zinc concentration		
		$Zn_1 = 0.005$ p.p.m.	$Zn_2 = 0.05$ p.p.m.	$Zn_3 = 0.5$ p.p.m.
Phosphorus concentration	$P_1 = 3$ p.p.m.	$\overline{Y}_{11} = 0.508215$	$\overline{Y}_{12} = 0.498785$	$\overline{Y}_{13} = 0.543687$
	$P_2 = 7.5$ p.p.m.	$\overline{Y}_{21} = 0.581597$	$\overline{Y}_{22} = 0.861794$	$\overline{Y}_{23} = 0.819579$
	$P_3 = 15$ p.p.m.	$\overline{Y}_{31} = 0.463414$	$\overline{Y}_{32} = 0.980882$	$\overline{Y}_{33} = 0.879430$

The analysis of variance is summarized in table 28.7.2.

Table 28.7.2
Analysis of variance of decimal logarithm of dry weight in g of two 66-day-old plants of clover (*Trifolium subterraneum* L.) of the Edenhope variety

Variation	SS	D. of F.	Variance	F	P
within-groups	0.380556	45	0.0084568		
phosphorus	0.739084	2	0.369542	43.6976	$P < 0.001$
zinc	0.737625	2	0.368812	43.6113	$P < 0.001$
interaction	0.445228	4	0.111307	13.1618	$P < 0.001$
total	2.30249	53	0.0434432		

Because the interaction is strongly significant from a statistical point of view ($P < 0.001$), it would be unadvisable to test the main effects (section 28.5), because the observed differences between cells could not be explained simply by additive actions of phosphorus and zinc. In order to describe the interaction between phosphorus and zinc in the data of Millikan (1963), a more complex statistical approach would be necessary, such as a *"response surface"* (Box, Hunter and Hunter, 1978, chapter 15; Tomassone, Dervin and Masson, 1993, pages 279-288; see also section 35.9).

The general nature of the interaction between phosphorus and zinc is visible in a stereogram of means (figure 28.7.1). In addition to being both necessary to the growth of clover, each of these two elements appears to influence the utilization of the other element. Thus, when there is a shortage of phosphorus, clover appears to be able to use less zinc, and vice versa. Conversely, when the concentrations of both elements are increased simultaneously, the growth of clover is increased more strongly than would be expected by adding up their separate effects. At high phosphorus concentrations ($P_2 = 7.5$ and $P_3 = 15$ p.p.m.), however, the mean decimal logarithm of dry weight is slightly smaller when zinc concentration is maximum ($Zn_3 = 0.5$ p.p.m.) than when it is lower ($Zn_2 = 0.05$ p.p.m.), which suggests that excessive zinc concentrations might have toxic effects (see also section 35.9).

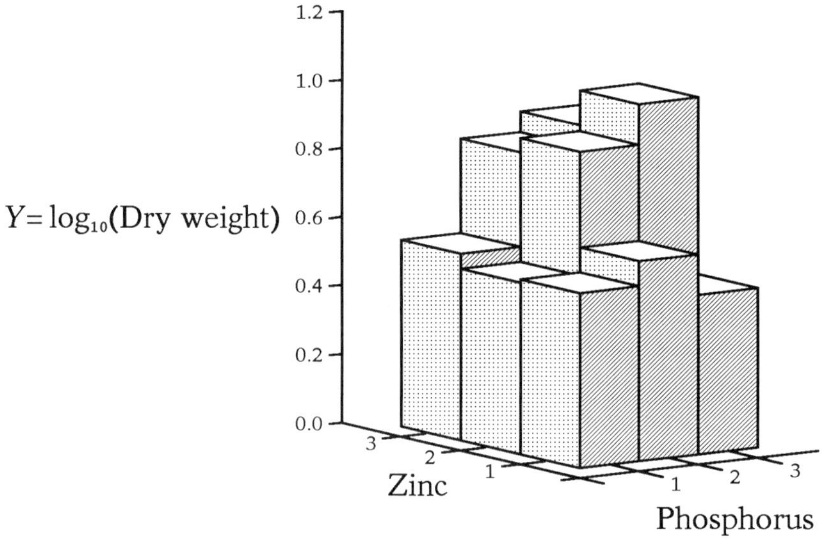

Figure 28.7.1
Stereogram of mean decimal logarithms of dry weight in g of two 66-day-old plants of clover (*Trifolium subterraneum* L.) at various zinc and phosphorus concentrations according to data from Millikan (1963)

A case where the simultaneous effects of two factors or experimental treatments are stronger than the sum of their separate effects, like the influence of phosphorus and zinc on the growth of clover, is referred to as an instance of *positive interaction* or *synergy*. On the contrary, a case where the joint effects of two factors are weaker than the sum of their separate effects is called a *negative interaction* or an *antagonism*.

Chapter 29
The multivariate normal distribution

Section 29.1: multivariate statistical methods

While partial and multiple regressions and correlations (chapters 23 and 25) belong to multivariate statistical methods in a broad sense, they are often treated separately because they are more widely known and have been used for a longer time. In fact, multiple regression goes back to Karl Friedrich Gauss (1777-1855). Moreover, multiple regression may be used even when the predictor variables are systematic and only the predicted variate is random. On the contrary, multivariate methods in a strict sense, discussed in chapters 29 to 34, have been developed mostly since 1930 and deal typically with the simultaneous analysis of several variates (random variables). For several decades, multivariate methods properly speaking were treated only at a highly technical level and were the object of very few monographs (Anderson, 1958; Kendall, 1957; Rao, 1952; Roy, 1957). However, many books have been published later about them (Anderson, 1984; Dagnelie, 1975; Dempster, 1969; Gnanadesikan, 1977; Kshirsagar, 1972; Lebart, Morineau and Piron, 1995; Morrison, 1967, 1990; Pontier, Dufour and Normand, 1990; Rao, 1965; Seal, 1964; Seber, 1984; Tomassone, Dervin and Masson, 1993). Moreover, some specialized discussions (Blackith and Reyment, 1971) and some reprint collections (Atchley and Bryant, 1975; Bryant and Atchley, 1975) have appeared in the literature.

Section 29.2: the multivariate normal probability density

When several variates $X_1, \dots, X_i, X_j, \dots, X_q$ have a joint multivariate normal distribution, the probability density of vector $\mathbf{X}=[X_1, \dots, X_i, X_j, \dots, X_q]$ may be concisely expressed by using vectors and matrices:

$$f(X_1, \dots, X_i, X_j, \dots, X_q) = \frac{1}{(2\pi)^{(q/2)}\,|\boldsymbol{\Sigma}|^{(1/2)}}\, e^{-(1/2)\,(\mathbf{X}-\boldsymbol{\mu})\,\boldsymbol{\Sigma}^{-1}(\tilde{\mathbf{X}}-\tilde{\boldsymbol{\mu}})},$$

where q is the number of elements in vector $\mathbf{X}$, that is the number of variates,

$\boldsymbol{\mu} = [\mu_1, \dots, \mu_i, \mu_j, \dots, \mu_q] = \mathscr{E}(\mathbf{X})$ is the vector of parametric means,

$$\boldsymbol{\Sigma} = \begin{bmatrix} \sigma_1^2 & \dots & \sigma_{1i} & \sigma_{1j} & \dots & \sigma_{1q} \\ \dots & \dots & \dots & \dots & \dots & \dots \\ \sigma_{1i} & \dots & \sigma_i^2 & \sigma_{ij} & \dots & \sigma_{iq} \\ \sigma_{1j} & \dots & \sigma_{ij} & \sigma_j^2 & \dots & \sigma_{jq} \\ \dots & \dots & \dots & \dots & \dots & \dots \\ \sigma_{1q} & \dots & \sigma_{iq} & \sigma_{jq} & \dots & \sigma_q^2 \end{bmatrix} = \mathscr{E}\left\{ \begin{bmatrix} X_1-\mu_1 \\ \dots \\ X_q-\mu_q \end{bmatrix} \begin{bmatrix} X_1-\mu_1 & \dots & X_q-\mu_q \end{bmatrix} \right\}$$

is the parametric covariance matrix,

$\pi = 3.141592653590\dots$, and $e = 2.718281828459\dots$.

The covariance matrix is represented by the upper case Greek letter sigma in bold type, $\boldsymbol{\Sigma}$, and should not be confused with the summation sign, which is represented by the same letter but in light type, Σ. For the normal distribution to be truly q-dimensional, the covariance matrix $\boldsymbol{\Sigma}$ must be regular (nonsingular) and its determinant $|\boldsymbol{\Sigma}|$ must be positive (section 24.18). The mean vector $\boldsymbol{\mu}$ determines the location of the center of the distribution and the covariance matrix $\boldsymbol{\Sigma}$ provides a joint numerical description of the extent of variation as well as of the presence or absence and of the strength of positive or negative straight-line relationships between observed variates $[X_1, \dots, X_q]$. However, the extent of variation may also be described by the standard deviations $(\sigma_1, \dots, \sigma_q)$ and relationships may be described separately by the covariance matrix of standardized variates $[(X_1 - \mu_1)/\sigma_1, \dots, (X_i - \mu_i)/\sigma_i, (X_j - \mu_j)/\sigma_j, \dots, (X_q - \mu_q)/\sigma_q]$, also known as the correlation matrix:

$$[\rho_{ij}] = \begin{bmatrix} 1 & \dots & \rho_{1i} & \rho_{1j} & \dots & \rho_{1q} \\ \dots & \dots & \dots & \dots & \dots & \dots \\ \rho_{1i} & \dots & 1 & \rho_{ij} & \dots & \rho_{iq} \\ \rho_{1j} & \dots & \rho_{ij} & 1 & \dots & \rho_{jq} \\ \dots & \dots & \dots & \dots & \dots & \dots \\ \rho_{1q} & \dots & \rho_{iq} & \rho_{jq} & \dots & 1 \end{bmatrix}.$$

The reader should note that the expression of the multivariate normal probability density is completely general. When the number q of variates is 1 or 2, in particular, the multivariate normal density gives back the univariate (section 5.2) or the bivariate (section 19.2) normal probability density. The argument of the exponential contains a quadratic form (section 24.24): $(\mathbf{X} - \boldsymbol{\mu})\boldsymbol{\Sigma}^{-1}(\tilde{\mathbf{X}} - \tilde{\boldsymbol{\mu}})$. When variation really spans q dimensions, this quadratic form is equivalent to a sum of squares of q standardized normal variates and follows a χ^2 distribution with q degrees of freedom. The equation

$$(\mathbf{X} - \boldsymbol{\mu})\,\boldsymbol{\Sigma}^{-1}(\tilde{\mathbf{X}} - \tilde{\boldsymbol{\mu}}) = \chi^2_{(1-\alpha;\, q)}$$

then delimits an ellipse (if $q = 2$), an ellipsoid (if $q = 3$) or a hyperellipsoid (if $q \geq 4$) within which each individual observation $\mathbf{X}$ coming from the multivariate normal distribution $\mathcal{N}(\boldsymbol{\mu}, \boldsymbol{\Sigma})$ has a probability $(1 - \alpha)$ of falling.

Section 29.3: calculating estimates of parameters

The point estimates (section 9.1) of the parameters of a multivariate normal distribution are generally computed through simple extensions of the procedures used in the case of univariate (sections 4.2, 4.5, 4.6, 4.7 and 4.10) and bivariate (sections 19.3 and 19.4) normal distributions. The present section is therefore limited to a discussion of the manner in which vectors and matrices may be used to carry out calculations in the case where there are several variates, a topic which has already been briefly mentioned concerning partial and multiple correlations and regressions (chapter 25). The first step is usually the computation of the *matrix of raw sums* (see also section 25.3) of sample data:

$$\left[\begin{array}{c|cccccc} N & \Sigma X_1 & \ldots & \Sigma X_i & \Sigma X_j & \ldots & \Sigma X_q \\ \hline \Sigma X_1 & \Sigma X_1^2 & \ldots & \Sigma X_1 X_i & \Sigma X_1 X_j & \ldots & \Sigma X_1 X_q \\ \ldots & \ldots & \ldots & \ldots & \ldots & \ldots & \ldots \\ \Sigma X_i & \Sigma X_1 X_i & \ldots & \Sigma X_i^2 & \Sigma X_i X_j & \ldots & \Sigma X_i X_q \\ \Sigma X_j & \Sigma X_1 X_j & \ldots & \Sigma X_i X_j & \Sigma X_j^2 & \ldots & \Sigma X_j X_q \\ \ldots & \ldots & \ldots & \ldots & \ldots & \ldots & \ldots \\ \Sigma X_q & \Sigma X_1 X_q & \ldots & \Sigma X_i X_q & \Sigma X_j X_q & \ldots & \Sigma X_q^2 \end{array}\right].$$

This matrix is of order $(q+1) \times (q+1)$, for it has one more row and one more column than the number of variates. The matrix of raw sums is subdivided above into four submatrices: the upper left submatrix has a single element, the sample size N; the upper right and lower left submatrices contain the sums of the q variates written as a row vector and as a column vector respectively; finally, the lower right submatrix, which is of order $(q \times q)$, contains the sums of squares and products of the variates.

The parametric mean vector $\boldsymbol{\mu}$ is generally estimated by multiplying the upper right submatrix by the inverse of sample size (section 24.10), which yields the sample mean vector:

$$\bar{\mathbf{X}} = [\bar{X}_1, \ldots, \bar{X}_i, \bar{X}_j, \ldots, \bar{X}_q] = (1/N)[\Sigma X_1, \ldots, \Sigma X_i, \Sigma X_j, \ldots, \Sigma X_q].$$

The *matrix of centered sums,* that is of the squares and products of the deviations from the means,

$$\begin{bmatrix} \Sigma(X_1-\bar{X}_1)^2 & \ldots & \Sigma(X_1-\bar{X}_1)\Sigma(X_i-\bar{X}_i) & \Sigma(X_1-\bar{X}_1)\Sigma(X_j-\bar{X}_j) & \ldots & \Sigma(X_1-\bar{X}_1)\Sigma(X_q-\bar{X}_q) \\ \ldots & \ldots & \ldots & \ldots & \ldots & \ldots \\ \Sigma(X_1-\bar{X}_1)\Sigma(X_i-\bar{X}_i) & \ldots & \Sigma(X_i-\bar{X}_i)^2 & \Sigma(X_i-\bar{X}_i)\Sigma(X_j-\bar{X}_j) & \ldots & \Sigma(X_i-\bar{X}_i)\Sigma(X_q-\bar{X}_q) \\ \Sigma(X_1-\bar{X}_1)\Sigma(X_j-\bar{X}_j) & \ldots & \Sigma(X_i-\bar{X}_i)\Sigma(X_j-\bar{X}_j) & \Sigma(X_j-\bar{X}_j)^2 & \ldots & \Sigma(X_j-\bar{X}_j)\Sigma(X_q-\bar{X}_q) \\ \ldots & \ldots & \ldots & \ldots & \ldots & \ldots \\ \Sigma(X_1-\bar{X}_1)\Sigma(X_q-\bar{X}_q) & \ldots & \Sigma(X_i-\bar{X}_i)\Sigma(X_q-\bar{X}_q) & \Sigma(X_j-\bar{X}_j)\Sigma(X_q-\bar{X}_q) & \ldots & \Sigma(X_q-\bar{X}_q)^2 \end{bmatrix},$$

is usually obtained next by subtracting the *matrix of centering terms,*

$$(1/N)\begin{bmatrix} \Sigma X_1 \\ \ldots \\ \Sigma X_i \\ \Sigma X_j \\ \ldots \\ \Sigma X_q \end{bmatrix}[\Sigma X_1 \ldots \Sigma X_i \; \Sigma X_j \ldots \Sigma X_q] = (1/N)\begin{bmatrix} (\Sigma X_1)^2 & \ldots & (\Sigma X_1)(\Sigma X_i) & (\Sigma X_1)(\Sigma X_j) & \ldots & (\Sigma X_1)(\Sigma X_q) \\ \ldots & \ldots & \ldots & \ldots & \ldots & \ldots \\ (\Sigma X_1)(\Sigma X_i) & \ldots & (\Sigma X_i)^2 & (\Sigma X_i)(\Sigma X_j) & \ldots & (\Sigma X_i)(\Sigma X_q) \\ (\Sigma X_1)(\Sigma X_j) & \ldots & (\Sigma X_i)(\Sigma X_j) & (\Sigma X_j)^2 & \ldots & (\Sigma X_j)(\Sigma X_q) \\ \ldots & \ldots & \ldots & \ldots & \ldots & \ldots \\ (\Sigma X_1)(\Sigma X_q) & \ldots & (\Sigma X_i)(\Sigma X_q) & (\Sigma X_j)(\Sigma X_q) & \ldots & (\Sigma X_q)^2 \end{bmatrix},$$

from the lower right submatrix of the matrix of raw sums:

$$[\Sigma(X_i-\bar{X}_i)(X_j-\bar{X}_j)] = [\Sigma X_i X_j] - (1/N)[(\Sigma X_i)(\Sigma X_j)].$$

The sample covariance matrix $\mathbf{S}$, which will be used as an estimate of the population covariance matrix $\boldsymbol{\Sigma}$, is then evaluated by multiplying the matrix of centered sums by the inverse of the number of degrees of freedom $(N-1)$:

$$\mathbf{S} = \begin{bmatrix} S_1^2 & \dots & S_{1i} & S_{1j} & \dots & S_{1q} \\ \dots & \dots & \dots & \dots & \dots & \dots \\ S_{1i} & \dots & S_i^2 & S_{ij} & \dots & S_{iq} \\ S_{1j} & \dots & S_{ij} & S_j^2 & \dots & S_{jq} \\ \dots & \dots & \dots & \dots & \dots & \dots \\ S_{1q} & \dots & S_{iq} & S_{jq} & \dots & S_q^2 \end{bmatrix} = (1/(N-1))\,[\Sigma(X_i - \overline{X}_i)(X_j - \overline{X}_j)]\,.$$

However, as emphasized in several earlier sections (4.6, 12.3, 13.6, 19.3, and 28.4), the utilization of centering terms occasionally entails numerical inaccuracies which may be prevented by calculating the centered sums directly by multiplying and squaring the deviations from the means, $(X_i - \overline{X}_i)$ and $(X_j - \overline{X}_j)$. The last step of computations is the evaluation of the sample correlation matrix $[r_{ij}]$,

$$[r_{ij}] = \begin{bmatrix} 1 & \dots & r_{1i} & r_{1j} & \dots & r_{1q} \\ \dots & \dots & \dots & \dots & \dots & \dots \\ r_{1i} & \dots & 1 & r_{ij} & \dots & r_{iq} \\ r_{1j} & \dots & r_{ij} & 1 & \dots & r_{jq} \\ \dots & \dots & \dots & \dots & \dots & \dots \\ r_{1q} & \dots & r_{iq} & r_{jq} & \dots & 1 \end{bmatrix},$$

which is used as an estimate of the population correlation matrix $[\rho_{ij}]$ (section 29.2). The sample correlation matrix $[r_{ij}]$ may be calculated by multiplying the ith row of the sample covariance matrix **S** by the inverse $(1/S_i)$ of the ith standard deviation and the jth column by the inverse $(1/S_j)$ of the jth standard deviation. The following box contains five statements which may be inserted in a *True BASIC™* computer program (section 24.26) in order to obtain the sample correlation matrix $\mathbf{R} = [r_{ij}]$ from the sample covariance matrix $\mathbf{S} = [S_{ij}]$:

```
FOR i = 1 TO q
  FOR j = 1 TO q
    LET R(i,j) = S(i,j)/SQR(S(i,i)*S(j,j))
  NEXT j
NEXT i
```

In *True BASIC*, like in several other computer programming languages, the subscripts i and j are placed between parentheses and separated by a comma. The asterisk indicates multiplication and the abbreviation SQR means that the square root of the expression following it between parentheses must be extracted. The division of each variate by its standard deviation constitutes a change of scale and may be done by premultiplying and postmultiplying the sample covariance matrix **S** by a diagonal matrix of which the ith diagonal element is $(1/S_i)$:

$$\mathbf{R} = [r_{ij}] = \begin{bmatrix} (1/S_1) & \dots & 0 & 0 & \dots & 0 \\ \dots & \dots & \dots & \dots & \dots & \dots \\ 0 & \dots & (1/S_i) & 0 & \dots & 0 \\ 0 & \dots & 0 & (1/S_j) & \dots & 0 \\ \dots & \dots & \dots & \dots & \dots & \dots \\ 0 & \dots & 0 & 0 & \dots & (1/S_q) \end{bmatrix} \mathbf{S} \begin{bmatrix} (1/S_1) & \dots & 0 & 0 & \dots & 0 \\ \dots & \dots & \dots & \dots & \dots & \dots \\ 0 & \dots & (1/S_i) & 0 & \dots & 0 \\ 0 & \dots & 0 & (1/S_j) & \dots & 0 \\ \dots & \dots & \dots & \dots & \dots & \dots \\ 0 & \dots & 0 & 0 & \dots & (1/S_q) \end{bmatrix}.$$

Section 29.4: testing the hypothesis that all correlations are null

The hypothesis that all variates $[X_1, \ldots, X_i, X_j, \ldots, X_q]$ are independent from each other implies that all correlations are null and that the parametric correlation matrix is a unit matrix: H_0: $[\rho_{ij}] = \mathbf{I}$. An approximate acceptance region of this hypothesis is

$$- [N - 1 - (2q + 5)/6] \log_e(|r_{ij}|) \leq \chi^2_{(1-\alpha;\, q(q-1)/2)},$$

where N is sample size, q is the number of variates included in the analysis, and $|r_{ij}|$ is the determinant of the sample correlation matrix (Bartlett, 1954). The above criterion may be used, for instance, for testing the hypothesis that the bilateral asymmetries of the humerus, radius, femur and tibia lengths of adult human skeletons (section 29.6) are all independent from each other. Even though most correlation coefficients in the matrices of table 29.6.1 are close to zero, the values of the criterion are $\chi^2_{(6)} = 22.7393$ for 117 male skeletons and $\chi^2_{(6)} = 41.7205$ for 110 female skeletons. The hypothesis that human limb bone asymmetries are uncorrelated must therefore be rejected ($P < 0.001$).

Section 29.5: means, variances and covariances of linear combinations

When a vector of several variates $\mathbf{X} = [X_1, \ldots, X_i, X_j, \ldots, X_q]$ has the mean vector $\boldsymbol{\mu} = \boldsymbol{\mu}_X = [\mu_1, \ldots, \mu_i, \mu_j, \ldots, \mu_q]$ and the covariance matrix

$$\boldsymbol{\Sigma} = \boldsymbol{\Sigma}_{XX} = \begin{bmatrix} \sigma_1^2 & \ldots & \sigma_{1i} & \sigma_{1j} & \ldots & \sigma_{1q} \\ \ldots & \ldots & \ldots & \ldots & \ldots & \ldots \\ \sigma_{1i} & \ldots & \sigma_i^2 & \sigma_{ij} & \ldots & \sigma_{iq} \\ \sigma_{1j} & \ldots & \sigma_{ij} & \sigma_j^2 & \ldots & \sigma_{jq} \\ \ldots & \ldots & \ldots & \ldots & \ldots & \ldots \\ \sigma_{1q} & \ldots & \sigma_{iq} & \sigma_{jq} & \ldots & \sigma_q^2 \end{bmatrix},$$

the linear combinations (section 24.22) of those variates, like

$$\tilde{\mathbf{Z}} = \mathbf{A}\tilde{\mathbf{X}} = \begin{bmatrix} Z_1 \\ \ldots \\ Z_k \end{bmatrix} = \begin{bmatrix} a_{11} & \ldots & a_{1q} \\ \ldots & \ldots & \ldots \\ a_{k1} & \ldots & a_{kq} \end{bmatrix} \begin{bmatrix} X_1 \\ \ldots \\ X_q \end{bmatrix},$$

have the mean vector

$$\mathscr{E}(\mathbf{Z}) = \tilde{\boldsymbol{\mu}}_Z = \mathbf{A}\tilde{\boldsymbol{\mu}} = \begin{bmatrix} a_{11} & \ldots & a_{1q} \\ \ldots & \ldots & \ldots \\ a_{k1} & \ldots & a_{kq} \end{bmatrix} \begin{bmatrix} \mu_1 \\ \ldots \\ \mu_q \end{bmatrix}$$

and the covariance matrix

$$\mathscr{E}[(\tilde{\mathbf{Z}} - \tilde{\boldsymbol{\mu}}_Z)(\mathbf{Z} - \boldsymbol{\mu}_Z)] = \boldsymbol{\Sigma}_{ZZ} = \mathbf{A}\,\boldsymbol{\Sigma}\,\tilde{\mathbf{A}} = \begin{bmatrix} a_{11} & \ldots & a_{1q} \\ \ldots & \ldots & \ldots \\ a_{k1} & \ldots & a_{kq} \end{bmatrix} \begin{bmatrix} \sigma_1^2 & \ldots & \sigma_{1q} \\ \ldots & \ldots & \ldots \\ \sigma_{1q} & \ldots & \sigma_q^2 \end{bmatrix} \begin{bmatrix} a_{11} & \ldots & a_{k1} \\ \ldots & \ldots & \ldots \\ a_{1q} & \ldots & a_{kq} \end{bmatrix},$$

provided the coefficients $[a_{ij}]$ are known *a priori,* which means that they are not estimated from the data.

In order for the linear combinations $Z_1, \dots, Z_k$ to occupy k dimensions, the number of combinations must not be greater than the number of original variates ($k \le q$), and the combinations must be linearly independent from each other, as for instance when the row vectors of the matrix **A** of the coefficients of the combinations are mutually orthogonal (section 24.6). Similar relationships hold at the sample level, so that the mean vector and the covariance matrix of linear combinations $\mathbf{Z} = [Z_1, \dots, Z_k]$ in a sample of size N are

$$\tilde{\bar{\mathbf{Z}}} = \mathbf{A}\tilde{\bar{\mathbf{X}}} \quad \text{and} \quad \mathbf{S}_{ZZ} = \mathbf{A}\,\mathbf{S}_{XX}\,\tilde{\mathbf{A}} .$$

When the coefficients of the combinations are known a priori, only the central matrix $\mathbf{S}_{XX}$ is random within the matrix product $\mathbf{A}\,\mathbf{S}_{XX}\,\tilde{\mathbf{A}}$, and the latter is an unbiased estimate of the parametric covariance matrix $\mathbf{\Sigma}_{ZZ}$ of the linear combinations.

Section 29.6: *a priori* linear combinations of normal variates

When the original variates $\mathbf{X} = [X_1, \dots, X_i, X_j, \dots, X_q]$ follow a multivariate normal distribution, the linear combinations $\mathbf{Z} = [Z_1, \dots, Z_k]$ discussed in the preceding section also have a multivariate normal distribution, and the methods based on the latter may be used. Thus, in a study of bilateral symmetry in the human skeleton based on data kindly provided by Professor Adolph H. Schultz, then Director of the Anthropological Institute of the University of Zürich, Jolicoeur (1963c) tried to verify whether there are relationships between the bilateral asymmetries of arm and leg bones, and considered the following dimensions in mm (appendix C):

X_1 = left humerus length, X_5 = left femur length,
X_2 = right humerus length, X_6 = right femur length,
X_3 = left radius length, X_7 = left tibia length,
X_4 = right radius length, X_8 = right tibia length.

In the case of linear dimensions of human adults, the normal distribution is often suitable, and the bilateral asymmetries of the proximal and distal segments of arms and legs may be described by the following four linear combinations:

$$\begin{bmatrix} \text{humerus asymmetry} \\ \text{radius asymmetry} \\ \text{femur asymmetry} \\ \text{tibia asymmetry} \end{bmatrix} = \begin{bmatrix} Z_1 \\ Z_2 \\ Z_3 \\ Z_4 \end{bmatrix} = \begin{bmatrix} -1 & +1 & 0 & 0 & 0 & 0 & 0 & 0 \\ 0 & 0 & -1 & +1 & 0 & 0 & 0 & 0 \\ 0 & 0 & 0 & 0 & -1 & +1 & 0 & 0 \\ 0 & 0 & 0 & 0 & 0 & 0 & -1 & +1 \end{bmatrix} \begin{bmatrix} X_1 \\ X_2 \\ X_3 \\ X_4 \\ X_5 \\ X_6 \\ X_7 \\ X_8 \end{bmatrix} .$$

According to the preliminary hypothesis, the parametric correlation between the Z_i and Z_j combinations should be null: H_0: $\rho_{ij} = 0$. The covariance matrix of linear combinations $[Z_1, Z_2, Z_3, Z_4]$ may easily be obtained by premultiplying and postmultiplying the covariance matrix of original variates $[X_1, \dots, X_8]$ by the matrix of coefficients and its transpose, respectively:

$$[s_{Z_iZ_j}] = \begin{bmatrix} -1 & +1 & 0 & 0 & 0 & 0 & 0 & 0 \\ 0 & 0 & -1 & +1 & 0 & 0 & 0 & 0 \\ 0 & 0 & 0 & 0 & -1 & +1 & 0 & 0 \\ 0 & 0 & 0 & 0 & 0 & 0 & -1 & +1 \end{bmatrix} [s_{X_iX_j}] \begin{bmatrix} -1 & 0 & 0 & 0 \\ +1 & 0 & 0 & 0 \\ 0 & -1 & 0 & 0 \\ 0 & +1 & 0 & 0 \\ 0 & 0 & -1 & 0 \\ 0 & 0 & +1 & 0 \\ 0 & 0 & 0 & -1 \\ 0 & 0 & 0 & +1 \end{bmatrix}.$$

The correlation matrix $[r_{Z_iZ_j}]$ of combinations may next be obtained from their covariance matrix following the procedure described in section 29.3. The correlation matrices of combinations $[Z_1, Z_2, Z_3, Z_4]$ for samples of human skeletons are given in table 29.6.1. Since the only two correlations which are negative are approximately null and are not statistically significant, we may accept the hypothesis that the asymmetries of one segment of a limb are not compensated for by those of another segment of the same limb. Only the positive correlation between humerus and radius bilateral asymmetries, reflected by combinations Z_1 and Z_2, is strongly significant in both sexes ($P < 0.001$). The other correlations do not differ significantly from 0 or barely reach the $\alpha = 0.05$ level.

Table 29.6.1
Correlation matrices of humerus, radius, femur and tibia asymmetries in adult human skeletons (Jolicoeur, 1963c)

117 men	110 women
$\begin{bmatrix} 1 & +0.361414 & +0.023747 & +0.166714 \\ & 1 & -0.047231 & +0.186796 \\ & & 1 & +0.084519 \\ & & & 1 \end{bmatrix}$	$\begin{bmatrix} 1 & +0.516997 & -0.004443 & +0.153873 \\ & 1 & +0.102485 & +0.070836 \\ & & 1 & +0.193036 \\ & & & 1 \end{bmatrix}$

The present analysis therefore shows that bilateral asymmetries tend to affect the proximal and distal segments of the arm jointly, but does not clearly disclose other trends.

Section 29.7: the multivariate lognormal distribution

The lognormal distribution is often more suitable than the normal distribution for the analysis of biological data, as already discussed in chapter 14 and in section 19.8 for the univariate and bivariate cases respectively. In the multivariate case also the lognormal distribution usually agrees better with the fact that most biological data are nonnegative and that their frequency distributions are often positively skewed (chapter 13) and heteroscedastic (sections 14.3 and 19.8). When the vector $\mathbf{X} = [X_1, \ldots, X_i, X_j, \ldots, X_q]$ of original variates follows a multivariate lognormal distribution, the vector of the logarithms of the data, $\mathbf{Y} = [Y_1, \ldots, Y_i, Y_j, \ldots, Y_q]$, where $Y_i = \log_e(X_i)$, follows a multivariate normal distribution (section 29.2) and the methods based on the latter may be applied to the transformed data. It must be noted that the mean vector and the covariance and correlation matrices are then those of the transformed variates, not of the original data. Once the statistical analysis is completed, however, its results, such as point or linear estimates or confidence, prediction or variation regions, may be reexpressed with respect to the original variates by using the antilogarithmic transformation (see figures 19.8.1 and 30.9.1).

Section 29.8: linear combinations of logarithms of lognormal variates

When the original variates $[X_1, \ldots, X_q]$ follow a lognormal distribution, the linear combinations of their logarithms $[Y_1, \ldots, Y_q]$, where $Y_i = \log_e(X_i)$, follow a multivariate normal distribution. Therefore, products and ratios of lognormal variates are lognormally distributed since they are equal to antilogarithms of linear combinations of normal variates. But biologists often have to deal with products or ratios of random variates (Jolicoeur, 1996a).

Thus, in their metabolic study of compulsive drinkers, Beerstecher and his coworkers (Beerstecher et al., 1950) determined the volume X_1 in ml and the creatinine concentration X_2 in mg/ml of complete morning specimens of the urine of eight healthy male control individuals. Since both variates are nonnegative, the logarithmic transformation may presumably make their distribution approximately normal as well as facilitate the exploration of linear combinations of logarithms.

In spite of small sample size ($N = 8$), the correlation coefficient between the logarithms of urine volume and creatinine concentration differs significantly from zero ($r = -0.782102$, $P < 0.05$), which indicates that, when the urine specimen is larger, it is simultaneously more dilute, a result which is physiologically somewhat trivial. The scatter diagram of logarithms Y_1 and Y_2 clearly reflects this negative correlation (figure 29.8.1, left). Because creatinine is currently used by physiologists as an indicator of renal function, it would be desirable to check whether, like creatinine concentration, the total amount of excreted creatinine is correlated with urinary volume or not. This may be done if the following linear combinations are defined:

$$\begin{bmatrix} Z_1 \\ Z_2 \end{bmatrix} = \begin{bmatrix} 1 & 0 \\ 1 & 1 \end{bmatrix} \begin{bmatrix} Y_1 \\ Y_2 \end{bmatrix}.$$

While the first linear combination Z_1 is equal simply to the logarithm Y_1 of urine volume, the second linear combination $Z_2 = Y_1 + Y_2$ is equal to the logarithm of the product of the volume of the urine specimen by its creatinine concentration, that is to the logarithm of the total amount of creatinine excreted. The covariance matrix of combinations Z_1 and Z_2 is easily obtained by premultiplying and postmultiplying the covariance matrix of transformed variates Y_1 and Y_2 by the matrix of the coefficients of the linear combinations and its transpose:

$$\begin{bmatrix} S^2_{Z_1} & S_{Z_1Z_2} \\ S_{Z_1Z_2} & S^2_{Z_2} \end{bmatrix} = \begin{bmatrix} 1 & 0 \\ 1 & 1 \end{bmatrix} \begin{bmatrix} S^2_{Y_1} & S_{Y_1Y_2} \\ S_{Y_1Y_2} & S^2_{Y_2} \end{bmatrix} \begin{bmatrix} 1 & 1 \\ 0 & 1 \end{bmatrix}.$$

The correlation coefficient of linear combinations Z_1 and Z_2 does not differ significantly from zero ($r = 0.054749$, $P > 0.05$), which confirms the graphical indications given by the corresponding scatter diagram (figure 29.8.1, right). The hypothesis that total creatinine is uncorrelated with urine volume may thus be accepted, which supports the utilization of creatinine as an indicator of renal function.

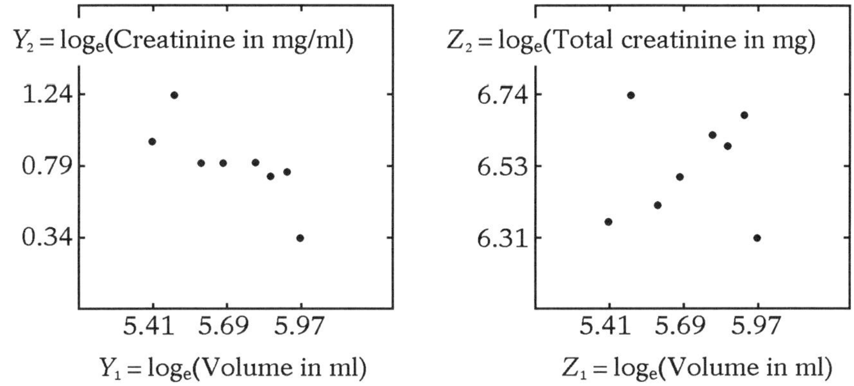

Figure 29.8.1
Left: negative correlation between the logarithms Y_1 of urine volume and Y_2 of creatinine concentration; right: lack of correlation between the logarithms Z_1 of urine volume and Z_2 of the total amount of creatinine excreted

A second example will be drawn from the field of morphometry. During a multivariate statistical study of skeletal variation in the North American marten (*Martes americana*), Jolicoeur (1963b) tried to check a hypothesis expressed earlier by an anatomist (Scott, 1957). According to that hypothesis, skeletal robustness is determined throughout the body by the activity of the subperiosteal tissue and is stimulated by muscular exercise and by sex hormones. Positive correlations would therefore be expected between the massiveness of various parts of the skeleton. Jolicoeur (1963b) thus measured the length and width in mm of several bones, the following variates being included in the present analysis:

X_1 = total skull length, X_5 = total humerus length,
X_2 = total skull width, X_6 = distal humerus width,
X_3 = total sacrum length, X_7 = total femur length,
X_4 = total sacrum width, X_8 = distal femur width.

Once variates have been submitted to the logarithmic transformation $Y_i = \log_e(X_i)$, four linear combinations may be defined which describe the relative width of each of the four structures:

$$\begin{bmatrix} Z_1 \\ Z_2 \\ Z_3 \\ Z_4 \end{bmatrix} = \begin{bmatrix} -1 & 1 & 0 & 0 & 0 & 0 & 0 & 0 \\ 0 & 0 & -1 & 1 & 0 & 0 & 0 & 0 \\ 0 & 0 & 0 & 0 & -1 & 1 & 0 & 0 \\ 0 & 0 & 0 & 0 & 0 & 0 & -1 & 1 \end{bmatrix} \begin{bmatrix} Y_1 \\ Y_2 \\ Y_3 \\ Y_4 \\ Y_5 \\ Y_6 \\ Y_7 \\ Y_8 \end{bmatrix} .$$

The correlation matrices of the four linear combinations in a sample of 52 males and in a sample of 33 females are given in table 29.8.1, and the scatter diagrams of linear combinations 1 and 2 on the one hand and 3 and 4 on the other hand are illustrated in figure 29.8.2, where prediction ellipses based on Hotelling's T^2 (chapter 30) at the $\alpha = 0.05$ significance level have been drawn.

Table 29.8.1
Correlation matrices of four linear combinations of logarithms of skeletal dimensions in the North American marten

52 males	33 females
$\begin{bmatrix} 1 & -0.114354 & +0.153447 & +0.074380 \\ & 1 & -0.103602 & +0.120005 \\ & & 1 & +0.610387 \\ & & & 1 \end{bmatrix}$	$\begin{bmatrix} 1 & -0.119540 & +0.058865 & +0.105141 \\ & 1 & +0.276222 & +0.345485 \\ & & 1 & +0.754863 \\ & & & 1 \end{bmatrix}$

The only correlation which is positive and highly conclusive in both samples is that between combinations Z_3 and Z_4 (table 29.8.1 and figure 29.8.2, right), which reflect the relative widths of the humerus and femur respectively ($P < 0.001$). Moreover, linear combinations 3 and 4 have higher numerical values (figure 29.8.2, right) and indicate that the humerus and the femur are more robust in males than in females. On the contrary, the other correlations, including that between the logarithms of the relative widths of the skull and of the sacrum (combinations Z_1 and Z_2, table 29.8.1, figure 29.8.2, left), are not statistically significant ($P > 0.05$) or barely significant ($P \leq 0.05$).

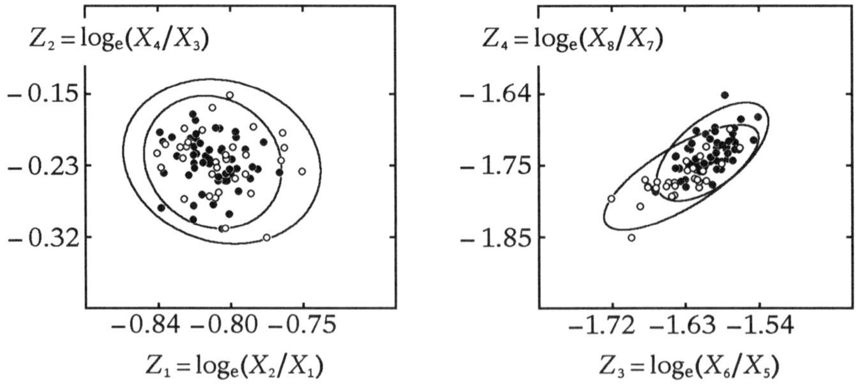

Figure 29.8.2
Left: lack of correlation of the logarithms of the relative widths of the skull (Z_1) and sacrum (Z_2) in the North American marten; right: the logarithms of the relative widths of the humerus (Z_3) and femur (Z_4) are positively correlated and larger in 52 males (black dots) than in 33 females (white dots); prediction ellipses at the $\alpha = 0.05$ significance level

The hypothesis of Scott (1957), according to which robustness is a joint property of several parts of the skeleton, appears to be confirmed in the case of the humerus and femur of the North American marten, of which the logarithms of relative widths are positively correlated. This may perhaps be explained by the fact that these long limb bones are involved mostly in locomotion and in supporting body weight. Moreover, since these bones have a rather simple shape, their robustness may be approximately proportional to their relative width. On the contrary, the relative widths of other skeletal components, like the skull and the sacrum, are not clearly correlated, which may perhaps be explained by the fact that these bones are not primarily involved in locomotion and are more complex, so that their robustness may not be a simple function of their relative width.

The examples discussed in section 29.6 and in the present section suggest that *a priori* linear combinations of normal variates or of logarithms of lognormal variates could facilitate the exploration of interesting correlations in various fields (physiology, morphology, ecology, quantitative genetics, etc.). The utilization of these linear combinations recalls the simplicity, the rigorousness and the power of the contrasts used in the analysis of variance (sections 26.3 to 26.6, 28.3), orthogonal contrasts in particular.

Section 29.9: the polynomial (also called multinomial) distribution

Like the binomial distribution, which becomes more and more closely similar to the univariate normal distribution when its exponent k increases (section 17.3), the polynomial distribution $(p_1 + \dots + p_q)^k = 1$ (section 17.12) becomes increasingly similar to the multivariate normal distribution. The mean vector $\boldsymbol{\mu}$ and the covariance matrix $\boldsymbol{\Sigma}$ of the polynomial variates $[X_1, \dots, X_i, X_j, \dots, X_q]$ have the following values:

$$\boldsymbol{\mu} = [kp_1, \dots, kp_i, kp_j, \dots, kp_q],$$

and

$$\boldsymbol{\Sigma} = \begin{bmatrix} kp_1(1-p_1) & \dots & -kp_1p_i & -kp_1p_j & \dots & -kp_1p_q \\ \dots & \dots & \dots & \dots & \dots & \dots \\ -kp_ip_1 & \dots & kp_i(1-p_i) & -kp_ip_j & \dots & -kp_ip_q \\ -kp_jp_1 & \dots & -kp_jp_i & kp_j(1-p_j) & \dots & -kp_jp_q \\ \dots & \dots & \dots & \dots & \dots & \dots \\ -kp_qp_1 & \dots & -kp_qp_i & -kp_qp_j & \dots & kp_q(1-p_q) \end{bmatrix}.$$

However, because of the constraint

$$X_1 + \dots + X_i + X_j + \dots + X_q = k,$$

the covariance matrix of the polynomial distribution is singular, because the variates $[X_1, \dots, X_i, X_j, \dots, X_q]$ are not linearly independent from each other since any one of them (X_q for instance) may be expressed as a function of the others:

$$X_q = k - (X_1 + \dots + X_i + X_j + \dots + X_{(q-1)}).$$

Nevertheless, if one of the variates is omitted (X_q for instance), a covariance matrix of order $(q-1)\times(q-1)$ corresponding to the other variates $[X_1, \dots, X_{(q-1)}]$ is obtained,

$$\mathbf{V} = \boldsymbol{\Sigma}_{((q-1)\times(q-1))} = \begin{bmatrix} kp_1(1-p_1) & \dots & -kp_1p_i & -kp_1p_j & \dots & -kp_1p_{(q-1)} \\ \dots & \dots & \dots & \dots & \dots & \dots \\ -kp_ip_1 & \dots & kp_i(1-p_i) & -kp_ip_j & \dots & -kp_ip_{(q-1)} \\ -kp_jp_1 & \dots & -kp_jp_i & kp_j(1-p_j) & \dots & -kp_jp_{(q-1)} \\ \dots & \dots & \dots & \dots & \dots & \dots \\ -kp_{(q-1)}p_1 & \dots & -kp_{(q-1)}p_i & -kp_{(q-1)}p_j & \dots & kp_{(q-1)}(1-p_{(q-1)}) \end{bmatrix},$$

and, provided none of the probabilities $[p_1, \dots, p_i, p_j, \dots, p_q]$ are null, this smaller covariance matrix $\boldsymbol{\Sigma}_{((q-1)\times(q-1))}$ is regular and its inverse $\mathbf{V}^{-1}$ has the value

$$(1/k)\begin{bmatrix} [(1/p_1)+(1/p_q)] & \dots & (1/p_q) & (1/p_q) & \dots & (1/p_q) \\ \dots & \dots & \dots & \dots & \dots & \dots \\ (1/p_q) & \dots & [(1/p_i)+(1/p_q)] & (1/p_q) & \dots & (1/p_q) \\ (1/p_q) & \dots & (1/p_q) & [(1/p_j)+(1/p_q)] & \dots & (1/p_q) \\ \dots & \dots & \dots & \dots & \dots & \dots \\ (1/p_q) & \dots & (1/p_q) & (1/p_q) & \dots & [(1/p_{(q-1)})+(1/p_q)] \end{bmatrix}.$$

The preceding facts may be used to explain why Pearson's criterion, used in chapters 15 and 16, follows the χ^2 distribution approximately. If the polynomial distribution is applied to a one-way frequency table comprising $R=q$ rows (section 15.2) for instance,

the total frequency N of the frequency table corresponds to the exponent k

of the polynomial distribution $\big(P(X_1), \dots, P(X_R)\big)^N = (p_1 + \dots + p_q)^k = 1$,

the observed frequencies $[F_1, \dots, F_R]$ of the frequency table correspond

to the values of the polynomial variates $[X_1, \dots, X_q]$,

and the expected frequencies $[E_1, \dots, E_R] = [NP(X_1), \dots, NP(X_R)]$ correspond

to the means $\boldsymbol{\mu} = [kp_1, \dots, kp_q]$.

On cursory examination, Pearson's criterion,

$$\sum_{i=1}^{R} (F_i - E_i)^2/E_i \;=\; \sum_{i=1}^{R} \big(F(X_i) - NP(X_i)\big)^2/NP(X_i) \;=\; \sum_{i=1}^{q} (X_i - kp_i)^2/kp_i\,,$$

does not seem equivalent to a sum of squares of standardized variates, because the denominator $E_i = NP(X_i) = kp_i$ of each of its q terms is equal to the mean $\mu_i = kp_i$ rather than the variance $\sigma_X^2 = kp_i(1-p_i)$ of the corresponding variate $F_i = X_i$. Moreover, it is not immediately obvious that, in the analysis of a one-way frequency table (section 15.2 and chapter 16) in which the probabilities are known theoretically, the number of degrees of freedom of Pearson's criterion is $(R-1)=(q-1)$ rather than $R=q$. However, both doubts may be cleared by an algebraic demonstration that the sum of the q terms of Pearson's criterion is equal to a quadratic form in only $(q-1)$ variates,

$$\big[(X_1 - kp_1) \quad \dots \quad (X_{(q-1)} - kp_{(q-1)})\big]\ \mathbf{V}^{-1} \begin{bmatrix} (X_1 - kp_1) \\ \dots \\ (X_{(q-1)} - kp_{(q-1)}) \end{bmatrix},$$

which possesses the same value whichever variate is omitted. This quadratic form is equivalent to a sum of squares of $(q-1)$ independent standardized variates and, since the polynomial distribution becomes more and more closely similar to the multivariate normal distribution when its exponent k increases, Pearson's criterion thus follows an approximate χ^2 distribution with $(q-1)=(R-1)$ degrees of freedom. When Karl Pearson (1857-1936) proposed, almost a century ago (Pearson, 1900), to apply the χ^2 distribution to the analysis of frequency tables, he made a pioneering discovery. The analysis of frequency tables with the χ^2 distribution is considered as one of the oldest among the so-called *nonparametric* methods.

Chapter 30

The distribution of Hotelling's T^2

Section 30.1: the multivariate generalization of Student's t

When several variates $X_1, \ldots, X_i, X_j, \ldots, X_q$ follow a multivariate normal distribution, the quadratic form $(\mathbf{X}-\boldsymbol{\mu})\,\boldsymbol{\Sigma}^{-1}(\tilde{\mathbf{X}}-\tilde{\boldsymbol{\mu}})$, which appears in the q-dimensional normal probability density, follows a χ^2 distribution with q degrees of freedom (section 29.1). In theory, the χ^2 distribution could thus be used to test hypotheses or to delimit confidence regions concerning the mean vector $\boldsymbol{\mu}$ if the parametric covariance matrix $\boldsymbol{\Sigma}$ were known. In practice, however, the population covariance matrix $\boldsymbol{\Sigma}$ is seldom known and is generally replaced by its estimate, the sample covariance matrix $\mathbf{S}$ (section 29.3).

Hotelling's T^2 distribution, which constitutes a generalization of Student's t distribution (chapter 6), must then be used instead of the χ^2 distribution. Discovered in 1931 by the American statistician Harold Hotelling (1895-1973), the T^2 distribution is related to the variance ratio (F) distribution (chapter 8) through the equation

$$T^2_{(P\,;\,q,\,\nu)} = [(\nu q)/(\nu - q + 1)]\, F_{(P\,;\,q,\,\nu-q+1)}\,,$$

where P is the cumulative probability, q is the number of dimensions occupied by variation and ν is the number of degrees of freedom of the estimated covariance matrix S used in the calculation of the T^2 criterion. In practice, the T^2 values necessary to determine confidence regions can thus be obtained from variance ratio tables through the above equation. Inversely, T^2 values obtained in significance tests can be transformed into variance ratios through the equation

$$F_{(P\,;\,q,\,\nu-q+1)} = [(\nu - q + 1)/(\nu q)]\, T^2_{(P\,;\,q,\,\nu)}\,.$$

Therefore, the utilization of Hotelling's T^2 does not require special tables. Hotelling's T^2 distribution is particularly easy to use when there are only $q = 2$ variates, since the cumulative probability P is then simply

$$P = 1 - [\nu/(\nu + T^2_{(P\,;\,2,\,\nu)})]^{(\nu-1)/2}\;;$$

inversely, $$T^2_{(P\,;\,2,\,\nu)} = \nu\,[1/(1-P)^{2/(\nu-1)} - 1]\,.$$

As for the limiting case where there is only one variate ($q = 1$), Hotelling's T^2 is then simply equal to a variance ratio or to the square of Student's t, and its number of degrees of freedom ν is the same as that of the corresponding variance ratio ($\nu - 1 + 1 = \nu$):

$$T^2_{(1,\,\nu)} = F_{(1,\,\nu)} = t^2_{(\nu)}\,.$$

When there are several variates ($q \geq 2$), however, the numbers of degrees of freedom of Hotelling's T^2 and of the denominator of the corresponding F are unequal, and differ occasionally from what would be expected intuitively (see section 30.7, for instance).

Contrary to Student's t, the square root of Hotelling's T^2 is not used because, when there are several variates, it would be impossible to define nonarbitrarily a unique direction opposing the positive and negative values of such a square root in multidimensional space.

Section 30.2: hypotheses concerning the mean vector μ

In the same manner as in the univariate case (section 9.5), when a vector of several variates $\mathbf{X} = [X_1, \ldots, X_q]$ follows a multivariate normal distribution of which the mean vector and covariance matrix are μ and Σ,

$$\mathbf{X} \leftarrow \mathcal{N}(\mu, \Sigma),$$

the mean vector $\bar{\mathbf{X}}$ of a sample of size N follows a multivariate normal distribution of which the mean vector is also μ but of which the parametric covariance matrix, $(1/N)\Sigma$, is N times smaller than that of individual observations:

$$\bar{\mathbf{X}} \leftarrow \mathcal{N}(\mu, (1/N)\Sigma).$$

The quadratic form appearing in the probability density of the distribution of $\bar{\mathbf{X}}$ is therefore $N(\bar{\mathbf{X}} - \mu)\,\Sigma^{-1}\,(\tilde{\bar{\mathbf{X}}} - \tilde{\mu})$ and, if the parametric covariance matrix Σ were known, the acceptance region of the preliminary hypothesis H_0: $\mu = \mu_0$ would be

$$N(\bar{\mathbf{X}} - \mu_0)\,\Sigma^{-1}\,(\tilde{\bar{\mathbf{X}}} - \tilde{\mu}_0) \leq \chi^2_{(1-\alpha;\,q)}.$$

In practice, the parametric covariance matrix Σ is generally unknown and is replaced by the sample covariance matrix $\mathbf{S}$, and the acceptance region of the preliminary hypothesis is

$$N(\bar{\mathbf{X}} - \mu_0)\mathbf{S}^{-1}(\tilde{\bar{\mathbf{X}}} - \tilde{\mu}_0) \leq T^2_{(1-\alpha;\,q,\,N-1)} = [((N-1)q)/(N-q)]\,F_{(1-\alpha;\,q,\,N-q)}.$$

The above criterion may be used, for instance, to test the hypothesis that the mean vector μ_Z of bilateral asymmetries of the human skeletons considered in sections 29.4 and 29.6 is null on the average:

$$H_0\colon \mu = \mu_0 = \mathbf{0} = [0, 0, 0, 0].$$

According to this hypothesis, even if the bilateral symmetry of the lengths of arm and leg bones is subject to individual fluctuations, it would be perfect on the average. The mean vectors $\bar{\mathbf{Z}}$ and covariance matrices $\mathbf{S}_{ZZ}$ of the samples of male and female skeletons are given in table 30.2.1 (next page). The values of Hotelling's T^2 fall in the rejection region of the preliminary hypothesis at all usual significance levels, since $T^2_{(4,\,116)} = 76.3039$ in men, which corresponds to the variance ratio $F_{(4,\,113)} = 18.5826$, and $T^2_{(4,\,109)} = 118.261$ in women, which corresponds to the variance ratio $F_{(4,\,106)} = 28.7516$. The hypothesis that arm and leg bones have the same length on the average on both sides of the body may thus be rejected conclusively ($P < 0.005$). It is interesting to note that the present test constitutes a generalization of the method of paired measurements (section 9.9).

Table 30.2.1
Mean vectors $\bar{\mathbf{Z}}$ and covariance matrices $\mathbf{S}_{ZZ}$ of bilateral asymmetries in mm of the humerus, radius, femur and tibia lengths of adult human skeletons (Jolicoeur, 1963c)

117 men	110 women
$[2.37607 \quad 1.96581 \quad -1.10256 \quad 0.042735]$	$[3.05455 \quad 2.57273 \quad -0.672727 \quad -0.036364]$
$\begin{bmatrix} 17.1160 & 4.71986 & 0.314766 & 2.49241 \\ & 9.96434 & -0.477675 & 2.13078 \\ & & 10.2653 & 0.978559 \\ & & & 13.0585 \end{bmatrix}$	$\begin{bmatrix} 13.5016 & 5.50976 & -0.045538 & 2.10292 \\ & 8.41209 & 0.829191 & 0.764137 \\ & & 7.78182 & 2.00284 \\ & & & 13.8335 \end{bmatrix}$

From a statistical viewpoint, in comparison with univariate methods, Hotelling's T^2 has the advantage of reducing the number of necessary comparisons as well as the risk of erroneously rejecting the preliminary hypothesis. This quality is analogous to that of the overall test of mean differences in the analysis of variance (section 12.1). Thus, if the average length differences of humeri, radii, femora and tibiae were tested using Student's t, four univariate tests would be necessary to replace each four-dimensional Hotelling's T^2 test. But the utilization of Hotelling's T^2 possesses an even greater asset: it allows positive or negative correlations between variates to be taken into account, which can affect conclusions drastically as shown in section 30.4.

Univariate methods should not be considered as obsolete for all that. Thus, if the null hypothesis is tested successively for each element of the mean vector using Student's t, it is found that average bilateral differences are conclusively positive for arm bones, conclusively negative for the femur, and not significant for the tibia. In the case of these data on the bilateral asymmetries of the human skeleton, univariate methods are therefore useful, because they show that the various limb bones do not depart from bilateral symmetry in the same way.

Section 30.3: confidence regions concerning the mean vector $\boldsymbol{\mu}$

When there is no preliminary hypothesis concerning the value of the population mean vector $\boldsymbol{\mu}$, the confidence region containing all likely values of that vector satisfies the following inequality:

$$N(\bar{\mathbf{X}}-\boldsymbol{\mu})\mathbf{S}^{-1}(\tilde{\bar{\mathbf{X}}}-\tilde{\boldsymbol{\mu}}) \le T^2_{(1-\alpha\,;\,q,\,N-1)} = [((N-1)q)/(N-q)]\,F_{(1-\alpha\,;\,q,\,N-q)}\,.$$

This confidence region has the shape of an ellipse when there are $q=2$ variates, of an ellipsoid when $q=3$, and of a hyperellipsoid when $q \ge 4$. While it is possible to represent a confidence ellipsoid in a three-dimensional diagram, the bivariate confidence ellipse is certainly the most easily drawn and interpreted and the most widely used version. The confidence ellipse at level $\alpha=0.05$ of the mean vector $\boldsymbol{\mu}=[\mu_1, \mu_2]$ of body weight X_1 in kg and skull length X_2 in mm of 60 adult female wolves (Jolicoeur, 1959) is illustrated in figure 30.3.1.

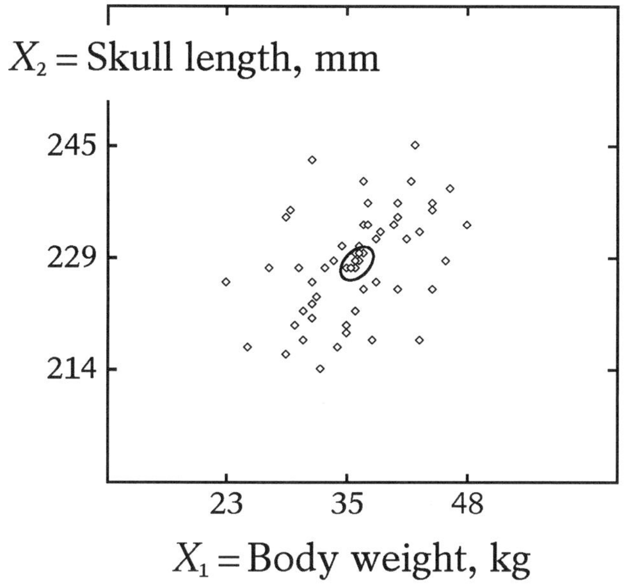

Figure 30.3.1
Confidence ellipse, at level $\alpha = 0.05$, of the mean vector $\boldsymbol{\mu} = [\mu_1, \mu_2]$ of body weight X_1 in kg and skull length X_2 in mm for 60 adult female wolves of the Canadian Northwest Territories (Jolicoeur, 1959)

The confidence ellipse of the population mean vector $\boldsymbol{\mu}$, which reflects the uncertainty of the sample mean vector $\overline{\mathbf{X}}$, has the same shape but a remarkably smaller size than the prediction ellipse calculated earlier (figure 19.7.1) at the same level, which reflected the extent of individual variation. The smallness of the confidence ellipse of $\boldsymbol{\mu}$ is due ultimately to the fact that the means $\overline{\mathbf{X}}$ of samples of size N are less variable than individual observations $\mathbf{X}$, a particular consequence of the law of large numbers already mentioned in section 9.5. The procedure for drawing a confidence ellipse using a computer will be discussed in detail at the same time as that for drawing prediction and variation ellipses in section 30.8.

Section 30.4: do two populations have equal mean vectors?

As in the univariate case (section 9.8), the vector $(\overline{\mathbf{X}}_B - \overline{\mathbf{X}}_A)$ of the difference between the mean vectors $\overline{\mathbf{X}}_A$ and $\overline{\mathbf{X}}_B$ of two samples of sizes N_A and N_B drawn from two multivariate normal populations $\mathcal{N}(\boldsymbol{\mu}_A, \boldsymbol{\Sigma}_A)$ and $\mathcal{N}(\boldsymbol{\mu}_B, \boldsymbol{\Sigma}_B)$ follows a multivariate normal distribution of which the parametric mean vector is $(\boldsymbol{\mu}_B - \boldsymbol{\mu}_A)$ and the parametric covariance matrix is $(1/N_A)\boldsymbol{\Sigma}_A + (1/N_B)\boldsymbol{\Sigma}_B$:

$$(\overline{\mathbf{X}}_B - \overline{\mathbf{X}}_A) \leftarrow \mathcal{N}\left(\boldsymbol{\mu}_B - \boldsymbol{\mu}_A,\ (1/N_A)\boldsymbol{\Sigma}_A + (1/N_B)\boldsymbol{\Sigma}_B\right).$$

Since the parametric covariance matrices $\boldsymbol{\Sigma}_A$ and $\boldsymbol{\Sigma}_B$ are generally unknown in practice, they are usually assumed to be equal, $\boldsymbol{\Sigma}_A = \boldsymbol{\Sigma}_B = \boldsymbol{\Sigma}$, which simplifies the problem of comparing means.

The pooled sample estimate $\mathbf{S}_{\text{within-groups}}$ of the covariance matrix $\boldsymbol{\Sigma}$ is then obtained by adding the matrices of centered sums of both samples and by dividing the resulting matrix by the total numbers of degrees of freedom, $(N_A - 1) + (N_B - 1) = (N_A + N_B - 2)$. If the sample covariance matrices of both groups, $\mathbf{S}_A$ and $\mathbf{S}_B$, have already been calculated, the pooled within-groups covariance matrix may also be evaluated as follows:

$$\mathbf{S}_{\text{within-groups}} = \mathbf{S} = [1/(N_A + N_B - 2)]\,[(N_A - 1)\,\mathbf{S}_A + (N_B - 1)\mathbf{S}_B] .$$

As for the covariance matrix of the vector $(\overline{\mathbf{X}}_B - \overline{\mathbf{X}}_A)$, its estimate is

$$[(1/N_A) + (1/N_B)]\,\mathbf{S}_{\text{within-groups}} = [(N_A + N_B)/(N_A\,N_B)]\,\mathbf{S} .$$

The acceptance region of the preliminary hypothesis that the vector of differences between parametric means is $(\boldsymbol{\mu}_B - \boldsymbol{\mu}_A)$ may then be expressed with respect to Hotelling's T^2:

$$\frac{(N_A\,N_B)}{(N_A + N_B)}\left((\overline{\mathbf{X}}_B - \overline{\mathbf{X}}_A) - (\boldsymbol{\mu}_B - \boldsymbol{\mu}_A)\right)\mathbf{S}^{-1}\left((\tilde{\overline{\mathbf{X}}}_B - \tilde{\overline{\mathbf{X}}}_A) - (\tilde{\boldsymbol{\mu}}_B - \tilde{\boldsymbol{\mu}}_A)\right) \leq T^2_{(1-\alpha;\,q,\,N_A + N_B - 2)} ,$$

where $\mathbf{S}^{-1}$ is the inverse of the within-groups covariance matrix

and $T^2_{(1-\alpha;\,q,\,N_A + N_B - 2)} = [((N_A + N_B - 2)q)/(N_A + N_B - q - 1)]\,F_{(1-\alpha\,;\,q,\,N_A + N_B - q - 1)}$.

In the particular case of the null hypothesis, the above expression becomes simpler because the vector $(\boldsymbol{\mu}_B - \boldsymbol{\mu}_A)$ vanishes:

$$\frac{(N_A\,N_B)}{(N_A + N_B)}(\overline{\mathbf{X}}_B - \overline{\mathbf{X}}_A)\,\mathbf{S}^{-1}(\tilde{\overline{\mathbf{X}}}_B - \tilde{\overline{\mathbf{X}}}_A) \leq T^2_{(1-\alpha;\,q,\,N_A + N_B - 2)} .$$

The comparison of the mean vectors of two samples through Hotelling's T^2 may be illustrated on data from a study (Ford, 1958) on the kidney development of the Pacific pink salmon, *Onchorynchus gorbuscha* (Walbaum). The renal glomeruli of bony fishes are used mostly by freshwater species to excrete excess water absorbed through osmosis. Therefore, Peter Ford, a British-born Canadian embryologist, tried to verify whether, at the time at which they would normally go back to sea, fry of Pacific pink salmon raised in a hatchery would slow down the rate of increase of renal glomeruli after being transferred to saltwater. Seven months after hatching, a group B of fry was thus transferred to salt water while another group A was kept in freshwater. During the following six weeks, a few fry of each group were sacrificed, their fork length X_1 was measured in mm and the number of glomeruli X_2 of their kidneys was counted. Ford based his analysis on over twenty specimens taken in each group after being maintained in saltwater (or in freshwater) from one to six weeks, but only two small samples obtained after five weeks and containing $N_A = N_B = 4$ specimens each are used here.

The mean vectors and the covariance matrices of both samples are given in table 30.4.1. If the means of the two samples are compared through univariate tests using Student's t distribution, one finds $t_{(6)} = +1.6397$ for fork length and $t_{(6)} = -0.595302$ for kidney glomeruli. Neither of these criteria is significant even at the least rigorous of usual levels of statistical significance ($\alpha = 0.05$). However, a bivariate comparison of the two sample mean vectors using Hotelling's T^2 yields

$$T^2_{(2,6)} = 29.1886, \text{ whence } F_{(2,5)} = 12.1619 > 8.4336 = F_{(0{,}975;\,2,\,5)},$$

which allows the null hypothesis to be rejected at the level $\alpha = 0.025$.

Table 30.4.1
Mean vectors $\overline{\mathbf{X}}$ and covariance matrices S of the fork length X_1 in mm and of the number X_2 of glomeruli of the kidney of Pacific pink salmon fry according to data from Ford (1958)

Group	A: 4 fry kept in fresh water	B: 4 fry transferred to salt water
$\overline{\mathbf{X}}$	[35.65 292]	[41.10 273]
S	$\begin{bmatrix} 6.810 & 105.667 \\ & 1894.000 \end{bmatrix}$	$\begin{bmatrix} 37.380 & 282.000 \\ & 2180.670 \end{bmatrix}$

The reasons for which Hotelling's T^2 test is more sensitive than the Student's t tests are made evident by diagrams of the acceptance region at level $\alpha = 0.05$ (figure 30.4.1, left) of the hypothesis that the vector $(\mu_B - \mu_A) = \mathbf{0}$ or of the confidence interval of the vector $(\mu_B - \mu_A)$ at the same level (figure 30.4.1, right). In the left side of figure 30.4.1, the ellipse centered on the hypothetical vector $(\mu_B - \mu_A) = \mathbf{0} = [0,\ 0]$ is the bivariate acceptance region based on Hotelling's T^2, while the rectangle is the intersection of the two univariate acceptance regions based on Student's t.

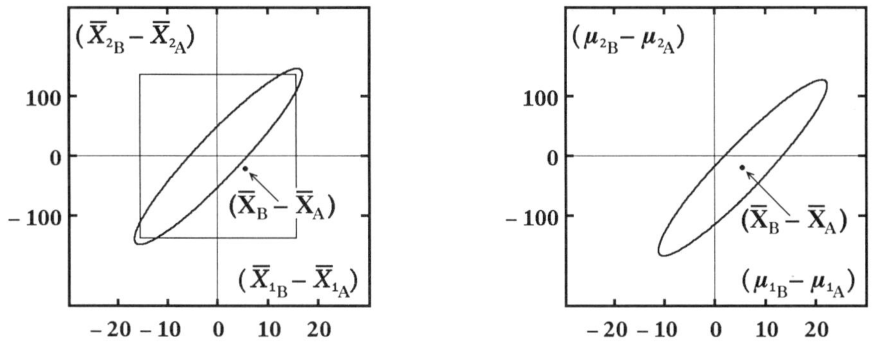

Figure 30.4.1
Fork length X_1 and number X_2 of the glomeruli of the kidney of Pacific pink salmon fry kept in fresh water (A) and transferred to salt water (B); left: acceptance region of the hypothesis that the vector $(\mu_B - \mu_A) = \mathbf{0}$; right: confidence region of the vector $(\mu_B - \mu_A)$

The vector $(\overline{\mathbf{X}}_B - \overline{\mathbf{X}}_A)$ of observed differences between sample means falls outside of the ellipse (figure 30.4.1, left), which takes into account the correlation between the two variates, but inside the rectangle, which neglects this correlation. Similarly, if a confidence region is used instead of a significance test, the confidence ellipse (figure 30.4.1, right), which is centered on the observed vector $(\overline{\mathbf{X}}_B - \overline{\mathbf{X}}_A)$, does not cover the hypothetical vector $(\mu_B - \mu_A) = \mathbf{0} = [0,\ 0]$.

The disagreement between the conclusions reached using Student's t and Hotelling's T^2 is due in part to the strong positive correlation between fork length and the number of glomeruli: $+0.930410$ in group A, $+0.987722$ in group B, and $+0.913587$ within groups $(0.01 < P < 0.02)$. Hotelling's T^2 test is more sensitive because it takes this strong positive correlation into account and combines the information coming from the two variates. In the present analysis, the simultaneous utilization of two univariate acceptance regions would be inefficient not only because it would be less sensitive than the bivariate elliptical region but also because it would entail a greater risk (approximately 2α) of erroneously rejecting the null hypothesis (see again section 26.1).

The fry transferred to saltwater thus appear to differ from those kept in freshwater by a smaller number of kidney glomeruli together with a greater fork length. In his original analysis, Ford (1958) used an analysis of covariance (chapter 21) to compare the estimation lines of the number of glomeruli from fork length, the latter serving somewhat as a measurement of *physiological age*. That approach is debatable because fork length too appears to be influenced by the transfer to saltwater. Even if growth rates may vary considerably among individual fry, it would undoubtedly have been better to use chronological age, which was known, as a predictor variate and to treat the other two variates as possibly affected by the experiment. The utilization of Hotelling's T^2 is nevertheless fruitful here, since the present analysis brings out a decrease in the number of renal glomeruli linked to an increase in fork length even though it is based only on a small part of Ford's original data.

Multivariate statistical methods strictly speaking, such as Hotelling's T^2 comparison of two sample mean vectors, have been used until now mostly in descriptive biology (anthropometry, quantitative taxonomy, observational ecology, etc.), and have seldom been applied to experimental data. Thanks to the influence of the classical works of Fisher (1960, 1970), most research workers are now aware of the necessity of designing experiments so as to test the effects and the interactions of many factors simultaneously, but many still do not realize that even a single experimental factor may have multivariate effects, for the analysis of which multivariate methods may be advisable.

Section 30.5: comparing an individual vector X with a sample mean vector $\overline{\mathbf{X}}$

Like in the univariate case (section 9.10), the comparison of an individual vector **X** with a sample is treated as a particular case of the comparison of the mean vectors $\overline{\mathbf{X}}_A$ and $\overline{\mathbf{X}}_B$ of two samples. The individual vector **X** is then treated as a sample of size $N_B = 1$, which allows the parametric mean vector $\boldsymbol{\mu}_B$ to be estimated but does not supply any information concerning the covariance matrix $\boldsymbol{\Sigma}_B$ of its population. The parametric covariance matrix $\boldsymbol{\Sigma}_A = \boldsymbol{\Sigma}$ is thus estimated exclusively from sample A by the covariance matrix $\mathbf{S}_A = \mathbf{S}$ of which the number of degrees of freedom is $(N_A - 1) = (N - 1)$. The Hotelling's T^2 criterion used to compare an individual vector **X** with the mean vector $\overline{\mathbf{X}}$ of a sample of size N is

$$\frac{N}{(N+1)}\left((\mathbf{X}-\overline{\mathbf{X}})-(\boldsymbol{\mu}_B-\boldsymbol{\mu}_A)\right)\mathbf{S}^{-1}\left((\tilde{\mathbf{X}}-\tilde{\overline{\mathbf{X}}})-(\tilde{\boldsymbol{\mu}}_B-\tilde{\boldsymbol{\mu}}_A)\right) \leq T^2_{(1-\alpha;\,q,\,N-1)} \ ,$$

where $T^2_{(1-\alpha;\,q,\,N-1)} = [((N-1)q)/(N-q)]\,F_{(1-\alpha\,;\,q,\,N-q)}$.

In the particular case of the null hypothesis H_0: $(\boldsymbol{\mu}_B - \boldsymbol{\mu}_A) = \mathbf{0}$, the acceptance region takes the simpler form

$$\frac{N}{(N+1)}\,(\mathbf{X}-\overline{\mathbf{X}})\,\mathbf{S}^{-1}(\tilde{\mathbf{X}}-\tilde{\overline{\mathbf{X}}}) \leq T^2_{(1-\alpha;\,q,\,N-1)} \ .$$

Section 30.6: prediction and variation regions

As in the univariate case (section 9.11), when a new individual vector **X** is assumed to come from the same population as a sample of which the mean vector is $\overline{\mathbf{X}}$, the last inequality of the preceding section can be used to delimit a prediction region in which the new individual vector is likely to fall. While it is possible to represent a trivariate prediction region graphically (see the ellipsoid in figure 23.2.1), the most frequently used version is certainly the bivariate prediction ellipse, which corresponds to the inequality

$$\frac{N}{(N+1)}\,(\mathbf{X}-\overline{\mathbf{X}})\,\mathbf{S}^{-1}(\tilde{\mathbf{X}}-\tilde{\overline{\mathbf{X}}}) \leq T^2_{(1-\alpha;\,2,\,N-1)} = (N-1)[1/\alpha^{2/(N-2)} - 1] \ .$$

In addition to being suitable for statistical inference, a prediction region is often useful in quantitative biology as an indicator of the observed or estimated extent of individual variation, provided sample size is large enough (let us say $N \geq 50$). Thus, when the variates are taxonomic characters largely determined by heredity, a prediction region provides a graphical description of the degree of genetic heterogeneity of the biological population under study. If the variates are characteristics of the habitat of a species, such as depth, salinity and water temperature for a marine organism, the prediction region delimits the set of environmental conditions which are suitable for the species and which may contribute to define its *ecological niche* (Green, 1971). Finally, if the variates are spatial (and temporal?) coordinates, the prediction region may reflect the extent of the *territory* defended or of the *home range* frequented by an individual organism.

In the case of a small sample, however, the prediction region is often excessively large in comparison with the observed extent of individual variation, because it is enlarged by the uncertainty of the sample mean vector $\overline{\mathbf{X}}$ and covariance matrix $\mathbf{S}$. Even though it is still suitable as a tool of statistical inference, a prediction region based on a small sample thus does not reflect the extent of individual variation satisfactorily. When what is really needed is a description of the observed or estimated extent of individual variation, a *variation region* (Jolicoeur and Heusner, 1986) should be used instead of a prediction region. A variation region is obtained simply by replacing the parameters $\boldsymbol{\mu}$ and $\boldsymbol{\Sigma}$ of the statistical population by the sample mean vector $\overline{\mathbf{X}}$ and covariance matrix $\mathbf{S}$ and by using the normal distribution instead of Student's t as well as the χ^2 distribution instead of Hotelling's T^2 or of the variance ratio F. A q-dimensional variation region is therefore defined by the inequality

$$(\mathbf{X}-\overline{\mathbf{X}})\,\mathbf{S}^{-1}(\widetilde{\mathbf{X}}-\widetilde{\overline{\mathbf{X}}}) \leq \chi^2_{(1-\alpha;q)}\ .$$

When there are $q=2$ variates, the variation ellipse corresponds to the inequality

$$(\mathbf{X}-\overline{\mathbf{X}})\,\mathbf{S}^{-1}(\widetilde{\mathbf{X}}-\widetilde{\overline{\mathbf{X}}}) \leq \chi^2_{(1-\alpha;2)}\ .$$

In the case of a small sample ($N=10$) like the one extracted from the data of Brown and Beerstecher (1951), already discussed in sections 19.1, 19.3 and 19.5, the variation ellipse is smaller and reflects the extent of individual variation better than the prediction ellipse (figure 30.6.1).

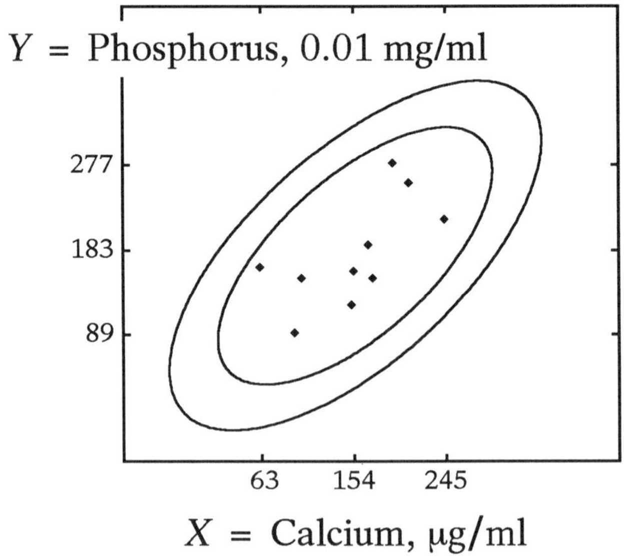

Figure 30.6.1
Scatter diagram of calcium (µg/ml) and phosphorus (0.01 mg/ml) concentrations in morning urine specimens from underweight adult men; prediction ellipse (large) and variation ellipse (small) at level $\alpha = 0.05$

Section 30.7: the vector of coefficients of a multiple regression

Hotelling's T^2 distribution may also be used to test hypotheses or to delimit confidence regions concerning the vector $\boldsymbol{\beta} = [\beta_1 \; ... \; \beta_k]$ of parametric coefficients of a multiple regression. In the case of the null hypothesis, H_0: $\boldsymbol{\beta} = [\beta_1 \; ... \; \beta_k] = \boldsymbol{\beta}_0 = \mathbf{0} = [0 \; ... \; 0]$, the variance ratio corresponding to Hotelling's T^2 is equal to the F value obtained when an analysis of variance is done on the multiple regression (section 25.6) or when the null hypothesis H_0: $\mathcal{R}_{(k+1).1...k} = 0$ is tested concerning the parametric multiple correlation coefficient (section 25.5):

$$F_{(k,\, N-1-k)} = (1/k)\,(\mathbf{b} - \boldsymbol{\beta}_0)\,[S_{b_i b_j}]^{-1}\,(\tilde{\mathbf{b}} - \tilde{\boldsymbol{\beta}}_0) = \frac{(N-1-k)\,R^2_{(k+1).1...k}}{k\,[1 - R^2_{(k+1).1...k}]} ,$$

where $\mathbf{b} = [b_1 \; ... \; b_k]$ is the vector of sample estimates of the coefficients of the multiple regression and $[S_{b_i b_j}]$ is the estimated covariance matrix of vector $\mathbf{b}$ (section 25.5). Moreover, the above variance ratio may also be used to test nonnull preliminary hypotheses or to delimit confidence regions concerning the parametric vector $\boldsymbol{\beta}$:

$$(1/k)\,(\mathbf{b} - \boldsymbol{\beta})\,[S_{b_i b_j}]^{-1}\,(\tilde{\mathbf{b}} - \tilde{\boldsymbol{\beta}}) \leq F_{(1-\alpha;\, k,\, N-1-k)} .$$

It is however disturbing to note that the corresponding Hotelling's T^2 value possesses $(N-2)$ degrees of freedom though the covariance matrix $[S_{b_i b_j}]$ of the multiple regression coefficients used in the computation of that T^2 value has $(N-1-k)$ degrees of freedom. This paradox is related to a duality question discussed by Kshirsagar (1972, pages 167-170), and reflects the theoretical complexity of multivariate analysis. The present formulation in terms of the variance ratio is more intuitive and will undoubtedly be preferred by most users.

The confidence regions of the coefficients of a multiple regression may be similar to those of the difference between two sample mean vectors (figure 30.4.1, right). In the particular case where there are two predictor variates, the correlation coefficient between the multiple regression coefficients b_1 and b_2 has the same absolute value as the correlation coefficient between the predictor variates, but the opposite sign. When two predictor variates are positively correlated and their multiple regression coefficients b_1 and b_2 have the same sign, the negative correlation between those coefficients may be considered as *compensatory*: when the value of the b_1 coefficient falls particularly low in random sampling, for instance, the value of the b_2 coefficient tends to fall high, which allows the X_2 predictor variate to compensate for the unusual predictive performance of variate X_1. Unlike total or partial correlations between biological variates (chapters 19, 23, 25 and 29), the correlations between the sample estimates of multiple regression coefficients must presumably be considered mostly as sampling artefacts for which biological interpretations should not be sought.

Section 30.8: plotting a confidence, a prediction, or a variation ellipse

Various methods have been used in the past for drawing confidence, prediction or variation ellipses. The first step of many of these methods was the calculation of the coordinates of the end points of the major and minor axes of the ellipse, which were plotted in the diagram. The ellipse was then delineated by using either a standard drafting technique or a special instrument known as an *ellipsograph* (available forty years ago from the Riefler company). Nowadays, it is easier to use a computer, and a short sequence of three changes of variables will now be described thanks to which an ellipse can be transformed into a circle (which is easier to draw) or vice versa (figure 30.8.1). In addition to providing an efficient method for plotting confidence, prediction and variation ellipses, these mathematical transformations constitute a simple bivariate example of principal component analysis, which will be treated in the next chapter (31).

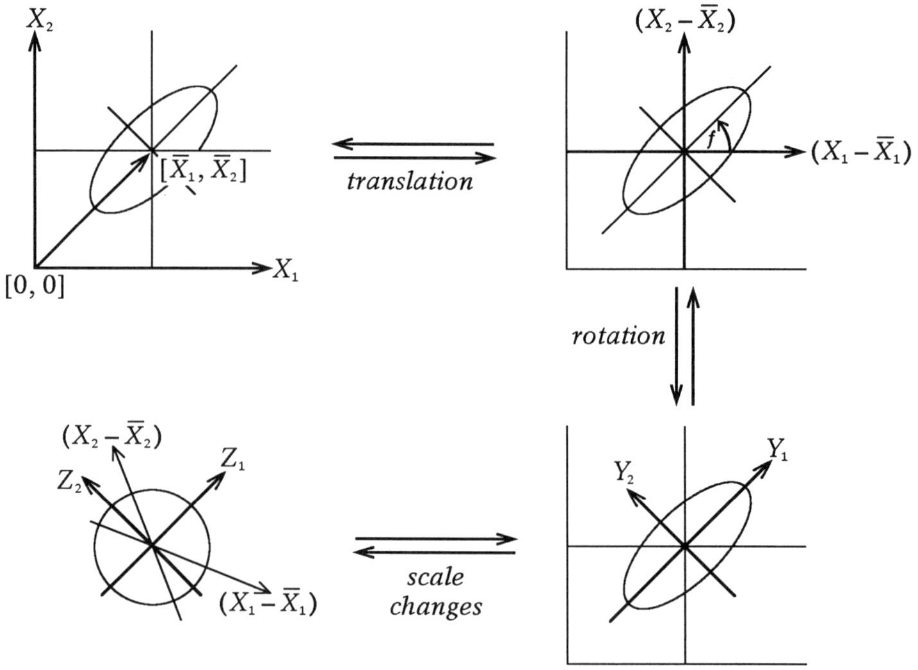

Figure 30.8.1
Translation, rotation and scale changes transforming the equation of an ellipse into that of a circle or vice versa

In order to draw a confidence, a prediction or a variation ellipse in a scatter diagram, the observed variates $\mathbf{X} = [X_1, \ X_2]$ may first be replaced by the centered variates $(\mathbf{X} - \bar{\mathbf{X}}) = [X_1 - \bar{X}_1, \ X_2 - \bar{X}_2]$, which is equivalent to moving the origin from the point $\mathbf{0} = [0, \ 0]$ to the sample mean vector $\bar{\mathbf{X}} = [\bar{X}_1, \bar{X}_2]$. Following this operation, known as a *translation,* the sample mean vector of the new (centered) variates is $\mathbf{0} = [0, \ 0]$. The coordinate axes may next be turned through an angle f, if necessary, so as to superpose the abscissa on the major (longest) axis of the ellipse by a *rotation.*

The angle f is chosen (section 22.2) so that

$$\tan(f) = \{(S_2^2 - S_1^2) + \sqrt{[(S_2^2 - S_1^2)^2 + 4S_{12}^2]}\}/(2S_{12}),$$

and the new variates are $\tilde{\mathbf{Y}} = \mathbf{U}(\tilde{\mathbf{X}} - \tilde{\bar{\mathbf{X}}})$, that is

$$\begin{bmatrix} Y_1 \\ Y_2 \end{bmatrix} = \begin{bmatrix} \cos(f) & \sin(f) \\ -\sin(f) & \cos(f) \end{bmatrix} \begin{bmatrix} (X_1 - \bar{X}_1) \\ (X_2 - \bar{X}_2) \end{bmatrix} = \begin{bmatrix} u_{11} & u_{12} \\ u_{21} & u_{22} \end{bmatrix} \begin{bmatrix} (X_1 - \bar{X}_1) \\ (X_2 - \bar{X}_2) \end{bmatrix},$$

where $\mathbf{U} = \begin{bmatrix} u_{11} & u_{12} \\ u_{21} & u_{22} \end{bmatrix} = \begin{bmatrix} \cos(f) & \sin(f) \\ -\sin(f) & \cos(f) \end{bmatrix}$ is the *rotation matrix.*

The mean vector of the new variates is $\bar{\mathbf{Y}} = [\bar{Y}_1, \bar{Y}_2] = [0, 0]$, their covariance is null, the variance is maximized in the direction of Y_1 and minimized in the direction of Y_2, and the covariance matrix is the diagonal matrix

$$\mathbf{D} = \begin{bmatrix} d_1 & 0 \\ 0 & d_2 \end{bmatrix} = \mathbf{U}\,\mathbf{S}\,\tilde{\mathbf{U}}.$$

The vectors $[u_{11}, u_{12}]$ and $[u_{21}, u_{22}]$ are the latent vectors and the scalars d_1 and d_2 are the latent roots (section 24.21) of the sample covariance matrix $\mathbf{S}$ of original variates X_1 and X_2. The transformation of the equation of the ellipse into that of a circle (section 24.24) can now be completed through a third and last change of variables:

$$\begin{bmatrix} Z_1 \\ Z_2 \end{bmatrix} = \begin{bmatrix} 1/\sqrt{d_1} & 0 \\ 0 & 1/\sqrt{d_2} \end{bmatrix} \begin{bmatrix} Y_1 \\ Y_2 \end{bmatrix} = \begin{bmatrix} 1/\sqrt{d_1} & 0 \\ 0 & 1/\sqrt{d_2} \end{bmatrix} \begin{bmatrix} u_{11} & u_{12} \\ u_{21} & u_{22} \end{bmatrix} \begin{bmatrix} (X_1 - \bar{X}_1) \\ (X_2 - \bar{X}_2) \end{bmatrix}.$$

The mean vector of the new variates is $\bar{\mathbf{Z}} = [\bar{Z}_1, \bar{Z}_2] = [0, 0]$ and their covariance matrix is a unit matrix $\mathbf{I}$. Therefore, the inequality defining the prediction region may be reexpressed as follows:

$$\frac{N}{(N+1)} (\mathbf{X} - \bar{\mathbf{X}})\,\mathbf{S}^{-1}(\tilde{\mathbf{X}} - \tilde{\bar{\mathbf{X}}}) = \frac{N}{(N+1)} \mathbf{Z}\,\mathbf{I}\,\tilde{\mathbf{Z}} = \frac{N}{(N+1)} (Z_1^2 + Z_2^2) \leq T^2_{(1-\alpha;\,2,\,N-1)}.$$

The division of the variates Y_1 and Y_2 by their respective standard deviations $\sqrt{d_1}$ and $\sqrt{d_2}$ constitutes a *change of scales.*

Once the coordinates of each one of the points of the circle have been determined with respect to the Z_1 and Z_2 variates by varying the angle between the abscissa and the radius from 0° to 360°, the coordinates of the corresponding point on the prediction ellipse of original variates X_1 and X_2 can be obtained from the triple inverse transformation

$$\begin{bmatrix} X_1 \\ X_2 \end{bmatrix} = \begin{bmatrix} \bar{X}_1 \\ \bar{X}_2 \end{bmatrix} + \begin{bmatrix} u_{11} & u_{21} \\ u_{12} & u_{22} \end{bmatrix} \begin{bmatrix} \sqrt{d_1} & 0 \\ 0 & \sqrt{d_2} \end{bmatrix} \begin{bmatrix} Z_1 \\ Z_2 \end{bmatrix}.$$

The following box contains a series of statements which may be inserted in a *True BASIC*™ computer program (see also sections 24.26 and 29.3 and appendices A and B) in order to draw a prediction ellipse in a scatter diagram. Earlier program statements (not given here) must be provided to calculate the sample mean vector $\overline{\mathbf{X}}$ and covariance matrix $\mathbf{S}$, to prepare the scatter diagram, and to choose the significance level α. In the fourth statement of the box, the expression SQR$[T^2*(N+1)/N]$ must be replaced by SQR(T^2/N) for the confidence ellipse of a mean vector $\boldsymbol{\mu}$ and by SQR$(\chi^2_{(1-\alpha;\,2)})$ for a variation ellipse.

```
! Prediction ellipse
OPTION ANGLE DEGREES
LET T2 = (N-1)*((1/alpha)^(2/(N-2))-1)
LET CircleRadius = SQR(T2*(N+1)/N)
LET D1 = ((S(1,1)+S(2,2)+SQR((S(1,1)-S(2,2))^2+4*S(1,2)^2))/2)
LET D2 = ((S(1,1)+S(2,2)-SQR((S(1,1)-S(2,2))^2+4*S(1,2)^2))/2)
LET U(1,1) = S(1,1)-S(2,2)+SQR((S(1,1)-S(2,2))^2+4*S(1,2)^2)
LET U(1,2) = 2*S(1,2)
LET SQ = SQR(U(1,1)^2+U(1,2)^2)
LET U(1,1),U(2,2) = U(1,1)/SQ
LET U(1,2) = U(1,2)/SQ
LET U(2,1) = -U(1,2)
FOR Angle = 0 TO 360
   LET Z(1) = COS(Angle)*CircleRadius
   LET Z(2) = SIN(Angle)*CircleRadius
   LET Y(1) = Z(1)*SQR(D1)
   LET Y(2) = Z(2)*SQR(D2)
   LET X(1) = M(1)+U(1,1)*Y(1)+U(2,1)*Y(2)
   LET X(2) = M(2)+U(1,2)*Y(1)+U(2,2)*Y(2)
   PLOT X(1),X(2);
NEXT Angle
PLOT X(1),X(2)
```

Section 30.9: antilogarithmic transform of an elliptical region

When the frequency distribution of the data resembles the lognormal (chapter 14, sections 19.8, 20.7, 22.5, 29.7 and 29.8) rather than the normal distribution, the confidence, prediction and variation regions obtained for the logarithms of the data may be reexpressed with respect to original variates. This inverse transformation has been illustrated in the case of anatomical data in figure 19.8.1, but it may be very useful too in the case of ecological data. The relationship between the dry weights of benthos and plankton in kg/ha in twelve Saskatchewan lakes studied by Rawson (1960) is represented in figure 30.9.1. Sample size is in fact $N = 13$ because two parts of one of the lakes were studied separately. The scatter diagram of original data (figure 30.9.1, right) shows high degrees of heteroscedasticity and skewness, which suggest that the bivariate normal distribution would be very ill-suited for the analysis of these data. On the contrary, logarithmically transformed data give no graphical indications of heteroscedasticity or skewness (figure 30.9.1, left). Moreover, while the correlation of original variates does not reach statistical significance (H_0: $\rho = 0$, H_1: $\rho \neq 0$; $r = +0.492140$, $\nu = 11$, $P > 0.05$), the correlation of logarithms is conclusive ($r = +0.710586$, $P < 0.01$).

The scatter of logarithmically transformed data has a clearly elliptical shape which agrees well with the variation ellipse (figure 30.9.1, left). On the contrary, the shape of the scatter of original data is very different from that of an ellipse (figure 30.9.1, right). The logarithmic transformation has drastic effects here because, when the abundance of plants or animals is measured, observed quantities often vary over several orders of magnitude. While the correlation coefficient of logarithmically transformed data is a remarkably efficient measurement of the relationship between the quantities of benthos and plankton, the antilogarithmic transform of the variation ellipse of logarithms provides a surprisingly good description of the heteroscedasticity and skewness of the frequency distribution of original data (figure 30.9.1, right). In conclusion, in the case of the present limnological data like in that of the anatomical data of figure 19.8.1, the logarithmic transformation allows the normal distribution to be applied satisfactorily to data for which it would never have been considered suitable otherwise. The study of these interesting data of Rawson (1960) will be carried on in sections 31.10 and 33.11.

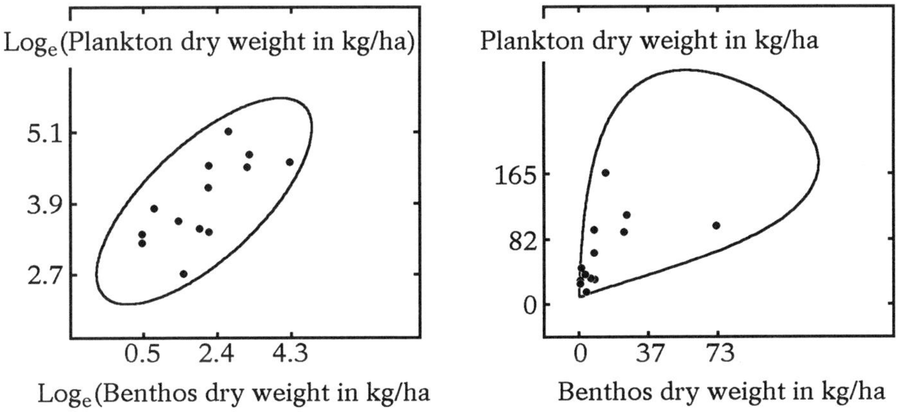

Figure 30.9.1
Logarithmic (left) and arithmetic (right) scatter diagrams of the quantities of benthos and plankton in kg/ha in Saskatchewan lakes according to data from Rawson (1960); variation ellipse (left) at level $\alpha = 0.05$ and antilogarithmic transform (right)

Chapter 31
Principal components or *principal axes*

Section 31.1: introduction

The expression *principal components* first appeared in the writings of the American statistician Harold Hotelling in 1933, but the technique was known earlier as *principal axes* and goes back to Karl Pearson (1901). In principle, principal components are appropriate for the analysis of variation within a sample coming from a single statistical population which follows the multivariate normal distribution (section 29.2). However, principal components are often used also in practice when the exact form of the probability distribution is uncertain or when data may come from several populations but the parent population of each individual observation is unknown. In principal component analysis, q linear combinations (sections 24.22 and 29.5) $\mathbf{Y} = [Y_1, \ldots Y_q]$ of the original variates $\mathbf{X} = [X_1, \ldots X_q]$ are determined in such a way that the new variates $\mathbf{Y} = [Y_1, \ldots Y_q]$ have zero covariances, that their coordinate axes are mutually orthogonal, and that their variances range from a maximum value for Y_1 to a minimum value for Y_q. The parametric (population) principal components correspond to the coordinates of observations on the principal axes of ellipses, ellipsoids or hyperellipsoids where the probability density is constant (sections 19.2, 23.1, 24.24 and 29.2): the first principal component corresponds to the major axis, the last principal component corresponds to the minor axis, and the other principal components correspond to successive intermediate axes. The simplest case is that of bivariate principal axes, which has already been discussed concerning the bivariate orthogonal estimation (regression) line (chapter 22) and the graphical representation of confidence, prediction or variation ellipses (section 30.8). The vectors of direction cosines (section 24.6) of principal axes are the latent vectors, and the variances of the new variates $[Y_1, \ldots Y_q]$ are the latent roots (sections 24.21 and 31.11) of the covariance matrix of original variates. The utilization of principal components has been discussed recently by Flury (1988) and Jackson (1991).

Section 31.2: the parametric (population) principal axes

Theoretically, in the ideal situation when the mean vector $\boldsymbol{\mu}$ and the covariance matrix $\boldsymbol{\Sigma}$ of the statistical population are known, the principal components of an individual observation $\mathbf{X} = [X_1, \ldots X_q]$ may be evaluated through the equation $\tilde{\mathbf{Y}} = \boldsymbol{\Upsilon}(\tilde{\mathbf{X}} - \tilde{\boldsymbol{\mu}})$, where $\boldsymbol{\Upsilon}$ denotes the (orthogonal) matrix of direction cosines of the parametric principal axes (or latent vectors):

$$\begin{bmatrix} Y_1 \\ \cdots \\ Y_i \\ \cdots \\ Y_q \end{bmatrix} = \begin{bmatrix} \upsilon_{11} & \cdots & \upsilon_{1j} & \cdots & \upsilon_{1q} \\ \cdots & \cdots & \cdots & \cdots & \cdots \\ \upsilon_{i1} & \cdots & \upsilon_{ij} & \cdots & \upsilon_{iq} \\ \cdots & \cdots & \cdots & \cdots & \cdots \\ \upsilon_{q1} & \cdots & \upsilon_{qj} & \cdots & \upsilon_{qq} \end{bmatrix} \begin{bmatrix} X_1 - \mu_1 \\ \cdots \\ X_j - \mu_j \\ \cdots \\ X_q - \mu_q \end{bmatrix} = \begin{bmatrix} \cos(\varphi_{11}) & \cdots & \cos(\varphi_{1j}) & \cdots & \cos(\varphi_{1q}) \\ \cdots & \cdots & \cdots & \cdots & \cdots \\ \cos(\varphi_{i1}) & \cdots & \cos(\varphi_{ij}) & \cdots & \cos(\varphi_{iq}) \\ \cdots & \cdots & \cdots & \cdots & \cdots \\ \cos(\varphi_{q1}) & \cdots & \cos(\varphi_{qj}) & \cdots & \cos(\varphi_{qq}) \end{bmatrix} \begin{bmatrix} X_1 - \mu_1 \\ \cdots \\ X_j - \mu_j \\ \cdots \\ X_q - \mu_q \end{bmatrix}.$$

$\boldsymbol{\Upsilon}$, υ and φ are the Greek letters upper-case upsilon, lower-case upsilon and phi, and correspond respectively to our upper-case **U**, and to our lower-case u and f. φ_{ij} is the angle between the coordinate axis of the ith population principal component Y_i and the coordinate axis of the jth original centered variate $(X_j - \mu_j)$.

The mean vector of principal components is $\boldsymbol{\mu}_Y = \mathbf{0} = [0, \ldots 0]$ and their covariance matrix is the diagonal matrix $\boldsymbol{\Delta}$:

$$\boldsymbol{\Delta} = \begin{bmatrix} \delta_1 & \ldots & 0 \\ \ldots & \ldots & \ldots \\ 0 & \ldots & \delta_q \end{bmatrix} = \mathscr{E}\left\{ \begin{bmatrix} v_{11} & \ldots & v_{1q} \\ \ldots & \ldots & \ldots \\ v_{q1} & \ldots & v_{qq} \end{bmatrix} \begin{bmatrix} X_1 - \mu_1 \\ \ldots \\ X_q - \mu_q \end{bmatrix} [X_1 - \mu_1 \quad \ldots \quad X_q - \mu_q] \begin{bmatrix} v_{11} & \ldots & v_{q1} \\ \ldots & \ldots & \ldots \\ v_{1q} & \ldots & v_{qq} \end{bmatrix} \right\},$$

$$\text{that is, } \boldsymbol{\Delta} = \begin{bmatrix} v_{11} & \ldots & v_{1q} \\ \ldots & \ldots & \ldots \\ v_{q1} & \ldots & v_{qq} \end{bmatrix} \begin{bmatrix} \sigma_1^2 & \ldots & \sigma_{1q} \\ \ldots & \ldots & \ldots \\ \sigma_{1q} & \ldots & \sigma_q^2 \end{bmatrix} \begin{bmatrix} v_{11} & \ldots & v_{q1} \\ \ldots & \ldots & \ldots \\ v_{1q} & \ldots & v_{qq} \end{bmatrix}.$$

While the covariance matrix $\boldsymbol{\Delta}$ of principal components $\mathbf{Y}$ may be obtained from the covariance matrix $\boldsymbol{\Sigma}$ of original variates $\mathbf{X}$ by the matrix product $\boldsymbol{\Upsilon}\,\boldsymbol{\Sigma}\,\tilde{\boldsymbol{\Upsilon}}$, the covariance matrix $\boldsymbol{\Sigma}$ of original variates may be reconstituted from its latent roots and latent vectors (section 24.25) by the product

$$\tilde{\boldsymbol{\Upsilon}}\,\boldsymbol{\Delta}\,\boldsymbol{\Upsilon} = \delta_1 \tilde{\boldsymbol{\Upsilon}}_1 \boldsymbol{\Upsilon}_1 + \delta_2 \tilde{\boldsymbol{\Upsilon}}_2 \boldsymbol{\Upsilon}_2 + \ldots + \delta_i \tilde{\boldsymbol{\Upsilon}}_i \boldsymbol{\Upsilon}_i + \ldots + \delta_q \tilde{\boldsymbol{\Upsilon}}_q \boldsymbol{\Upsilon}_q = \boldsymbol{\Sigma},$$

where $\boldsymbol{\Upsilon}_1, \boldsymbol{\Upsilon}_2, \ldots \boldsymbol{\Upsilon}_i, \ldots \boldsymbol{\Upsilon}_q$ are the successive row vectors of matrix $\boldsymbol{\Upsilon}$.

Section 31.3: the estimated (sample) principal axes

Since the principal components of the statistical population are generally unknown in practice, they are estimated from the sample mean vector $\overline{\mathbf{X}}$ and covariance matrix $\mathbf{S}$. The matrix $\mathbf{U}$ of direction cosines and the diagonal covariance matrix $\mathbf{D}$ of the sample principal axes correspond to the latent vectors and latent roots of the sample covariance matrix $\mathbf{S}$. Once matrices $\mathbf{U}$ and $\mathbf{D}$ have been evaluated, the coordinates of observations on the sample principal axes can be determined through the equation $\tilde{\mathbf{Y}} = \mathbf{U}(\tilde{\mathbf{X}} - \tilde{\overline{\mathbf{X}}})$, where f_{ij} is the angle between the ith sample principal axis and the coordinate axis of the jth centered original variate $(X_j - \overline{X}_j)$:

$$\begin{bmatrix} Y_1 \\ \ldots \\ Y_i \\ \ldots \\ Y_q \end{bmatrix} = \begin{bmatrix} u_{11} & \ldots & u_{1j} & \ldots & u_{1q} \\ \ldots & \ldots & \ldots & \ldots & \ldots \\ u_{i1} & \ldots & u_{ij} & \ldots & u_{iq} \\ \ldots & \ldots & \ldots & \ldots & \ldots \\ u_{q1} & \ldots & u_{qj} & \ldots & u_{qq} \end{bmatrix} \begin{bmatrix} X_1 - \overline{X}_1 \\ \ldots \\ X_j - \overline{X}_j \\ \ldots \\ X_q - \overline{X}_q \end{bmatrix} = \begin{bmatrix} \cos(f_{11}) & \ldots & \cos(f_{1j}) & \ldots & \cos(f_{1q}) \\ \ldots & \ldots & \ldots & \ldots & \ldots \\ \cos(f_{i1}) & \ldots & \cos(f_{ij}) & \ldots & \cos(f_{iq}) \\ \ldots & \ldots & \ldots & \ldots & \ldots \\ \cos(f_{q1}) & \ldots & \cos(f_{qj}) & \ldots & \cos(f_{qq}) \end{bmatrix} \begin{bmatrix} X_1 - \overline{X}_1 \\ \ldots \\ X_j - \overline{X}_j \\ \ldots \\ X_q - \overline{X}_q \end{bmatrix}.$$

The numerical evaluation of matrices $\mathbf{U}$ and $\mathbf{D}$ from the sample covariance matrix $\mathbf{S}$ has been discussed in detail for the bivariate case in section 30.8. In the multivariate case, matrices $\mathbf{U}$ and $\mathbf{D}$ may be obtained through various algorithms for which programs are available for most computers. The diagonalization method of the German mathematician Carl Gustav Jacobi (1804-1851), which has been described and discussed by White (1958), is rather simple and generally satisfactory. A short computer program of Jacobi's method in *True BASIC*™ is given in Appendix A. Jacobi's diagonalization method constitutes a generalization of the bivariate procedure described in section 30.8 and consists in a sequence of two-dimensional rotations of which each one reduces to zero one of the covariances of matrix $\mathbf{S}$ or of one of the matrices succeeding it during the computations. Such a method is said to be *iterative* (an adjective derived from the Latin *iter,* meaning road), because the solution is reached gradually through a repetition of similar steps called *iterations*.

First, a search is made for the covariance having the largest absolute value, let us say S_{ij}. The angle $f = 0.5 \operatorname{arc tan}[2S_{ij}/(S_{ii} - S_{jj})]$ is then obtained, where $\operatorname{arc tan} = \tan^{-1}$ denotes the inverse function of the trigonometric tangent function. Next, a matrix $\mathbf{U}_1$ is formed by replacing the ii^{th}, ij^{th}, ji^{th} and jj^{th} elements of a unit matrix by $\cos(f)$, $\sin(f)$, $-\sin(f)$ and $\cos(f)$ respectively:

$$\mathbf{U}_1 = \begin{bmatrix} 1 & 0 & \dots & 0 & \dots & 0 & \dots & 0 & 0 \\ 0 & 1 & \dots & 0 & \dots & 0 & \dots & 0 & 0 \\ \dots & \dots & \dots & \dots & \dots & \dots & \dots & \dots & \dots \\ 0 & 0 & \dots & u_{ii} & \dots & u_{ij} & \dots & 0 & 0 \\ \dots & \dots & \dots & \dots & \dots & \dots & \dots & \dots & \dots \\ 0 & 0 & \dots & u_{ji} & \dots & u_{jj} & \dots & 0 & 0 \\ \dots & \dots & \dots & \dots & \dots & \dots & \dots & \dots & \dots \\ 0 & 0 & \dots & 0 & \dots & 0 & \dots & 1 & 0 \\ 0 & 0 & \dots & 0 & \dots & 0 & \dots & 0 & 1 \end{bmatrix} = \begin{bmatrix} 1 & 0 & \dots & 0 & \dots & 0 & \dots & 0 & 0 \\ 0 & 1 & \dots & 0 & \dots & 0 & \dots & 0 & 0 \\ \dots & \dots & \dots & \dots & \dots & \dots & \dots & \dots & \dots \\ 0 & 0 & \dots & \cos(f) & \dots & \sin(f) & \dots & 0 & 0 \\ \dots & \dots & \dots & \dots & \dots & \dots & \dots & \dots & \dots \\ 0 & 0 & \dots & -\sin(f) & \dots & \cos(f) & \dots & 0 & 0 \\ \dots & \dots & \dots & \dots & \dots & \dots & \dots & \dots & \dots \\ 0 & 0 & \dots & 0 & \dots & 0 & \dots & 1 & 0 \\ 0 & 0 & \dots & 0 & \dots & 0 & \dots & 0 & 1 \end{bmatrix}.$$

If the matrix product $\mathbf{S}_1 = \mathbf{U}_1 \mathbf{S} \tilde{\mathbf{U}}_1$ is now evaluated, its ij^{th} and ji^{th} elements are found to be null. The operations described from the top of the present page down to here constitute the first iteration. The second iteration is similar to the first one except that matrix $\mathbf{S}_1$ is substituted for matrix $\mathbf{S}$, a second rotation matrix $\mathbf{U}_2$ is found, and a second matrix product $\mathbf{S}_2 = \mathbf{U}_2 \mathbf{S}_1 \tilde{\mathbf{U}}_2$ is obtained. Subsequent iterations are similar to the second one, and computations are carried out until the largest covariance (in absolute value) in matrix $\mathbf{S}_k = \mathbf{U}_k \mathbf{S}_{(k-1)} \tilde{\mathbf{U}}_k$ is close enough to zero. The matrix $\mathbf{U}$ of the direction cosines of the principal axes (latent vectors) and the diagonal form $\mathbf{D}$ of the sample covariance matrix $\mathbf{S}$ are then obtained from the matrix products

$$\mathbf{U} = \mathbf{U}_k \mathbf{U}_{(k-1)} \dots \mathbf{U}_2 \mathbf{U}_1 \quad \text{and} \quad \mathbf{D} = \mathbf{U} \mathbf{S} \tilde{\mathbf{U}}.$$

Section 31.4: interpreting principal axes

In order to disclose the nature of the variation corresponding to each principal axis, the first equation of the preceding section may be inverted, original variates $\tilde{\mathbf{X}}$ may be reexpressed in terms of principal components $\tilde{\mathbf{Y}}$,

$$\begin{bmatrix} X_1 \\ \dots \\ X_q \end{bmatrix} = \begin{bmatrix} \overline{X}_1 \\ \dots \\ \overline{X}_q \end{bmatrix} + \begin{bmatrix} u_{11} & \dots & u_{q1} \\ \dots & \dots & \dots \\ u_{1q} & \dots & u_{qq} \end{bmatrix} \begin{bmatrix} Y_1 \\ \dots \\ Y_q \end{bmatrix},$$

and only one of the latter may be varied at a time. Thus, if all principal components except the first one are kept equal to zero, the equation of the first principal component is obtained:

$$\begin{bmatrix} X_1 \\ \dots \\ X_q \end{bmatrix} = \begin{bmatrix} \overline{X}_1 \\ \dots \\ \overline{X}_q \end{bmatrix} + \begin{bmatrix} u_{11} \\ \dots \\ u_{1q} \end{bmatrix} Y_1,$$

that is
$$\frac{X_1 - \overline{X}_1}{u_{11}} = \dots = \frac{X_j - \overline{X}_j}{u_{1j}} = \dots = \frac{X_q - \overline{X}_q}{u_{1q}} = Y_1.$$

It must be emphasized that, when principal component analysis is applied to data collected for other purposes, it is often difficult and it may even be almost impossible to give clear biological interpretations to principal components, except possibly the first one.

When the original variates **X** are bodily dimensions of living organisms differing from each other with respect to age or size, the covariances in matrix **S** are often all positive and the direction cosines $[u_{11} \ldots u_{1q}]$ of the first principal axis usually all have the same sign (and may be made all positive by changing the sense of the corresponding vector if necessary). On the contrary, subsequent principal axes have direction cosines which differ in sign. For instance, figure 31.4.1 represents the principal components of a sample of 24 female painted turtles (*Chrysemys picta marginata*) caught within a single day (August 2, 1956) in a small stagnant pond of the St. Lawrence valley, near Montreal (Jolicoeur and Mosimann, 1960). The three variates considered here are

$$X_1 = \log_e(\text{carapace length in mm}),$$

$$X_2 = \log_e(\text{carapace width in mm}),$$

$$X_3 = \log_e(\text{carapace height in mm}).$$

The mean vector $\overline{\mathbf{X}}$, the covariance matrix **S**, the matrix **U** of direction cosines and the diagonal covariance matrix **D** of principal axes are given in table 31.4.1 (next page). Matrices **S** and **D** are multiplied by 10^4 for easier reading.

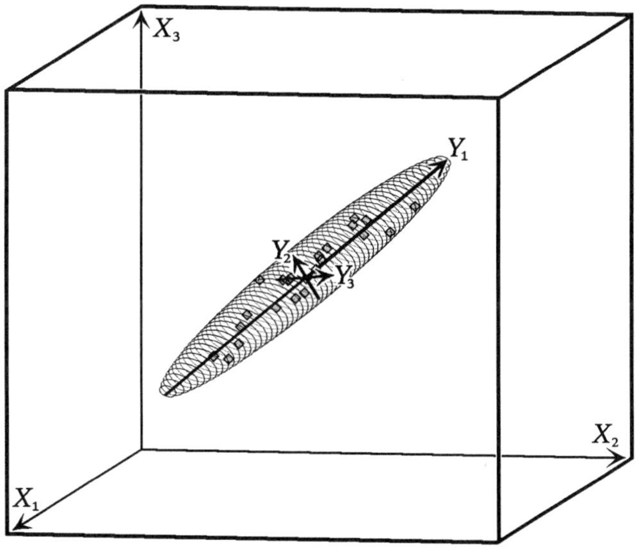

Figure 31.4.1
Principal components of logarithms of carapace length, width and height in mm of 24 female painted turtles (Jolicoeur and Mosimann, 1960)

Since the direction cosines of the major axis, [0.622264 0.484071 0.615193], are all positive, this axis corresponds to a trend of joint increase (or of joint decrease) of the three variates, and the corresponding variance, $d_1 = 671.996/10^4$, may be interpreted broadly speaking as a measurement of individual *size* variation. However, the second direction cosine of the major axis of these turtles has a lower numerical value than the other two, which appears to indicate that carapace width increases less rapidly than carapace length and height.

Table 31.4.1
Mean vector $\overline{\mathbf{X}}$, covariance matrix S, matrix U of direction cosines and diagonal covariance matrix D of principal axes of logarithms of carapace dimensions of 24 female painted turtles

$\overline{\mathbf{X}} = [4.90036 \quad 4.62291 \quad 3.93825]$	$10^4\mathbf{S} = \begin{bmatrix} 263.910 & 201.240 & 254.429 \\ 201.240 & 161.904 & 197.819 \\ 254.429 & 197.819 & 258.986 \end{bmatrix}$
$\mathbf{U} = \begin{bmatrix} 0.622264 & 0.484071 & 0.615193 \\ -0.455179 & -0.415619 & 0.787447 \\ -0.636866 & 0.770023 & -0.038286 \end{bmatrix}$	$10^4\mathbf{D} = \begin{bmatrix} 671.996 & 0 & 0 \\ 0 & 7.50452 & 0 \\ 0 & 0 & 5.30001 \end{bmatrix}$

As for the intermediate and minor axes, since their direction cosines do not all have the same sign, they correspond to trends of variation in the direction of which some variates increase but others decrease. Thus the intermediate axis contrasts carapace height on the one hand with carapace length and width on the other hand, while the minor axis, of which the vector of direction cosines is very close to $[-1/\sqrt{2},\ +1/\sqrt{2},\ 0]$, opposes carapace length and width. The variances $d_2 = 7.50452/10^4$ and $d_3 = 5.30001/10^4$ can thus be interpreted as measurements of *shape* variation (in a broad sense).

The direction cosines of each principal axis thus tend to disclose the nature of variation, while the corresponding latent root reflects the extent of variation. The data of the present example will be considered further in sections 31.6 and 31.7, and the question of the analysis of *size and shape variation* will be discussed in greater detail in sections 31.5, 31.8 and 31.9.

In ecology, original variates are often measurements of the abundance of several species and, when the direction cosines of the first principal axis are all positive (which is not always the case), the first principal component may be interpreted as a measurement of the general abundance of all species. As for the other principal axes, they then correspond to trends of variation in the direction of which some species become abundant while others become scarce.

Section 31.5: analysis of the total variation of several variates

Because the diagonalization of a covariance matrix does not change its trace (section 24.21), the sum $(d_1 + \ldots + d_q)$ of the variances of the principal components $[Y_1, \ldots Y_q]$ is equal to the sum $(S_1^2 + \ldots + S_q^2)$ of the variances of original variates $[X_1, \ldots X_q]$. The trace of the covariance matrix **S** (which is equal to that of the diagonal matrix **D**) may thus be considered as a measure of the total variation of all q variates jointly. However, while the total variation of principal components is equal to that of original variates, it breaks down differently, since the variance d_1 of the first principal axis makes up a maximal part of the total variation, the variance d_2 of the second principal axis makes up a maximal part of the remainder once d_1 is subtracted, and so on down to the variance d_q of the last principal axis, which is minimal.

In some cases, the principal components of the correlation matrix, which is the covariance matrix of standardized variates (section 24.16), can be interpreted more easily than those of the covariance matrix of original variates. This follows from the fact that standardized variates are on an equal footing, so to speak, at the start of a principal component analysis, since their variances are all equal to unity and the trace of the correlation matrix is equal to the total number q of variates.

The data of Schultz on the left and right humerus, radius, femur and tibia lengths of 117 men and 110 women, already considered in sections 29.4, 29.6 and 30.2, provide an excellent example for the analysis of the trace of correlation or covariance matrices, because the intensity of bilateral symmetry in human limbs should make for clear relationships. The correlation matrices are given in table 31.5.1. The highest correlations ($r \geq 0.96$) are those between bilateral homologues, that is corresponding left and right bones such as left and right humeri, etc.. Among other correlations, some appear to be particularly high between bones belonging to the same limb (arm versus leg) or which are situated similarly close (proximal) or far (distal) from the body. However, it is obviously difficult to draw clear conclusions from such a mass of numerical data.

Table 31.5.1
Correlations of left and right humerus, radius, femur and tibia lengths in 117 men (above main diagonal) and 110 women (below main diagonal)

Left humerus	Right humerus	Left radius	Right radius	Left femur	Right femur	Left tibia	Right tibia
1	0.964826	0.817164	0.792966	0.794824	0.800620	0.756549	0.773786
0.971165	1	0.804800	0.806282	0.782573	0.789134	0.725603	0.750664
0.810960	0.810033	1	0.967167	0.762217	0.749690	0.802115	0.802965
0.780147	0.807019	0.974459	1	0.764900	0.750241	0.791873	0.802095
0.846822	0.834010	0.790756	0.775075	1	0.987566	0.862919	0.872100
0.842514	0.829157	0.783361	0.770302	0.992904	1	0.859202	0.870762
0.787713	0.762344	0.830505	0.799288	0.911178	0.903724	1	0.982497
0.790529	0.770537	0.845622	0.816391	0.899287	0.895226	0.986596	1

One may wonder whether principal component analysis would allow more explicit information to be extracted from such a set of variable quantities. The direction cosines of principal axes are given in table 31.5.2. As in the preceding section, the direction cosines of the major axis are all positive, and the variance d_1 may again be interpreted as a measurement of size variation in a broad sense. Moreover, the direction cosines of the first principal axis are all approximately equal to a common value $1/\sqrt{q} = 1/\sqrt{8} = 0.353553$:

$$[+1/\sqrt{8},\ +1/\sqrt{8},\ +1/\sqrt{8},\ +1/\sqrt{8},\ +1/\sqrt{8},\ +1/\sqrt{8},\ +1/\sqrt{8},\ +1/\sqrt{8}].$$

Since principal component analysis is applied here to correlation matrices, the order of magnitude of variates should not have much influence, and the present results suggest that the lengths of the eight limb bones tend to increase (or to decrease) jointly at relative rates which are approximately equal.

While the direction cosines of the other seven principal axes are not all positive, axes 2, 3 and 4 differ in an important manner from axes 5, 6, 7 and 8: except in the case of a few cosines which are approximately null, the direction cosines of bilateral homologues (like the left and right humeri, etc.) have similar signs on axes 2, 3 and 4 but opposite signs on the last four axes. This shows that axes 2, 3 and 4 correspond to contrasts between proximal and distal bones of upper and lower limbs but do not oppose left and right bones. On the contrary, axes 5, 6, 7 and 8 contrast left with right bones and therefore reflect the fluctuations of bilateral symmetry. The vectors of direction cosines of some principal axes are strikingly similar to the vectors of coefficients of some *a priori* linear combinations considered in section 29.6. Thus the vector of direction cosines of the eighth principal axis of each sample (117 men and 110 women) is closely similar to the vector

$$[0,\ 0,\ 0,\ 0,\ -1/\sqrt{2},\ +1/\sqrt{2},\ 0,\ 0],$$

which opposes the lengths of left and right femora and is proportional to the vector $[0,\ 0,\ 0,\ 0,\ -1,\ +1,\ 0,\ 0]$ of the coefficients of the third *a priori* linear combination used in section 29.6. As for the vectors of direction cosines of axes 5, 6 and 7, they are rather clearly made up of linear combinations of the vectors of coefficients of combinations 1, 2 and 4 of section 29.6.

Finally, the second, third and fourth principal axes also correspond to trends of variation which can easily be identified: for instance, the second principal axis of each sample is very close to the vector

$$[+1/\sqrt{8},\ +1/\sqrt{8},\ +1/\sqrt{8},\ +1/\sqrt{8},\ -1/\sqrt{8},\ -1/\sqrt{8},\ -1/\sqrt{8},\ -1/\sqrt{8}],$$

which contrasts the lengths of arm bones with those of leg bones, while axes 3 and 4 contrast proximal bones (humeri and femora) with distal bones (radii and tibiae).

Table 31.5.2
Matrices of direction cosines of principal axes of correlation matrices of left and right humerus, radius, femur and tibia lengths

117 men

+0.350761	+0.346666	+0.351103	+0.349499	+0.358041	+0.356999	+0.355640	+0.359505
+0.352474	+0.406032	+0.339489	+0.329675	−0.345730	−0.350557	−0.361432	−0.337504
+0.413679	+0.429498	−0.441781	−0.457548	+0.234029	+0.281892	−0.252312	−0.204910
+0.279321	+0.192554	−0.211790	−0.285560	−0.447763	−0.409401	+0.447841	+0.435708
−0.490694	+0.505353	−0.504406	+0.455018	−0.029699	−0.023728	−0.088169	+0.182080
−0.511235	+0.484002	+0.472896	−0.482014	−0.050194	+0.061445	+0.159126	−0.129604
+0.024126	+0.032075	−0.198286	+0.166798	+0.173878	−0.172300	+0.650966	−0.669008
+0.029231	−0.052423	−0.063835	+0.105217	−0.680721	+0.684517	+0.147692	−0.165815

110 women

+0.349394	+0.346955	+0.350053	+0.343682	+0.361376	+0.359703	+0.357853	+0.358962
+0.254420	+0.336594	+0.391742	+0.440798	−0.343060	−0.351118	−0.366826	−0.314127
+0.497756	+0.487043	−0.402834	−0.419979	+0.161144	+0.169151	−0.232384	−0.260364
+0.284321	+0.154363	−0.086930	−0.276560	−0.449008	−0.485666	+0.410291	+0.453291
−0.506791	+0.551927	−0.511391	+0.386011	−0.055229	−0.034038	+0.026778	+0.151951
−0.344269	+0.307743	+0.434843	−0.444508	−0.011718	+0.087588	−0.461579	+0.423360
−0.321233	+0.313506	+0.308312	−0.285781	+0.238238	−0.250052	+0.498687	−0.503810
−0.077756	+0.087600	+0.092214	−0.080308	−0.681320	+0.641475	+0.229301	−0.207512

To complement the examination of the direction cosines, the values of the variances of principal axes are given in table 31.5.3, which summarizes the nature and the relative importance of the corresponding trends of variation. Surprisingly, even though the fluctuations of bilateral symmetry occupy half (4) the total number (8) of available dimensions, the sum of the corresponding variances makes up only approximately 1% of the trace of correlation matrices, that is of the total variation of all variates. This reflects the small extent of the fluctuations of bilateral symmetry in human limb bones. Other trends of shape variation are much more important for, although they occupy only three dimensions, they make up from 13% to 14% of the trace. Finally, the strongest trend of variation in these samples of adult individuals corresponds to size which, while it occupies only one dimension, makes up approximately 85% of the trace.

Table 31.5.3
Nature and relative importance of the trends of variation corresponding to the principal axes of the correlation matrices of the lengths of human arm and leg bones (Jolicoeur, 1963c)

<table>
<tr><th rowspan="2">Principal axis</th><th colspan="2">117 men</th><th colspan="2">110 women</th><th rowspan="2" colspan="2">Nature of variation</th></tr>
<tr><th>Variance</th><th>Percentage of trace</th><th>Variance</th><th>Percentage of trace</th></tr>
<tr><td>1</td><td>6.74796</td><td>84.35</td><td>6.90646</td><td>86.33</td><td colspan="2">size variation</td></tr>
<tr><td>2</td><td>0.547582</td><td>6.84</td><td>0.468795</td><td>5.86</td><td>contrast between arms and legs</td><td rowspan="3">general shape variation (symmetry excluded)</td></tr>
<tr><td>3</td><td>0.402484</td><td>5.03</td><td>0.414908</td><td>5.19</td><td rowspan="2">proximodistal contrasts</td></tr>
<tr><td>4</td><td>0.207246</td><td>2.59</td><td>0.141036</td><td>1.76</td></tr>
<tr><td>5</td><td>0.046670</td><td>0.58</td><td>0.036667</td><td>0.46</td><td rowspan="4">contrasts between bilateral homologues</td><td rowspan="4">fluctuations of bilateral symmetry</td></tr>
<tr><td>6</td><td>0.020821</td><td>0.26</td><td>0.013312</td><td>0.17</td></tr>
<tr><td>7</td><td>0.015715</td><td>0.20</td><td>0.012677</td><td>0.16</td></tr>
<tr><td>8</td><td>0.011522</td><td>0.14</td><td>0.006147</td><td>0.08</td></tr>
</table>

While the simple examination of correlation matrices done at the beginning of this section indicated mostly that the highest correlations were those between bilateral homologues, principal component analysis does reveal the multivariate structure of variation in a considerably more detailed and more accurate manner.

Section 31.6: the principal axes of logarithmically transformed data, and the multivariate generalization of the allometry equation

The lengths of human limb bones analyzed in sections 29.6 and 30.2 and in the preceding section (31.5) were not submitted to the logarithmic transformation because, when relative variation is slight, the normality hypothesis can occasionally be disproved neither for the untransformed variate nor for its logarithm (end of section 16.4). This can happen in particular for human stature (height) or other linear measurements of adult individuals belonging to species in which somatic growth ends long before death.

For the sake of simplicity, the normal distribution may then be applied to untransformed data. Nevertheless, the frequency distribution of biological data is generally more similar to the lognormal than to the normal distribution (chapter 14, and sections 4.3, 4.9, 12.7, 16.4, 19.8, 20.6, 20.7, 22.5, 28.7, 29.7, 29.8 and 30.9). Moreover, the logarithmic transformation may be desirable in order to prevent the occurrence of negative confidence limits for original variates, to straighten exponential or allometric relationships, or to simplify the study of relationships between ratios or products of variates using *a priori* linear combinations of logarithms (section 29.8).

The principal components of logarithmically transformed data may be interpreted in the same general manner as those of raw (untransformed) data, except that the equation of the multidimensional straight line corresponding to each principal axis applies to logarithms instead of original variates X_i. Thus the equation of the major axis of logarithms $Y_i = \log_e(X_i)$ is

$$\frac{Y_1 - \overline{Y}_1}{u_{11}} = \dots = \frac{Y_i - \overline{Y}_i}{u_{1i}} = \frac{Y_j - \overline{Y}_j}{u_{1j}} = \dots = \frac{Y_q - \overline{Y}_q}{u_{1q}} = Z_1 ,$$

where Z_1 denotes the value of the first principal component. When the preceding equation is expressed in terms of original variates, one gets

$$\left(\frac{X_1}{G_1}\right)^{\frac{1}{u_{11}}} = \dots = \left(\frac{X_i}{G_i}\right)^{\frac{1}{u_{1i}}} = \left(\frac{X_j}{G_j}\right)^{\frac{1}{u_{1j}}} = \dots = \left(\frac{X_q}{G_q}\right)^{\frac{1}{u_{1q}}} = e^{Z_1},$$

where $G_i = \exp(\overline{Y}_i)$ is the geometric mean of the ith original variate X_i. Pairing any two of the first q members of this multidimensional equation yields a bivariate power function, that is an allometry relationship (sections 20.7 and 22.5), of which the *allometry exponent* is $k = u_{1j}/u_{1i}$:

$$\left(\frac{X_j}{G_j}\right) = \left(\frac{X_i}{G_i}\right)^{\frac{u_{1j}}{u_{1i}}}$$

Because samples of living organisms captured in nature generally contain individuals differing from each other with respect to age and size, the most pronounced trend of variation of bodily dimensions often reflects size, which then corresponds approximately to the first principal component of data (here transformed into logarithms). It is therefore intuitively reasonable to use the equation of the first principal axis of logarithmically transformed data as a multivariate generalization of the allometry equation.

From a historical viewpoint, the first research worker who attempted to develop a multivariate version of the allometry equation was the French biometrician Georges Teissier (1900-1972), who initially proposed a simplified multivariate version of the standardized major axis (section 22.6) of logarithmically transformed data (Teissier, 1948):

$$\frac{Y_1 - \overline{Y}_1}{s_{Y_1}} = \dots = \frac{Y_i - \overline{Y}_i}{s_{Y_i}} = \frac{Y_j - \overline{Y}_j}{s_{Y_j}} = \dots = \frac{Y_q - \overline{Y}_q}{s_{Y_q}} .$$

Determining the standardized major axis required only the calculation of the mean and of the standard deviation of each variate, clearly a convenient solution before computers became widely available. However, in addition to the statistical weaknesses of the bivariate version, which were discovered many years later (section 22.6; Jolicoeur, 1975, 1990; Kimura, 1992), Teissier's multivariate version of the standardized major axis had the drawback that it did not take into account correlations between variates. In later works, Teissier (1955, 1960, 1961) thus proposed using either a multivariate version of the bivariate structural relationship (section 22.7) or the major axis of the correlation matrix of logarithmically transformed data. However, the multivariate structural relationship, which is related to the so-called *factor analysis* used in statistical psychology, is plagued by the frequent occurrence of negative estimates for residual variances (the Heywood case), even though each variate is assumed theoretically to possess its own residual variance. As for the major axis of the correlation matrix, the major reason for using it is to prevent the influence of unequal scale changes of original variates, but this is not necessary in the case of logarithmically transformed data, for the covariance matrix of logarithms is already *dimensionless* (section 22.5). Moreover, the probability distribution of principal components is simpler and better known for the covariance than for the correlation matrix. For the preceding reasons, Jolicoeur (1963a) finally concluded that the most straightforward generalization of the allometry equation was the major axis of the covariance matrix of logarithmically transformed data.

If the equation of the major axis of the logarithms $Y_i = \log_e(X_i)$ of the carapace length, width and height measurements of the 24 female painted turtles studied in section 31.4 is expressed with respect to original variates, one gets:

$$\left(\frac{X_1}{134.338}\right)^{\frac{1}{0.622264}} = \left(\frac{X_2}{101.790}\right)^{\frac{1}{0.484071}} = \left(\frac{X_3}{51.3287}\right)^{\frac{1}{0.615193}} = e^{Z_1} .$$

If a large (old ?) turtle is compared with a small (young ?) one, carapace length X_1 thus appears to increase more rapidly than carapace width X_2, for it is proportional to the power $0.622264/0.484071 = 1.28548$ of the latter, but it seems to increase at the same speed as carapace height X_3 with respect to which its allometry exponent is $0.622264/0.615193 = 1.01149$. This example will be considered again in sections 31.7 and 31.9.

Section 31.7: hypotheses concerning the directions of principal axes

For the proportions of the turtles discussed in the last section to remain constant as their size increases, the direction cosines of the major axis of the covariance matrix of the logarithmically transformed data should all be equal to a common value, $1/\sqrt{q} = 1/\sqrt{3}$. It should be noted that, since the preceding statement constitutes a hypothesis, it concerns the statistical population rather than the sample, and it must be expressed with respect to parametric principal axes (section 31.2) rather than estimated principal axes. This *isometry* hypothesis (section 20.7) may thus be formulated with respect to all variates jointly as follows:

$$H_0\colon\ [\upsilon_{11}, \upsilon_{12}, \upsilon_{13}] = [1/\sqrt{q}, \ldots, 1/\sqrt{q}] = [1/\sqrt{3}, 1/\sqrt{3}, 1/\sqrt{3}] .$$

According to the isometry hypothesis, all allometry exponents like (v_{1j}/v_{1i}) would be equal to unity and all original variates would vary in direct proportion to their geometric means, following the equation of the parametric (population) major axis:

$$\left(\frac{X_1}{\Gamma_1}\right) = \left(\frac{X_2}{\Gamma_2}\right) = \left(\frac{X_3}{\Gamma_3}\right) ,$$

where Γ_i is the (parametric) geometric mean of the ith original variate X_i.

Other hypotheses may also have to be tested concerning other principal axes. Assume, for instance, that there would be theoretical reasons to believe that the ratio (X_2/X_1) of the width to the length of these turtles would be the least pronounced trend of variation. The variate $\log_e(X_2/X_1) = (Y_2 - Y_1)$ might then be expected to have a particularly small variance. But this variate is proportional to the coordinate of an observation on the minor axis of logarithmically transformed data if the vector of direction cosines of this axis is $[-1/\sqrt{2}, +1/\sqrt{2}, 0]$. The following hypothesis should then be tested:

$$\text{H}_0\text{: } [v_{31}, v_{32}, v_{33}] = [-1/\sqrt{2}, +1/\sqrt{2}, 0] .$$

More generally, the equation of the kth hypothetical principal axis may be expressed in terms of original variates X_i when the latter are submitted to the logarithmic transformation before the analysis:

$$\left(\frac{X_1}{\Gamma_1}\right)^{\frac{1}{v_{k1}}} = \dots = \left(\frac{X_i}{\Gamma_i}\right)^{\frac{1}{v_{ki}}} = \left(\frac{X_j}{\Gamma_j}\right)^{\frac{1}{v_{kj}}} = \dots = \left(\frac{X_q}{\Gamma_q}\right)^{\frac{1}{v_{kq}}} = \mathrm{e}^{Z_k} .$$

When the analysis is done on untransformed original variates X_i, however, the equation of the kth hypothetical principal axis is

$$\frac{X_1 - \mu_1}{v_{k1}} = \dots = \frac{X_i - \mu_i}{v_{ki}} = \frac{X_j - \mu_j}{v_{kj}} = \dots = \frac{X_q - \mu_q}{v_{kq}} .$$

In order to test hypotheses concerning the direction of a principal axis, asymptotic criteria, which may be used when sample size is large (let us say $N \geq 30$), were developed by Anderson (1963), Kshirsagar (1961) and Mallows (1961). The asymptotic criterion which is most frequently used in practice is the one of Anderson (1963), which follows the χ^2 distribution approximately:

$$(N-1)\left(d_i \mathbf{v}_i \mathbf{S}^{-1} \tilde{\mathbf{v}}_i + \mathbf{v}_i \mathbf{S} \tilde{\mathbf{v}}_i / d_i - 2\right) \doteq \chi^2_{(q-1)} ,$$

where $\mathbf{v}_i$ is the vector of direction cosines of the ith hypothetical principal axis,

$\mathbf{S}^{-1}$ is the inverse of the sample covariance matrix $\mathbf{S}$,

and d_i is the ith latent root of matrix $\mathbf{S}$.

The preliminary hypothesis that the vector of direction cosines of the ith principal axis is $\mathbf{v}_i$ may be accepted if

$$(N-1)\left(d_i \mathbf{v}_i \mathbf{S}^{-1} \tilde{\mathbf{v}}_i + \mathbf{v}_i \mathbf{S} \tilde{\mathbf{v}}_i / d_i - 2\right) \leq \chi^2_{(1-\alpha;\, q-1)} .$$

First applied by Jolicoeur (1963b) in the context of allometry, Anderson's criterion was subsequently used as the basis of an asymptotic (large-sample) confidence interval for the slope of the major axis of a bivariate normal distribution (Jolicoeur, 1965).

A small-sample bivariate confidence interval was later based (Jolicoeur, 1968; Jolicoeur and Mosimann, 1968) on the principle that the sample correlation between the coordinates of observations on the hypothetical major and minor axes should not differ significantly from zero (section 22.3). The multivariate version of the criterion corresponding to this bivariate confidence interval was finally discovered independently by Vaughton (1970, unpublished Master's thesis) and by Jolicoeur (1984), the latter having shown that

$$(N-q)\left(\mathbf{v}_i\mathbf{S}\tilde{\mathbf{v}}_i\ \mathbf{v}_i\mathbf{S}^{-1}\tilde{\mathbf{v}}_i - 1\right)/(q-1) = F_{(q-1,\,N-q)}\,.$$

This small-sample criterion corresponds to a test of the hypothesis that there is a null parametric multiple correlation between the coordinates of observations on the hypothetical principal axis and the $(q-1)$ other principal axes (which do not have to be specified). The preliminary hypothesis that $\mathbf{v}_i$ is the vector of direction cosines of one of the principal axes may be accepted if

$$(N-q)\left(\mathbf{v}_i\mathbf{S}\tilde{\mathbf{v}}_i\ \mathbf{v}_i\mathbf{S}^{-1}\tilde{\mathbf{v}}_i - 1\right)/(q-1) \leq F_{(1-\alpha;\,q-1,\,N-q)}\,.$$

Unlike the asymptotic criterion of Anderson (1963), the above small-sample criterion does not link the hypothetical vector $\mathbf{v}_i$ with any latent root, d_i, of the sample covariance matrix, and thus does not allow the number of the hypothetical principal axis to be specified. In practice, however, it is often easy to see which sample principal axis $\mathbf{u}_i$ is most closely similar to the hypothetical principal axis $\mathbf{v}_i$ by evaluating the cosine of the angle f between them,

$$\cos(f) = \mathbf{v}_i\,\tilde{\mathbf{u}}_i\,,$$

and by making sure that this angle is not too large (let us say $f \leq 20°$). Because of their differences, the asymptotic criterion of Anderson (1963) and the small-sample criterion may both be used within a single analysis, for they complement rather than replace each other. James (1977) and Tyler (1981) have developed more complex methods for testing hypotheses concerning several principal axes jointly, but these will perhaps be less often useful to biologists. When all principal axes are considered simultaneously, the hypothesis that their correlation matrix is a unit matrix may be tested (section 29.4).

In the case of the 24 female painted turtles already considered in sections 31.4 and 31.6, the isometry hypothesis, according to which the vector of direction cosines of the major axis would be $\mathbf{v}_1 = [1/\sqrt{3}, 1/\sqrt{3}, 1/\sqrt{3}]$, must be rejected according to the small-sample criterion ($F_{(2,\,21)} = 14.7864$, $P < 0.0005$) as well as according to Anderson's criterion ($\chi^2_{(2)} = 32.7863$, $P < 0.0005$). The shape of female painted turtles thus appears to change as their size increases. As for the hypothesis that $\mathbf{v}_3 = [-1/\sqrt{2}, +1/\sqrt{2}, 0]$ is the vector of direction cosines of the minor axis, it must also be rejected ($F_{(2,\,21)} = 12.3911$, $P < 0.0005$; $\chi^2_{(2)} = 27.4105$, $P < 0.0005$). Therefore, the data do not support the conjecture that the ratio of carapace width to carapace length corresponds to the least pronounced trend of variation.

A second example may be drawn from a study of a sample of 68 male North American martens (appendix D; Jolicoeur, 1984). The variates included in the present analysis are

$$Y_1 = \log_e(X_1) = \log_e(\text{skull length in mm}),$$

$$Y_2 = \log_e(X_2) = \log_e(\text{humerus length in mm}),$$

$$Y_3 = \log_e(X_3) = \log_e(\text{femur length in mm}).$$

The principal components are illustrated in figure 31.7.1, while table 31.7.1 contains the mean vector $\overline{\mathbf{Y}}$, the covariance matrix **S**, the matrix **U** of direction cosines of the principal axes and the diagonal form **D** of the covariance matrix.

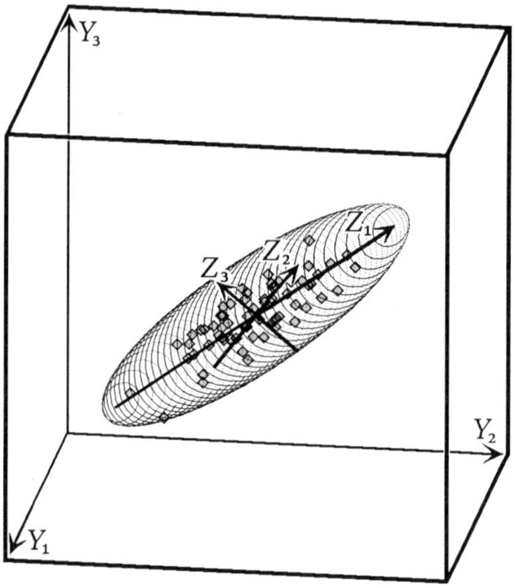

Figure 31.7.1
Principal components of logarithms of skull, humerus and femur lengths in mm in a sample of 68 male North American martens (Jolicoeur, 1984)

Table 31.7.1
Mean vector $\overline{\mathbf{Y}}$, covariance matrix S, matrix U of direction cosines and diagonal covariance matrix D of principal axes of logarithmically transformed data of a sample of 68 male North American martens

$\overline{\mathbf{Y}} = [4.36919 \quad 4.16006 \quad 4.25933]$	$10^4\mathbf{S} = \begin{bmatrix} 5.28200 & 4.03648 & 3.98936 \\ 4.03648 & 6.95118 & 6.19409 \\ 3.98936 & 6.19409 & 6.99730 \end{bmatrix}$
$\mathbf{U} = \begin{bmatrix} 0.463356 & 0.626388 & 0.626848 \\ -0.885872 & 0.309013 & 0.346036 \\ 0.023049 & -0.715646 & 0.698083 \end{bmatrix}$	$10^4\mathbf{D} = \begin{bmatrix} 16.1357 & 0 & 0 \\ 0 & 2.31568 & 0 \\ 0 & 0 & 0.779092 \end{bmatrix}$

The isometry hypothesis could barely be rejected here with Anderson's criterion ($\chi^2_{(2)} = 6.09488$, $P \leq 0.05$) but can be accepted with the small-sample criterion ($F_{(2,\ 65)} = 2.90394$, $P > 0.05$). Hence, this analysis does not disclose conclusive shape differences between small and large male martens.

As for the minor axis, there are *a priori* reasons to consider the hypothesis

$$H_0:\ \boldsymbol{v}_3 = [\ 0,\ -1/\sqrt{2},\ +1/\sqrt{2}\ ],$$

because the ratio of femur length to humerus length should be more critical than the ratio of either of these two variates to skull length with respect to locomotor efficiency. In the course of evolution, natural selection should thus have decreased the variability of the variate $\log_e(X_3/X_2) = (Y_3 - Y_2)$, which is proportional to the coordinate of an observation on the hypothetical vector $\boldsymbol{v}_3$. But the sample minor axis,

$$\mathbf{u}_3 = [0.023049,\ -0.715646,\ +0.698083\ ],$$

is closely similar to the hypothetical minor axis $\boldsymbol{v}_3$ with which it makes a very small angle, $f = 1.50025°$, and the hypothesis H_0 can be accepted with Anderson's criterion ($\chi^2_{(2)} = 0.060228$, $P > 0.05$) as well as with the small-sample criterion ($F_{(2,\ 65)} = 0.029195$, $P > 0.05$). The hypothesis according to which the ratio of femur length to humerus length would have a minimal variability is therefore supported by the present data.

Section 31.8: the analysis of size and shape variation

The analysis of size and shape variation has been discussed by many authors, including Mosimann (1970), Sprent (1972), and Darroch and Mosimann (1985). When the direction cosines of the major axis of the sample covariance matrix **S** of logarithmically transformed data are all positive, the coordinate Z_1 of an observation on that axis may be interpreted as a *size variable* (or *size variate*):

$$Z_1 = u_{11}(Y_1 - \overline{Y}_1) + \ldots + u_{1q}(Y_q - \overline{Y}_q) = \log_e[(X_1/G_1)^{u_{11}} \times \ldots \times (X_q/G_q)^{u_{1q}}]\,.$$

When, moreover, the isometry hypothesis $H_0:\ \boldsymbol{v}_1 = [1/\sqrt{q}\,,\ \ldots\,,\ 1/\sqrt{q}]$ is acceptable, the size variate may be defined with respect to the parameters of the statistical population as follows:

$$Z_1 = (Y_1 - \mu_1)/\sqrt{q} + \ldots + (Y_q - \mu_q)/\sqrt{q} = \log_e[(X_1/\Gamma_1) \times \ldots \times (X_q/\Gamma_q)]^{(1/\sqrt{q})}\,.$$

However, the preceding definition refers to the geometric means $\Gamma_1,\ \ldots\ \Gamma_q$ of the population, which are usually unknown and may be replaced by sample means $G_1\ ,\ \ldots\ G_q$:

$$Z_1 = (Y_1 - \overline{Y}_1)/\sqrt{q} + \ldots + (Y_q - \overline{Y}_q)/\sqrt{q} = \log_e[(X_1/G_1) \times \ldots \times (X_q/G_q)]^{(1/\sqrt{q})}\,.$$

While the size variate Z_1 defined above is related to the geometric mean of original variates considered by Mosimann (1970),

$$\sqrt[q]{\prod_{i=1}^{q} X_i} = \exp[\log_e(X_1 \times \ldots \times X_q)^{(1/q)}]\,,$$

it differs from it in three important respects:

(1) the size variate Z_1 is expressed on a logarithmic rather than on an arithmetic scale, in order to take into account the fact that the frequency distribution of biological data is more often similar to the lognormal than to the normal distribution;
(2) because each original variate X_i is divided by its geometric mean G_i or Γ_i in the calculation of the size variate Z_1, the latter is a dimensionless number, unaffected by linear scale changes; original variates may thus be measured in different units even though they must still be of the same kind (Mosimann, 1970, page 931; lengths, surfaces and volumes or masses should not be mixed);
(3) the exponent of each original variate X_i or of their product is $(1/\sqrt{q})$ instead of $(1/q)$ as in Mosimann's geometric mean, which makes $\mathbf{v}_1 = [1/\sqrt{q}, \ldots, 1/\sqrt{q}]$ a unit vector, allowing the same scale to be used for the size variate Z_1 as for the shape variates $[Z_2, \ldots Z_q]$ which are orthogonal to it; therefore, the variance of the size variate Z_1 is readily comparable to those of the shape variates $[Z_2, \ldots Z_q]$.

While the major axis of the covariance matrix of logarithmically transformed data may still be the best multivariate generalization of the allometry equation in practice (see the discussion of Shea, 1985), the reader should not believe that it is always easy to analyze size and shape variation in a satisfactory manner. In his important paper of 1970, Mosimann emphasized that size may be defined and measured in many different manners and that, theoretically, shape can be independent of at most one size variate. To be sure, the various possible size variates include the Z_1 variate defined above as well as Mosimann's related geometric mean of original variates, but they also include the sum $(X_1 + \ldots + X_q)$ of all original variates or of part of them, the square root of the sum of their squares, $\sqrt{(X_1^2 + \ldots + X_q^2)}$, any one of the original variates, X_i, or its logarithm, Y_i, etc., etc. . Anyone of those size variates may be preferable depending on the nature of the study being done. However, as noted by Mosimann (1970, pages 931-932), even though, in theory, shape may be independent of only one size variate, a relationship of shape with size is certainly less debatable when it can be demonstrated using several different size variates.

In order to make the present discussion easier to grasp, two puzzling examples are illustrated in figures 31.8.1 and 31.8.2. Figure 31.8.1 represents the case where the major axis coincides with the *isometry line* (also called *equiangular line* because it makes equal angles with the coordinate axes of all variates), shown as a thin line. The isometry hypothesis is thus accepted if the coordinate Z_1 on the major axis is used as a size variate. On the contrary, when the correlation is imperfect and the sample size is large enough, the isometry hypothesis would be rejected if either of the transformed variates Y_1 or Y_2 were used as a size variate, since this is equivalent to using the $Y_2.Y_1$ estimation (regression) line of Y_2 from Y_1 or the $Y_1.Y_2$ estimation line of Y_1 from Y_2. Moreover, the conclusion drawn by using Y_1 as a size variate (Y_2 negatively allometric) would contradict the conclusion drawn by using Y_2 (Y_2 positively allometric)! In such a case, where observed variates Y_1 and Y_2 show a clear positive correlation, there is little doubt that the coordinate Z_1 on the major axis is more satisfactory as a size variate than either Y_1 or Y_2 .

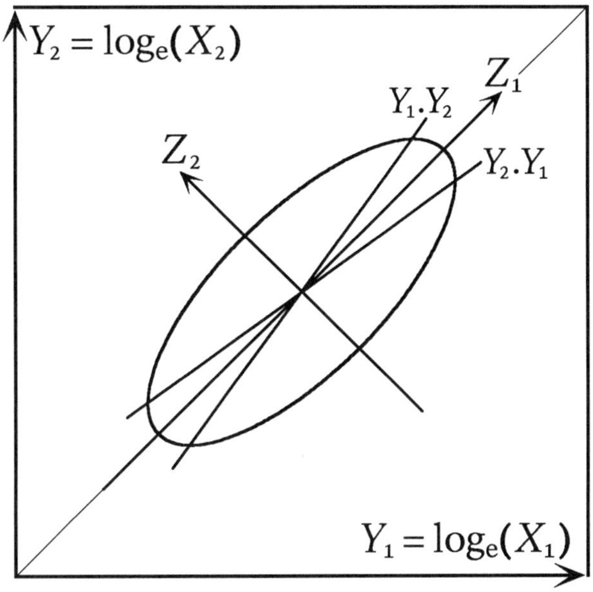

Figure 31.8.1
The isometry hypothesis may be acceptable if the major axis is used but unacceptable if ordinary regression lines ($Y_2.Y_1$ or $Y_1.Y_2$) are used, which is equivalent to using the logarithm of either variate as a size variate

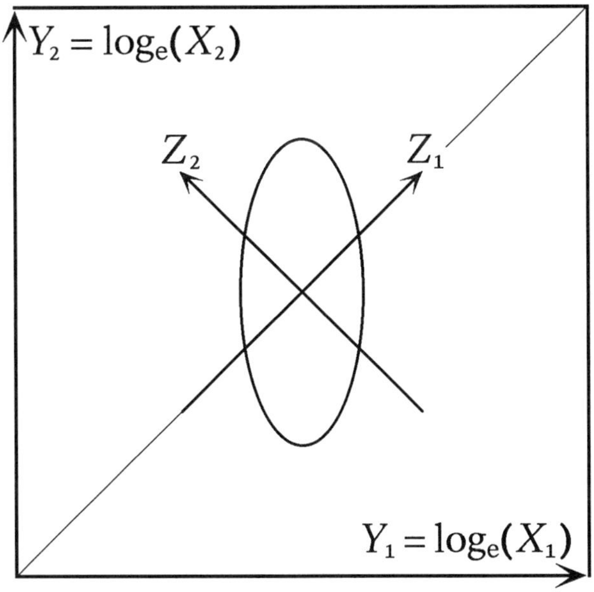

Figure 31.8.2
Even when there is no correlation between the logarithms $Y_1 = \log_e(X_1)$ and $Y_2 = \log_e(X_2)$ of original variates, there may be a strong correlation between the size variate $Z_1 = (Y_1 + Y_2)/\sqrt{2}$ and the shape variate $Z_2 = (Y_2 - Y_1)/\sqrt{2}$

As for figure 31.8.2, it illustrates the case where there is no correlation between transformed variates Y_1 and Y_2 but where one of these two variates (Y_2) varies much more than the other one. If the coordinate $Z_1 = (Y_1 + Y_2)/\sqrt{2}$ on the isometry line were used as a size variate and the coordinate $Z_2 = (Y_2 - Y_1)/\sqrt{2}$ on the perpendicular axis as a shape variate, size and shape would appear to be positively correlated. Yet, since observed variates Y_1 and Y_2 are uncorrelated, considering variate Y_2 as positively allometric would seem obviously incorrect! – Of course, in such an extreme case, a biometrician using the major axis would avoid concluding that there is allometry, since the hypothesis that the major axis is vertical could be accepted, meaning that there is no relationship between observed variates. Nevertheless, if the coordinate Z_1 on the isometry line is used as a size variate, an observed variate Y_2 which has a large variance but which is not strongly related to the other observed variates might tend to be mistakenly interpreted as positively allometric.

Section 31.9: the method of correlations between size and shape variates

When a need is felt to confirm the allometry indications provided by the major axis of the covariance matrix of logarithmically transformed data, it is possible to check whether the relationship between size and shape remains statistically significant even if different size variates are used. In the case of the 24 female painted turtles considered in sections 31.4, 31.6 and 31.7, for instance, the correlation matrix of the following variates may be calculated (table 31.9.1):

$$Y_1 = \log_e(X_1) = \log_e(\text{carapace length in mm}),$$

$$Y_2 = \log_e(X_2) = \log_e(\text{carapace width in mm}),$$

$$Y_3 = \log_e(X_3) = \log_e(\text{carapace height in mm}),$$

$$Y_4 = \log_e(X_2 - X_1) = \log_e(\text{width/length}),$$

$$Y_5 = \log_e(X_3 - X_1) = \log_e(\text{height/length}).$$

Table 31.9.1
Correlation matrix of three size variates (Y_1, Y_2 and Y_3) and two shape variates (Y_4 and Y_5) in 24 female painted turtles

$$[r_{ij}] = \begin{bmatrix} 1 & +0.973545 & +0.973198 & -0.798595 & -0.155765 \\ +0.973545 & 1 & +0.966050 & -0.639943 & -0.071756 \\ +0.973198 & +0.966050 & 1 & -0.728201 & +0.075573 \\ -0.798595 & -0.639943 & -0.728201 & 1 & +0.334823 \\ -0.155765 & -0.071756 & +0.075573 & +0.334823 & 1 \end{bmatrix}$$

The variates Y_1, Y_2 and Y_3 are the logarithms of carapace dimensions, of which the principal components have already been analyzed, but which will be considered here as three possible alternative size variates. The variates Y_4 and Y_5 are logarithms of ratios of carapace width and carapace height to carapace length, two shape variates.

Whichever size variate is used (Y_1, Y_2 or Y_3), its correlation with Y_4 is negative and differs significantly from 0 ($P<0.001$), but the contrary is true with Y_5 ($P>0.05$). The allometry indications yielded by the major axis in sections 31.4, 31.6 and 31.7, according to which carapace width appears to increase less rapidly than carapace length and height, are thus confirmed.

The present approach, which may be called the *method of correlations between size and shape variates,* may also be applied to the data on 68 male North American martens previously considered in section 31.7. The correlations between the following five variates are given in table 31.9.2:

$$Y_1 = \log_e(X_1) = \log_e(\text{skull length in mm}),$$

$$Y_2 = \log_e(X_2) = \log_e(\text{humerus length in mm}),$$

$$Y_3 = \log_e(X_3) = \log_e(\text{femur length in mm}),$$

$$Y_4 = \log_e(X_2 - X_1) = \log_e(\text{humerus length/skull length}),$$

$$Y_5 = \log_e(X_3 - X_1) = \log_e(\text{femur length/skull length}).$$

Table 31.9.2
Correlation matrix of three size variates (Y_1, Y_2 and Y_3) and two shape variates (Y_4 and Y_5) in 68 male North American martens

$$[r_{ij}] = \begin{bmatrix} 1 & +0.666154 & +0.656204 & -0.265701 & -0.271216 \\ +0.666154 & 1 & +0.888144 & +0.542008 & +0.394621 \\ +0.656204 & +0.888144 & 1 & +0.408632 & +0.548327 \\ -0.265701 & +0.542008 & +0.408632 & 1 & +0.815698 \\ -0.271216 & +0.394621 & +0.548327 & +0.815698 & 1 \end{bmatrix}$$

The variates Y_1, Y_2 and Y_3 are the logarithms of the lengths of the skull, humerus and femur, which will be used here as size variates, while Y_4 and Y_5 are logarithms of ratios of humerus and femur length to skull length, two shape variates. Unlike what was found in turtles, shape variates are significantly correlated with all size variates (at levels ranging from $\alpha=0.05$ to $\alpha=0.001$), but these correlations are negative when the size variate is the logarithm of skull length while they are positive when the size variate is the logarithm of humerus or femur length. The lengths of limb bones thus appear positively or negatively allometric depending on the size variate chosen. Because of these disagreements, the relationship between size and shape is doubtful, and it seems safer to retain the isometry hypothesis until concordant indications are obtained. Therefore, in the case of male North American martens also, the *method of correlations between size and shape variates* confirms the conclusion drawn from principal component analysis, since the isometry hypothesis was similarly accepted when the small-sample criterion was applied to the major axis of the same data in section 31.7.

Section 31.10: principal axes of heterogeneous data

While the theory of principal components implies that a single multivariate normal population is involved, principal component analysis is frequently used in practice as a substitute for more complicated approaches even when the nature of the probability distribution is not known accurately or when observations may come from several unknown populations. Scatter diagrams of the first three principal components of the limnological data of Rawson (1960), already considered in section 30.9, are represented in figure 31.10.1.

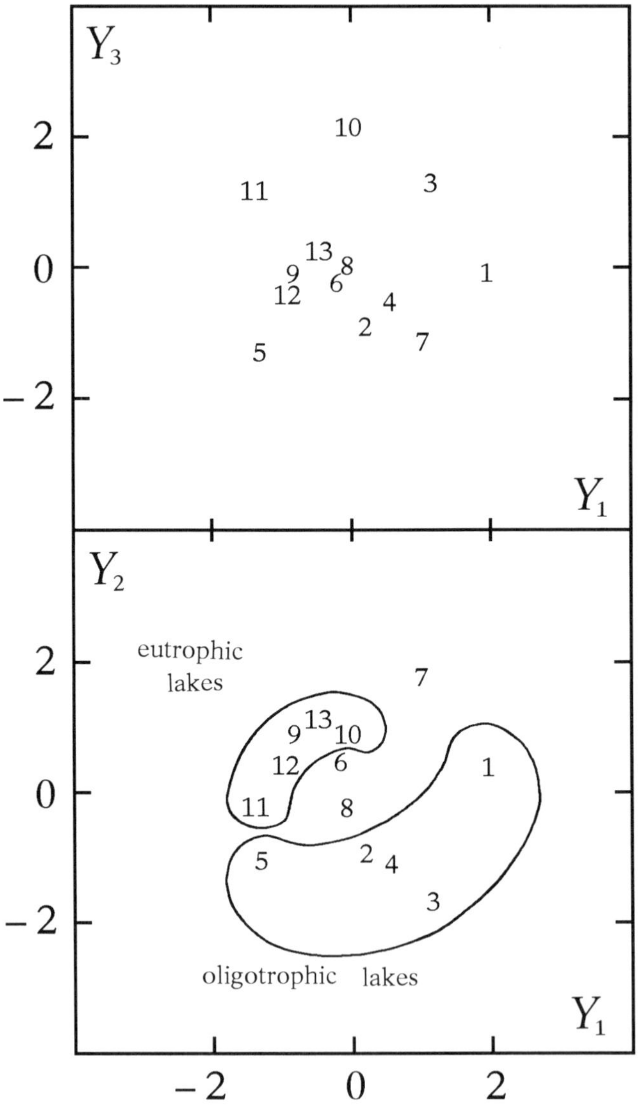

Figure 31.10.1
Scatter diagram of coordinates on the first three principal axes of the Saskatchewan lakes studied by Rawson (1960)

The geographical location of thirteen lakes (or lake sections) studied in the province of Saskatchewan by Rawson (1960) is summarily described in table 31.10.1. Among the physical, chemical and biological characteristics determined by Rawson (1960), the following variates are included in the present analysis:

X_1 = average depth of lake in m,

X_2 = average temperature in °C of surface waters (from 0 to 10 m) in July and August,

X_3 = $\log_e$(dry weight of plankton in kg/ha),

X_4 = $\log_e$(dry weight of benthos in kg/ha),

X_5 = $\log_e$(average weight of fish caught per standard net, in kg/net).

Table 31.10.1
Geographical location of the thirteen Saskatchewan lakes (or lake sections) studied by Rawson (1960)

Number	Name	Longitude	Latitude	Geological substratum
1	Athabasca	110° west	59° north	Precambrian rocky outcroppings of *Canadian Shield*
2	Cree	107° west	57° north	
3	Wollaston	103° west	58° north	
4	Reindeer	102° west	57° north	
5	Frobisher	108° west	57° north	
6	La Ronge	105° west	55° north	border between *Canadian Shield* and glacial deposits
7	Hunter Bay	104° west	55° north	
8	Amisk	102° west	54° north	
9	Churchill	108° west	56° north	glacial deposits south of *Canadian Shield*
10	Big Peter Pond	109° west	56° north	
11	Little Peter Pond	108° west	56° north	
12	Île à la Crosse	108° west	56° north	
13	Waskesiu	107° west	54° north	

Even though lakes 1 to 5 are located on rocky outcroppings of the *Canadian Shield,* and lakes 9 to 13 on glacial deposits south of the *Canadian Shield,* it would not necessarily be appropriate to classify these thirteen bodies of water into distinct categories right from the start of the analysis, because the geological substratum is heterogeneous even on the *Canadian Shield.* Thus, Rawson (1960) noted that, while lakes Wollaston and Reindeer are surrounded by granitic ridges, lake Cree is located on a sandstone plateau. Moreover, while the north shore of lake Athabasca is granitic, its south shore is sandy. It may therefore be preferable to consider this sample of lakes initially as coming from a continuum. If these various lakes belong in fact to well-defined categories, *between-groups* variation should show through even if the sample is treated as if it came from a single homogeneous population.

In order to prevent possible disparities of variances d_1, d_2, ... d_q from excessively distorting the scatter diagrams of principal components, the latter are represented in standardized form, $\mathbf{Y} = \mathbf{D}^{-1/2}\,\mathbf{U}(\mathbf{X}-\overline{\mathbf{X}})$, that is

$$\begin{bmatrix} Y_1 \\ Y_2 \\ \cdots \\ Y_q \end{bmatrix} = \begin{bmatrix} 1/\sqrt{d_1} & 0 & \cdots & 0 \\ 0 & 1/\sqrt{d_2} & \cdots & 0 \\ \cdots & \cdots & \cdots & \cdots \\ 0 & 0 & \cdots & 1/\sqrt{d_q} \end{bmatrix} \begin{bmatrix} u_{11} & u_{12} & \cdots & u_{1q} \\ u_{21} & u_{22} & \cdots & u_{2q} \\ \cdots & \cdots & \cdots & \cdots \\ u_{q1} & u_{q2} & \cdots & u_{qq} \end{bmatrix} \begin{bmatrix} (X_1 - \bar{X}_1) \\ (X_2 - \bar{X}_2) \\ \cdots \\ (X_q - \bar{X}_q) \end{bmatrix} .$$

In the scatter diagram of principal components Y_1 and Y_2 (figure 31.10.1, bottom), lakes 1 to 5 are situated below and to the right of lakes 9 to 13, while lakes 6 to 8 have an intermediate position. However, the three groups of lakes are not so clearly separated from each other in the scatter diagram of principal components Y_1 and Y_3 (figure 31.10.1, top). The scatter diagram of the first two principal components (figure 31.10.1, bottom) is interesting because it provides an overall view of the data based on all original variates $[X_1, X_2, \ldots X_5]$ jointly.

Nevertheless, in the case of Rawson's data, some of the scatter diagrams of original variates are at least as informative and more easily interpreted than those of principal components. Thus the scatter diagram of the average depth X_1 and of the average summer temperature X_2 of surface waters shows that, though lake Frobisher (No. 5) is located on the *Canadian Shield,* it has a small average depth and warm surface waters in the summer like the lakes situated on glacial deposits (figure 31.10.2). Yet the scatter diagram of the logarithms of the dry weights per hectare of benthos (X_3) and plankton (X_4) indicates that lake Frobisher (No. 5) has a low productivity like the other lakes located on the *Canadian Shield* (figure 31.10.3). Unlike lakes 1 to 4, however, lake Frobisher may owe its low productivity to the lack of transparency of its waters (Rawson, 1960, pages 202-203).

The strong negative correlation between average depth and the average summer temperature of surface waters (figure 31.10.2; $r = -0.834638$, $P < 0.001$) is interesting because it reflects the fact that surface waters (the *epilimnion*) are generally more difficult to warm up in a deep lake, not only because its lower waters (the *hypolimnion*) store a greater amount of cold but also because its thermal stratification (the *thermocline*) tends to be more stable. According to limnological theory, a deeper lake also tends to be less productive because the minerals released by the bacterial decomposition of plant and animal fragments tend to remain in deep waters most of the year and are less available to fertilize surface waters. To sum up, the results of the present analysis appear to be consonant with the theoretical basis of the limnological classification of *oligotrophic, mesotrophic* and *eutrophic* lakes, which have respectively low, average and high productivities.

From a methodological point of view, it should be noted that the three groups of lakes overlap to a smaller extent in the scatter diagram of original variates X_3 and X_4 (figure 31.10.3) than with respect to the first two principal components (figure 31.10.1, bottom). It will be indicated in chapters 32 and 33 that, when there are *a priori* reasons to suspect that two or several groups of multivariate observations come from different statistical populations, discriminant functions generally bring out differences between groups more efficiently than principal components.

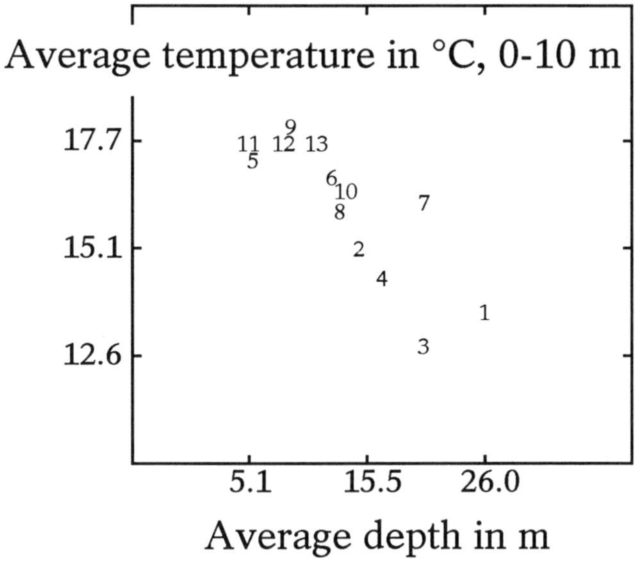

Figure 31.10.2
Scatter diagram of average depth in m and average summer temperature in °C of surface waters (from 0 to 10 m) in the Saskatchewan lakes studied by Rawson (1960)

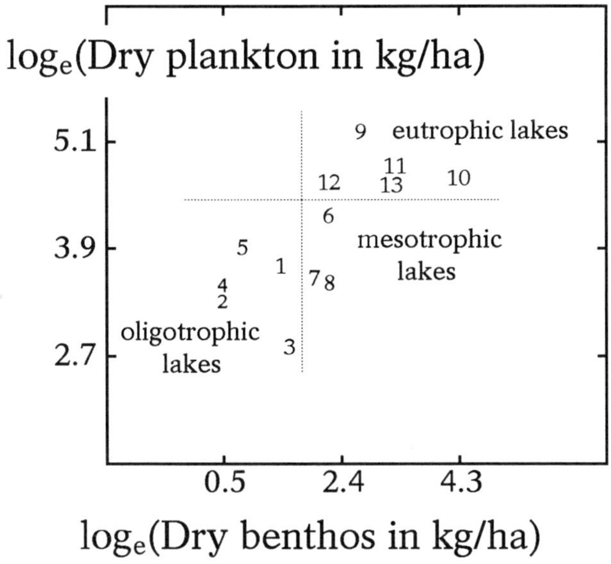

Figure 31.10.3
Scatter diagram of logarithms of average dry weight per hectare of benthos and plankton in the Saskatchewan lakes studied by Rawson (1960)

In a study of a heterogeneous collection of fishes belonging to the genus *Etheostoma* and coming from the region of Ottawa, McAllister, Jolicoeur and Tsuyuki (1972) showed that the frequency distribution of the first principal component of a set of variates comprising qualitative and meristic characters (counts of pores, fin rays, etc.) was significantly trimodal. This suggested that the sampled population contained natural hybrids in addition to specimens belonging to the parent species *Etheostoma nigrum* and *Etheostoma olmstedi*.

Section 31.11: mathematical aspects

The linear combinations of original variates possessing the largest and the smallest variances correspond to the latent vectors and roots of the sample covariance matrix $\mathbf{S}$, as may be shown by maximizing or minimizing the quadratic form $\mathbf{U}_i\mathbf{S}\,\tilde{\mathbf{U}}_i$ while respecting the condition $\mathbf{U}_i\mathbf{I}\tilde{\mathbf{U}}_i = \mathbf{U}_i\tilde{\mathbf{U}}_i = 1$, according to which $\mathbf{U}_i$ is a unit vector (section 24.6). This conditional maximization or minimization is generally carried out by using the method of *undetermined multipliers* of the French mathematician Joseph Louis de Lagrange (1736-1813), according to which

$$\frac{\partial}{\partial \mathbf{U}_i}[\mathbf{U}_i\mathbf{S}\tilde{\mathbf{U}}_i - d_i(\mathbf{U}_i\mathbf{I}\tilde{\mathbf{U}}_i - 1)] = 2\mathbf{U}_i\mathbf{S} - 2d_i\mathbf{U}_i\mathbf{I} = \mathbf{0}\,,$$

whence $$\mathbf{U}_i\mathbf{S} = d_i\mathbf{U}_i\mathbf{I} = d_i\mathbf{U}_i\,.$$

The product of matrix $\mathbf{S}$ by vector $\mathbf{U}_i$ is proportional to that vector, which shows that $\mathbf{U}_i$ is a latent vector and that d_i is the corresponding latent root of matrix $\mathbf{S}$ (section 24.21). A covariance matrix of order $(q \times q)$ possesses q latent vectors and q latent roots, and the q scalar equations $\mathbf{U}_i\mathbf{S} = d_i\mathbf{U}_i$ may be assembled into a single matrix equation,

$$\mathbf{US} = \mathbf{DUI} = \mathbf{DU}\,,$$

where $\mathbf{U}$ is the orthogonal matrix of latent vectors and $\mathbf{D}$ is the diagonal matrix of latent roots. A similar demonstration can be made also for the latent vectors $\boldsymbol{\Upsilon}$ and the latent roots $\boldsymbol{\Delta}$ of the population covariance matrix $\boldsymbol{\Sigma}$ (section 31.2):

$$\boldsymbol{\Upsilon\Sigma} = \boldsymbol{\Delta\Upsilon}\,.$$

However, the preceding relationships may also be demonstrated without using Lagrange's undetermined multipliers, since

$$\frac{\partial}{\partial \mathbf{U}_i}\mathbf{U}_i\mathbf{S}\tilde{\mathbf{U}}_i/\mathbf{U}_i\tilde{\mathbf{U}}_i = \frac{\partial}{\partial \mathbf{U}_i}\mathbf{U}_i\mathbf{S}\tilde{\mathbf{U}}_i/\mathbf{U}_i\mathbf{I}\tilde{\mathbf{U}}_i = [2\mathbf{U}_i\mathbf{S}(\mathbf{U}_i\mathbf{I}\tilde{\mathbf{U}}_i) - 2(\mathbf{U}_i\mathbf{S}\tilde{\mathbf{U}}_i)\mathbf{U}_i\mathbf{I}]/(\mathbf{U}_i\mathbf{I}\tilde{\mathbf{U}}_i)^2.$$

Therefore, in order to get $$\frac{\partial}{\partial \mathbf{U}_i}\mathbf{U}_i\mathbf{S}\tilde{\mathbf{U}}_i/\mathbf{U}_i\tilde{\mathbf{U}}_i = \mathbf{0}\,,$$

one must have $$\mathbf{U}_i\mathbf{S} = [(\mathbf{U}_i\mathbf{S}\tilde{\mathbf{U}}_i)/(\mathbf{U}_i\mathbf{I}\tilde{\mathbf{U}}_i)]\mathbf{U}_i\mathbf{I} = d_i\mathbf{U}_i\,.$$

The reader who is not familiar with the differentiation of vectors and of their functions, like quadratic forms, should consult Graybill (1969, chapter 10).

Chapter 32
Fisher's linear discriminant function

Section 32.1: introduction

When Hotelling's T^2 distribution (section 30.4) indicates that one should reject the hypothesis that two multivariate normal distributions have equal mean vectors, it may be useful to determine the linear combination of original variates,

$$Y = b_1 X_1 + b_2 X_2 + \dots + b_q X_q ,$$

which brings out differences between observations belonging to the two populations as clearly as possible, and which may be used to estimate from which population a new individual observation,

$$\mathbf{X} = [X_1, X_2, \dots X_q] ,$$

has come. This linear combination is called a *discriminant function* and was developed by Fisher (1936), whose attention was drawn to the problem by Edgar Anderson (see Anderson, 1954), an American botanist, during a study of iris species of the Gaspé Peninsula, Québec. In his paper, which he illustrated with data provided by Anderson, Fisher (1936) obtained his discriminant function by maximizing the ratio of the difference between the means of two groups to the within-groups standard deviation. Fisher also showed that the same function could be obtained by estimating the coefficients

$$\mathbf{b} = [b_1, b_2, \dots b_q]$$

through multiple regression (sections 23.3, 25.3 and 25.5) after attributing arbitrary Y values to individual observations, such arbitrary values being constant within each group but different from one group to the other. Fisher's linear discriminant function can be useful not only in taxonomy but also in medical diagnosis, where the variates $\mathbf{X} = [X_1, X_2, \dots X_q]$ may represent symptoms, and in various other fields where a conclusion must be drawn or a decision must be made on the basis of many quantitative data jointly.

Section 32.2: choosing between two populations known *a priori*

In order to introduce the problem in the simplest possible manner, let us assume at first that an individual observation $\mathbf{X} = [X_1, X_2, \dots X_q]$ has been randomly obtained from one of two multivariate normal populations of which the parameters are already known: $\mathcal{N}(\mu_A, \Sigma_A)$ and $\mathcal{N}(\mu_B, \Sigma_B)$. On the basis of the numerical values of the elements of vector $\mathbf{X}$, from which of the two populations should the individual observation be deemed to have come? – It seems reasonable to assign the individual observation $\mathbf{X}$ to the population having the highest probability of containing a random observation differing as little as $\mathbf{X}$ (or less) from the average. But this probability may be evaluated for each population by using the quadratic form $(\mathbf{X} - \mu)\, \Sigma^{-1} (\tilde{\mathbf{X}} - \tilde{\mu})$, which follows the χ^2 distribution with q degrees of freedom (section 29.2). The larger the numerical value of this quadratic form, the smaller the probability that the individual observation $\mathbf{X}$ has come from that population.

The individual observation **X** should therefore be attributed to population A

if $$(\mathbf{X}-\mu_A)\,\Sigma_A^{-1}(\tilde{\mathbf{X}}-\tilde{\mu}_A) < (\mathbf{X}-\mu_B)\,\Sigma_B^{-1}(\tilde{\mathbf{X}}-\tilde{\mu}_B),$$

that is if $$0 < Z = (\mathbf{X}-\mu_B)\,\Sigma_B^{-1}(\tilde{\mathbf{X}}-\tilde{\mu}_B) - (\mathbf{X}-\mu_A)\,\Sigma_A^{-1}(\tilde{\mathbf{X}}-\tilde{\mu}_A),$$

and to population B otherwise. When the covariance matrices Σ_A and Σ_B of the two populations are equal, $\Sigma_A = \Sigma_B = \Sigma$, the preceding inequality boils down to

$$\mathbf{X}\,\Sigma^{-1}(\tilde{\mu}_B-\tilde{\mu}_A) < (1/2)\,(\mu_B+\mu_A)\,\Sigma^{-1}(\tilde{\mu}_B-\tilde{\mu}_A)$$

where $$\Sigma^{-1}(\tilde{\mu}_B-\tilde{\mu}_A) = \tilde{\beta}$$

is the column vector of the parametric coefficients of the *linear discriminant function* Y. The inequality may thus be expressed as

$$Y = \mathbf{X}\tilde{\beta} < (1/2)\,(\mu_B+\mu_A)\,\tilde{\beta},$$

which shows that the individual observation $\mathbf{X}=[X_1, X_2, \ldots X_q]$ should be attributed to population A if the value $Y=\mathbf{X}\tilde{\beta}$ of its discriminant function lies closer to the discriminant function $\mu_A\tilde{\beta}$ of the average μ_A of population A than the midpoint $(1/2)(\mu_B+\mu_A)\tilde{\beta}$ between the discriminant functions of the averages of the two populations. In other words, the individual observation **X** should be attributed to population A if its discriminant function Y lies closer to the discriminant function of the average of population A than to the discriminant function of the average of population B, a readily understandable result.

When the *a priori* probabilities $P(\mathrm{A})$ and $P(\mathrm{B})=1-P(\mathrm{A})$, that the individual observation **X** comes from population A or B respectively, are different, or when the two kinds $c(1|2)$ and $c(2|1)$ of misclassification costs are unequal, the preceding procedure may be modified in order to reduce the probability or the total cost of classification errors (see Anderson, 1984, chapter 6 and Morrison, 1990, chapter 6). Finally, when the covariance matrices Σ_A and Σ_B of the two populations differ, the inequality at the top of this page delimits a curved boundary between the regions in which the individual observation **X** is attributed to population A or B, and the Z variate is referred to as a *quadratic discriminant function.*

Section 32.3: two populations known through samples

When the parameters of populations A and B are estimated from samples instead of being known a priori, a basically similar procedure is followed except that parameters are replaced by sample estimates (see section 30.4) and the individual observation **X** is attributed to population A if

$$\mathbf{X}\tilde{\mathbf{b}} < (1/2)\,(\overline{\mathbf{X}}_B+\overline{\mathbf{X}}_A)\,\tilde{\mathbf{b}},$$

where $$\tilde{\mathbf{b}} = \mathbf{S}^{-1}(\tilde{\overline{\mathbf{X}}}_B - \tilde{\overline{\mathbf{X}}}_A)$$

and $$\mathbf{S} = [1/(N_A+N_B-2)]\,[(N_A-1)\,\mathbf{S}_A+(N_B-1)\,\mathbf{S}_B] = \mathbf{S}_{\text{within-groups}}\ .$$

Of course, attributing an individual observation to one of two populations, A or B, is even more uncertain when parameters are estimated from limited-size samples than when they are known a priori.

Section 32.4: Fisher's linear discriminant function and Hotelling's T^2

While the T^2 distribution of Hotelling (1931) and the linear discriminant function of Fisher (1936) were developed independently and for different purposes, they are closely related. For instance, in the particular case of the test of the hypothesis that two populations have equal mean vectors,

$$H_0: \mu_B = \mu_A \text{ , that is } H_0: \mu_B - \mu_A = 0 \, ,$$

the value of Hotelling's T^2 is

$$\frac{(N_A N_B)}{(N_A + N_B)} (\bar{X}_B - \bar{X}_A) \, S^{-1} (\tilde{\bar{X}}_B - \tilde{\bar{X}}_A) = \frac{(N_A N_B)}{(N_A + N_B)} (\bar{X}_B - \bar{X}_A) \, \tilde{b} = T^2_{(q, \, N_A + N_B - 2)} \, .$$

where $\tilde{b}$ is the column vector of the coefficients of Fisher's linear discriminant function.

Section 32.5: Mahalanobis' generalized distance

Moreover, Hotelling's T^2 and Fisher's discriminant function are related to the *generalized distance* proposed in 1936 by the Indian statistician P. C. Mahalanobis (1893-1972) as a measure of the relative importance of the difference between the mean vectors of two multivariate normal populations. Theoretically, the generalized distance between the means of two populations A and B is defined as the square root Δ of the quadratic form

$$\Delta^2 = (\mu_B - \mu_A) \, \Sigma^{-1} (\tilde{\mu}_B - \tilde{\mu}_A) \, .$$

In comparison with the ordinary distance of Euclidean geometry,

$$\sqrt{((\mu_B - \mu_A) \, I \, (\tilde{\mu}_B - \tilde{\mu}_A))} \, ,$$

Mahalanobis' generalized distance Δ has the advantage of using within-groups variation as a yardstick for differences between means and of taking into account the fact that the within-groups standard deviation is greater in some directions than others. In fact, because of correlations between variates, within-groups variation usually has an elliptical (or ellipsoidal) rather than a circular (or spherical) shape. By using Mahalanobis' generalized distance, one is applying an elliptical standard to original variates or, equivalently, one is working with new variates with respect to which within-groups variation is circular (see figures 30.8.1 and 33.3.1).

When population parameters are estimated from samples, the generalized distance is estimated by the square root D of the quadratic form

$$D^2 = (\bar{X}_B - \bar{X}_A) \, S^{-1} (\tilde{\bar{X}}_B - \tilde{\bar{X}}_A) \, .$$

The square of the generalized distance between sample means, D^2, thus has merely to be multiplied by $(N_A N_B)/(N_A + N_B)$ in order to get the value of Hotelling's T^2 used to test the null hypothesis $H_0: \mu_B - \mu_A = 0$ (section 30.4).

When samples coming from more than two populations are compared, generalized distances may be calculated after pairing these groups in all possible manners. An approximate three-dimensional model of differences between group means could then be built using balls and sticks as chemists used to do to represent molecular structures. However, the methods discussed in the next chapter (33) are more efficient for the analysis of multivariate differences between many groups of observations.

Section 32.6: a taxonomical example

Fisher's linear discriminant function is illustrated here on data about two butterfly species already mentioned in section 26.6. Each species is represented by a sample of 10 male specimens coming from British Columbia in the case of *Papilio glaucus* (A) and from Colorado in the case of *Papilio rutulus* (B). In order to allow relationships between original variates X_1 and X_2 and the discriminant function $Y = b_1 X_1 + b_2 X_2$ to be easily seen, only two variates are included in the present analysis:

X_1 = the length of the superuncus in mm, multiplied by 8,

X_2 = the length of the right genital valva, in mm.

The mean vectors $\overline{\mathbf{X}}$ and the covariance matrices $\mathbf{S}$ of both samples are given in table 32.6.1.

Table 32.6.1
Statistics of two samples of male butterflies; data from Brower (1959)

Species	*Papilio glaucus*	*Papilio rutulus*
Origin	British Columbia	Colorado
Sample size	$N_A = 10$	$N_B = 10$
Mean vector $\overline{\mathbf{X}}$	[9.000 4.752]	[10.95 5.540]
Covariance matrix $\mathbf{S}$	$\begin{bmatrix} 0.3333333 & 0.0411111 \\ 0.0411111 & 0.0625067 \end{bmatrix}$	$\begin{bmatrix} 0.1916667 & 0.0650000 \\ 0.0650000 & 0.0695556 \end{bmatrix}$

The estimated coefficients of the discriminant function (see section 32.3) are

$$\mathbf{b} = (\overline{\mathbf{X}}_B - \overline{\mathbf{X}}_A)\mathbf{S}^{-1} = [5.98920 \quad 7.12148],$$

the value of Hotelling's T^2 (sections 30.4 and 32.4) is

$$T^2_{(2,\,18)} = 86.4534, \text{ whence } F_{(2,\,17)} = 40.8252 \text{ and } P < 0.001,$$

and Mahalanobis' generalized distance D (section 32.5) is

$$D = \sqrt{D^2} = \sqrt{17.2907} = 4.15821.$$

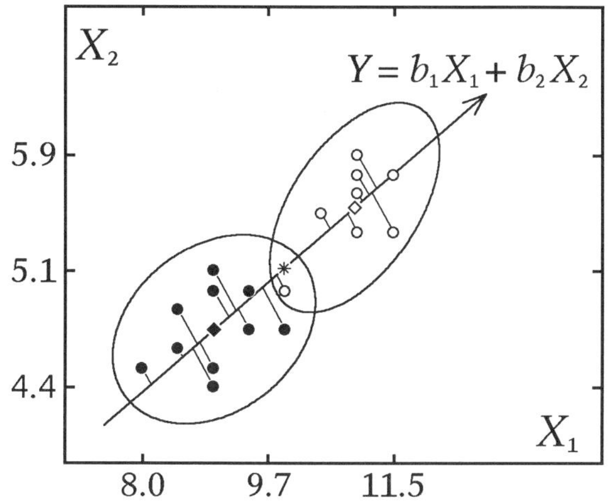

Figure 32.6.1
Scatter diagram of X_1 = (length of superuncus) × 8 and of X_2 = length of right genital valva, in mm, in male butterflies of the species *Papilio glaucus* (●, mean : ◆) and *Papilio rutulus* (○, mean : ◇); individual observations are projected (obliquely) on the coordinate axis of Fisher's linear discriminant function $Y = b_1X_1 + b_2X_2$; the general mean is represented by an asterisk (✳)

The hypothesis that the mean vectors $\boldsymbol{\mu}_A$ and $\boldsymbol{\mu}_B$ of the two species are equal must therefore be rejected, and the estimate of Mahalanobis' generalized distance between the means of the two populations is equal to approximately four times the within-groups standard deviation (D = 4.15821). Moreover, if a tentative identification were required for an additional specimen presumed to belong to one of the two species, the discriminant function Y could be used. Variation ellipses drawn at the level $\alpha = 0.05$ as well as the projections of individual observations on the coordinate axis of Fisher's linear discriminant function $Y = b_1X_1 + b_2X_2$ are represented in figure 32.6.1. Because sample sizes are equal here ($N_A = N_B = 10$), the general mean (✳) coincides with the midpoint between the means $\bar{\mathbf{X}}_A$ (◆) and $\bar{\mathbf{X}}_B$ (◇) of the two groups. Only one of the individual observations of the two samples, the lowest one of species *Papilio rutulus* (○), would be misclassified using the discriminant function, since it lies lower than the midpoint (✳), and is therefore located closer to the mean of the other species, $\bar{\mathbf{X}}_A$ (◆), than to the mean of its own species, $\bar{\mathbf{X}}_B$ (◇).

It is noteworthy that the within-groups variance of the discriminant function $Y = \mathbf{b}\tilde{\mathbf{X}}$ is (see section 29.5)

$$S_Y^2 = \mathbf{b}\mathbf{S}\tilde{\mathbf{b}} = (\bar{\mathbf{X}}_B - \bar{\mathbf{X}}_A)\mathbf{S}^{-1}\mathbf{S}\,\mathbf{S}^{-1}(\tilde{\bar{\mathbf{X}}}_B - \tilde{\bar{\mathbf{X}}}_A) = (\bar{\mathbf{X}}_B - \bar{\mathbf{X}}_A)\mathbf{S}^{-1}(\tilde{\bar{\mathbf{X}}}_B - \tilde{\bar{\mathbf{X}}}_A) = D^2.$$

The within-groups variance of the discriminant function of individual observations is thus equal to the square of the generalized distance between sample means, a rather surprising result!

The coefficients $\mathbf{b} = [b_1 \;\; b_2]$ are often adjusted so that the within-groups variance of the discriminant function of individual observations is equal to unity, which agrees with the scale in which the generalized distance D is expressed. This is done by redefining the coefficients of the discriminant function as follows:

$$\mathbf{b} = (1/D)(\overline{\mathbf{X}}_B - \overline{\mathbf{X}}_A)\mathbf{S}^{-1}.$$

In the case of the present data, this yields:

$$\mathbf{b} = (1/4.15821)[5.98920 \quad 7.12148] = [1.44033 \quad 1.71263].$$

Chapter 33
Multiple discriminant analysis

Section 33.1: introduction

While multivariate differences between several groups of observations could be described by pairing these groups in all possible manners and by evaluating Mahalanobis' generalized distance between their means (section 32.5), a much better performing method known as *multiple discriminant analysis* will be discussed in this chapter. Multiple discriminant analysis is also called *canonical variate analysis,* but the latter expression is less adequate because canonical variates may be used even when there is only one group of observations (section 30.8, chapter 31, chapter 34). Multiple discriminant analysis may be considered as a principal component analysis (chapter 31) in which the principal axes of between-groups variation are determined after within-groups variation has been taken as a yardstick (sections 33.3 and 33.12). In the particular case where only two groups are compared, multiple discriminant analysis reduces to Fisher's linear discriminant function (chapter 32).

Section 33.2: multivariate within-groups and between-groups variation

According to the preliminary hypothesis H_0, on the one hand, the samples to be compared in a multiple discriminant analysis have come from a single multivariate normal population $\mathcal{N}(\boldsymbol{\mu}, \boldsymbol{\Sigma})$ or, equivalently, from several multivariate normal populations having identical mean vectors and covariance matrices (figure 33.2.1, left). According to the alternative hypothesis H_1, on the other hand, the samples have come from dissimilar multivariate normal populations (figure 33.2.1, right). When the preliminary hypothesis $H_0 : \boldsymbol{\mu}_1 = ... = \boldsymbol{\mu}_C = \boldsymbol{\mu}$ is true, the mean vector $\overline{\mathbf{X}}_g$ of the gth sample follows a multivariate normal distribution (sections 29.2 and 30.2) of which the mean vector is $\boldsymbol{\mu}$ but of which the covariance matrix $(1/N_g)\boldsymbol{\Sigma}$ is N_g times smaller than the covariance matrix $\boldsymbol{\Sigma}$ of the individual observations $\mathbf{X}_{gh}$. From a graphical viewpoint, this is reflected

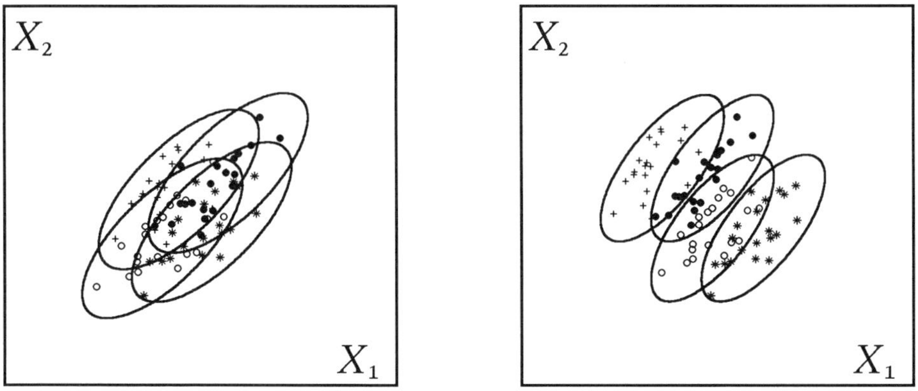

Figure 33.2.1
Individual observations and prediction ellipses for samples coming from the same population or from identical populations (left) and from dissimilar populations (right)

by the fact that the groups of dots show much overlap and that their centers (means) are dispersed within an elliptical (or ellipsoidal) region having the same shape but a smaller size than the region of individual (within-groups) variation (figure 33.2.1, left). When the alternative hypothesis is true, on the contrary, clusters of dots tend to show less overlap and their centers (means) are dispersed within an elliptical (or ellipsoidal) region which may have a different shape and a larger size than the region of individual variation (figures 33.2.1 and 33.2.2, right). It should be noted that, like in several earlier chapters (sections 9.8, 12.7, 30.4 and 32.2), the problem of comparing means is made as simple as possible by assuming that the populations compared have equal covariance matrices,

$$\mathbf{\Sigma}_1 = \dots = \mathbf{\Sigma}_g = \dots = \mathbf{\Sigma}_C = \mathbf{\Sigma},$$

even when the alternative hypothesis H_1: $\boldsymbol{\mu}_i \neq \boldsymbol{\mu}_j$ is true. However, even when covariance matrices are unequal or when the data do not follow the multivariate normal distribution exactly, multiple discriminant analysis provides a simplified solution which may often be useful for the comparison of several group means with respect to many variates.

In multiple discriminant analysis as in the analysis of variance (chapter 12), it is by comparing between-groups and within-groups variation (figure 33.2.2) that it may be decided whether the preliminary hypothesis H_0 should be accepted or rejected in favor of the alternative hypothesis H_1. Yet within-groups and between-groups variations are described here by matrices of sums of (squares and) products of the deviations of the q variates $[X_1 \dots X_q]$ rather than by simple sums of squares of a single variate as in the univariate analysis of variance. The matrix $\mathbf{W}$ of *"within-groups sums of products"* is obtained by adding up the sums of products of deviations obtained within each sample with respect to the mean vector $\overline{\mathbf{X}}_g$ of that sample, which is equivalent to doing as if the mean vectors of the various samples did not differ from each other (figure 33.2.2, left). The matrix $\mathbf{B}$ of *"between-groups sums of products"* is equal to the difference $\mathbf{B} = \mathbf{T} - \mathbf{W}$, where $\mathbf{T}$ denotes the *"total sums of products"* obtained by adding up the sums of products of deviations with respect to the general mean vector $\overline{\mathbf{X}}$.. Matrix $\mathbf{B}$ reflects the dispersion of the sample mean vectors $\overline{\mathbf{X}}_g$ about the general mean vector $\overline{\mathbf{X}}$. (figure 33.2.2, right).

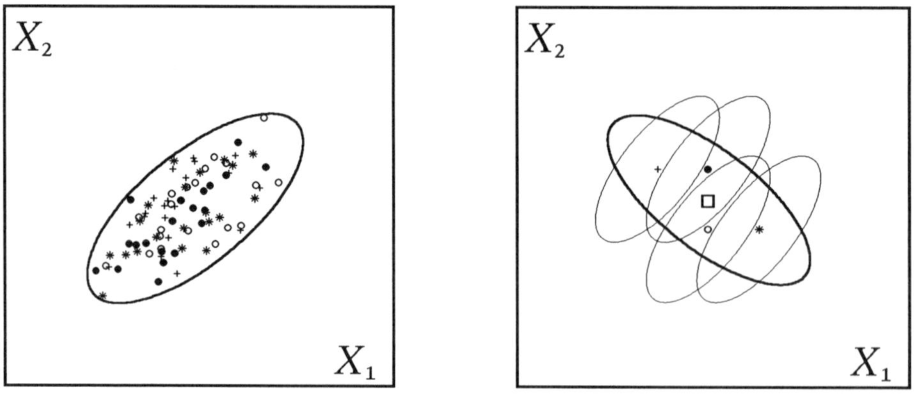

Figure 33.2.2
Within-groups variation of individual observations about the mean of each sample (left), and between-groups variation of sample means about the general mean (□, right)

Multiple discriminant analysis is intimately related to the multivariate analysis of variance ("MANOVA"), and it may also be combined with the analysis of covariance (chapter 21; see also Baron and Jolicoeur, 1980) and with the two-way (chapter 28) or multi-way analysis of variance, but the present account is limited to the simplest case, where the groups compared may be classified with respect to only one criterion. The notation and the computing equations used for the within-groups and between-groups matrices of sums of products **W** and **B** are summarized in table 33.2.1.

Table 33.2.1
Notation and computing equations for the within-groups and between-groups matrices W and B of sums of squares and products of deviations

X_{ghi} = the numerical value of the ith variate of the hth individual observation of the gth group (sample); the ith element of vector $\mathbf{X}_{gh}$;
$i = 1, \ldots q$, and $j = 1, \ldots q$, are subscripts of any two different variates;
$q =$ the total number of variates;
$h = 1, \ldots N_g$, is the subscript of one individual observation;
$N_g =$ the sample size of the gth group (sample);
$g = 1, \ldots C$, is the subscript of one particular group (sample);
$C =$ is the total number of groups (samples) compared;
$N. = N_1 + \ldots + N_g + \ldots + N_C = \sum_{g=1}^{C} N_g$ = the general (total) sample size;
$\overline{\mathbf{X}}_g = [\overline{X}_{g1} \ldots \overline{X}_{gq}]$ is the mean vector of the gth sample; $\overline{\mathbf{X}}. = [\overline{X}._1 \ldots \overline{X}._q]$ is the general mean vector of all observations, also known as the overall mean or grand mean vector;
$\mathbf{T} = [\sum_{g=1}^{C} \sum_{h=1}^{N_g} (X_{ghi} - \overline{X}._i)(X_{ghj} - \overline{X}._j)]$ = the matrix of total sums of products;
$\mathbf{W} = [\sum_{g=1}^{C} \sum_{h=1}^{N_g} (X_{ghi} - \overline{X}_{gi})(X_{ghj} - \overline{X}_{gj})]$ = the matrix of within-groups sums of products;
$\mathbf{B} = \mathbf{T} - \mathbf{W} = [\sum_{g=1}^{C} N_g(\overline{X}_{gi} - \overline{X}._i)(\overline{X}_{gj} - \overline{X}._j)]$ = the matrix of between-groups sums of products.

Theoretically, when the preliminary hypothesis H_0: $\mu_1 = \mu_2 = \ldots = \mu_C = \mu$ is true, the individual observation vectors $\mathbf{X}_{gh} = [X_{ghi}]$ are dispersed about the mean vector μ with the covariance matrix $\boldsymbol{\Sigma}$, while the mean vector $\overline{\mathbf{X}}_g$ of the gth sample is dispersed about μ with a covariance matrix $(1/N_g)\boldsymbol{\Sigma}$, but the covariance matrix $\boldsymbol{\Sigma}$ could be estimated either by the within-groups estimate,

$$\mathbf{S}_{\text{within-groups}} = \mathbf{S} = [(1/(N. - C)]\,\mathbf{W}, \text{ which has } (N. - C) \text{ degrees of freedom,}$$

or by the between-groups estimate,

$$\mathbf{S}_{\text{between-groups}} = [(1/(C-1)]\,\mathbf{B}, \text{ which has } (C-1) \text{ degrees of freedom.}$$

When the preliminary hypothesis is not true, the population mean vectors $\mu_1, \ldots, \mu_g, \ldots, \mu_C$ do not coincide. Even when the mean vectors do not coincide, however, they may not completely "fill" the total number of available dimensions. For instance, let us assume that three groups of living organisms belonging to the same genus are compared with respect to three quantitative characters, and that the individual specimens of the first group are hybrids between closely related species represented by the second and third groups. If the hybrid specimens are exactly intermediate (with respect to all characters included in the study) between the two species crossed, their mean should be located on the same straight line as the means of the two species and at half distance between them. In this context, the means of the three groups would then be expected to be *"collinear"*, that is to lie on a single straight line within a three-dimensional space. When *"collinearity"* occurs in this way, three mean vectors may be concluded to differ less than if they had occupied two dimensions as they could have done, since three points can determine a plane. Similarly, if the mean vectors of four groups were contained within a plane (two dimensions) instead of filling three dimensions as they could have done, they would be said to be *"coplanar"*.

During a multiple discriminant analysis, a whole sequence of hypotheses may thus be considered: first, the preliminary hypothesis H_0 according to which mean vectors do not differ at all and occupy a single point (0 dimensions); second, the hypothesis H_1 according to which mean vectors are collinear and occupy only one dimension; third, the hypothesis H_2 according to which mean vectors are coplanar and occupy only two dimensions; ... and so on up to the hypothesis that mean vectors occupy the full number of available dimensions, which is equal to minimum$(q, C-1)$ where q is the number of variates and C is the number of groups. The next section will describe a series of changes of variables through which multivariate differences between several groups of observations can be brought out as clearly as possible.

Section 33.3: geometrical aspects of multiple discriminant analysis

The changes of variables involved in multiple discriminant analysis are similar to those already described concerning the delineation of a confidence, a prediction or a variation ellipse in section 30.8 except that, in this case, there are generally more than two variates, and several groups of observations instead of only one. In the first change of variables, the original variates $\mathbf{X} = [X_1 \ldots X_q]$, represented by the column vector $\tilde{\mathbf{X}}$ at the top of figure 33.3.1, are replaced by centered variates $(\tilde{\mathbf{X}} - \tilde{\bar{\mathbf{X}}}.)$, on the second line of figure 33.3.1. This is equivalent to using the general mean vector $\bar{\mathbf{X}}. = [\bar{X}._1 \ldots \bar{X}._q]$ as an origin instead of the point $\mathbf{0} = [0 \ldots 0]$ and is called a *translation*. In the second change, centered variates $(\tilde{\mathbf{X}} - \tilde{\bar{\mathbf{X}}}.)$ are replaced by the coordinates $\tilde{\mathbf{Y}} = \mathbf{U}(\tilde{\mathbf{X}} - \tilde{\bar{\mathbf{X}}}.)$ of observations on the principal axes $\mathbf{U}$ of the within-groups covariance matrix (third line of figure 33.3.1). This is a *rotation*. In the third change, the coordinates $\mathbf{Y} = [Y_1 \ldots Y_q]$ of observations on the principal axes of within-groups variation are scaled in such a way that each coordinate Y_i is divided by its standard deviation, which equals the square root $\sqrt{d_i}$ of the corresponding latent root (fourth line of figure 33.3.1). This set of scale changes is carried out by premultiplying the column vector $\tilde{\mathbf{Y}}$ by the inverse square root of the diagonal form $\mathbf{D}$ of the within-groups covariance matrix. It must be emphasized that the matrix $\mathbf{D}$, denoted by a bold-type D, should not be confused with Mahalanobis' generalized distance D, denoted by light-type italic in chapter 32. Following these scale changes, the new variates are $\mathbf{D}^{-1/2}\,\mathbf{U}\,(\tilde{\mathbf{X}} - \tilde{\bar{\mathbf{X}}}.)$ and within-groups variation now has a circular, spherical or hyperspherical shape depending on the number of variates (2, 3 or more than 3). At this stage of computations, the within-groups covariance matrix **S** has become a unit matrix,

$$\mathbf{D}^{-1/2}\,\mathbf{U}\,\mathbf{S}\,\tilde{\mathbf{U}}\,\mathbf{D}^{-1/2} = \mathbf{D}^{-1/2}\,\mathbf{D}\,\mathbf{D}^{-1/2} = \mathbf{I}\,,$$

while the matrix **B** of between-groups sums of products has become $\mathbf{D}^{-1/2}\,\mathbf{U}\,\mathbf{B}\,\tilde{\mathbf{U}}\,\mathbf{D}^{-1/2}$.

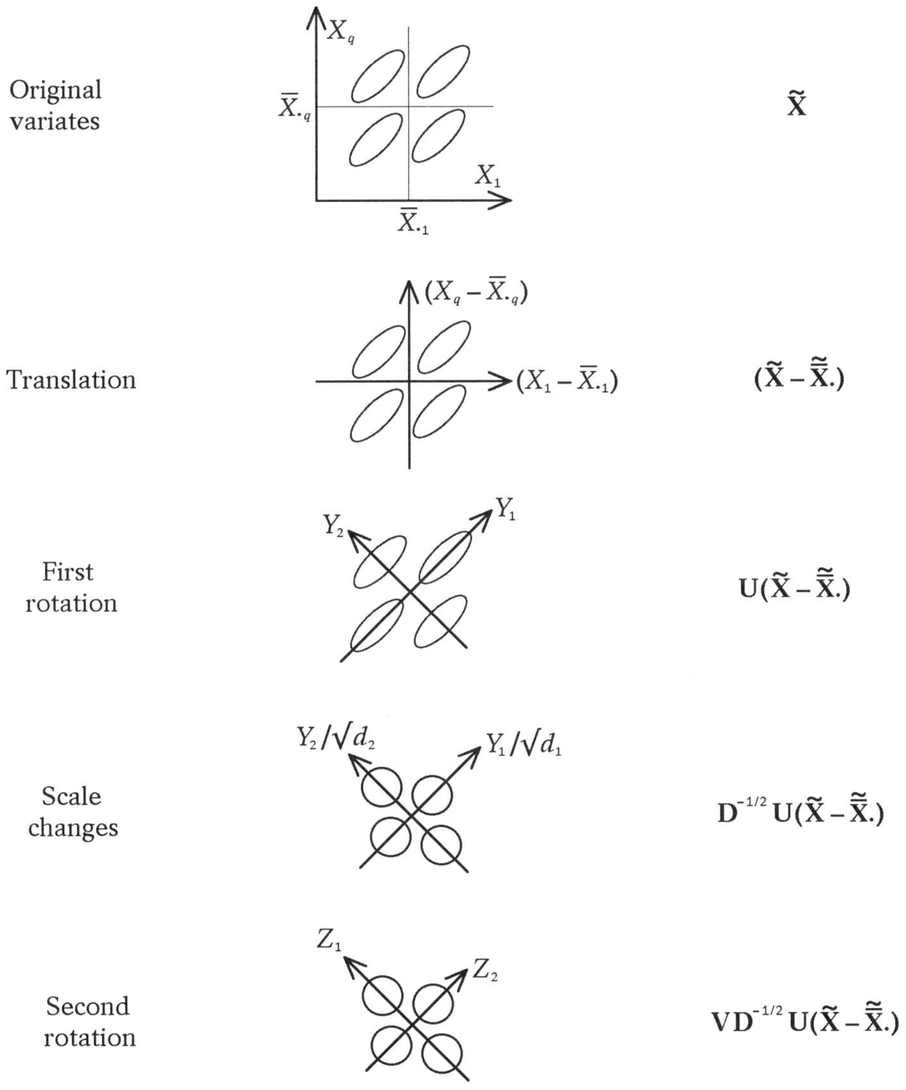

Figure 33.3.1
Geometrical interpretation of changes of variables in multiple discriminant analysis

Under its new form $\mathbf{D}^{-1/2}\mathbf{U}\,\mathbf{B}\,\tilde{\mathbf{U}}\,\mathbf{D}^{-1/2}$, the matrix of between-groups sums of products still reflects the dispersion of sample mean vectors $\overline{\mathbf{X}}_g = [\,\overline{X}_{g1}\ \dots\ \overline{X}_{gq}]$ about the general mean vector $\overline{\mathbf{X}}. = [\,\overline{X}._1 \dots \overline{X}._q]$, but this dispersion is expressed with respect to the standardized coordinates $[\,Y_1/\sqrt{d_1}\ \dots\ Y_q/\sqrt{d_q}\,] = \mathbf{D}^{-1/2}\,\mathbf{U}\,(\tilde{\mathbf{X}} - \tilde{\overline{\mathbf{X}}}.)$ on the principal axes of within-groups variation. The new form $\mathbf{D}^{-1/2}\mathbf{U}\,\mathbf{B}\,\tilde{\mathbf{U}}\,\mathbf{D}^{-1/2}$ of the matrix of between-groups sums of products is usually not yet diagonal, because the principal axes of between-groups variation generally differ from those of within-groups variation. Therefore, a second rotation, involving another orthogonal matrix $\mathbf{V}$, must be carried out in order to express observations with respect to the principal axes of between-groups variation (last line of figure 33.3.1),

$$\tilde{\mathbf{Z}} = \mathbf{V}\,\mathbf{D}^{-1/2}\,\mathbf{U}\,(\tilde{\mathbf{X}} - \tilde{\overline{\mathbf{X}}}.) = \mathbf{K}\,(\tilde{\mathbf{X}} - \tilde{\overline{\mathbf{X}}}.)\,,$$

and the matrix product $\mathbf{K} = \mathbf{V}\,\mathbf{D}^{-1/2}\mathbf{U}$ contains the coefficients of the discriminant functions (also called discriminant axes) $\tilde{\mathbf{Z}}$. Following this last change of variables, the matrix of between-groups sums of products takes the diagonal form

$$\mathbf{K}\,\mathbf{B}\,\tilde{\mathbf{K}} = \mathbf{V}\,\mathbf{D}^{-1/2}\,\mathbf{U}\mathbf{B}\tilde{\mathbf{U}}\,\mathbf{D}^{-1/2}\,\tilde{\mathbf{V}} = \mathbf{E}\,,$$

of which the diagonal elements e_j (an instance of latent roots; section 24.21) may be arranged according to decreasing magnitude by reordering the row vectors of matrix $\mathbf{K}$ if necessary. The numerical values of the latent roots e_j reflect the importance of the differences between sample mean vectors in the direction of each discriminant axis, the maximum number of nonnull latent roots e_j (within the limits of rounding errors) being $\min(q, C-1)$. When the preliminary hypothesis H_0: $\mu_1 = \mu_2 = \dots = \mu_C = \mu$ is true, the sum of the latent roots e_j of which the subscript $j \leq \min(q, C-1)$ follows the χ^2 distribution with $q(C-1)$ degrees of freedom approximately. It must be noted that, unlike matrices $\mathbf{U}$ and $\mathbf{V}$, the matrix $\mathbf{K}$ of the coefficients of discriminant functions is not orthogonal. Consequently, if the values of original variates must ever be recalculated from those of discriminant functions, the following equation must be used:

$$\tilde{\mathbf{X}} = \tilde{\overline{\mathbf{X}}}. + \mathbf{K}^{-1}\,\tilde{\mathbf{Z}} = \tilde{\overline{\mathbf{X}}}. + \tilde{\mathbf{U}}\,\mathbf{D}^{+1/2}\,\tilde{\mathbf{V}}\,\tilde{\mathbf{Z}}\,.$$

Section 33.4: hypotheses about the dimensionality of between-groups variation

Hypotheses can be tested about the number of dimensions occupied by the population mean vectors μ_g through an approximate method discovered by Bartlett and described by Rao (1952, page 373; 1965, page 474). This method is summarized in table 33.4.1. While the sum of the latent roots e_j follows the χ^2 distribution approximately, Bartlett (1954) found that the approximation can be improved if each latent root e_j is replaced by the expression $a \times \log_e[1 + e_j/(N. - C)]$, where $a = [(N. - 1) - (1/2)(q + C)]$. The χ^2 approximation is then so satisfactory that it is almost exact, as noted by Kshirsagar (1972, pages 301-302).

Table 33.4.1

Tests of hypotheses about the dimensionality of between-groups variation

Latent root	χ^2 component	Degrees of freedom	Sum of χ^2 components	Cumulated degrees of freedom	Tested hypothesis
e_1	$a \times \log_e[1 + e_1/(N. - C)]$	$(q + C - 2)$	$\sum_{i=1}^{\min(q, C-1)} a \times \log_e[1 + e_i/(N. - C)]$	$q(C-1)$	H_0: 0 dimensions
e_2	$a \times \log_e[1 + e_2/(N. - C)]$	$(q + C - 4)$	$\sum_{i=2}^{\min(q, C-1)} a \times \log_e[1 + e_i/(N. - C)]$	$q(C-1) - (q + C - 2)$	H_1: 1 dimension
. . .	. . .	. . .	. . .	. . .	. . .
e_j	$a \times \log_e[1 + e_j/(N. - C)]$	$(q + C - 2j)$	$\sum_{i=j}^{\min(q, C-1)} a \times \log_e[1 + e_i/(N. - C)]$	$(q - j + 1)(C - j)$	$H_{(j-1)}$: $(j-1)$ dimensions
. . .	. . .	. . .	. . .	. . .	. . .
$e_{\min(q, C-1)}$	$a \times \log_e[1 + e_{\min(q, C-1)}/(N. - C)]$	$(C - q)$ if $q \leq (C-1)$, $(q - C + 2)$ if $(C-1) < q$	$a \times \log_e[1 + e_{\min(q, C-1)}/(N. - C)]$	$(C - q)$ if $q \leq (C-1)$, $(q - C + 2)$ if $(C-1) < q$	$H_{\min(q-1, C-2)}$: $\min(q-1, C-2)$ dimensions

Note: Bartlett's factor $a = [(N. - 1) - (1/2)(q + C)]$.

However, Bartlett found that the χ^2 approximation would fail if one of the latent roots e_j were excluded in a direction in which the parametric mean vectors $\boldsymbol{\mu}_g$ do not differ from each other. In order to prevent this possibility in practice, the χ^2 components and the corresponding numbers of degrees of freedom are cumulated down to the bottom of table 33.4.1 for each significance test (fourth and fifth columns).

The first significance test bears on the preliminary hypothesis H_0, according to which the parametric mean vectors $\boldsymbol{\mu}_g$ would not differ at all from each other and would occupy zero dimensions. The χ^2 components $[(N. - 1) - (1/2)(q + C)] \log_e[1 + e_j/(N. - C)]$ corresponding to all observed latent roots from e_1 down to $e_{\min(q,\ C-1)}$ are then summed up. If this sum, recorded in the fourth column of table 33.4.1, is greater than $\chi^2_{(1-\alpha,\ q(C-1))}$, the preliminary hypothesis H_0 is rejected and the alternative hypothesis H_1, according to which the parametric mean vectors $\boldsymbol{\mu}_g$ would occupy (at least) one dimension, is accepted.

The second significance test bears on hypothesis H_1, which is now treated as the preliminary hypothesis. The χ^2 components (in the second column of table 33.4.1) are summed up from the second latent root e_2 down to $e_{\min(q,\ C-1)}$, which means once again down to the bottom line of the table. If this sum, recorded in the fourth column of table 33.4.1, is greater than $\chi^2_{(1-\alpha,\ q(C-1)-(q+C-2))}$, the hypothesis H_1 is rejected and the alternative hypothesis H_2, according to which the parametric mean vectors $\boldsymbol{\mu}_g$ would occupy (at least) two dimensions, is accepted.

Similar significance tests are thus carried out toward the bottom of the table as long as the sum of χ^2 components (in the fourth column of table 33.4.1) on each line is greater than $\chi^2_{(1-\alpha)}$ with the number of degrees of freedom recorded in the fifth column, but significance tests are stopped as soon as one of the hypotheses indicated in the sixth column can be accepted. For instance, if hypothesis H_3 were rejected but hypothesis H_4 were accepted, the final conclusion of the analysis would be that the data show that the number of dimensions occupied by the parametric mean vectors $\boldsymbol{\mu}_g$ is equal to 4, but do not show the number to be larger than 4.

If a multiple discriminant analysis indicates that all hypotheses mentioned in the sixth column of the table must be rejected, including the last one, $H_{\min(q-1,\ C-2)}$, the final conclusion would be that the parametric mean vectors $\boldsymbol{\mu}_g$ occupy all available dimensions, which corresponds to hypothesis $H_{\min(q,\ C-1)}$. However, the latter hypothesis does not appear in table 33.4.1 and is never tested, because it is the ultimate alternative hypothesis, which does not set any restrictions on the number of dimensions occupied by the parametric mean vectors $\boldsymbol{\mu}_g$, and which is accepted when all preceding hypotheses have to be rejected.

Section 33.5: summary of computing steps

Because the computations required by multiple discriminant analysis are long and complicated, the major steps are summarized in table 33.5.1. While all computational steps have already been discussed, it may be useful to note that, in scalar algebra, the ith discriminant function $\bar{Z}_{gi}$ of the mean vector $\bar{\mathbf{X}}_g = [\bar{X}_{g1} \ldots \bar{X}_{gq}]$ of the gth sample is

$$\bar{Z}_{gi} = k_{i1}(\bar{X}_{g1} - \bar{X}_{\cdot 1}) + \ldots + k_{iq}(\bar{X}_{gq} - \bar{X}_{\cdot q}) ,$$

where $\bar{\mathbf{X}}_{\cdot} = [\bar{X}_{\cdot 1} \ldots \bar{X}_{\cdot q}]$ is the general mean vector of all sample data (table 33.2.1). Similarly, the discriminant function of an individual specimen $\mathbf{X} = [X_1 \ldots X_q]$ is

$$Z_i = k_{i1}(X_1 - \bar{X}_{\cdot 1}) + \ldots + k_{iq}(X_q - \bar{X}_{\cdot q}) .$$

Table 33.5.1
Summary of computing steps

(1)	the matrices $\mathbf{W}$ and $\mathbf{B}$ of within-groups and between-groups sums of products are computed in the manner indicated in section 33.2;
(2)	the within-groups covariance matrix $\mathbf{S} = \mathbf{S}_{\text{within-groups}} = [1/(N. - C)]\,\mathbf{W}$ is computed;
(3)	matrix $\mathbf{S}$ is diagonalized in the manner indicated in section 31.3, which yields the orthogonal matrix $\mathbf{U}$ and the diagonal matrix $\mathbf{D}$;
(4)	the matrix product $\mathbf{D}^{-1/2}\,\mathbf{U}\,\mathbf{B}\,\breve{\mathbf{U}}\,\mathbf{D}^{-1/2}$ is evaluated;
(5)	the matrix product $\mathbf{D}^{-1/2}\,\mathbf{U}\,\mathbf{B}\,\breve{\mathbf{U}}\,\mathbf{D}^{-1/2}$ is diagonalized in the manner indicated in section 31.3, which yields the orthogonal matrix $\mathbf{V}$ and the diagonal matrix $\mathbf{E}$;
(6)	the matrix $\mathbf{K} = \mathbf{V}\mathbf{D}^{-1/2}\mathbf{U}$ of the coefficients of the discriminant functions is evaluated;
(7)	the preceding calculations are verified by checking if $\mathbf{K}\mathbf{S}\breve{\mathbf{K}} = \mathbf{I}$ and if $\mathbf{K}\mathbf{B}\breve{\mathbf{K}} = \mathbf{E}$;
(8)	hypotheses H_0, H_1, H_2, ... etc. are tested in the manner described in section 33.4 and table 33.4.1;
(9)	numerical values are computed for the discriminant functions corresponding to latent roots (e_1, e_2, ... $e_{\min(q,\ C-1)}$) which are statistically significant; discriminant functions may be evaluated for individual observations, for sample mean vectors, as well as for specimens of which the origin is doubtful and should be estimated; the values of these discriminant functions may be represented graphically in scatter diagrams (see sections 33.6 to 33.11).

The relationships between the within-groups covariance matrix $\mathbf{S}$, the between-groups matrix $\mathbf{B}$ of sums of (squares and) products, the matrix $\mathbf{K}$ of the coefficients of discriminant functions and the diagonal matrix $\mathbf{E}$ of the corresponding latent roots may be summarized overall (see section 33.12) by the equation $\mathbf{KB} = \mathbf{EKS}$.

Section 33.6: graphical representation of results

What makes multiple discriminant analysis particularly interesting is the fact that its results may be represented graphically and constitute two-dimensional or three-dimensional views of the groups of multivariate data to be compared (see figures 33.8.1, 33.9.1, 33.10.1 and 33.11.1). When the total number of variates is larger than two or three, discriminant functions usually allow most of the information relative to differences between group means to be expressed using a minimum number of linear combinations (section 24.22) of original variates. In principle, the only discriminant functions which should be represented graphically are those which the methods of section 33.4 have shown to be statistically significant. Nevertheless, it may occasionally be useful to represent the first nonsignificant discriminant function (the third one, for instance, when only the first and second discriminant functions reach statistical significance), provided it is made clear that the nonsignificant discriminant function does not bring out conclusive differences.

The manner in which scatter diagrams of discriminant functions should be used depends on the nature of the information sought. When differences between means are of paramount importance, a confidence region of the discriminant functions $\mathbf{K}(\tilde{\boldsymbol{\mu}}_g - \tilde{\bar{\bar{\mathbf{X}}}}.)$ of the mean vector $\boldsymbol{\mu}_g$ of each population is generally represented. If the assumption that population covariance matrices are equal is reasonable, a two-dimensional confidence region of the mean vector is a circle of which the radius is

$$\sqrt{(\chi^2_{(1-\alpha;2)}/N_g)}\,,$$

or, preferably,

$$\sqrt{(T^2_{(1-\alpha;2,N.-C)}/N_g)}\,,$$

and of which the center has the discriminant functions $\tilde{\bar{\mathbf{Z}}}_g = \mathbf{K}(\tilde{\bar{\mathbf{X}}}_g - \tilde{\bar{\bar{\mathbf{X}}}}.)$ of the sample mean vector $\bar{\mathbf{X}}_g$ as coordinates.

When the assumption that population covariance matrices are equal is doubtful, a more satisfactory confidence region may be obtained for the mean vector $\boldsymbol{\mu}_g$ of each population by replacing the confidence circle discussed in the preceding paragraph by a confidence ellipse. The latter likewise has its center located at the discriminant functions of the sample mean vector $\bar{\mathbf{X}}_g$ but is based on $T^2_{(1-\alpha;\,2,\,N_g-1)}$ and on elements of the estimate $\mathbf{K}\,\mathbf{S}_g\,\tilde{\mathbf{K}} \neq \mathbf{I}$ of the covariance matrix of discriminant functions obtained by using the covariance matrix $\mathbf{S}_g$ of each sample instead of the pooled within-groups covariance matrix $\mathbf{S} = \mathbf{S}_{\text{within-groups}}$.

In some studies, it must be determined whether several groups of living organisms overlap to a large extent or not. In taxonomy, for instance, the absence of overlap implies that doubtful specimens could eventually be identified with little risk of error. In quantitative genetics and in evolutionary studies, the absence of overlap with respect to many characters might indicate that populations have become different enough to be reproductively incompatible (morphological differences do not always imply reproductive incompatibility, however). In such studies, it would be desirable not only to represent group means but also to depict the extent of individual variation in each group, either by representing all individual observations as dots or by drawing a prediction circle (figure 33.9.1), having the radius

$$\sqrt{[T^2_{(1-\alpha;2,N.-C)}\,(N_g+1)/N_g]}\,,$$

or a prediction ellipse based on $T^2_{(1-\alpha;\,2,\,N_g-1)}$ and on the appropriate elements of matrix

$\mathbf{K S_g \tilde{K}}$ (figure 33.8.1). When samples are very small, however, prediction circles or ellipses may be too large in comparison with the observed or estimated extent of individual variation (section 30.6) and may be replaced by variation circles of radius $\sqrt{\chi^2_{(1-\alpha; 2)}}$ (figure 33.10.1) or by variation ellipses.

When samples are large, all or almost all discriminant functions are often statistically significant even when observed differences between means are slight. However, most of the variation between means lies usually in the direction of the first few discriminant axes. Even though none of the significant discriminant axes should be ignored, the first few axes are nevertheless the most important ones for obtaining a broad outline of between-groups variation.

Therefore, the three-dimensional scatter diagram of the first three discriminant functions is often represented, sometimes by using a pair of stereoscopic views, for instance. Following a simpler approach, two bivariate scatter diagrams in which the abscissas represent the same discriminant function with the same scale, may be joined in such a way that two perpendicular views of the same trivariate scatter diagram are seen next to each other (figures 33.9.1 and 33.10.1). Moreover, the three-dimensional interpretation of such a pair of diagrams may be made easier by examining a copy folded at right angle along the abscissa of the upper part (figure 33.6.1), since the two bivariate scatter diagrams may then be perceived as projections upon the front face (in grey) and the upper face(in white) of a "box" containing the trivariate diagram (in black).

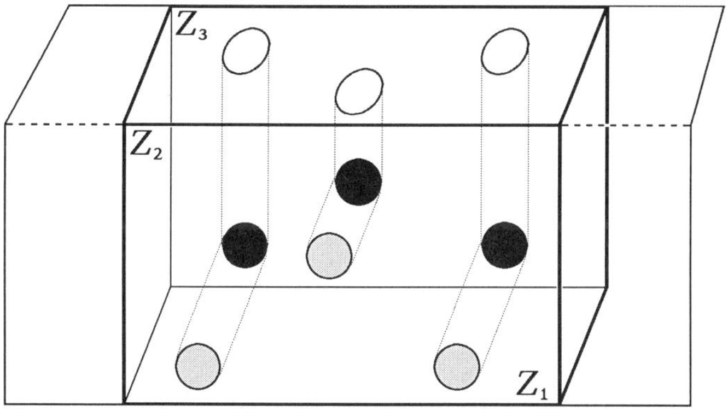

Figure 33.6.1
Three-dimensional interpretation of a joined pair of scatter diagrams

Section 33.7: confidence circles and the problem of axis reflections

The sense of the coordinate axis of a discriminant function is arbitrary, and the signs of all coefficients may change when a different computing algorithm is used or when a different set of data is analyzed. This kind of phenomenon may occur not only for discriminant functions but also for principal components (chapter 31) or for canonical variates (chapter 34). While this phenomenon, which corresponds geometrically to the reflection of a coordinate axis, is somewhat trifling, it could conceal the similarity of the results obtained in related analyses, but it may be easily avoided in practice by changing if necessary the signs of all elements of a vector of coefficients and by defining arbitrarily as positive (or negative) the element which has the largest absolute value.

Preventing axis reflections is particularly important when attempts are made to check the performance of confidence regions. For instance, Krzanowski (1989) claimed that the probability of coverage of confidence circles for the means of discriminant functions was very inaccurate, sometimes less than half the desired confidence level of 95%. However, the frequencies obtained by Krzanowski (1989) using the *"bootstrap"* resampling method were wrong, in part because he did not take axis reflections into account. Krzanowski's error has been alluded to by Ringrose and Krzanowski (1991), but in a periodical which is still relatively new and not widely available, and many users of discriminant functions may still be unaware that the earlier claim of Krzanowski (1989) was unfounded. Moreover, some statisticians keep trying to replace confidence circles of the means of discriminant functions by nonparametric methods, which would not likely be more accurate.

It may be shown by a sampling experiment that the confidence circles of group means are approximately correct in multiple discriminant analysis. In order to make the present results as comparable as possible with those of Krzanowski (1989), the parameters used in this sampling experiment were made equal to the sample mean vectors and covariance matrices of the data published by Lubischew (1962) on three Russian species of beetles belonging to the genus *Chaetocnema,* data on which Krzanowski based one of his examples (Krzanowski, 1989, page 115, figure 1). Moreover, the sample sizes used in the present sampling experiment were the same as those of Lubischew (1962): $N_1 = 21$ specimens for the first species (*Chaetocnema concinna*), $N_2 = 31$ specimens for the second species (*Chaetocnema heikertingeri*) and $N_3 = 22$ specimens for the third species (*Chaetocnema heptapotamica*). Normal random numbers were produced by subtracting 24 from the sum of 48 uniform random numbers and dividing the difference by 2. These independent random numbers were then transformed into correlated numbers by using the latent roots and vectors of the covariance matrix of each species.

The results of three simulated multiple discriminant analyses of the data of Lubischew (1962) are illustrated in figure 33.7.1. The confidence circle of the parametric mean vector $\boldsymbol{\mu}_g = [\mu_{g1} \ldots \mu_{gq}]$ of each group, based on Hotelling's T^2 at level $\alpha = 0.05$, is shown in the scatter diagram of the first two discriminant functions Z_1 and Z_2. Each circle bears a label indicating the group number (1, 2 or 3) and the simulation number (a, b or c). Moreover, the discriminant function $\mathbf{K}(\tilde{\boldsymbol{\mu}}_g - \tilde{\tilde{\mathbf{X}}}.)$ of the parametric mean of each group is represented by a dot (• for simulation a, ○ for simulation b and $*$ for simulation c) showing whether the projection of the parametric group mean is covered or not by the corresponding confidence circle. Even though the confidence circles of each group mean move about the scatter diagram from one simulation to the next, each of the nine confidence circles in figure 33.7.1 covers the projection of the corresponding parametric mean. Therefore, the present results thus far do not seem to support the claim made by Krzanowski in 1989.

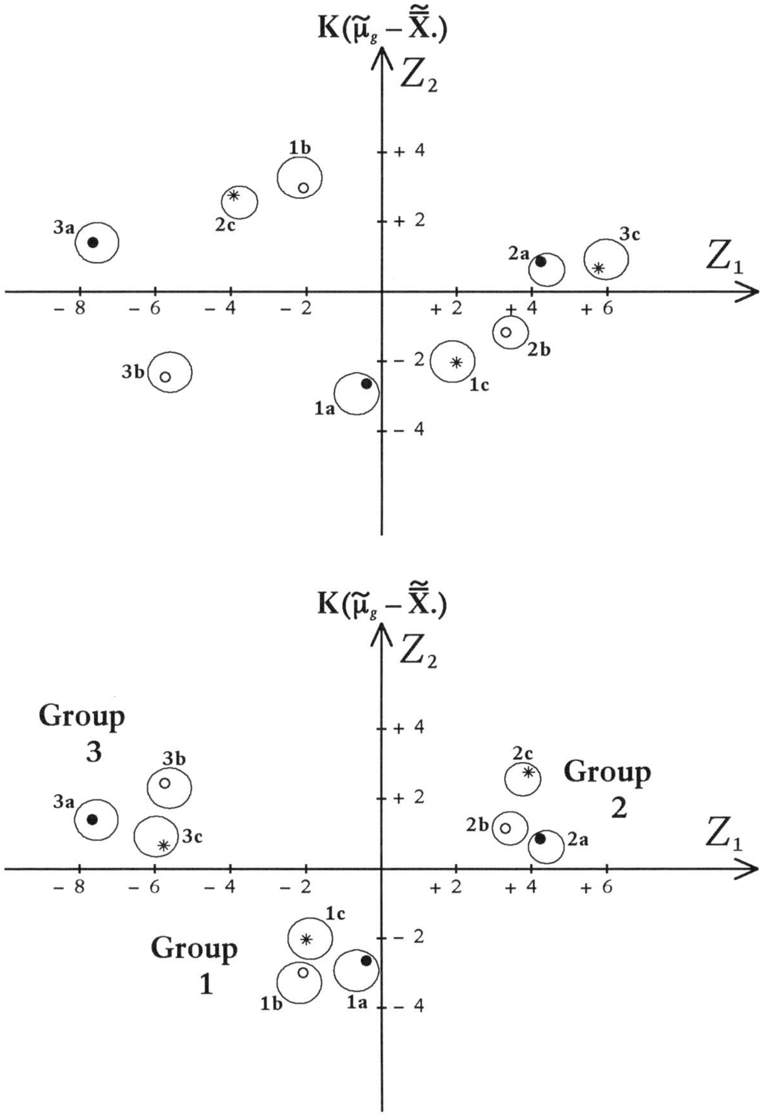

Figure 33.7.1
Scatter diagrams of confidence circles ($1-\alpha=0.95$) of group means in the plane of the first two discriminant functions Z_1 and Z_2 when axis reflections are allowed (top) and prevented (bottom); each label indicates group number (1, 2 or 3) and simulation number (a, b, or c); dots (●, ○ or ∗) represent parametric group means μ_g and are covered by the corresponding confidence circle in all cases illustrated

Axis reflections have been allowed in the upper half but prevented in the lower half of figure 33.7.1. The upper half of figure 33.7.1. shows that the coordinate axis of the second discriminant function is reflected from simulation "a" to simulation "b", and that the coordinate axis of the first function is reflected from simulations "a" and "b" to simulation "c". Therefore the confidence circles of the mean of each group are dispersed in three different quadrants and exhibit no clear pattern. Once axis reflections have been prevented, however, the means of groups 1, 2 and 3 fall respectively in the lower central, upper right and upper left regions of the scatter diagram (lower half of figure 33.7.1), and the pattern is obvious.

A longer series of twenty random simulations is illustrated in figure 33.7.2, where the projection of the parametric mean $\boldsymbol{\mu}_g$ of each group is represented by a black dot. The upper halves of figures 33.7.1 and 33.7.2 indicate that the various kinds of axis reflections do not occur with the same frequency. Thus, the confidence circle of the mean of the second group falls in the lower right quadrant approximately twice as often as in the upper right quadrant, it seldom falls in the upper left quadrant, and it never fell in the lower left quadrant during the present series of simulations.

Even when axis reflections are prevented, the confidence circle of the mean of each group still moves about the scatter diagram of discriminant functions to a considerable extent (figures 33.7.1 and 33.7.2, bottom), but it keeps covering the projection of the corresponding parametric mean most of the time, which shows that both move about in a correlated manner. The coordinates of the sample mean vector $\bar{\mathbf{X}}_g$ and of the parametric mean vector $\boldsymbol{\mu}_g$ of the gth group are thus subject to strong fluctuations, but the numerical values of those coordinates have little interest in themselves, because what is most important in multiple discriminant analysis is the pattern and the extent of between-groups dispersion.

As for the performance of the confidence circles of group means, what really matters is the distance between the projection $\mathbf{K}\tilde{\bar{\mathbf{X}}}_g$ of the sample mean vector of the gth group and thc projcction $\mathbf{K}\tilde{\boldsymbol{\mu}}_g$ of the corresponding parametric mean vector, rather than the distance from either to the projection $\mathbf{K}\tilde{\bar{\mathbf{X}}}.$ of the general mean $\bar{\mathbf{X}}.$ of all data. Consequently, when the performance of confidence circles of group means must be evaluated, the difference $\mathbf{K}(\tilde{\boldsymbol{\mu}}_g - \tilde{\bar{\mathbf{X}}}.) - \mathbf{K}(\tilde{\bar{\mathbf{X}}}_g - \tilde{\bar{\mathbf{X}}}.) = \mathbf{K}(\tilde{\boldsymbol{\mu}}_g - \tilde{\bar{\mathbf{X}}}_g)$ must be considered rather than the values $\mathbf{K}(\tilde{\bar{\mathbf{X}}}_g - \tilde{\bar{\mathbf{X}}}.)$ and $\mathbf{K}(\tilde{\boldsymbol{\mu}}_g - \tilde{\bar{\mathbf{X}}}.)$ of the discriminant functions as usually calculated. Scatter diagrams of the differences $\mathbf{K}(\tilde{\boldsymbol{\mu}}_g - \tilde{\bar{\mathbf{X}}}_g)$ have been inset in the upper left, upper right and lower left corners of the lower half of figure 33.7.2, and the projection of each parametric mean vector lies outside of the corresponding confidence circle in only two cases out of 60, and lies close to the circle in a third case. Since these confidence circles have been based on Hotelling's T^2 distribution at the $\alpha = 0.05$ level, the frequencies observed here agree very well with expectations.

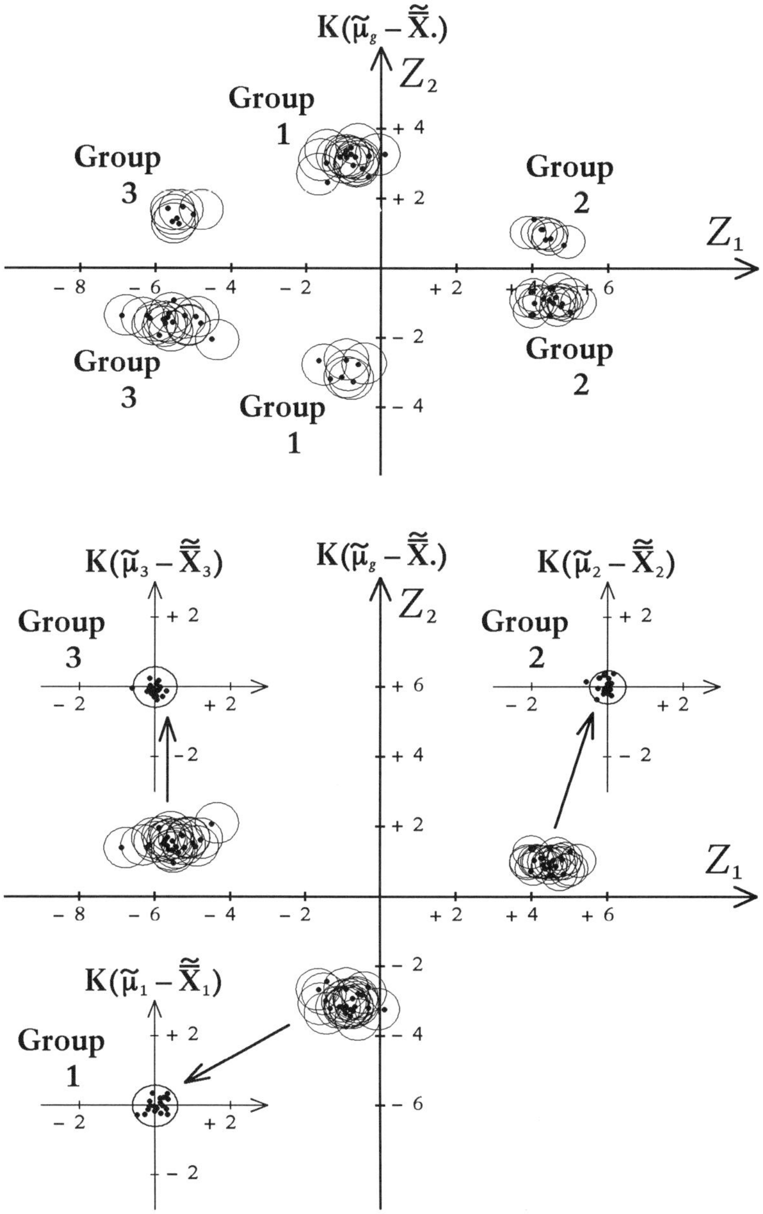

Figure 33.7.2

Scatter diagrams of confidence circles ($1 - \alpha = 0.95$) of group means in the plane of the first two discriminant functions Z_1 and Z_2 when axis reflections are allowed (top) and prevented (bottom); the projection of each parametric group mean μ_g in each simulation is represented by a black dot; the smaller diagrams inset in the corners of the lower diagram are centered on the mean of each sample in each simulation instead of the general mean

The results of a much larger sampling experiment, comprising 10,000 simulations, are summarized in table 33.7.1. Frequencies of coverage have been determined at levels $\alpha = 0.05$ and $\alpha = 0.01$ for three methods. According to the first method, referred to as "traditional" by Krzanowski (1989), confidence circles were determined using the χ^2 distribution with two degrees of freedom and the pooled within-groups covariance matrix $\mathbf{S} = \mathbf{S}_{\text{within-groups}}$, which possesses $(N. - C)$ degrees of freedom, where C and $N.$ denote respectively the number of groups and the general (total) sample size. According to the second method, confidence circles were again based on the pooled within-groups covariance matrix but the χ^2 distribution was replaced by Hotelling's T^2 distribution. According to the third method, finally, Hotelling's T^2 distribution was again used but the assumption that parametric covariance matrices are equal was given up and the confidence circles of group means were replaced by ellipses, the covariance matrix of discriminant functions in the gth group being estimated by the matrix product $\mathbf{K}\mathbf{S}_g\mathbf{\tilde{K}}$ with $(N_g - 1)$ degrees of freedom, where $\mathbf{S}_g$ is the sample covariance matrix of the gth group. It must be emphasized that the present confidence ellipses are much smaller and less elongated that the (utterly incorrect) ellipses proposed by Krzanowski in 1989.

Table 33.7.1
Relative frequencies of coverage of the projections of parametric means by the corresponding confidence regions in 10,000 simulations

Kind of region	confidence circles		confidence circles		confidence ellipses	
Distribution	$\chi^2_{(1-\alpha;\,2)}$		Hotelling's $T^2_{(1-\alpha;\,2,\,N.-C)}$		Hotelling's $T^2_{(1-\alpha;\,2,\,N_g-1)}$	
Covariance matrix	pooled within-groups		pooled within-groups		covariance matrix $\mathbf{S}_g$ of each sample	
$(1-\alpha)$	0.95	0.99	0.95	0.99	0.95	0.99
Group 1	0.9394	0.9843	0.9494	0.9889	0.9342	0.9859
Group 2	0.8511	0.9431	0.8667	0.9542	0.9272	0.9831
Group 3	0.9613	0.9911	0.9669	0.9941	0.9331	0.9826
Average	0.9173	0.9728	0.9277	0.9791	0.9315	0.9839

Even though the average relative frequencies of coverage of the three groups, on the last line of table 33.7.1, are all lower than the desired confidence level, they approximate $(1-\alpha)$ more closely when the χ^2 distribution is replaced by Hotelling's T^2 distribution and when the pooled within-groups covariance matrix is replaced by the covariance matrix $\mathbf{S}_g$ of each sample. Frequencies of coverage are clearly lower for group 2 than for groups 1 and 3, particularly when the pooled covariance matrix is used, but the reason for this becomes clear if the scatter diagrams of the original paper of Lubischew (1962) are reexamined, since dispersion is visibly greater in species 2 than in species 1 and 3. Therefore, when confidence circles are based on the pooled within-groups covariance matrix, they tend to be too small for species 2 and too large for species 1 and 3.

The coverage frequencies of the three groups are more homogeneous when confidence circles are replaced by ellipses and when the sample covariance matrix of discriminant functions is estimated separately for each group by the matrix product $\mathbf{K}\mathbf{S}_g\tilde{\mathbf{K}}$, because this procedure allows the dispersion of each group to have its own extent and its own principal directions. Even in this way, however, coverage frequencies are still slightly lower than $(1-\alpha)$, and cautious users will no doubt avoid using the $\alpha = 0.05$ level for drawing conclusions regarding differences between means. Moreover, some investigators might prefer treating the comparison of several (C) group means as a multiple estimation problem and using enlarged Bonferroni confidence regions (section 26.2) with a significance level of $(1 - \alpha/C)$ for each group.

Anyhow, before detailed conclusions are drawn regarding differences between group means, the preliminary hypothesis H_0 should first be tested in an overall manner through the methods discussed in section 33.4. Once the general statistical significance of between-groups variation has been established, however, the scatter diagrams of sample discriminant functions should be considered as the best view of differences between groups obtainable from the data, since these diagrams are based on maximum-likelihood estimates.

Section 33.8: the differences between three species of irises

Multiple discriminant analysis is illustrated first (figure 33.8.1) in the case of famous data on three North-American species of irises (Appendix E). These data were obtained by the well-known American botanist Edgar Anderson, who drew Ronald A. Fisher's attention to the need of developing statistical methods for distinguishing closely related species (see Anderson, 1954). It is in order to analyze these data that Fisher developed his linear discriminant function (chapter 32), which is suitable for the comparison of two groups (Fisher, 1936). In addition to being published by Fisher in 1936 in his original paper, these data have been reprinted many times (Andrews and Herzberg, 1985; Atchley and Bryant, 1975; Morrison, 1990; Pontier, Dufour and Normand, 1990). Summary information concerning the three iris species is given in table 33.8.1.

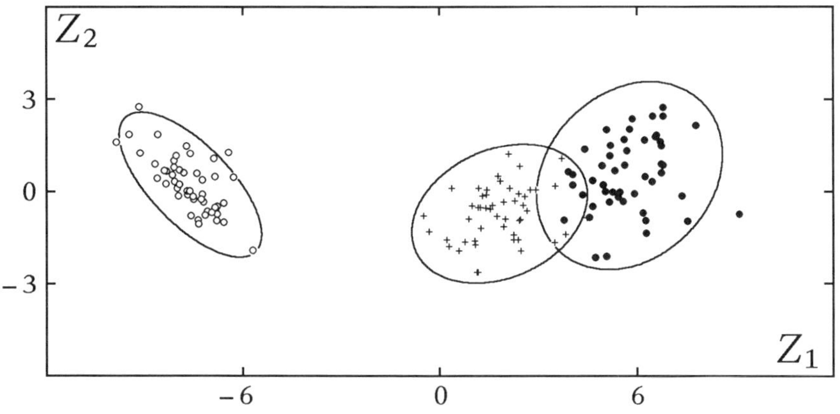

Figure 33.8.1
Scatter diagram of the two discriminant functions of the three iris species *Iris setosa* (○), *Iris versicolor* (+) and *Iris virginica* (●) according to data published by Fisher (1936); prediction ellipses at level $\alpha = 0.05$

Table 33.8.1
Characteristics of three North-American species of irises

Species	*Iris setosa canadensis* (*Iris Hookeri*)	*Iris versicolor*	*Iris virginica*
Range	southeastern Quebec and Canada, northeastern U. S. A.	southeastern Quebec and Canada, northeastern U. S. A.	southwestern Quebec, Canada and U. S. A.
Habitat	upper beaches and meadows, dunes, rocky slopes	wet shores and meadows, marshes	wet shores and meadows, marshes
Chromosome number	38	108	70

The subspecies *Iris setosa canadensis* (also called *Iris Hookeri* in the United States of America) is particularly interesting because it is related to a species occurring in Eastern Asia and in Alaska, from which it is thought to have been separated by Pleistocene ice sheets. The data on *Iris setosa canadensis* and *Iris versicolor* analyzed by Fisher were collected by Edgar Anderson in the Gaspé peninsula, in the meadows bordering the south shore of the St. Lawrence River between l'Île Verte and Trois-Pistoles, where these two species grow together. The third species, *Iris virginica*, does not occur in the Gaspé peninsula, but it was considered by Anderson and Fisher in order to test the hypothesis that the "hexaploid" species *Iris versicolor* would go back to natural hybrids between the "diploid" species *Iris setosa* and the "tetraploid" species *Iris virginica*. The quantitative characters measured by Anderson on fifty specimens of each species and analyzed by Fisher were the following:

$$X_1 = \text{sepal length in mm,}$$
$$X_2 = \text{sepal width in mm,}$$
$$X_3 = \text{petal length in mm,}$$
$$X_4 = \text{petal width in mm.}$$

Some authors have noted that discriminant functions were not really necessary in the comparison of species *Iris setosa canadensis* and *Iris versicolor* because the petals of the former are tiny and their dimensions do not overlap with those of the latter. This lack of overlap is evident in the scatter diagram of the discriminant functions (figure 33.8.1), where the two species are completely separated. However, *Iris versicolor* differs much less from *Iris virginica*, with which it shows partial overlap (figure 33.8.1). The range of individual variation has been outlined here by prediction ellipses rather than prediction circles because the extent and the principal directions of dispersion in *Iris setosa canadensis* differ visibly very much from those in *Iris versicolor* and in *Iris virginica*.

If the species *Iris versicolor* descended from natural hybrids between *Iris setosa canadensis* and *Iris virginica*, and if there were no dominance or epistasis phenomena, the quantitative characters of *Iris versicolor* would be expected to be intermediate between those of *Iris setosa canadensis* and *Iris virginica* and the parametric mean vector of *Iris versicolor* should lie on the same straight line as the mean vectors of the other two species. The preceding assumptions are apparently not all justified, because Fisher noted that the sepal width (X_2) of *Iris versicolor* is not intermediate between those of the other two species. Moreover, multiple discriminant analysis shows that the hypothesis that the three parametric mean vectors occupy a single dimension must be rejected (table 33.8.2, $\chi^2_{(3)} = 36.5297$, $P = 0.5786/10^7$). Nevertheless, the greater phenotypical similarity between *Iris virginica* and *Iris versicolor* than between the latter and *Iris setosa canadensis* (figure 33.8.1) could be explained (Fisher, 1936) by the fact that *Iris virginica*, being a tetraploid, would have contributed twice as much genetic material as *Iris setosa canadensis* to the genotype of *Iris versicolor*.

Table 33.8.2
Multiple discriminant analysis of the data published by Fisher (1936) on the three species *Iris setosa*, *Iris versicolor* and *Iris virginica*

Latent root	χ^2 component	Degrees of freedom	Sum of χ^2 components	Cumulated degrees of freedom	Probability
4732.21	509.586	5	546.115	8	0
41.9525	36.5297	3	36.5297	3	$0.5786/10^7$

In conclusion, the fact that between-groups variation is significantly two-dimensional does not completely rule out the hypothesis according to which *Iris versicolor* would descend from natural hybrids between *Iris setosa canadensis* and *Iris virginica*. The fact that the sepal width of *Iris versicolor* is not intermediate could indeed be due to dominance or epistasis phenomena. Moreover, even if *Iris versicolor* originated from natural hybrids between the other two species, this could have happened thousands of years ago, and the three species could have evolved further since then.

Section 33.9: the differences between four species of butterflies

A second example may be found in the interesting data of Brower (1959) on male butterflies of the genus *Papilio*, already considered in several earlier chapters. The specific designations and the geographical origin of specimens are summarized in table 33.9.1.

Table 33.9.1
Data on male butterflies belonging to four species of the genus *Papilio* (Brower, 1959)

Species	Source	Sample size
Papilio glaucus	British Columbia	$N_1 = 10$
Papilio rutulus	Colorado	$N_2 = 10$
Papilio eurymedon	Colorado	$N_3 = 10$
Papilio multicaudatus	Colorado	$N_4 = 10$

The following variates are considered in the present analysis:

X_1 = the length of the eighth tergite in mm, multiplied by 8,
X_2 = the length of the superuncus in mm, multiplied by 8,
X_3 = the length of the right valve in mm,
X_4 = the length of the right forewing in mm.

The multiple discriminant analysis is summarized in table 33.9.2 and, while the first two discriminant axes are highly significant, the third one is not ($\chi^2_{(2)} = 4.54663$, $P = 0.10297$). However, since sample sizes are small, the third discriminant axis might possibly turn out to be significant if larger samples were available.

Table 33.9.2
Multiple discriminant analysis of data from Brower (1959) on the four species *Papilio glaucus, P. rutulus, P. eurymedon* and *P. multicaudatus*

Latent root	χ^2 component	Degrees of freedom	Sum of χ^2 components	Cumulated degrees of freedom	Probability
539.174	96.9904	6	117.027	12	0
20.0418	15.4903	4	20.0369	6	0.0027278
4.99387	4.54663	2	4.54663	2	0.10297

The scatter diagrams of the three discriminant functions are illustrated in figure 33.9.1. The lack of statistical significance of the third discriminant axis is reflected graphically by the fact that the four prediction circles have practically the same vertical position in the scatter diagram of functions Z_1 and Z_3 (upper half of figure 33.9.1). With respect to the characters included in the present analysis, *Papilio multicaudatus* differs markedly from the other three species, but the latter overlap to a considerable extent and are difficult to distinguish from each other.

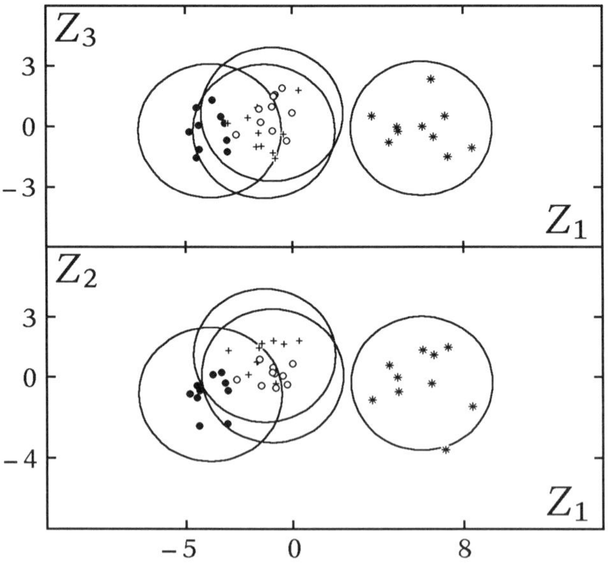

Figure 33.9.1
Scatter diagrams of the three discriminant functions of male specimens of the four butterfly species *Papilio glaucus* (●), *Papilio rutulus* (○), *Papilio eurymedon* (+) and Papilio *multicaudatus* (∗) with respect to four quantitative characters; data from Brower (1959); prediction circles at level $\alpha = 0.05$

It should be noted that expert taxonomists of this group of butterflies take into account many other characters which could not readily be included in a single discriminant analysis, such as habitat and food preferences, reproductive behavior and larval characteristics. From a biogeographical viewpoint, *Papilio glaucus* ranges over most of North America, excluding western United States where it is replaced by the other three species (as well as by a fifth species in Mexico, *Papilio alexiares*). It is currently believed that Pleistocene ice sheets isolated *Papilio glaucus* in the East from the other species in the West and in the South. After the retreat of glaciers, the different species reextended their ranges and came into contact with each other, but they appear to be differentiated enough to remain distinct since natural hybrids are rare and not very fertile.

Section 33.10: the evolutionary divergence of several mammalian orders

As shown in earlier sections (33.8 and 33.9), multiple discriminant analysis can be applied to the study of slight differences between closely related biological populations (microevolution of races, subspecies or allied species), but it can be useful also for the analysis of major differences between higher taxonomic categories (macroevolution of genera, families, orders, etc.), for it can yield integrated descriptions of evolutionary differences with respect to many quantitative characters jointly. When major taxonomic groups are compared in this way through multiple discriminant analysis, each specimen usually represents a whole species or genus, rather than just an individual organism as in microevolutionary studies.

Jolicoeur, Pirlot, Baron and Stephan (1984) compared common (base 10) logarithms of the body weight in g and of the volumes in mm^3 of eleven major brain regions in five groups of mammals. The data used in their study are summarily described in table 33.10.1, and the results of the multiple discriminant analysis are recorded in table 33.10.2.

Table 33.10.1
Data of Jolicoeur, Pirlot, Baron and Stephan (1984) on the evolutionary divergence of the brain in Insectivora, Chiroptera and Primates

Group	Sample size
basal Insectivora	$N_1 = 12$
progressive Insectivora	$N_2 = 16$
Chiroptera	$N_3 = 63$
Prosimians	$N_4 = 21$
Simians	$N_5 = 27$

The four discriminant axes are all statistically significant, but 93.10% of the sum of the latent roots (2297.84) falls on the first two axes and 98.12% on the first three axes. Little information (1.88%) is therefore neglected is only the first three discriminant functions are included in the scatter diagram (figure 33.10.1).

Table 33.10.2
Multiple discriminant analysis of data on five groups of mammals (Jolicoeur, Pirlot, Baron and Stephan, 1984)

Latent root	χ^2 component	Degrees of freedom	Sum of χ^2 components	Cumulated degrees of freedom	Probability
1836.19	348.102	15	617.845	48	0
303.03	153.09	13	269.743	33	0
115.391	80.4431	11	116.653	20	0
43.2304	36.2096	9	36.2096	9	$3.63911/10^5$

The absence of overlap between Insectivora and Prosimians on the one hand, and between the latter and Simians on the other hand, is remarkable and reflects large differences in the size of the brain, of which the relative volume is almost negligible in primitive Insectivora but important in Simians. The exceptional position of man (1) is conspicuous. With respect to the characters included in the study, Chiroptera appear to be intermediate between progressive Insectivora and Prosimians. More exhaustive discussions and conclusions may be found in the paper of Jolicoeur, Pirlot, Baron and Stephan (1984), where the present data were examined in considerably greater detail.

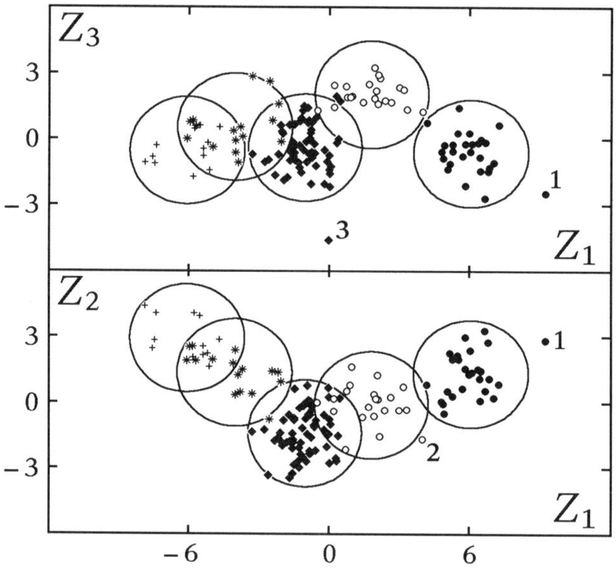

Figure 33.10.1
Scatter diagrams of the first three discriminant functions of five groups of mammals [basal Insectivora (+), progressive Insectivora (∗), Chiroptera (◆), Prosimians (○) and Simians (●)] with respect to the common logarithms of body weight and of of the volumes of eleven major brain regions, according to data from Jolicoeur, Pirlot, Baron and Stephan (1984); variation circles ($\alpha = 0.05$); outliers: 1 : *Homo sapiens,* 2: *Tarsius* sp., 3: *Hipposideros armiger* (a bat)

Section 33.11: an ecological and limnological example

In a fourth example, multiple discriminant analysis is applied to the interesting data of Rawson (1960) on three groups of lakes from Saskatchewan, which have already been considered in sections 30.9 and 31.10. The same variates are included and the lakes are numbered the same way as in section 31.10. The scatter diagram of discriminant functions Z_1 and Z_2 is illustrated in figure 33.11.1. While the lakes situated on the Canadian Shield (1 to 5) fall clearly to the left, the lakes situated south of the Shield (9 to 13) fall clearly to the right, and the lakes straddling the edge of the Shield (6, 7 and 8) fall in the center.

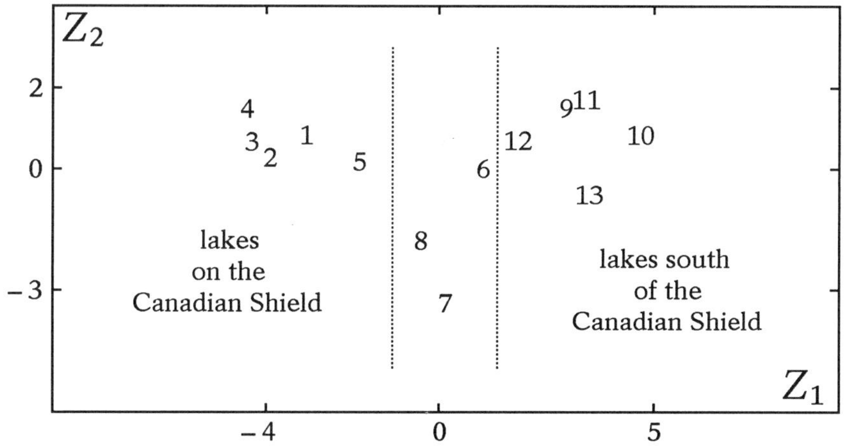

Figure 33.11.1
Scatter diagram of the two discriminant functions of three groups of lakes from Saskatchewan according to data from Rawson (1960); the lakes are numbered as in section 31.10

Comparing figure 33.11.1 with figures 31.10.1, 31.10.2 and 31.10.3 shows that the three groups of lakes are separated more distinctly by multiple discriminant analysis than by other methods. The multiple discriminant analysis is summarized in table 33.11.1, which indicates that the second discriminant axis is not statistically significant ($\chi^2_{(4)} = 6.95326$, $P = 0.138379$). Therefore, even though lakes 6, 7 and 8 appear to fall lower than the other lakes, this is not conclusive, and the three groups of lakes may be considered to follow a straight-line relationship.

Table 33.11.1
Multiple discriminant analysis of the data of Rawson (1960) on three groups of lakes from Saskatchewan

Latent root	χ^2 component	Degrees of freedom	Sum of χ^2 components	Cumulated degrees of freedom	Probability
115.583	20.2431	6	27.1963	10	0.00242438
13.849	6.95326	4	6.95326	4	0.138379

A precise description of the preceding straight-line relationship may be obtained by reexpressing original variates in terms of discriminant functions (as also done for principal components in section 31.4). This yields, like at the end of section 33.3,

$$\tilde{\mathbf{X}} = \tilde{\bar{\mathbf{X}}}. + \mathbf{K}^{-1}\,\tilde{\mathbf{Z}} = \tilde{\bar{\mathbf{X}}}. + \tilde{\mathbf{U}}\,\mathbf{D}^{+1/2}\,\tilde{\mathbf{V}}\,\tilde{\mathbf{Z}}.$$

The first discriminant function Z_1, which is the only one to be statistically significant, is then allowed to vary, while the second discriminant function Z_2 is kept constant and equal to 0, whence

$$(\mathbf{X}-\bar{\mathbf{X}}.) = Z_1[-1.06533 \quad +0.419305 \quad +0.196160 \quad +0.323173 \quad +0.078120].$$

An increase of the first discriminant function Z_1 thus corresponds to a decrease of the average depth of lakes in m but to an increase of the last four variates, which are positively related to eutrophy:

X_2 = average temperature in °C of surface waters (from 0 to 10 m) in July and August,

X_3 = $\log_e$ (dry weight of plankton in kg/ha),

X_4 = $\log_e$ (dry weight of benthos in kg/ha),

X_5 = $\log_e$ (average weight of fish caught per standard net, in kg/net).

The lakes on the Canadian Shield may therefore be considered as oligotrophic, those straddling the limit of the Shield as mesotrophic, and those south of the Shield as eutrophic. Multiple discriminant analysis thus confirms the conclusions drawn in section 31.10.

When the first discriminant axis is the only one to be statistically significant, as here, its equation may be used as a description of the multivariate linear relationship between all variates. This relationship is said to be *functional* if the function Z_1 is deterministic but *structural* if Z_1 is random (see also section 22.7). The data of each sample may be considered as *replicates* allowing residual variances and covariances with respect to the relationship to be estimated. The expression *generalized least squares* is used in this context (Sprent, 1969).

Section 33.12: final remarks

Many biological examples of multiple discriminant analysis have been published, including those of Jolicoeur (1959, 1975), Green (1971), Baron and Jolicoeur (1980) and Jolicoeur, Pirlot, Baron and Stephan (1984). In many cases, when many groups are compared with respect to many correlated variates, the probability of detecting slight average differences is considerably enhanced by using multiple discriminant analysis.

Moreover, even when no significant differences are found, the utilization of multiple discriminant analysis remains worthwile because it decreases the risk that the lack of significant results might be due to the use of less sensitive (powerful) methods. Finally, Pontier, Jolicoeur and Pernin (1987) have noted that, when the number q of variates is greater than $(C-1)$, where C is the number of groups compared, it may be useful to consider, in addition to the $(C-1)$ discriminant functions, $[q-(C-1)]$ additional linear combinations with respect to which group mean vectors do not differ. Examples of such so-called *antidiscriminant functions* have been drawn from the biometry of sportsmen and sportswomen by Pontier, Dufour and Normand (1990) and by Touabti-Mimouni (1996). In multiple discriminant analysis as in principal component analysis [section 31.7; see also Jolicoeur (1963d) and Jolicoeur (1984)], the linear combinations which vary little or which do not vary at all reflect characteristics with respect to which the organism tends to remain constant.

From a mathematical viewpoint, each row vector $\mathbf{K}_i$ of the matrix $\mathbf{K}$ of the coefficients of discriminant functions may be obtained by maximizing the between-groups sum of squares $\mathbf{K}_i \mathbf{B} \tilde{\mathbf{K}}_i$ under the condition that $\mathbf{K}_i \mathbf{S} \tilde{\mathbf{K}}_i = 1$, according to which the within-groups variance of the ith discriminant function is equal to unity. Like in the case of principal components (section 31.11), this maximization can be carried out by using the method of *Lagrange's undetermined multipliers*, according to which

$$\frac{\partial}{\partial \mathbf{K}_i}[\mathbf{K}_i \mathbf{B} \tilde{\mathbf{K}}_i - e_i(\mathbf{K}_i \mathbf{S} \tilde{\mathbf{K}}_i - 1)] = 2\mathbf{K}_i \mathbf{B} - 2e_i \mathbf{K}_i \mathbf{S} = \mathbf{0},$$

whence

$$\mathbf{K}_i \mathbf{B} = e_i \mathbf{K}_i \mathbf{S}.$$

The equations concerning all discriminant functions, of which the number is equal to minimum$(q, C-1)$, may then be arranged into a single matrix equation (section 33.5),

$$\mathbf{KB} = \mathbf{EKS},$$

which may also be expressed as

$$\mathbf{KBS}^{-1} = \mathbf{EK}$$

if each member is postmultiplied by the inverse of the within-groups covariance matrix $\mathbf{S}$. This last equation shows that the row vectors of matrix $\mathbf{K}$ are the latent vectors and that the elements of the main diagonal of the diagonal matrix $\mathbf{E}$ are the latent roots of the matrix $\mathbf{BS}^{-1}$. Since the latter is asymmetrical, however, its left latent vectors differ from its right latent vectors, which may give rise to confusion. Consequently, computations are often carried out in the symmetrical manner described in sections 33.3 and 33.5.

The equation $\mathbf{KB} = \mathbf{EKS}$ may also be obtained without using Lagrange's undetermined multipliers, since

$$\frac{\partial}{\partial \mathbf{K}_i} \mathbf{K}_i \mathbf{B} \tilde{\mathbf{K}}_i / \mathbf{K}_i \mathbf{S} \tilde{\mathbf{K}}_i = [2\mathbf{K}_i \mathbf{B}(\mathbf{K}_i \mathbf{S} \tilde{\mathbf{K}}_i) - 2(\mathbf{K}_i \mathbf{B} \tilde{\mathbf{K}}_i)\mathbf{K}_i \mathbf{S}]/(\mathbf{K}_i \mathbf{S} \tilde{\mathbf{K}}_i)^2 = \mathbf{0},$$

whence

$$\mathbf{K}_i \mathbf{B} = [(\mathbf{K}_i \mathbf{B} \tilde{\mathbf{K}}_i)/(\mathbf{K}_i \mathbf{S} \tilde{\mathbf{K}}_i)]\mathbf{K}_i \mathbf{S} = e_i \mathbf{K}_i \mathbf{S}.$$

Multiple discriminant analysis is related to the analysis of canonical correlations (chapter 34), of which it may be considered a particular case.

Chapter 34
Canonical correlations

Section 34.1: introduction

Harold Hotelling proposed the method of canonical correlations in 1936 to describe linear relationships between two subsets of variates as simply and as efficiently as possible. The vector of the complete set of all variables, $\mathbf{X}=[X_1, \ldots X_q]$, is subdivided into two subvectors, $\mathbf{X}_1$ and $\mathbf{X}_2$,

$$\mathbf{X} = [\mathbf{X}_1 | \mathbf{X}_2] = [X_{11}, \ldots X_{1q_1} | X_{21}, \ldots X_{2q_2}],$$

of which the first has q_1 and the second has q_2 elements. Conventionally, the variates are ordered so that $q_1 \leq q_2$. The mean vector $\boldsymbol{\mu}$ and the covariance matrix $\boldsymbol{\Sigma}$ of the $q = q_1 + q_2$ variates are subdivided in conformable fashion:

$$\boldsymbol{\mu} = [\boldsymbol{\mu}_1 | \boldsymbol{\mu}_2] \qquad \text{and} \qquad \boldsymbol{\Sigma} = \begin{bmatrix} \boldsymbol{\Sigma}_{11} & \boldsymbol{\Sigma}_{12} \\ \boldsymbol{\Sigma}_{21} & \boldsymbol{\Sigma}_{22} \end{bmatrix}.$$

Canonical correlations possess algebraic properties which are unaffected by the nature of the probability distribution of variates $\mathbf{X}=[\mathbf{X}_1 | \mathbf{X}_2]$, but current methods for testing hypotheses assume that the multivariate normal distribution applies to at least one of the two subvectors $\mathbf{X}_1$ and $\mathbf{X}_2$, the variates having possibly been submitted to prior transformations if necessary.

Canonical correlation analysis seeks q_1 pairs of linear combinations,

$$\mathfrak{A}_{11} X_{11} + \ldots + \mathfrak{A}_{1q_1} X_{1q_1} \qquad \text{and} \qquad \mathfrak{B}_{11} X_{21} + \ldots + \mathfrak{B}_{1q_2} X_{2q_2},$$

$$\ldots$$

$$\mathfrak{A}_{q_1 1} X_{11} + \ldots + \mathfrak{A}_{q_1 q_1} X_{1q_1} \qquad \text{and} \qquad \mathfrak{B}_{q_1 1} X_{21} + \ldots + \mathfrak{B}_{q_1 q_2} X_{2q_2},$$

that is to say,

$$\mathfrak{A}\tilde{\mathbf{X}}_1 \qquad \text{and} \qquad \mathfrak{B}\tilde{\mathbf{X}}_2,$$

where

$$\mathfrak{A}_{(q_1 \times q_1)} = \begin{bmatrix} \mathfrak{A}_{11} & \ldots & \mathfrak{A}_{1q_1} \\ \ldots & \ldots & \ldots \\ \mathfrak{A}_{q_1 1} & \ldots & \mathfrak{A}_{q_1 q_1} \end{bmatrix} \qquad \text{and} \qquad \mathfrak{B}_{(q_1 \times q_2)} = \begin{bmatrix} \mathfrak{B}_{11} & \ldots & \mathfrak{B}_{1q_2} \\ \ldots & \ldots & \ldots \\ \mathfrak{B}_{q_1 1} & \ldots & \mathfrak{B}_{q_1 q_2} \end{bmatrix}.$$

German style A and B letters are used here because the coefficients of these linear combinations are population parameters. The linear combinations $\mathfrak{A}\tilde{\mathbf{X}}_1$ and $\mathfrak{B}\tilde{\mathbf{X}}_2$ are chosen so that the variance of each combination is equal to unity (section 34.2), that the two combinations of each pair have maximum correlation, and that all other correlations are null. Matrices $\mathfrak{A}$ and $\mathfrak{B}$ must therefore satisfy the following conditions:

$$[\mathfrak{A} \; \mathfrak{B}] \begin{bmatrix} \boldsymbol{\Sigma}_{11} & \boldsymbol{\Sigma}_{12} \\ \boldsymbol{\Sigma}_{21} & \boldsymbol{\Sigma}_{22} \end{bmatrix} \begin{bmatrix} \tilde{\mathfrak{A}} \\ \tilde{\mathfrak{B}} \end{bmatrix} = \begin{bmatrix} \mathbf{I} & \mathbf{P} \\ \mathbf{P} & \mathbf{I} \end{bmatrix},$$

where $\mathbf{I}$ is a unit (identity) matrix of order $(q_1 \times q_1)$ and $\mathbf{P}$ (the letter $\mathbf{P}$ being the Greek upper case rho) is a square diagonal matrix (section 24.17) of order $(q_1 \times q_1)$ of which the diagonal elements ρ_i are the canonical correlations and other elements are null:

$$\mathbf{P} = \begin{bmatrix} \rho_1 & \dots & 0 \\ \dots & \dots & \dots \\ 0 & \dots & \rho_{q_1} \end{bmatrix} .$$

Each of the linear combinations $\mathfrak{A}\tilde{\mathbf{X}}_1$ of the variates of the first subset is thus correlated with one, and only one, of the linear combinations $\mathfrak{B}\tilde{\mathbf{X}}_2$ of the variates of the second subset (the one having the same subscript). Conventionally, pairs of linear combinations are ordered according to the decreasing magnitude of the corresponding canonical correlations, so that

$$\rho_1 \geqq \rho_2 \geqq \dots \geqq \rho_i \geqq \dots \geqq \rho_{q_1} .$$

Like multiple correlations (section 23.5), canonical correlations are nonnegative, but some may be null. The major aims of the analysis are to discover the number and the magnitude of nonnull canonical correlations in the statistical population as well as the manner in which the original variates of the two subsets are combined in functions $\mathfrak{A}\tilde{\mathbf{X}}_1$ and $\mathfrak{B}\tilde{\mathbf{X}}_2$.

The sample estimates $\mathbf{A}$ and $\mathbf{B}$ of matrices $\mathfrak{A}$ and $\mathfrak{B}$ are obtained by replacing the parametric covariance matrix $\mathbf{\Sigma}$ by the sample covariance matrix $\mathbf{S}$, which yields

$$[\mathbf{A}\ \mathbf{B}] \begin{bmatrix} \mathbf{S}_{11} & \mathbf{S}_{12} \\ \mathbf{S}_{21} & \mathbf{S}_{22} \end{bmatrix} \begin{bmatrix} \tilde{\mathbf{A}} \\ \tilde{\mathbf{B}} \end{bmatrix} = \begin{bmatrix} \mathbf{I} & \mathbf{R} \\ \mathbf{R} & \mathbf{I} \end{bmatrix} ,$$

where $\mathbf{R}$ is the matrix of sample canonical correlations:

$$\mathbf{R} = \begin{bmatrix} r_1 & \dots & 0 \\ \dots & \dots & \dots \\ 0 & \dots & r_{q_1} \end{bmatrix} .$$

Section 34.2: computational principles and techniques

The goal is to maximize the correlation r_i of the ith pair $\mathbf{A}_i\tilde{\mathbf{X}}_1$ and $\mathbf{B}_i\tilde{\mathbf{X}}_2$ of linear combinations. As in the cases of principal components (section 31.11) and of multiple discriminant analysis (section 33.12), this maximization can be carried out by using undetermined Lagrange multipliers. Maximizing the correlation

$$r_i = \mathbf{A}_i\mathbf{S}_{12}\tilde{\mathbf{B}}_i \,/ \sqrt{[(\mathbf{A}_i\mathbf{S}_{11}\tilde{\mathbf{A}}_i)\,(\mathbf{B}_i\mathbf{S}_{22}\tilde{\mathbf{B}}_i)]}$$

amounts to the same as maximizing the covariance $\mathbf{A}_i\mathbf{S}_{12}\tilde{\mathbf{B}}_i$ while respecting the conditions $\mathbf{A}_i\mathbf{S}_{11}\tilde{\mathbf{A}}_i = 1$ and $\mathbf{B}_i\mathbf{S}_{22}\tilde{\mathbf{B}}_i = 1$. One may thus write

$$\frac{\partial}{\partial \mathbf{A}_i}[\mathbf{A}_i\mathbf{S}_{12}\tilde{\mathbf{B}}_i - (l_i/2)(\mathbf{A}_i\mathbf{S}_{11}\tilde{\mathbf{A}}_i - 1) - (m_i/2)(\mathbf{B}_i\mathbf{S}_{22}\tilde{\mathbf{B}}_i - 1)] = \mathbf{S}_{12}\tilde{\mathbf{B}}_i - l_i\mathbf{S}_{11}\tilde{\mathbf{A}}_i = \mathbf{0} ,$$

and

$$\frac{\partial}{\partial \mathbf{B}_i}[\mathbf{A}_i\mathbf{S}_{12}\tilde{\mathbf{B}}_i - (l_i/2)(\mathbf{A}_i\mathbf{S}_{11}\tilde{\mathbf{A}}_i - 1) - (m_i/2)(\mathbf{B}_i\mathbf{S}_{22}\tilde{\mathbf{B}}_i - 1)] = \mathbf{A}_i\mathbf{S}_{12} - m_i\mathbf{B}_i\mathbf{S}_{22} = \mathbf{0} ,$$

where l_i and m_i, as well as $(l_i/2)$ and $(m_i/2)$, are undetermined Lagrange multipliers.

Transposing the matrix products $\mathbf{S}_{12}\tilde{\mathbf{B}}_i$ and $\mathbf{S}_{11}\tilde{\mathbf{A}}_i$ in the first of the two preceding equations yields

$$\mathbf{B}_i\mathbf{S}_{21} = l_i\mathbf{A}_i\mathbf{S}_{11} \tag{1}$$

and

$$\mathbf{A}_i\mathbf{S}_{12} = m_i\mathbf{B}_i\mathbf{S}_{22}\,. \tag{2}$$

If both members of equation (1) are then postmultiplied by $\tilde{\mathbf{A}}_i$ and both members of equation (2) by $\tilde{\mathbf{B}}_i$, one has

$$\mathbf{B}_i\mathbf{S}_{21}\tilde{\mathbf{A}}_i = l_i\mathbf{A}_i\mathbf{S}_{11}\tilde{\mathbf{A}}_i = l_i \times 1 = l_i$$

and

$$\mathbf{A}_i\mathbf{S}_{12}\tilde{\mathbf{B}}_i = m_i\mathbf{B}_i\mathbf{S}_{22}\tilde{\mathbf{B}}_i = m_i \times 1 = m_i\,.$$

But the bilinear form (section 24.23) $\mathbf{B}_i\mathbf{S}_{21}\tilde{\mathbf{A}}_i$ is equal to $\mathbf{A}_i\mathbf{S}_{12}\tilde{\mathbf{B}}_i$ since it is simply its transpose, which shows that $l_i = m_i$; equation (2) can thus be rewritten as:

$$\mathbf{A}_i\mathbf{S}_{12} = l_i\mathbf{B}_i\mathbf{S}_{22}\,. \tag{3}$$

Once the vector $\mathbf{A}_i$ of coefficients of the linear combination of the variates of the subvector $\mathbf{X}_1$ will have been evaluated, it will therefore be possible to obtain the value of the vector $\mathbf{B}_i$ of coefficients of the linear combination of the variates of the subvector $\mathbf{X}_2$ from equation (3) as follows:

$$\mathbf{B}_i = (1/l_i)\,\mathbf{A}_i\mathbf{S}_{12}\mathbf{S}_{22}^{-1}\,. \tag{4}$$

For the time being, the value of $\mathbf{B}_i$ given by (4) may be substituted in equation (1), which yields

$$(1/l_i)\,\mathbf{A}_i\mathbf{S}_{12}\mathbf{S}_{22}^{-1}\mathbf{S}_{21} = l_i\mathbf{A}_i\mathbf{S}_{11}\,,$$

whence

$$\mathbf{A}_i\mathbf{S}_{12}\mathbf{S}_{22}^{-1}\mathbf{S}_{21} = l_i^2\mathbf{A}_i\mathbf{S}_{11}\,. \tag{5}$$

The equations corresponding to the q_1 pairs of linear combinations may finally be expressed in a single matrix equation,

$$\mathbf{A}\mathbf{S}_{12}\mathbf{S}_{22}^{-1}\mathbf{S}_{21} = \mathbf{L}^2\mathbf{A}\mathbf{S}_{11}\,, \tag{6}$$

where the elements of the main diagonal of the diagonal matrix $\mathbf{L}$ are the undetermined multipliers $l_1, \dots\, l_i, \dots\, l_{q_1}$, which are square roots of latent roots (section (24.21).

Canonical correlations may also be evaluated without using undetermined Lagrange multipliers, if the vectors of partial derivatives with respect to vectors $\mathbf{A}_i$ and $\mathbf{B}_i$ of the square of the ith canonical correlation,

$$r_i^2 = (\mathbf{A}_i\mathbf{S}_{12}\tilde{\mathbf{B}}_i)^2 / [(\mathbf{A}_i\mathbf{S}_{11}\tilde{\mathbf{A}}_i)(\mathbf{B}_i\mathbf{S}_{22}\tilde{\mathbf{B}}_i)]\,,$$

are set equal to null vectors. As soon as these derivatives have been obtained, the resulting equations may be considerably simplified if $\mathbf{A}_i\mathbf{S}_{11}\tilde{\mathbf{A}}_i$ and $\mathbf{B}_i\mathbf{S}_{22}\tilde{\mathbf{B}}_i$ are replaced by 1.0 and $\mathbf{A}_i\mathbf{S}_{12}\tilde{\mathbf{B}}_i$ is replaced by r_i :

$$\mathbf{B}_i\mathbf{S}_{21} = r_i\mathbf{A}_i\mathbf{S}_{11} \tag{7}$$

and

$$\mathbf{A}_i\mathbf{S}_{12} = r_i\mathbf{B}_i\mathbf{S}_{22}\,. \tag{8}$$

Equations (7) and (8) are similar to equations (1) and (2) except that the canonical correlation r_i is substituted for the undetermined multipliers l_i and m_i, which shows that the latter are in fact equal to the desired canonical correlation. Equations (6) and (4) may therefore be reexpressed more explicitly as:

$$\mathbf{A}\,\mathbf{S}_{12}\,\mathbf{S}_{22}^{-1}\,\mathbf{S}_{21} = \mathbf{R}^2\mathbf{A}\,\mathbf{S}_{11}\,, \tag{9}$$

and

$$\mathbf{B} = \mathbf{R}^{-1}\mathbf{A}\,\mathbf{S}_{12}\,\mathbf{S}_{22}^{-1}\,. \tag{10}$$

Equations (9) and (10) may also easily be deduced (particularly when $q_1 = q_2$) from the algebraic properties discussed in section 34.1 without using differential calculus. Equation (9) is visibly similar to the equation $\mathbf{KB} = \mathbf{EKS}$ obtained in multiple discriminant analysis (section 33.12), which indicates that a similar computing method (sections 33.3 and 33.5) can be used, the following matrix substitutions being made:

Canonical correlations	Multiple discriminant analysis
$\mathbf{A}$	$\mathbf{K}$
$\mathbf{S}_{12}\,\mathbf{S}_{22}^{-1}\,\mathbf{S}_{21}$	$\mathbf{B}$
$\mathbf{R}^2$	$\mathbf{E}$
$\mathbf{S}_{11}$	$\mathbf{S}$

However, it must be noted that matrix **B** does not have the same meaning in canonical correlation analysis as in multiple discriminant analysis, where it denotes the between-groups sums of squares and products.

Section 34.3: hypotheses about the dimensionality of correlations

The fact that multiple discriminant analysis and canonical correlation analysis are related is also evident from the nature of the hypotheses made and from the way in which they are tested. While multiple discriminant analysis brings out the dimensionality of between-group variation (section 33.4), canonical correlations reveal the dimensionality of relationships between two subsets of variates. Hypothesis testing tables for multiple discriminant analysis (table 33.4.1) and for canonical correlations (table 34.3.2) are very similar (see equivalences in table 34.3.1) and are used in the same way (see section 33.4).

Table 34.3.1
Equivalences between multiple discriminant and canonical correlation analysis

	Discriminant functions	Canonical correlations
Number of variates of first subset	q if $q \le (C-1)$, $(C-1)$ if $(C-1) < q$	q_1
Number of variates of second subset	$(C-1)$ if $q \le (C-1)$, q if $(C-1) < q$	q_2
j^{th} χ^2 component	$a \times \log_e[1 + e_j/(N. - C)]$	$-a \times \log_e(1 - r_j^2)$

Table 34.3.2
Tests of hypotheses about the dimensionality of correlations

Canonical correlation	χ^2 component	Degrees of freedom	Sum of χ^2 components	Cumulated degrees of freedom	Tested hypothesis
r_1	$-a \times \log_e(1-r_1^2)$	(q_1+q_2-1)	$\sum_{i=1}^{q_1}[-a \times \log_e(1-r_i^2)]$	$q_1 q_2$	H_0: 0 dimensions
r_2	$-a \times \log_e(1-r_2^2)$	(q_1+q_2-3)	$\sum_{i=2}^{q_1}[-a \times \log_e(1-r_i^2)]$	$(q_1-1)(q_2-1)$	H_1: 1 dimension
...	...	...	...	...	...
r_j	$-a \times \log_e(1-r_j^2)$	(q_1+q_2+1-2j)	$\sum_{i=j}^{q_1}[-a \times \log_e(1-r_i^2)]$	$(q_1+1-j)(q_2+1-j)$	$H_{(j-1)}$: $(j-1)$ dimensions
...	...	...	...	...	...
r_{q_1}	$-a \times \log_e(1-r_{q_1}^2)$	(q_2+1-q_1)	$-a \times \log_e(1-r_{q_1}^2)$	(q_2+1-q_1)	$H_{(q_1-1)}$: (q_1-1) dimensions

Note 1 : Bartlett's factor $a = [(N-1)-(1/2)(q_1+q_2+1)]$; Note 2 : conventionally, $q_1 \leq q_2$.

Section 34.4: a classical example from Rao (1952) using data from Frets (1921)

While multiple discriminant analysis (chapter 33) has been used extensively, there are few convincing examples of canonical correlations, and the practical utility of the latter has been questioned (see Kendall and Stuart, 1966, page 305, for instance). The data on the head length and breadth of the first and second adult sons of 25 Dutch families, extracted by Rao (1952, page 245) from Frets' monograph (1921), have therefore been reutilized by other authors, including Anderson (1958, page 58; 1984, page 97) and Seber (1984, page 263). This example may be considered a classic, and has been briefly reanalyzed in this section. However, it is not fully satisfactory because of small sample size ($N=25$), and different results will be obtained in the next section (34.5) on a new and more substantial sample of Frets' data. Table 34.4.1 contains the means, covariance matrix, standard deviations and correlation matrix of the small sample of Frets' data drawn by Rao (1952). The individual data are reproduced in appendix F.

Table 34.4.1
Means, covariance matrix, standard deviations and correlations of the head length and breadth in mm of the first two adult sons in 25 Dutch families, based on data extracted by Rao (1952) from Frets' monograph (1921)

	first son		second son	
	length	breadth	length	breadth
Means	185.72	151.12	183.84	149.24
Covariance matrix	95.2933	52.8683	69.6617	46.1117
		54.3600	51.3117	35.0533
			100.807	56.5400
				45.0233
Standard deviations	9.76183	7.37292	10.0403	6.70994
Correlation matrix	1	0.734556	0.710752	0.703981
		1	0.693157	0.708550
			1	0.839252
				1

What is being sought is the linear combinations of the head dimensions of the first adult son which have the highest correlations with the corresponding linear combinations of the head dimensions of the second son. The numerical values of the coefficients of those linear combinations, which are the elements of matrices **A** and **B**, are given in table 34.4.2, and the canonical correlation analysis is summarized in table 34.4.3.

Table 34.4.2
Coefficients of the linear combinations of the head length and breadth in mm of the first two adult sons in 25 Dutch families, based on data extracted by Rao (1952) from Frets' monograph (1921)

	first son		second son	
	length	breadth	length	breadth
1st combination	+0.0565662	+0.0707368	+0.0502426	+0.0802224
2nd combination	−0.139971	+0.186950	−0.176148	+0.262084

The linear combinations of the first pair reflect head size, since their coefficients are all positive. Moreover, the corresponding canonical correlation (the first one) is highly significant ($r_1 = 0.788508$, $P = 0.00032189$), which indicates that the first and second adult sons of each family resemble each other with respect to head size. On the contrary, the second canonical correlation does not reach statistical significance ($r_2 = 0.053740$, $P = 0.803082$), and the corresponding linear combinations reflect the relative breadth (shape) of the head, since their coefficients are negative for head length but positive for head breadth. Thus, these data do not indicate that the first and second adult sons of each family tend to have similar head shapes. This lack of likeness in relative head breadth is surprising, because siblings would be expected to show hereditary similarities in shape as well as in size. This question will be reexamined in the next section on the basis of a more substantial sample of Frets' data (1921).

Table 34.4.3
Canonical correlation analysis of the head length and breadth in mm of the first two adult sons in 25 Dutch families, based on data extracted by Rao (1952) from Frets' monograph (1921)

Canonical correlation	χ^2 component	Degrees of freedom	Sum of χ^2 components	Cumulated degrees of freedom	Probability
0.788508	20.9020	3	20.9642	4	0.00032189
0.053740	0.062181	1	0.062181	1	0.803082

A comment should be made here concerning numerical accuracy. Twenty or thirty years ago, it was difficult to compute canonical correlations and their associated vectors of coefficients with accuracy, especially when the numbers of variates q_1 and q_2 were large. Nowadays, thanks to the considerable accuracy of computers and software, there is little excuse for publishing inaccurate figures. For instance, if the numerical results of the preceding analysis are checked entirely within the computer by calculating the correlation matrix of linear combinations,

$$[\mathbf{A}\ \mathbf{B}]\begin{bmatrix}\mathbf{S}_{11} & \mathbf{S}_{12}\\ \mathbf{S}_{21} & \mathbf{S}_{22}\end{bmatrix}\begin{bmatrix}\tilde{\mathbf{A}}\\ \tilde{\mathbf{B}}\end{bmatrix} = \begin{bmatrix}\mathbf{I} & \mathbf{R}\\ \mathbf{R} & \mathbf{I}\end{bmatrix},$$

the elements of this matrix which should be equal to 0 or to 1 do so with a numerical accuracy reaching 15 digits. The numerical results obtained several decades ago on the same data by Anderson (1958, pages 303-305; 1984, pages 500-502) agree with the present results and have from 3 to 6 accurate digits. However, the results obtained more recently by Seber (1984, pages 262-264), still on the same data, have only from 3 to 4 accurate digits in canonical correlations and from 1 to 2 in the coefficients of linear combinations. The major difficulty at the present time may be the poor quality and the unsatisfactory documentation of some currently used commercial software packages. The concluding remarks of Lefebvre and Tassencourt (1968) are still worthy of attention: "It is worrisome to find that the importance of verifications is neglected in most current computer programs. The considerable increase in the power of computers seems to have intoxicated users and made them oblivious of the fact that computing speed is useless when there is a risk that numerical results may be wrong."

Section 34.5: a more substantial sample of Frets' data (1921)

It is unfortunately difficult to draw a satisfactory random sample of Frets' data (1921) because that author, as was currently done at that time, split his data into numerous tables according to the type of hereditary phenomenon he was trying to detect. Moreover, the data pertaining to some families are repeated in several tables. Finally, some of the repeated data differ in a few instances, presumably because of typesetting errors. In order to obtain a sample of Frets' data as free from bias and from errors as possible, we have therefore extracted from Frets' monograph (1921) the data concerning all families possessing at least one adult son and one adult daughter and exhibiting no numerical inconsistencies. The statistics concerning these 73 families are summarized in table 34.5.1.

Table 34.5.1
Means, covariance matrix, standard deviations and correlations of the head length and breadth in mm of the father, mother, first adult son and first adult daughter in 73 Dutch families, based on data extracted from Frets' monograph (1921)

father		mother		son		daughter	
length	breadth	length	breadth	length	breadth	length	breadth
			means				
194.507	154.685	184.096	149.329	192.890	153.863	183.260	147.822
			covariance matrix				
39.5312	16.2452	5.50628	7.58105	15.9174	16.0982	13.9774	3.71651
	28.6632	3.73896	3.24391	12.7567	12.1785	9.87481	9.16533
		24.0323	10.5930	13.2884	5.69387	11.6275	4.96176
			30.9460	5.60597	14.6429	5.91324	10.1566
				44.7100	13.7487	16.3761	5.95244
					32.0643	5.55004	10.9753
						31.2785	8.81088
							19.1484
			standard deviations				
6.28738	5.35381	4.90228	5.56291	6.68656	5.66254	5.59272	4.37589
			correlation matrix				
1	0.482607	0.178645	0.216749	0.378617	0.452164	0.397495	0.135083
	1	0.142459	0.108919	0.356346	0.401716	0.329794	0.391219
		1	0.388437	0.405390	0.205116	0.424096	0.231298
			1	0.150712	0.464851	0.190064	0.417234
				1	0.363117	0.437911	0.203435
					1	0.175252	0.442933
						1	0.360023
							1

During the analysis of the small sample of size $N=25$ extracted from Frets' monograph (1921) by Rao, in section 34.4, the lack of correlation between linear combinations reflecting head shape was deemed surprising. A similar question may be considered in the case of the first adult son and first adult daughter in the larger sample analyzed in this section. Table 34.5.2 contains the coefficients of linear combinations and the analysis of canonical correlation is summarized in table 34.5.3.

Table 34.5.2
Coefficients of the linear combinations of the head length and breadth in mm of the first adult son and first adult daughter in 73 Dutch families, based on data extracted from Frets' monograph (1921)

	son		daughter	
	length	breadth	length	breadth
1st combination	+0.0878041	+0.110184	+0.0993871	+0.149665
2nd combination	−0.134364	+0.154219	−0.163872	+0.193910

As in the preceding section, the first and second pairs of linear combinations reflect head size and head shape respectively but, this time, both canonical correlations are substantial (0.462898 and 0.393440) and statistically significant ($P<0.001$). Unlike the small sample extracted from Frets' monograph (1921) by Rao (1952), this new larger sample thus shows that hereditary similarities between siblings affect head shape as well as head size. Analyses based on other data from table 34.5.1 bring out substantial and conclusive canonical correlations also between the head dimensions of children and those of their parents.

Table 34.5.3
Canonical correlation analysis of the head length and breadth in mm of the first adult son and first adult daughter in 73 Dutch families, based on data extracted from Frets' monograph (1921)

Canonical correlation	χ^2 component	Degrees of freedom	Sum of χ^2 components	Cumulated degrees of freedom	Probability
0.462898	16.7598	3	28.4481	4	0.0000101
0.393440	11.6883	1	11.6883	1	0.0006290

One may wonder whether the father and mother too might tend to resemble each other with respect to head dimensions, since an analysis of Galton's data by Karl Pearson and Alice Lee (1900) indicated eye color similarities between husband and wife, which Pearson interpreted (perhaps incorrectly) as due to mating preferences (see section 15.5). The coefficients of linear combinations of head dimensions of the father and mother are given in table 34.5.4 and the analysis of canonical correlations is summarized in table 34.5.5. In this case, however, neither of the canonical correlations is significant ($P>0.05$). Moreover, the coefficients of the linear combinations of the second pair have opposite signs in the father and mother. The present analysis thus does not reveal any tendency to similarity in head dimensions between husband and wife.

Table 34.5.4
Coefficients of the linear combinations of the head length and breadth in mm of the father and mother in 73 Dutch families, based on data extracted from Frets' monograph (1921)

	father		mother	
	length	breadth	length	breadth
1st combination	+0.144513	+0.0312045	+0.101711	+0.124848
2nd combination	−0.109968	+0.210967	+0.196620	−0.149898

Table 34.5.5
Canonical correlation analysis of the head length and breadth in mm of the father and mother in 73 Dutch families, based on data extracted from Frets' monograph (1921)

Canonical correlation	χ^2 component	Degrees of freedom	Sum of χ^2 components	Cumulated degrees of freedom	Probability
0.242218	4.20204	3	4.43963	4	0.349766
0.058419	0.237596	1	0.237596	1	0.625948

Altogether, the conclusions drawn from the analysis of this new larger sample of Frets' data (1921) agree better with current knowledge concerning the hereditary control of quantitative characters. From a methodological viewpoint, the descriptive interpretations of the linear combinations corresponding to canonical correlations could easily be confirmed using *a priori* linear combinations (sections 29.6 and 29.8) defined in such a way as to reflect head size and relative breadth exactly.

Section 34.6: possibilities and limitations of canonical correlations

In general, the analysis of canonical correlations is exploratory in nature, because the corresponding linear combinations can seldom be interpreted as easily as in the case of the data of Frets (1921) used in the examples of this chapter. For all that, one should not think that canonical correlations can be useful only in quantitative genetics.

In human biology, for instance, one could consider correlations, within a group of individuals, between a set of physiological and a set of morphometric characteristics. In developmental studies, possible relationships could be sought between the features of two very different life-forms assumed successively by the individuals of a single species during their life cycle, like the larvae and adults of anuran amphibians (frogs and toads) or of Lepidoptera (butterflies).

In neurobiology, one could try to see whether there are relationships between the quantitative degree of development of various sensory or motor centers of the central nervous system and measurements of the success of the organism in various activities or when facing different kinds of challenges. In psychology or in the study of animal behavior, one could attempt to relate a set of quantitative components of a complex stimulus with a set of quantitative characteristics of an equally complex response of the organism. In ecology, it might be interesting to correlate a set of physicochemical conditions in the environment with the biological characteristics of a community.

The analysis of canonical correlations could thus be applied to a broad variety of problems, but it must be emphasized that interpreting the results of such analyses typically requires as much effort and as much care as the numerical procedures themselves. Trying to confirm the suggestions gleaned from the analysis of canonical correlations with the help of *a priori* linear combinations (sections 29.6 and 29.8) may frequently be advisable. Moreover, in highly multivariate situations, it may be better to work initially with relatively limited subsets of variates, as in the preceding sections of this chapter, and try only later to get a general overview of relationships between all variates.

Is the method of canonical correlations a tool of analysis or synthesis? – Paradoxically, it may be both.

Chapter 35

Growth curves and other nonlinear relationships

Section 35.1: introduction

The words *linear* and *nonlinear* have special technical meanings in statistics. Thus, on the one hand, a relationship (also called a *model*) $Y=f(A, B, C, \ldots ; X)$ is said to be *linear* if the partial derivative of the function $f(A, B, C, \ldots ; X)$ with respect to each of its coefficients (parameters) $A, B, C, \ldots$ etc. does not involve either the same coefficient or another coefficient. On the other hand, the relationship $Y=f(A, B, C, \ldots ; X)$ is said to be nonlinear if the partial derivative of the function $f(A, B, C, \ldots ; X)$ with respect to any of its coefficients involves either the same or another coefficient. Let us consider for instance the relationship between the oxygen consumption Y of guinea pigs in g per kg of body weight per hour and the environmental temperature $X=T$ in °C (figure 35.1.1).

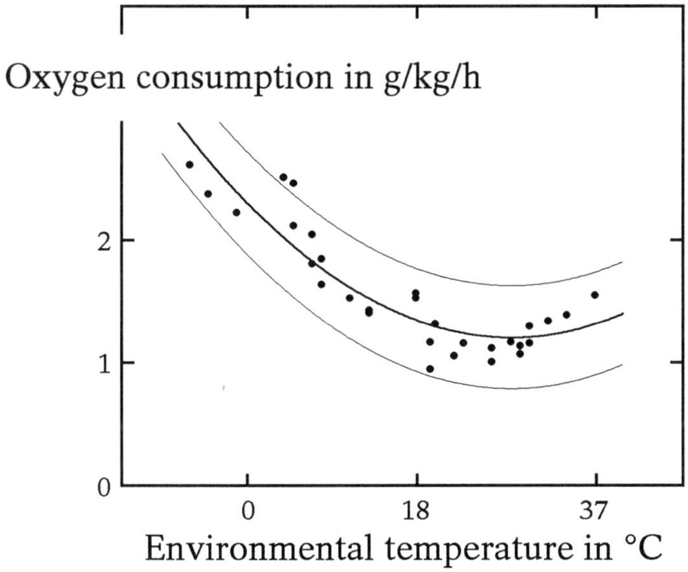

Figure 35.1.1
Curvilinear relationship between the oxygen consumption Y in g/kg/h of guinea pigs and the environmental temperature T in °C, according to data from Terroine and Trautmann (1927); variation belt (thin lines) at level $\alpha = 0.05$

While this relationship is clearly not straight (*rectilinear*), it may be described using the polynomial

$$Y = a+bX+cX^2 = a+bT+cT^2,$$

which, even though it is not linear with respect to the predictor variate $X=T$, is linear with respect to each of its coefficients a, b, and c, since $\partial Y/\partial a=1$, $\partial Y/\partial b=T$, and $\partial Y/\partial c=T^2$. Consequently the coefficients a, b, and c may be estimated simply by using the multiple regression (chapters 23 and 25) of the predicted variate Y on the first and second powers, T and T^2, of the predictor variate.

However, the preceding relationship may also be described using another function,

$$Y = A\{1+[(T-B)/C]^2\},$$

which is nonlinear, since $\partial Y/\partial A = \{1+[(T-B)/C]^2\}$, etc., but which may be easier to interpret in a biologically interesting manner,

for A = an estimate of the average minimum oxygen consumption,
B = the environmental temperature at which the average minimum oxygen consumption A is reached,
and C = the temperature difference, below or above B, at which the average oxygen consumption is equal to $2A$.

Point and interval estimates of the coefficients of a nonlinear relationship like $Y=A\{1+[(T-B)/C]^2\}$ are generally much more difficult to obtain (sections 35.2, 35.3, 35.4 and 35.5) than those of a linear relationship like $Y=a+bT+cT^2$. Consequently, many statisticians prefer using linear models as often as possible, rather than nonlinear models. From a biological viewpoint, however, a thoughtfully conceived nonlinear model is often more explicitly meaningful and easier to interpret than a linear model, as shown by the present example.

In the particular case of the guinea pig data of Terroine and Trautmann (1927), the nonlinear model $Y=A\{1+[(T-B)/C]^2\}$ can be transformed into the linear model $Y=a+bT+cT^2$ and vice versa through a simple reparameterization. The nonlinear model may indeed be written out in expanded form as

$$Y = A + AB^2/C^2 - 2ABT/C^2 + AT^2/C^2,$$

whence

$$a = A + AB^2/C^2,$$

$$b = -2AB/C^2$$

and

$$c = A/C^2.$$

The coefficients a, b, and c of the linear model $Y=a+bT+cT^2$ would obviously have no explicitly useful biological meaning, since they are only related in a complicated manner to the coefficients A, B, and C of the nonlinear model, which are themselves easily interpretable as indicated above.

Inversely,

$$A = a - (b^2/4c),$$

$$B = -b/(2c),$$

and

$$C = \sqrt{\{(a/c) - [b^2/(4c^2)]\}}.$$

However, many other nonlinear models cannot be reexpressed in a linear form through a reparameterization and are then said to be *intrinsically nonlinear*. Some intrinsically nonlinear relationships may be linearized through a transformation of the predicted variate Y (sections 20.6 and 20.7), but this procedure is advisable mostly when the transformed variate has a constant variance and follows a reasonably well-known probability distribution.

While the coefficients of a linear model are often denoted by lower case letters (a, b, c, etc.), those of a nonlinear model are often denoted by upper case letters (A, B, C, etc.). As for the corresponding statistical parameters, upper case Greek letters are not generally used because some of them are identical to the corresponding Roman letters while others have no clear equivalents. Some authors thus use upper case letters (A, B, C, etc.) to denote parameters and the same letters topped by a circumflex accent (a "hat") to denote sample estimates or estimators ($\hat{A}$, $\hat{B}$, $\hat{C}$, etc.). Following another satisfactory notation, which is used in this chapter, sample estimates are denoted by unaccented upper case Roman letters (A, B, C, etc.) and population parameters by the corresponding German upper case letters ($\mathfrak{A}$, $\mathfrak{B}$, $\mathfrak{C}$, etc.).

In the case of a nonlinear model, estimating the residual variance by dividing the residual sum of squares (RSS) by the number of degrees of freedom ($N-q$), where q denotes the number of parameters, is only approximate, but a sampling experiment (Jolicoeur, 1995) has shown that this may nevertheless be satisfactory in practice. Several textbooks are devoted in part or completely to nonlinear models (Bates and Watts, 1988; Draper and Smith, 1981; Jolivet, 1983; Lebreton, Millier, Jolivet, Pavé and Vila, 1982; Ratkowsky, 1983, 1990; Ross, 1990; Seber and Wild, 1989; Tomassone, Audrain, Lesquoy-de Turckheim and Millier, 1992; Tomassone, Dervin and Masson, 1993). However, the mathematical level of some of these textbooks is not suitable for every biologist.

Section 35.2: iterative minimization of the residual sum of squares (RSS)

The coefficients A, B, C, ... of a nonlinear model $\hat{Y}=f(A, B, C, \dots ; X)$ appear nonlinearly not only in the function $f(A, B, C, \dots ; X)$ but also in the partial derivatives with respect to these coefficients of the residual sum of squares,

$$\mathrm{RSS} = \sum_{h=1}^{N} (Y_h - \hat{Y})^2.$$

Consequently, the equations obtained by setting the preceding derivatives equal to zero, following the *principle of least squares,* generally do not yield explicit estimators A, B, C, ... of parameters $\mathfrak{A}$, $\mathfrak{B}$, $\mathfrak{C}$, ... Unlike in the case of linear models (sections 20.1, 22.2, 23.3, 25.3 and 25.5), nonlinear least-squares estimates are thus usually obtained by a trial-and-error process instead of through direct calculations, repetitive steps being called *iterations* (see also section 31.3).

Several iterative procedures are available to minimize the residual sum of squares [or to maximize the likelihood (section 4.10)] with respect to the values of the coefficients A, B, C, ... Most of these algorithms require either the derivatives of the residual sum of squares (RSS) or numerical estimates of those derivatives. However, the *simplex* method of Nelder and Mead (1965) does not require derivatives or numerical estimates of derivatives and is therefore easy to program for computers (and to modify if necessary). Nelder and Mead's simplex method is not to be confused with a similarly named method due to George B. Dantzig and used in operations research and linear programming (Vajda, 1960). In comparison with other current minimization methods, Nelder and Mead's algorithm is sometimes slower, but it is robust and it does work even when the fitted model and its residual sum of squares do not possess derivatives at some values of the predictor variate, as in the case of segmented models (section 35.8).

Moreover, with Nelder and Mead's simplex method, estimates of parameters can easily be submitted to constraints (to be nonnegative or nonpositive, for instance) when desirable (see sections 35.13, 35.14, 35.17). The gradual recognition of the merits of Nelder and Mead's procedure in various fields of applied science has been summarily described by Olsson and Nelson (1975). The present writer has used Nelder and Mead's algorithm successfully in his own research for over fifteen years. A *True BASIC*™ computer program enabling the user to fit a nonlinear model through Nelder and Mead's method is given in Appendix B. Unlike earlier implementations of Nelder and Mead's algorithm, including the FORTRAN version by O'Neill (1971), the present program uses two convergence criteria, and puts an end to iterations not only when the decrease of the residual sum of squares (RSS) becomes negligible but also when point estimates of the parameters practically no longer change, which does not always happen exactly at the same time.

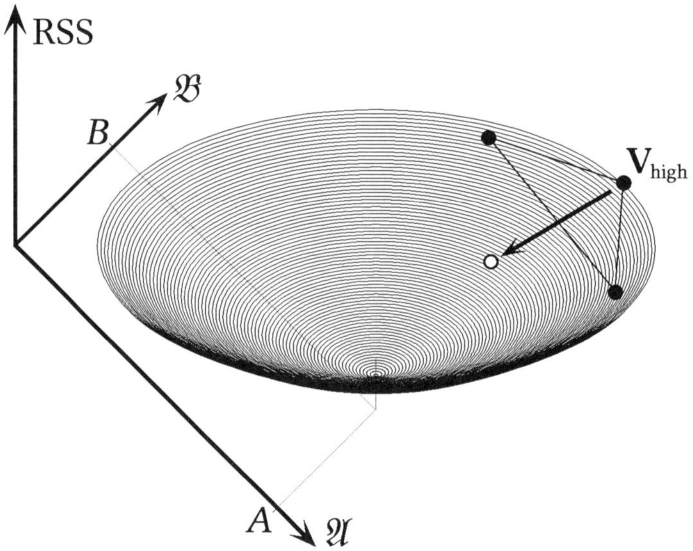

Figure 35.2.1
During a typical iteration of Nelder and Mead's method (1965), the highest point of the simplex, represented here by a triangle of black dots, is replaced by its reflection (○) with respect to the midpoint of the other dots

Since the simplex method has already been described several times in the literature (Nelder and Mead, 1965; O'Neill, 1971; Olsson and Nelson, 1975), the present section is limited to a brief discussion of the program given in Appendix B, which includes, by the way, several explanatory comments. The initial vector of trial estimates (or *starting point*) is used to generate a set of q linearly independent vectors, where q denotes the number of parameters of the fitted model. Together with the starting point, these vectors constitute a matrix $\mathbf{V}$ of order $(q+1)\times(q+1)$ and can be represented geometrically (figure 35.2.1) by a group of points (also called a *simplex*) in a space where the $(q+1)^{\text{th}}$ dimension corresponds to the residual sum of squares (RSS).

In the case where point estimates A and B are sought for two parameters $\mathfrak{A}$ and $\mathfrak{B}$, for instance, the values of the residual sum of squares corresponding to various possible values of coefficients A and B are represented by a bowl-shaped surface having a circular or elliptical cross-section if the relationship is linear, but an irregular shape if the relationship is nonlinear. The estimates A and B are the horizontal coordinates of the lowest point of this bowl (figure 35.2.1).

During a typical iteration of Nelder and Mead's minimization method (1965), the point $\mathbf{V}_{\text{high}}$ having the greatest height (RSS) is replaced by its reflection $\mathbf{V}_1$ (the white dot in figure 35.2.1) with respect to the midpoint $\mathbf{V}_{\text{mid}}$ of the other points of the simplex. When the reflected point $\mathbf{V}_1$ is lower than the lowest point $\mathbf{V}_{\text{low}}$, on the one hand, an attempt is made to go even lower to point $\mathbf{V}_2$, which would expand the simplex. On the other hand, when the reflected point $\mathbf{V}_1$ is higher than all other points (except possibly $\mathbf{V}_{\text{high}}$), the simplex is contracted toward its lowest point. Unless there are special difficulties, the simplex moves gradually toward the lowest point of the bowl and shrinks down around it.

It must be emphasized, however, that every minimization algorithm is liable to fail occasionally. The most frequent problem is that, in the case of a nonlinear model, the surface of the residual sum of squares may possess not just one but several local ("relative") minima situated at different heights and more or less remote from each other. Hence, the algorithm may converge on a secondary minimum instead of the lowest (overall or "absolute") minimum. Since the initial simplex of each series of iterations usually spans a broad region around the initial trial vector, an attempt can be made to check whether the end point of a first series of iterations is the overall minimum by using it as the starting point of a second series of iterations.

Nelder and Mead (1965) stressed that their method is adaptive, inasmuch as their simplex expands when there is an easy way down but contracts when a crooked or almost horizontal surface slows down numerical convergence. A biologist watching a simplex moving down gradually toward the lowest point of the surface of a residual sum of squares may sometimes have the illusion that he is looking at an amoeba! – Nelder and Mead's method may also be used to maximize a function: the sign of the function merely has to be changed before minimization.

Section 35.3: the so-called multiplicative least squares

Even though, in the first two sections of this chapter, it has been assumed for the sake of simplicity that the predicted variate Y followed the normal distribution, the lognormal distribution is usually more suitable for biological variates, which are generally nonnegative, heteroscedastic and often have positively skewed distributions (chapter 14, and sections 16.4, 19.8, 20.6, 20.7, 22.5, 28.7, 29.7, 29.8, 30.9, 31.6, and 31.8). Assuming that the variate predicted through a nonlinear model follows the lognormal distribution does not make the analysis more difficult. One only has to use the so-called *multiplicative least squares,* according to which the residual sum of squares (RSS) is defined on a logarithmic scale:

$$\text{RSS} = \sum_{h=1}^{N} [\log_e(Y_h) - \log_e(\hat{Y})]^2 = \sum_{h=1}^{N} [\log_e(Y_h/\hat{Y})]^2 .$$

However, the value $\hat{Y}$ predicted from the model must then be interpreted as a geometric mean (sections 4.3 and 14.2) instead of an (ordinary) arithmetic mean. Moreover, the residual standard deviation $\sqrt{[\mathrm{RSS}/(N-q)]}$, where N and q denote the sample size and the number of parameters respectively, pertains to the logarithmically transformed variate $Z_h = \log_e(Y_h)$. Confidence, prediction or variation limits must therefore be determined for logarithms and then be submitted to the antilogarithmic transformation $Y = \exp(Z)$ before being used or represented in the scatter diagram of original data (section 35.6). While demographers and econometricians often prefer to keep using the ordinary arithmetic mean for objective reasons even when the distribution of their variates is positively skewed, biologists usually do not object to using the geometric mean. Figure 35.3.1 illustrates the result of using multiplicative least squares to fit the model $Y = A\{1 + [(T-B)/C]^2\}$, which had been fitted earlier (figure 35.1.1) through ordinary additive least squares.

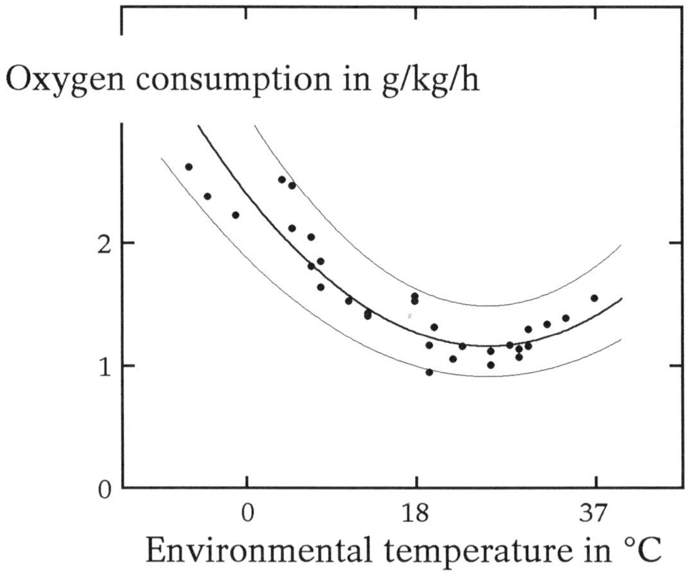

Figure 35.3.1
Relationship between the oxygen consumption Y in g/kg/h of guinea pigs and the environmental temperature T in °C, fitted through multiplicative least squares; according to data from Terroine and Trautmann (1927); variation belt (thin lines) at level $\alpha = 0.05$

If figures 35.3.1 and 35.1.1 are compared, the variation belts (section 35.6) obtained through multiplicative least squares are seen to be narrower for low oxygen consumptions but broader for high oxygen consumptions. Therefore, the fit obtained through multiplicative least squares is more accurate in the neighbourhood of the average minimum oxygen consumption, of which the estimate A tends to be more precise than when using additive least squares. Moreover, the lower variation belt is visibly narrower than the upper variation belt, as it should be for a positively skewed distribution.

Section 35.4: multivariate confidence regions of parameters

When least-squares estimates $[A, B, C, \ldots]$ and the minimum value of the residual sum of squares, RSS$(A, B, C, \ldots)$, have been found for a nonlinear model,

$$\hat{Y} = f(A, B, C, \ldots ; X),$$

the preliminary hypothesis H_0 that the vector of parameters $[\mathfrak{A}, \mathfrak{B}, \mathfrak{C}, \ldots]$ has a theoretical value $[\mathfrak{A}_0, \mathfrak{B}_0, \mathfrak{C}_0, \ldots]$ can be submitted to an approximate likelihood-ratio test using the variance ratio criterion

$$\{[\mathrm{RSS}(\mathfrak{A}_0, \mathfrak{B}_0, \mathfrak{C}_0, \ldots) - \mathrm{RSS}(A, B, C, \ldots)]/q\} / \{[\mathrm{RSS}(A, B, C, \ldots)]/(N-q)\},$$

where RSS$(\mathfrak{A}_0, \mathfrak{B}_0, \mathfrak{C}_0, \ldots)$ is the value of the residual sum of squares computed using the hypothetical values $[\mathfrak{A}_0, \mathfrak{B}_0, \mathfrak{C}_0, \ldots]$ of the parameters instead of their least-squares estimates $[A, B, C, \ldots]$. The hypothesis

$$H_0 : [\mathfrak{A}, \mathfrak{B}, \mathfrak{C}, \ldots] = [\mathfrak{A}_0, \mathfrak{B}_0, \mathfrak{C}_0, \ldots]$$

can be accepted if the above variance ratio $\leq F_{(1-\alpha;\, q,\, N-q)}$, but must be rejected otherwise.

Inversely, when no hypothetical values $[\mathfrak{A}_0, \mathfrak{B}_0, \mathfrak{C}_0, \ldots]$ are available, the set of all vectors satisfying

$$\mathrm{RSS}(\mathfrak{A}, \mathfrak{B}, \mathfrak{C}, \ldots) \leq \mathrm{RSS}(A, B, C, \ldots) \times \{1 + [q \times F_{(1-\alpha;\, q,\, N-q)}]/(N-q)\}$$

constitutes a q-dimensional confidence region for the vector $[\mathfrak{A}, \mathfrak{B}, \mathfrak{C}, \ldots]$. This confidence region is elliptical, ellipsoidal or hyperellipsoidal in the case of a linear model but more or less warped in the case of a nonlinear model.

These q-dimensional confidence regions are mentioned in many textbooks on nonlinear models (Bates and Watts, 1988, page 200; Draper and Smith, 1981, page 472; Ross, 1990, page 9) and are often considered as the best summary of relationships between the estimates of all parameters. In practice, however, a priori hypothetical values are seldom available concerning several or all parameters jointly. Moreover, since these q-dimensional confidence regions are difficult to use as such, their exploration is often limited to two-dimensional slices or projections (Bates and Watts, 1988, pages 204-216; Miller, 1994). Finally, when one wishes to find out whether a particular parameter should be kept in a model or could be eliminated, a univariate significance test or confidence interval (section 35.5) generally provides a more direct answer.

Section 35.5: univariate confidence limits of parameters

For several decades, statisticians have obtained approximate univariate confidence limits for the parameters of nonlinear models by treating them as if they were linear and by using the inverse of the *information matrix* as their covariance matrix (Ratkowsky, 1983, page 16; Ross, 1990, page 7). Incidentally, these *"linearized"* confidence limits are still practically the only ones offered in the nonlinear regression computer programs presently available on the market. However, Donaldson and Schnabel (1987) have shown that these linearized limits are often very inaccurate while more accurate limits can be obtained iteratively by using the likelihood-ratio criterion and taking into account the nonlinearity of the fitted model.

The determination of marginal confidence limits for one of the parameters of a nonlinear model has been outlined by Ratkowsky (1990, page 40). The confidence interval $[B_1, B_2]$ of parameter $\mathfrak{B}$, for instance, includes all values of this parameter which satisfy

$$\mathrm{RSS}(\mathfrak{B}) \leq \mathrm{RSS}(A, B, C, \ldots) \times \{1 + [1 \times F_{(1-\alpha;\, 1,\, N-q)}]/(N-q)\},$$

where $\mathrm{RSS}(\mathfrak{B})$ denotes the residual sum of squares obtained by fixing the value of parameter $\mathfrak{B}$ arbitrarily but by estimating all other parameters ($\mathfrak{A}$, $\mathfrak{C}$, ...) according to the least-squares principle. It is noteworthy that the numerator of the variance ratio involved in the univariate confidence interval above has only one degree of freedom, unlike the numerator of the variance ratio in the q-dimensional confidence region (section 35.4), which had q degrees of freedom. It has been shown in a sampling experiment (Jolicoeur, 1995) that these univariate confidence intervals cover the corresponding parameters with a relative frequency very close to the desired confidence level $(1-\alpha)$ even when multiplicative least squares are used (section 35.3) and when parameters are constrained to be nonnegative.

The numerical evaluation of the upper marginal confidence limit of one of the parameters of a nonlinear model having $q=2$ parameters is illustrated in Figure 35.5.1. Each iteration begins with the attribution of a trial value, represented by a vertical arrow, to the parameter of which the limits are sought, and goes on with the iterative minimization of the residual sum of squares with respect to all other parameters, represented by a horizontal arrow. The *starting point* of each iteration may be either the vector $[A, B, C, \ldots]$ of general least-squares estimates (figure 35.5.1, left) or the end point of the preceding iteration (figure 35.5.1, right).

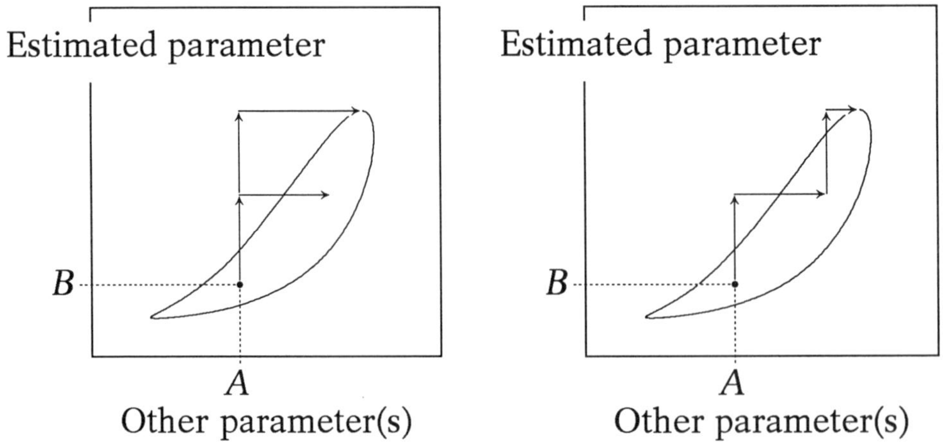

Figure 35.5.1
During the computation of a marginal confidence limit of a parameter ($\mathfrak{B}$ for instance), the *starting point* of each iteration may be either the vector $[A, B]$ of general least-squares estimates (left) or the end point of the preceding iteration (right)

The numerical evaluation of the marginal confidence limits of the parameters of a nonlinear model is difficult and time-consuming because it requires repeated minimizations, of which many converge far from the vector [A, B, C, ...] of general least-squares estimates. It would therefore be tempting to compute conditional confidence limits instead of marginal confidence limits of the parameters, since the computation of conditional confidence limits does not require repeated minimizations and is much faster. Moreover, conditional confidence intervals would be narrower (and thus putatively more accurate) because they take correlations between parameters into account. The numerical evaluation of conditional confidence limits is done in the same manner as that of marginal confidence limits, except that parameters other than the parameter estimated remain equal to specified values, such as their own general least-squares estimates for instance (figure 35.5.2).

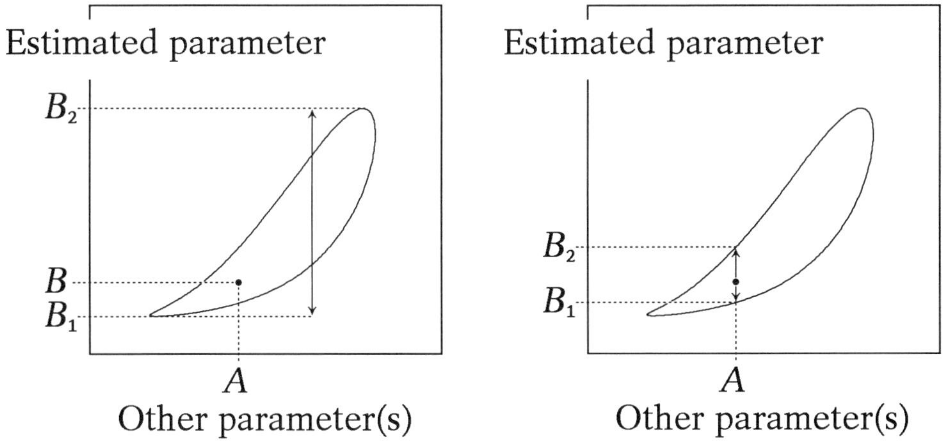

Figure 35.5.2
Parameters other than the estimated parameter are allowed to vary in the case of marginal confidence limits (left), but remain equal to a specified value, such as their own general least-squares estimates, in the case of conditional confidence limits (right)

One might think that conditional confidence intervals of the parameters of a nonlinear model (figure 35.5.2, right), narrower and perhaps more precise than marginal confidence intervals (figure 35.5.2, left), would have the same kind of advantage as estimation (regression) lines (chapter 20) or the T^2 distribution (chapter 30), which similarly take correlations into account. When one must decide whether a parameter should be retained or eliminated from a model, however, marginal confidence limits are preferable since, when a parameter is eliminated, the simplified model is refitted by minimizing the residual sum of squares with respect to the other parameters. In the case of the comparison of nested models (section 35.7), the conclusions drawn from analyses of variance therefore correspond to the indications provided by the marginal confidence limits of the parameters of the full model rather than the conditional limits.

Moreover, because the estimates of parameters are correlated, the conditional confidence limits of one parameter ($\mathfrak{B}$ for instance) are not invariant when the model is reparameterized with respect to another parameter (let us say $\mathfrak{A}$), while marginal confidence limits are unaffected. The conditional confidence limits of one of the parameters of a nonlinear model thus have the serious drawback of not being unique, since they depend on the way in which the other parameters are chosen. In the case of nonlinear models as in that of multiple regression (section 30.7), the correlations between the estimates of parameters should presumably be considered as sampling effects which should not be given biological interpretations. It is noteworthy that, in the case of the coefficients of multiple regressions too, the currently used confidence limits are marginal limits (sections 23.4 and 25.7) rather than conditional limits. The utilization of the marginal confidence limits of the parameters of a nonlinear model will be illustrated in section 35.7 (table 35.7.3).

Section 35.6: variation belts

In this chapter, the estimated extent of individual variation about nonlinear relationships is described by *variation belts.* These variation belts, discussed by Jolicoeur and Heusner (1986), are obtained by substituting the estimated coefficients $[A, B, C, \ldots]$ for the parameters $[\mathfrak{A}, \mathfrak{B}, \mathfrak{C}, \ldots]$ of the model and by using the normal distribution as well as the estimated residual standard deviation $S_{Y.X} = \sqrt{[\mathrm{RSS}/(N-q)]}$, where N and q denote sample size and the number of parameters. The variation belt comprises all values of the predicted variate Y satisfying

$$f(A, B, C, \ldots; X_0) + t_{(\alpha/2;\, \infty)} S_{Y.X} \leq Y \leq f(A, B, C, \ldots; X_0) + t_{(1-\alpha/2;\, \infty)} S_{Y.X}$$

for every particular value X_0 of the predictor variate. When multiplicative least squares are used, as indicated in section 35.3, the limits of the variation belt must be determined with respect to the logarithmically transformed variates at first and then be submitted to the antilogarithmic transformation, which yields

$$f(A, B, C, \ldots; X_0) \times \exp[t_{(\alpha/2;\, \infty)} S_{Y.X}] \leq Y \leq f(A, B, C, \ldots; X_0) \times \exp[t_{(1-\alpha/2;\, \infty)} S_{Y.X}].$$

The variation belt is the set of all variation intervals (section 9.11) of the predicted variate Y corresponding to all possible successive values of the predictor variate.

Section 35.7: analyses of variance of *nested models*

In statistics, the expression *nested models* is applied to models of which some may be considered as particular cases of the others. Let us reconsider the guinea pig data of Terroine and Trautmann (1927) on the oxygen consumption Y in g/kg/h as related to the environmental temperature T in °C (sections 35.1 and 35.3, figures 35.1.1 and 35.3.1). First, let us note that the three-parameter model used in sections 35.1 and 35.3, $Y = A\{1 + [(T-B)/C]^2\}$, can be reexpressed as

$$Y = A\{1 + |(T-B)/C|^2\},$$

since taking the absolute value $|(T-B)/C|$ does not affect the function Y because the square of the quantity $[(T-B)/C]$ is positive anyhow.

The second power was chosen arbitrarily, and one could wonder whether a model involving another value of the exponent might not fit Terroine and Trautmann's guinea pig data better. This question may be answered by comparing the preceding model with a more general one,

$$Y = A\{1+|(T-B)/C|^D\},$$

where the arbitrary exponent 2 is replaced by a fourth parameter $\mathfrak{D}$ (D) to be estimated from the data. It may be said that the two preceding models are *nested*, or that the model $Y=A\{1+|(T-B)/C|^2\}$ is *nested within* the model $Y=A\{1+|(T-B)/C|^D\}$, because the first model can be obtained from the second one by fixing the value of parameter $\mathfrak{D}$ arbitrarily at 2. It may also be said, using the same terms as in section 35.4, that the first model is a *simplified version* of the *full model.*

The point estimates of the parameters and the residual sums of squares obtained through multiplicative least squares for the two models are given in table 35.7.1. The residual sum of squares (RSS) is clearly smaller for the full model (0.313434) than for the simplified model (0.425075).

Table 35.7.1
Multiplicative least-squares estimates of parameters, residual sums of squares (RSS) and standard deviations with two nested models for the guinea pig data of Terroine and Trautmann (1927); the sample size $N = 30$

		Model	
		$Y=A\{1+\|(T-B)/C\|^2\}$	$Y=A\{1+\|(T-B)/C\|^D\}$
Number of parameters		3	4
Parameters	$\mathfrak{A}$ (A)	1.15551 g/kg/h	1.01434 g/kg/h
	$\mathfrak{B}$ (B)	25.6480 °C	25.0373 °C
	$\mathfrak{C}$ (C)	24.6730 °C	20.3544 °C
	$\mathfrak{D}$ (D)	Hypothesized equal to 2.0	1.20387 (dimensionless)
RSS		0.425075	0.313434
Standard deviations		0.125473	0.109796

The comparison of the two models through an approximate analysis of variance is summarized in table 35.7.2.

Table 35.7.2
Comparison through an approximate analysis of variance of two nested models for the guinea pig data of Terroine and Trautmann (1927)

Nature of variation	RSS	Degrees of freedom	Observed variance	Variance ratio	Probability
Complete model	0.313434	$(N-4)=26$	0.0120552		
Extra parameter	0.111641	1	0.111641	$F_{(1, 26)} = 9.26081$	$P = 0.0052986$
Simplified model	0.425075	$(N-3)=27$	0.0157435		

Even though the preceding analysis of variance is approximate, the hypothesis that parameter $\mathfrak{D}$ (corresponding to the estimated coefficient D) is equal to 2 can thus be rejected conclusively, for the increase of the residual sum of squares due to fixing the fourth parameter is too large to be attributed to random. The full model is represented in figure 35.7.1, and the smallness of the residual sum of squares is reflected by the narrowness of variation belts (compare figure 35.7.1 with figure 35.3.1).

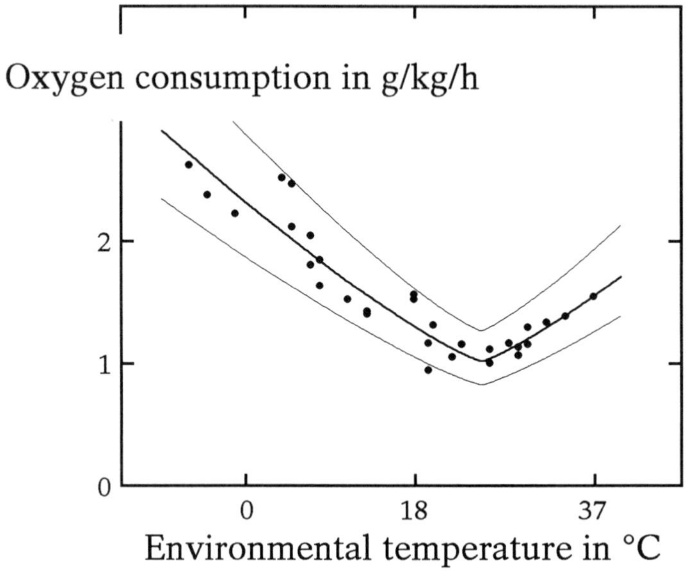

Figure 35.7.1
Relationship between the oxygen consumption Y in g/kg/h of guinea pigs and the environmental temperature T in °C, described by the model $Y=A\{1+|(T-B)/C|^D\}$; variation belt at level $\alpha=0.05$

The conclusion drawn from the comparison of the two models through the analysis of variance agrees with the indications given by the marginal confidence limits of the parameters of the full model (section 35.5 and table 35.7.3). The confidence interval of exponent $\mathfrak{D}$ does not cover 2, not even at significance level $\alpha=0.01$. Using the simplified model in which the exponent equals 2 is thus not justified for the guinea pig data of Terroine and Trautmann (1927). While the confidence interval covers a whole range of likely noninteger values, it does contain the integer value $\mathfrak{D}_0=1.0$. Moreover, the point estimate $D=1.20387$ (table 35.7.1) is not far from unity. Consequently the simplified model

$$Y = A\{1+|(T-B)/C|^1\} = A\{1+|(T-B)/C|\}$$

would be expected to fit Terroine and Trautmann's guinea pig data satisfactorily. Now this model happens to be an instance of the piecewise-linear models discussed in the next section (35.8).

Table 35.7.3
Marginal confidence limits of the model $Y = A\{1 + |(T-B)/C|^D\}$ applied to the guinea pig data of Terroine and Trautmann (1927)

α	0.05		0.01	
Limit	lower	upper	lower	upper
$\mathfrak{A}$ (A)	0.822630	1.14466	0.719393	1.18419
$\mathfrak{B}$ (B)	23.6781	26.8032	23.1990	27.6575
$\mathfrak{C}$ (C)	12.4947	26.1771	8.36807	28.3152
$\mathfrak{D}$ (D)	0.790053	1.70856	0.656723	1.91784

In general, when one wishes to simplify a nonlinear model, the parameters which one could consider eliminating are additive constants of which the confidence intervals cover 0 as well as multiplicative constants and exponents of which the confidence intervals cover unity. However, one should refrain from eliminating parameters which are necessary to maintain the dimensional homogeneity of a model (Bridgman, 1931). Thus, in the model

$$Y = A\{1 + |(T-B)/C|\},$$

the parameters $\mathfrak{A}(A)$ and $\mathfrak{C}(C)$ should be retained even if their confidence intervals covered unity, for $\mathfrak{A}(A)$ is necessary to ensure that the right member of the equation is expressed in the same unit (g/kg/h) as the predicted variate Y, while $\mathfrak{C}(C)$ is necessary to make sure that the value of the ratio $(T-B)/C$ remains unchanged if the temperature T is measured in tenths of degrees instead of degrees. However, one could consider eliminating parameter $\mathfrak{B}$ (B) if its confidence interval covered 0, because this would not impair the dimensional homogeneity of the equation. Nevertheless, the nullity of parameter $\mathfrak{B}$ (B) would imply that the average oxygen consumption of guinea pigs would be minimum when the environmental temperature $T = 0°C$, an hypothesis which is neither physiologically realistic nor acceptable according to the confidence intervals based on Terroine and Trautmann's data (table 35.7.3).

In practice, new users of nonlinear models should avoid beginning an analysis by using a complicated model with the intention of eliminating superfluous parameters later, for the point and interval estimation of the parameters of an excessively complicated nonlinear model is numerically difficult or impossible. The difficulties are analogous to those occurring in multiple regression when there are too many highly intercorrelated predictor variates (section 25.9), but the problems are even worse for nonlinear models. It is therefore advisable to begin using a reasonably simple model, in accord with the *principle of parsimony* (see also section 9.2), and to check subsequently whether the observed relationship could be described better through a slightly more complex model.

Section 35.8: segmented models (*"piecewise models"*)

The simplest and best-known segmented models are the so-called *piecewise-linear models,* which are made up of consecutive straight-line segments, the end point of each segment coinciding generally with the starting point of the next segment. However, all or part of the segments of a *piecewise* model may also be curvilinear (see section 35.15).

For several decades, breaking down an overall relationship into successive segments has not been done much in biology, undoubtedly because computers have facilitated the fitting of complex curvilinear relationships, but also perhaps in part because it may be philosophically more satisfying to describe natural phenomena in a gradual and unified manner. Some biological relationships nevertheless appear to show that sudden changes may occur at some values of the predictor variate, and segmented models may then be suitable. Moreover, Prunty (1983) has proposed using smooth functions which are *piecewise-linear in the limit,* and in which additional parameters allow curved transitions from one segment to the next.

The model $Y = A\{1 + |(T - B)/C|\}$, mentioned in the preceding section, is a particularly simple example of piecewise-linear model. The results obtained by fitting this model through multiplicative least squares to the data of Terroine and Trautmann (1927) on the oxygen consumption of guinea pigs and environmental temperature is illustrated in figure 35.8.1. Even though this model possesses one parameter less than the model $Y = A\{1 + |(T - B)/C|^D\}$, in which it is nested, its residual sum of squares (RSS) is barely higher, and its residual standard deviation is slightly lower (table 35.8.1).

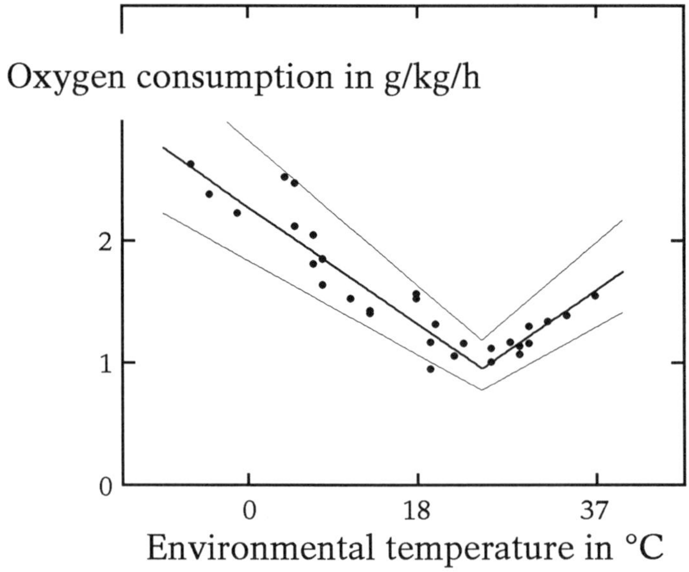

Figure 35.8.1
Piecewise-linear model $Y = A\{1 + |(T - B)/C|\}$ fitted through multiplicative least squares to the relationship between the oxygen consumption Y in g/kg/h of guinea pigs and the environmental temperature T in °C; variation belt (thin lines) at level $\alpha = 0.05$

Moreover, the marginal confidence intervals obtained for the parameters $\mathfrak{A}$ (A), $\mathfrak{B}$ (B) and $\mathfrak{C}$ (C) are somewhat narrower (tables 35.7.3 and 35.8.2) for the simplified model $Y = A\{1 + |(T - B)/C|\}$ than for the complete model $Y = A\{1 + |(T - B)/C|^D\}$. The parameters of the simpler model can therefore be estimated with more accuracy than those of the full model.

Table 35.8.1
Multiplicative least-squares estimates of parameters, residual sums of squares (RSS) and standard deviations with two nested models, including the piecewise-linear model $Y = A\{1 + |(T - B)/C|^1\}$, for the guinea pig data of Terroine and Trautmann (1927); the sample size $N = 30$

		Model	
		$Y = A\{1 + \|(T - B)/C\|^1\}$	$Y = A\{1 + \|(T - B)/C\|^D\}$
Number of parameters		3	4
Parameters	𝔄 (A)	0.946898 g/kg/h	1.01434 g/kg/h
	𝔅 (B)	24.9392 °C	25.0373 °C
	ℭ (C)	17.8458 °C	20.3544 °C
	𝔇 (D)	Hypothesized equal to 1.0	1.20387 (dimensionless)
RSS		0.324820	0.313434
Standard deviations		0.109683	0.109796

To sum up, even though the model $Y = A\{1 + |(T - B)/C|\}$ implies that the oxygen consumption Y of guinea pigs stops decreasing and suddenly starts increasing when the environmental temperature T crosses the point of thermal neutrality B, this simplified model appears to provide a satisfactory description of the phenomenon.

Table 35.8.2
Marginal confidence limits of the parameters of the model $Y = A\{1 + |(T - B)/C|\}$ applied to the guinea pig data of Terroine and Trautmann (1927)

α	0.05		0.01	
Limit	lower	upper	lower	upper
𝔄 (A)	0.858597	1.03965	0.828676	1.07325
𝔅 (B)	23.8056	26.4493	23.4285	27.0800
ℭ (C)	13.8781	23.5691	12.7638	26.2167

A strikingly different relationship is found (figure 35.8.2, next page) between the oxygen consumption Y and the environmental temperature T in another mammal studied by Terroine and Trautmann (1927), the rabbit. Unlike the guinea pig, of which the oxygen consumption rises as soon as the environmental temperature deviates from the point of thermal neutrality, the rabbit keeps its oxygen consumption practically constant within a broad temperature range going approximately from − 4°C to +33°C ! A model comprising 3 segments is therefore appropriate in this case.

The five-parameter model $Y = A\{1 + \max[0, (B - T)/C] + \max[0, (T - D)/E]\}$ has been used here because the slope of the third segment (figure 35.8.2, right) has a visibly larger absolute value than the slope of the first segment (figure 35.8.2, left). The two "max" functions allow the physiological reactions of the rabbit to be treated differently at the two ends of the *range of thermal neutrality*. Multiplicative least-squares estimates of the five parameters are given in table 35.8.3 (next page).

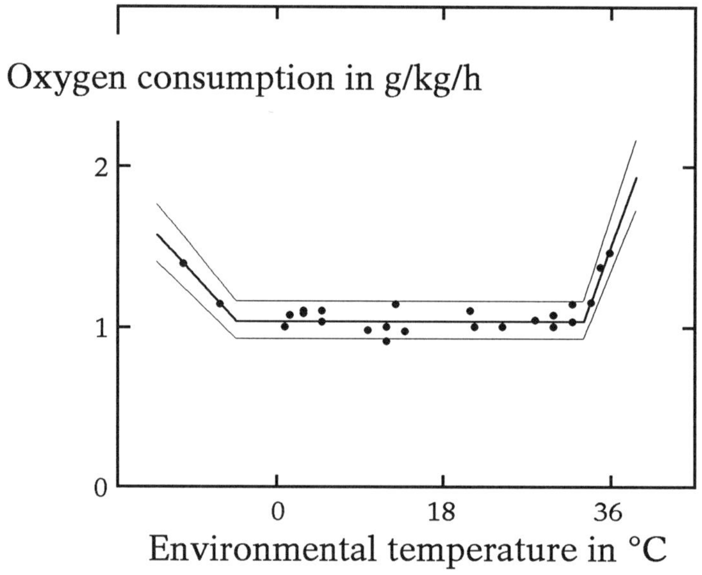

Figure 35.8.2
Piecewise-linear model $Y = A\{1 + \max[0, (B - T)/C] + \max[0, (T - D)/E]\}$ fitted to the relationship between the oxygen consumption Y in g/kg/h of rabbits and the environmental temperature T in °C according to data from Terroine and Trautmann (1927); variation belt (thin lines) at level $\alpha = 0.05$

The average minimum oxygen consumption of the rabbit would be doubled if the environmental temperature reached $C = 16.5291$ °C below the lower limit $B = -4.28907$ °C or $E = 6.45260$ °C above the upper limit $D = 33.1671$ °C of the range of thermal neutrality. The oxygen consumption increases more than twice as fast when the environmental temperature rises above $D = 33.1671$ °C as when it decreases below $B = -4.28907$ °C, because the rabbit must fight more actively against high temperatures, which could rapidly be lethal.

Table 35.8.3
Multiplicative least-squares estimates of parameters, residual sum of squares (RSS) and standard deviation for a piecewise-linear model of the rabbit data of Terroine and Trautmann (1927), $Y = A\{1 + \max[0, (B - T)/C] + \max[0, (T - D)/E]\}$; the sample size $N = 26$

Parameters	$\mathfrak{A}$ (A)	1.03307 g/kg/h
	$\mathfrak{B}$ (B)	−4.28907 °C
	$\mathfrak{C}$ (C)	16.5291 °C
	$\mathfrak{D}$ (D)	33.1671 °C
	$\mathfrak{E}$ (E)	6.45260 °C
RSS		0.0704539
Residual standard deviation		0.0579219

Numerical difficulties unfortunately occur if attempts are made to compute the confidence limits of this five-parameter model, because Terroine and Trautmann (1927) obtained too few observations at very low as well as at very high temperatures. Scrutiny of the scatter diagram (figure 35.8.2) shows indeed that there are only two observations in the zone of the first segment (at left) and only three in the zone of the third segment (at right). This example illustrates the importance, when one tries to describe a quantitative relationship, to have enough data for all parts of the relationship.

The occurrence of two very different relationships (figures 35.8.1 and 35.8.2) for the same kind of phenomenon in two species belonging to the same class raises an important question. The guinea pig and the rabbit [at least of the variety studied by Terroine and Trautmann (1927)] obviously have very dissimilar strategies with respect to thermal acclimatization. Like many other small mammals, the guinea pig does not have a strongly insulating fur and escapes very low and very high environmental temperatures by seeking refuge in its burrow. Nevertheless, as soon as the external temperature deviates from the optimum, the guinea pig must fight to prevent its body temperature from increasing or decreasing. On the contrary, the rabbit has a strongly insulating fur and can remain exposed to cold winter temperatures many hours without raising its metabolic activities. When the environmental temperature is high, however, the rabbit must cool down, which it does thanks to its large ears, which are strongly vascularized and sparsely furred. Moreover, the rabbit gets rid of its thick fur coat just before the warm season by molting!

From a methodological viewpoint, the necessity of using two different models to describe the same kind of phenomenon in two species belonging to the same class shows that, in order to choose a statistical model correctly, a biometrician must mandatorily have a good knowledge and a good understanding of the biological context. Even the way in which a statistical model is parameterized is biologically important. Thus, in a segmented model, the coordinates of the join-points between successive segments may have major interest, as for instance the *point of thermal neutrality* in guinea pigs or the limits of the *range of thermal neutrality* in rabbits. Ratkowsky's claim (1983) that nonlinear models should be parameterized mostly according to statistical criteria is therefore not completely acceptable.

Section 35.9: optimization problems

In living organisms, optimization consists basically either in minimizing energy expenditures or in maximizing returns, such as somatic (bodily) growth or reproduction (see also Ferron, 1987, concerning the optimization of behavior). The relationship between oxygen consumption and environmental temperature, considered in the preceding sections of this chapter, is a good example of minimization, since oxygen consumption is related to the metabolic activities and energy expenditures of living organisms. The second kind of optimization, according to which the production of biomass is maximized, is illustrated in figure 35.9.1 (next page). The average weight at the age of 66 days of two plants of clover (*Trifolium subterraneum* L.) of the *Edenhope* variety is to be maximized with respect to the zinc concentration of the applied fertilizer. The data come from the study of Millikan (1963) already cited in sections 28.1 and 28.7.

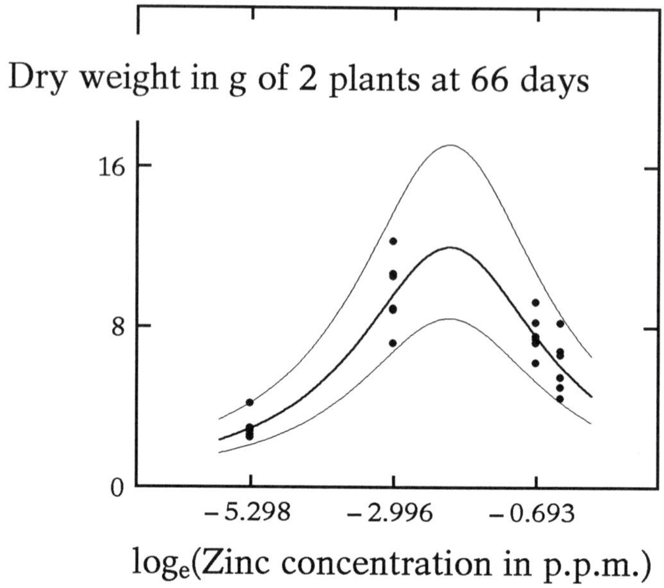

Figure 35.9.1
Maximisation of the dry weight in g of two plants of clover (*Trifolium subterraneum* L.) with respect to the natural logarithm X of the zinc concentration in p.p.m., described using the model $Y = A/\{1+[(X-B)/C]^2\}$; variation belt at level $\alpha = 0.05$; data from Millikan (1963)

Because the first three zinc concentrations form a geometric progression, ($Zn_1 = 0.005$ p.p.m., $Zn_2 = 0.05$ p.p.m., $Zn_3 = 0.5$ p.p.m. and $Zn_4 = 0.75$ p.p.m.), the natural logarithm X of zinc concentration has been used in order to space the values of the predictor variate as regularly as possible. The fertilizer also contained 15 p.p.m. of phosphorus. The maximization model applied here,

$$Y = A/\{1+[(X-B)/C]^2\},$$

is closely related to the nonlinear minimization model fitted in sections 35.1 and 35.3, except that the coefficient A is divided instead of being multiplied by the expression $\{1+[(X-B)/C]^2\}$. Multiplicative least-squares estimates of the three parameters are given in table 35.9.1.

Table 35.9.1
Multiplicative least-squares estimates of parameters, residual sum of squares (RSS) and standard deviation for the model $Y = A/\{1+[(X-B)/C]^2\}$; clover data from Millikan (1963); the sample size $N = 24$

Parameters	$\mathfrak{A}$ (A)	11.9711 g
	$\mathfrak{B}$ (B)	- 2.09355
	$\mathfrak{C}$ (C)	1.81214
RSS		0.693860
Residual standard deviation		0.181772

The point estimates of thc parameters thus indicate that the average maximum weight at 66 days of two plants of clover of the Edenhope variety is

$$A = 11.9711 \text{ grams},$$

that this maximum weight is reached when the zinc concentration of the fertilizer is

$$\exp(B) = \exp(-2.09355) = 0.123249 \text{ p.p.m},$$

and that the average weight is divided by half and equal to $A/2 = 5.98555$ grams when the zinc concentration of the fertilizer is

$$\exp(1.81214) = 6.12354 \text{ times}$$

higher or lower than the optimum concentration (0.123249 p.p.m.). The present model is a simple example of a *response curve.* In order to describe the effects of zinc and phosphorus simultaneously, a more complex model involving two predictor variates and corresponding to a *response surface* would have to be used (section 28.7; Box, Hunter and Hunter, 1978, chapter 15; Lebreton, Millier, Jolivet, Pavé and Vila, 1982, chapter 5; Tomassone, Dervin and Masson, 1993, pages 279-288).

Section 35.10: the growth of biological populations

After the exponential relationship (section 20.6), the best-known and most widely used nonlinear model for the growth of biological populations is undoubtedly the *logistic* (also called *autocatalytic*) curve, originally conceived in 1838 by the Belgian mathematician Verhulst. The equation of this curve may be written as follows:

$$Y = A/\{1 + \exp[(B - t)/C]\},$$

where Y denotes the number of individuals in the population at time t;

A is the upper asymptote of the curve, the average maximum (expected) number of individuals in the population when the time $t = +\infty$; in ecology, the upper asymptote A is usually interpreted as the *carrying capacity* of the environment;

B is the abscissa of the inflection point $[t, Y] = [B, A/2]$ of the curve; it is the time at which the number Y of individuals is half the *carrying capacity;*

and C determines the time $t = (B - C)$ at which the number Y of individuals is still only equal to $A/(1 + e) = 0.268941\,A$; in population ecology, the inverse $r = 1/C$ of coefficient C is called the *intrinsic rate of natural increase* of a population.

The three coefficients A, B are C are positive in the case of growth, but C would be negative in the case of a logistic decrease (section 21.4). The logistic curve also has a lower asymptote, equal to 0 at time $t = -\infty$. The derivative of the logistic function with respect to time t may be expressed as

$$dY/dt = Y \times (1/C) \times [1 - (Y/A)],$$

which shows that the rate of increase of the number Y of individuals equals the product of a factor proportional to present population size, $Y \times (1/C)$, by a so-called *braking function,* $[1 - (Y/A)]$, which is equal to unity at first but decreases toward 0 as population size Y increases toward the upper asymptote A.

Very similar to exponential growth (section 20.6) at the beginning, logistic growth thus differs more and more from it and slows down as the number of individuals gets near the carrying capacity of the environment, as reflected by the *sigmoid* shape (approximately similar to an elongated and slanted S) of the corresponding curve. Because the logistic curve simulates the nutritional and spatial constraints faced by living organisms as they grow in size and in numbers, it provides a more realistic model of biological growth than the exponential curve. Figure 35.10.1 illustrates a logistic curve fitted through multiplicative least squares to data published in a classical work by the Russian ecologist Gause (1934). The original data are included in "DATA" statements at the end of appendix B, and the point and interval estimates of the parameters have already been published (Jolicoeur, 1995).

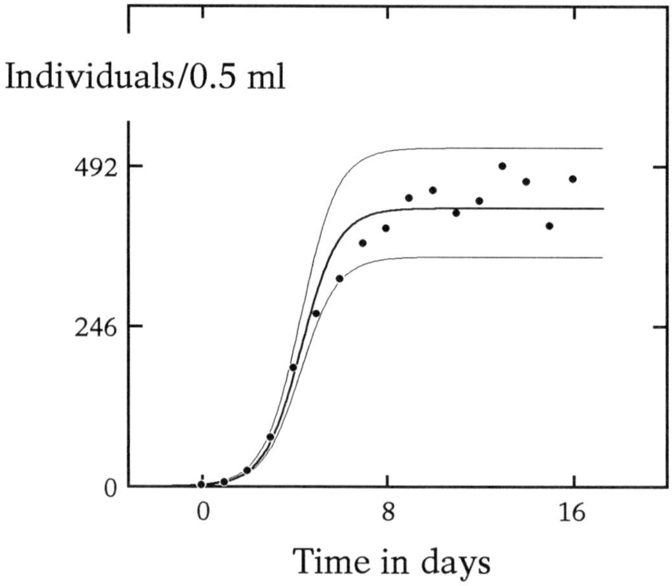

Figure 35.10.1
Growth of an experimental population of the species *Paramecium aurelia* described using the logistic curve fitted through multiplicative least squares; data from Gause (1934); variation belt at level $\alpha = 0.05$

The logistic curve is just one among many models which can be applied to population growth, including some models where time is treated as a discrete variable and where matrix algebra is used (one of the early papers in that vein being that of Leslie, 1945). The study of those models would require several chapters or even a whole monograph and is outside of the scope of the present textbook. Interested readers should consult works on population dynamics.

Section 35.11: somatic (bodily) growth

Applied at first to the growth of biological populations (section 35.10), the logistic curve has also been used to describe the somatic (bodily) growth of individual organisms. Similarly, the curve of Gompertz (section 35.12), initially proposed in 1825 to facilitate the construction of human survival tables, has also been applied later to somatic growth (Winsor, 1932). However, the logistic curve and the Gompertz curve both possess a lower asymptote in addition to their upper asymptote, which means that their ordinate is only null at time $t = -\infty$ and that they are less suitable for somatic than for population growth. In fact, while every biological population goes back ultimately to remote ancestors, each individual organism only begins precisely at the time of fecundation. Moreover, excepting reserve matters, the fertilized egg is generally much smaller than the adult, with the result that a somatic growth curve should practically have a null ordinate at time 0, and pass approximately through the origin. Finally, since it is not always easy to get adequate numbers of very young organisms in growth studies, curves having a lower asymptote often tend to pass far from the origin, a deficiency against which curves passing through the origin, like those of Hill (1913) and Pütter (1924), are evidently protected.

Proposed initially in 1913 by the British physiologist Hill to model the saturation of haemoglobin by oxygen and by carbon monoxide, Hill's curve may also be used (Jolicoeur, 1985) to describe the growth with time t of a somatic dimension Y :

$$Y = A/[1+(C/t)^B],$$

where $Y =$ the somatic dimension of the individual organism at time t,

$A =$ the upper asymptote of the curve, that is the average maximum value of the dimension Y at time $t = +\infty$;

$B =$ a dimensionless number determining the shape of the curve;

and $C =$ the time at which the somatic dimension $Y = A/2$.

Even though Hill's curve has only three parameters, which are all positive, it may be either sigmoid (when $B > 1$) or nonsigmoid (when $B \leq 1$), and it has an inflection point of which the height $(B-1)A/(2B)$ varies from 0 when $B \leq 1$ to $(A/2)$ when $B = +\infty$. Hill's curve is thus more flexible than the logistic and Gompertz curves, of which the inflection points have heights equal to fixed percentages of the upper asymptote A. The statement of Winsor (1932), according to which a curve should have at least four parameters in order to have a flexible shape, is thus incorrect.

Somatic growth may also be analyzed using a modified version (Jolicoeur, 1985; Lebeau, Jolicoeur, Pageau and Crossman, 1986),

$$Y = A[1-\exp(-t/C)]^B,$$

of a curve originally proposed by Pütter (1924) but subsequently modified (section 35.14) and popularized by Bertalanffy (1951). Similarly to Hill's curve, Pütter's curve may be either sigmoid (when $B > 1$) or nonsigmoid (when $B \leq 1$) and has an inflection point of which the height $A[(B-1)/B]^B$ varies from 0 to (A/e). In contrast with Hill's curve, Pütter's curve has the advantage of being algebraically compatible with the presence of simple allometry between several somatic dimensions (sections 20.7, 31.6 and 35.12), but it has the disadvantage of being difficult to evaluate with accuracy when t is small and C is large, because the numerical value of the expression $\exp(-t/C)$ is then very close to unity from which it is subtracted.

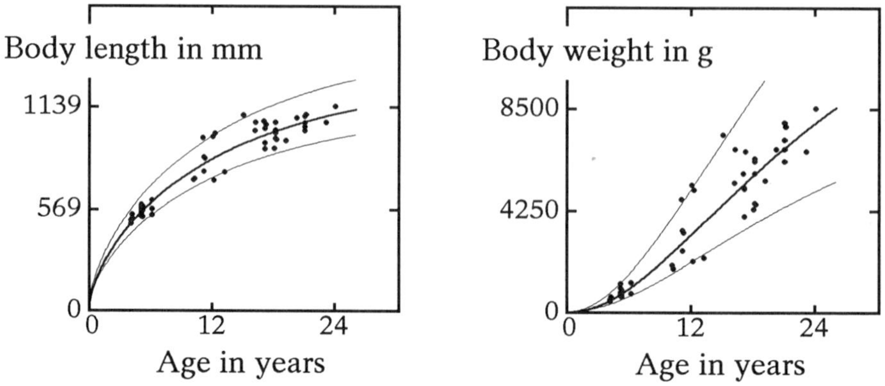

Figure 35.11.1
Growth in length and in weight of 59 female muskellunges (*Esox masquinongy*), described using the curve $Y = A[1 - \exp(-t/C)]^B$ fitted through multiplicative least squares to data from Lebeau (1984); variation belts at level $\alpha = 0.05$

However, numerical inaccuracies are seldom a problem when computations are done with high precision and, when Pütter's curve fits data satisfactorily, it is particularly interesting because of its algebraic compatibility with simple allometry. Figure 35.11.1 shows Pütter's curve fitted through multiplicative least squares to the growth in length and weight of 59 females of the muskellunge (*Esox masquinongy*), an especially large species of North-American pike. The data were kindly provided by Dr. Bernard Lebeau (1984), an ichthyologist. The point estimates of parameters given in table 35.11.1 differ slightly from those published by Lebeau, Jolicoeur, Pageau and Crossman (1986), because the latter were obtained under the assumption that the $\mathfrak{C}(C)$ parameter had the same numerical value for all variates. Marginal confidence limits are given in table 35.11.2 for body length and in table 35.11.3 for body weight. In the case of body length (table 35.11.2) as well as body weight (table 35.11.3), the upper confidence limits of the asymptote $\mathfrak{A}(A)$ and of the time-scale parameter $\mathfrak{C}(C)$ are finite at significance level $\alpha = 0.05$ but infinite at $\alpha = 0.01$.

Table 35.11.1
Multiplicative least-squares estimates of parameters, residual sums of squares (RSS) and standard deviations for the model $Y = A[1 - \exp(-t/C)]^B$ fitted to the body length in cm and body weight in kg of 59 female muskellunges, according to data from Lebeau (1984)

		Somatic dimension	
		body length in cm	body weight in kg
Parameters	$\mathfrak{A}(A)$	136.788 cm	17.9303 kg
	$\mathfrak{B}(B)$	0.490988 (dimensionless)	1.78395 (dimensionless)
	$\mathfrak{C}(C)$	8.44025 years	6.96365 years
RSS		0.104958	1.94658
Standard deviations		0.0432926	0.186441

Table 35.11.2

Marginal confidence limits of the parameters of the model $Y = A[1 - \exp(-t/C)]^B$ applied to the body length in cm of 59 female muskellunges, according to data from Lebeau (1984)

α	0.05		0.01	
Limit	lower	upper	lower	upper
$\mathfrak{A}(A)$	128.053	179.687	126.193	$+\infty$
$\mathfrak{B}(B)$	0.281213	0.836228	0.231133	1.00096
$\mathfrak{C}(C)$	4.76097	45.458	4.14007	$+\infty$

The hypothesis that the muskellunge has an unlimited growth pattern can thus be rejected at significance level $\alpha = 0.05$ but not at the more rigorous $\alpha = 0.01$ level. As for the sigmoid pattern of growth in body weight, it is not established beyond all doubt, since the confidence interval of the exponent $\mathfrak{B}(B)$ covers unity at significance level $\alpha = 0.05$ as well as at $\alpha = 0.01$ (table 35.11.3). In conclusion, Bernard Lebeau's data supply provisional evidence ($\alpha = 0.05$) that the growth pattern of the muskellunge is limited (asymptotic), but additional data would be necessary to make this conclusion more definitive.

Table 35.11.3

Marginal confidence limits of the parameters of the model $Y = A[1 - \exp(-t/C)]^B$ applied to the body weight in kg of 59 female muskellunges, according to data from Lebeau (1984)

α	0.05		0.01	
Limit	lower	upper	lower	upper
$\mathfrak{A}(A)$	14.4950	62.3438	13.8108	$+\infty$
$\mathfrak{B}(B)$	0.835359	3.67404	0.692086	4.71536
$\mathfrak{C}(C)$	3.72545	75.6338	3.19469	$+\infty$

The reader should note that, theoretically, it may not be impossible for some living organisms to possess an unlimited growth pattern because, even if such organisms keep growing until they die, their death rates do not allow them to go on forever. Unlimited growth patterns may often be described satisfactorily by using a simple power function of time,

$$Y = At^B ,$$

where A denotes the value of dimension Y when time $t = 1$. This simple two-parameter nonasymptotic chronological growth model could be referred to as *time allometry*. The analysis by Jolicoeur (1985) and by Jolicoeur and Heusner (1986) of the data of Laws (1953) on the southern elephant seal (*Mirounga leonina* L.) suggests that the males of this species may possess an unlimited growth pattern, natural selection favoring large males able to secure a harem. Female elephant seals, however, only reach a modest size and have a clearly limited growth pattern. Somatic growth is clearly limited in many bird and mammal species, including man (section 35.13). The growth of plants has been studied by many workers, including Causton and Venus (1981) and Hunt (1982).

Section 35.12: chronological growth and multivariate allometry

Research workers interested in *morphometrics,* that is in the quantitative study of animal and plant morphology, have long wondered whether models of somatic growth with respect to time (also called *chronological growth,* section 35.11) are compatible with the presence of simple allometry (sections 20.7 and 31.6) between the various body dimensions of living organisms belonging to the same or to different species. The question has been reviewed by Lebeau, Jolicoeur, Pageau and Crossman (1986) and by Jean (1987, chapter 5). When the growth pattern of several somatic dimensions $[Y_1, \ldots, Y_i, Y_j, \ldots Y_q]$ is unlimited and may be described by power functions of time t,

$$Y_1 = A_1 t^{B_1}, \ldots, Y_i = A_i t^{B_i}, Y_j = A_j t^{B_j}, \ldots, Y_q = A_q t^{B_q},$$

these q two-dimensional equations may be rearranged into a single $(q+1)$-dimensional equation,

$$(Y_1/A_1)^{1/B_1} = \ldots = (Y_i/A_i)^{1/B_i} = (Y_j/A_j)^{1/B_j} = \ldots = (Y_q/A_q)^{1/B_q} = t\,.$$

The coefficients $[A_1, \ldots, A_i, A_j, \ldots A_q]$ above denote the numerical values of the somatic dimensions $[Y_1, \ldots, Y_i, Y_j, \ldots Y_q]$ when time $t=1$. However, each member of the preceding equation (including time t) can be made dimensionless if it is divided by its geometric mean:

$$(Y_1/G_1)^{1/B_1} = \ldots = (Y_i/G_i)^{1/B_i} = (Y_j/G_j)^{1/B_j} = \ldots = (Y_q/G_q)^{1/B_q} = t/G_t\,.$$

The first q members of the latter are analogous to the multivariate generalization of the allometry equation (section 31.6), and pairing any two of these members yields a power function of which the exponent (B_j/B_i) is equal to the ratio of the allometry exponents with respect to time of the two somatic dimensions involved,

$$Y_j = A_j (Y_i/A_i)^{B_j/B_i} = G_j (Y_i/G_i)^{B_j/B_i}\,.$$

An unlimited chronological growth pattern according to which each somatic dimension is equal to a power function of age is therefore fully compatible algebraically with the presence of simple allometry between these dimensions, and the same would be true if somatic growth with respect to time were exponential (section 20.6).

Relationships between simple allometry and limited (asymptotic) sigmoid patterns of somatic growth are less evident. Lumer (1937) claimed that simple allometry is not fully compatible with sigmoid growth, but Kavanagh and Richards (1942) noted that somatic dimensions may be allometric with respect to each other while being related to time through sigmoid functions which, however, may not necessarily be identical in form. When several somatic dimensions are considered simultaneously, the Gompertz curve,

$$Y_i = A_i \exp\{-\exp[(B_i - t)/C_i]\}\,,$$

where A_i denotes the upper asymptote, may be expressed in multidimensional form,

$$\ldots = B_i - C_i \log_e[\log_e(A_i/Y_i)] = B_j - C_j \log_e[\log_e(A_j/Y_j)] = \ldots = t\,,$$

but the value of each the first q members of the latter is only defined if the corresponding variate does not exceed its asymptote: $Y_i \leq A_i$. Pairing any two of the first q members of this equation shows that

$$Y_j = A_j \exp\{-\exp[(B_j - B_i)/C_j)] \times [\log_e(A_i/Y_i)]^{(C_i/C_j)}\}.$$

For all somatic dimensions to be related to each other through simple allometry, the time-scale coefficients $[C_1, \ldots, C_i, C_j, \ldots C_q]$ must all have a common value C, for one then has a power function,

$$Y_j = A_j(Y_i/A_i)^{\exp[(B_j - B_i)/C]}.$$

Similarly, Pütter's curve (section 35.11),

$$Y_i = A_i[1 - \exp(-t/C_i)]^{B_i},$$

may be put in multidimensional form,

$$\ldots = -C_i \log_e[1 - (Y_i/A_i)^{1/B_i}] = -C_j \log_e[1 - (Y_j/A_j)^{1/B_j}] = \ldots = t,$$

and pairing the ith and j^{th} member shows that

$$Y_j = A_j\{1 - [1 - (Y_i/A_i)^{1/B_i}]^{(C_i/C_j)}\}^{B_j}.$$

Here also, for all of these functions to be simple power functions, all time-scale coefficients $[C_1, \ldots, C_i, C_j, \ldots C_q]$ must be equal to a common value C, since one then has

$$Y_j = A_j(Y_i/A_i)^{(B_j/B_i)},$$

and the multidimensional version of the curve may then be expressed more simply as

$$(Y_1/A_1)^{1/B_1} = \ldots = (Y_i/A_i)^{1/B_i} = (Y_j/A_j)^{1/B_j} = \ldots = (Y_q/A_q)^{1/B_q} = [1 - \exp(-t/C)].$$

In conclusion, some asymptotic growth models, the curves of Gompertz and Pütter in particular, are algebraically compatible with the presence of simple allometry between somatic dimensions provided the time-scale coefficients of these dimensions are equal to each other. In the case of the body length, body weight and egg number of female muskellunges, Lebeau, Jolicoeur, Pageau and Crossman (1986) found that the estimates of allometry exponents obtained from the above multivariate version of Pütter's curve were approximately similar to those obtained from principal component analysis (section 31.6) or from ordinary bivariate estimation (regression) lines (section 20.7).

Section 35.13: human growth and growth spurts

The somatic growth models discussed in the preceding sections are not satisfactory for human stature and body weight, because of the occurrence of the *pubertal growth spurt*. After slowing down at the end of early childhood, human growth in stature and body weight speeds up at the time of puberty, in part because of the production of growth hormone by the hypophysis (also known as the pituitary gland). In order to adequately reflect the prepubertal slowdown and the pubertal acceleration of human growth, a growth model must have at least two phases.

Proposed a couple of decades ago, the two best-known and most widely used models of human growth are perhaps the triple logistic curve of Bock and Thissen (1976) and the so-called "No. 2" curve of Preece and Baines (1978). Even though a simplified version possessing only eight parameters is sometimes used, the triple logistic curve involves basically nine parameters and is obtained by summing up three logistic functions,

$$Y = A_1/\{1+\exp[(B_1-t)/C_1]\}+A_2/\{1+\exp[(B_2-t)/C_2]\}+A_3/\{1+\exp[(B_3-t)/C_3]\}\,,$$

while the curve of Preece and Baines (1978),

$$Y = A-2(A-B)/\{\exp[(t-D)/C_1]+\exp[(t-D)/C_2]\}\,,$$

has only five parameters, which makes it easier to fit. In principle, the triple logistic curve of Bock and Thissen (1976) and the curve of Preece and Baines (1978) suffer from the disadvantage that they do not pass through the origin, and they tend not to provide a good fit to data obtained on very young children. In order to solve this problem, Jolicoeur, Pontier, Pernin and Sempé (1988) proposed a new human growth model involving seven parameters and currently referred to by the acronym JPPS:

$$Y = A\{1-1/[(t/C_1)^{B_1}+(t/C_2)^{B_2}+(t/C_3)^{B_3}]\}^k.$$

The JPPS model passes through the origin $[t, Y]=[0, 0]$ and is expressed as a function of total age (measured from the time of fecundation), which may be estimated if necessary by adding the average duration of pregnancy (39 weeks $=0.75$ year) to ordinary postnatal age. The coefficient k is given the value $k=1$ when the model is applied to stature and $k=3$ in the case of body weight. A is an estimate of adult size. Unlike the model of Preece and Baines (1978), the JPPS model fits very young children as satisfactorily as older ones. Moreover, since it involves only seven parameters, it is easier and faster to fit than the triple logistic curve of Bock and Thissen (1976). The coefficients of the JPPS model must satisfy the constraints $A>0$, $B_i\geqq 0$, $C_i>0$, which is ensured by applying the simplex method (section 35.2) of Nelder and Mead (1965) to the logarithms of the coefficients instead of the coefficients themselves.

In a later work, Jolicoeur, Pontier and Abidi (1992) have proposed a more elaborate version of the JPPS model, referred to by the acronym JPA-2, which is expressed as a function of (ordinary) postnatal age and in which the constraint of passing through the origin is relaxed by introducing an eighth parameter $\mathfrak{D}(D)\geqq 0$:

$$Y = A\{1-1/[[(t+D)/C_1]^{B_1}+[(t+D)/C_2]^{B_2}+[(t+D)/C_3]^{B_3}]\}^k.$$

The eighth parameter $\mathfrak{D}(D)$ of model JPA-2 denotes the estimated duration of prenatal development. In fact, the duration of pregnancy is not always equal to its average length of 39 weeks. Moreover, the development of some characters may progress actively only once a certain preliminary organization (circulatory system, etc.) is in place and allows the fetus to profit fully from the nutrients absorbed through the placenta. Finally, estimating parameter $\mathfrak{D}(D)$ from the data compensates automatically for any error or uncertainty concerning the exact date of fecundation. Figure 35.13.1 illustrates model JPA-2 fitted to data on 13 boys from the French auxological survey (see appendix G). Multiplicative least-squares estimates are given in table 35.13.1.

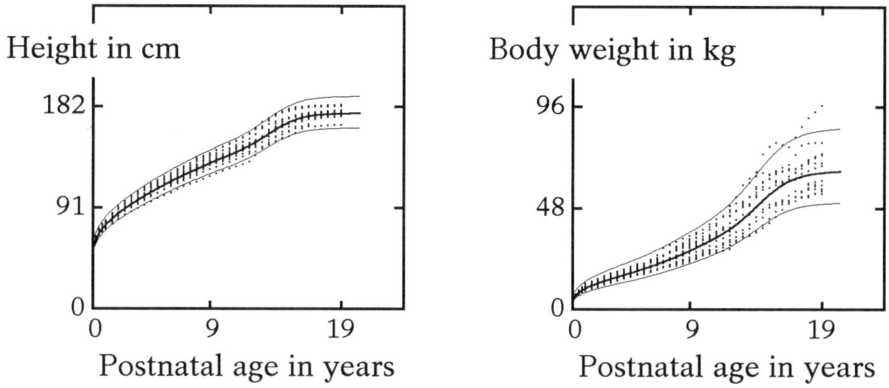

Figure 35.13.1
Cross-sectional growth of the height (left) and body weight (right) of 13 boys from the French auxological survey (Jolicoeur, Pontier, Pernin and Sempé, 1988); the JPA-2 growth curve (Jolicoeur, Pontier and Abidi, 1992; Jolicoeur and Pontier, 1993), $Y = A\{1 - 1/[[(t+D)/C_1]^{B_1} + [(t+D)/C_2]^{B_2} + [(t+D)/C_3]^{B_3}]\}^k$, is fitted through multiplicative least squares; variation belts at level $\alpha = 0.05$

The data used here are longitudinal, which means that a whole series of consecutive measurements (41 measurements from the age of 4 weeks to 19 years, in this case) have been taken on each child. Moreover, the growth models discussed in this section were originally conceived for the analysis of longitudinal data. Nevertheless, the present analysis has been carried out as if the data were cross-sectional: the association between the measurements obtained on the same child has been disregarded, in order to check whether model JPA-2 works satisfactorily also when a cross-sectional study is simulated (Jolicoeur and Pontier, 1993).

Table 35.13.1
Multiplicative least-squares estimates of parameters, residual sums of squares (RSS) and standard deviations for the JPA-2 growth curve fitted to 13 boys, the data being treated as if they were cross-sectional

		height in cm	body weight in kg
Parameters	$\mathfrak{A}$ (A)	175.666 cm	65.7518 kg
	$\mathfrak{B}_1$ (B_1)	0.358447	0.255682
	$\mathfrak{B}_2$ (B_2)	2.35008	2.88564
	$\mathfrak{B}_3$ (B_3)	14.1509	12.6349
	$\mathfrak{C}_1$ (C_1)	2.97728 years	0.691983 years
	$\mathfrak{C}_2$ (C_2)	8.03954 years	8.92748 years
	$\mathfrak{C}_3$ (C_3)	12.6333 years	12.7071 years
	$\mathfrak{D}$ (D)	0.223400 years	0.0499509 years
RSS		0.898578	9.63642
Standard deviations		0.423929	1.38827

Even though cross-sectional studies, in which many children of different ages are measured simultaneously but each child is measured only once, are generally considered as less desirable, such studies are nevertheless useful because they can be done much more rapidly than longitudinal studies, which require a generation!

After a detailed comparison of eight asymptotic models of human growth, including the JPPS and JPA-2 models, the triple logistic curve of Bock and Thissen (1976), the curve of Preece and Baines (1978), and the models of Shohoji and Sasaki (1987) and of Kanefuji and Shohoji (1990), Jolicoeur, Pontier and Abidi (1992) finally concluded that the JPA-2 model presently provides the best fit for the growth of human stature.

Because all of these models are empirical, and since several of them already provide a practically perfect fit, in particular when applied to stature data, it is perhaps futile to keep searching for models possessing an even greater goodness of fit. Some authors have found that prepubertal growth may be cyclical in nature (Butler, McKie and Ratcliffe, 1990), and the existence of a *mid-growth spurt* may still be considered as controversial, but one may wonder whether such minor trends should necessarily be incorporated into the deterministic part of a growth model or should rather be treated as mere fluctuations. Hauspie (1989) has done an excellent review of mathematical methods available for the study of human growth.

Section 35.14: complex allometry

Following the publication of the classical monograph of Huxley (1932), the study of allometry has been pursued actively by many biologists (sections 20.7, 31.6, 35.12), but the cases where the power function (simple allometry) is not adequate have tended to be considered as exceptional, difficult to analyze, and of little interest. While admittedly much more difficult to analyze than simple allometry, complex allometry should certainly not be viewed as uncommon or of little interest.

Thus, the first example in this section deals with a supposedly well-known species, the white laboratory rat (*Rattus norvegicus*). Body length in mm, body weight in g and brain weight in mg have been measured on four specimens at birth, and every 10 days up to the age of 100 days. The original data have been published by Jolicoeur and Pirlot (1988) and are reproduced in appendix H thanks to the kind permission of the Editor of the periodical *Growth, Development & Aging*. Scatter diagrams of somatic dimensions as related to *total age* are illustrated in figure 35.14.1. Total age is determined by adding the average duration of pregnancy (24 days in the white rat) to the ordinary postnatal age. The total age of a newborn white rat is thus $t=(0+24)=24$ days. The first attempts to analyze the data showed that Pütter's curve (sections 35.11 and 35.12),

$$Y_i = A_i[1-\exp(-t/C_i)]^{B_i}$$

fitted body length and body weight very well but brain weight very poorly, because the latter begins growing actively later than the former (figure 35.14.1, a, b and c), many days after fecundation. In order to take this delay of brain growth into account, total age t could be replaced in the equation of Pütter's curve by the difference $(t-t_{0i})$ or $(t-D_i)$, following Bertalanffy (1951), where $t_{0i}=D_i$ is the initial delay after which the dimension Y_i begins growing actively.

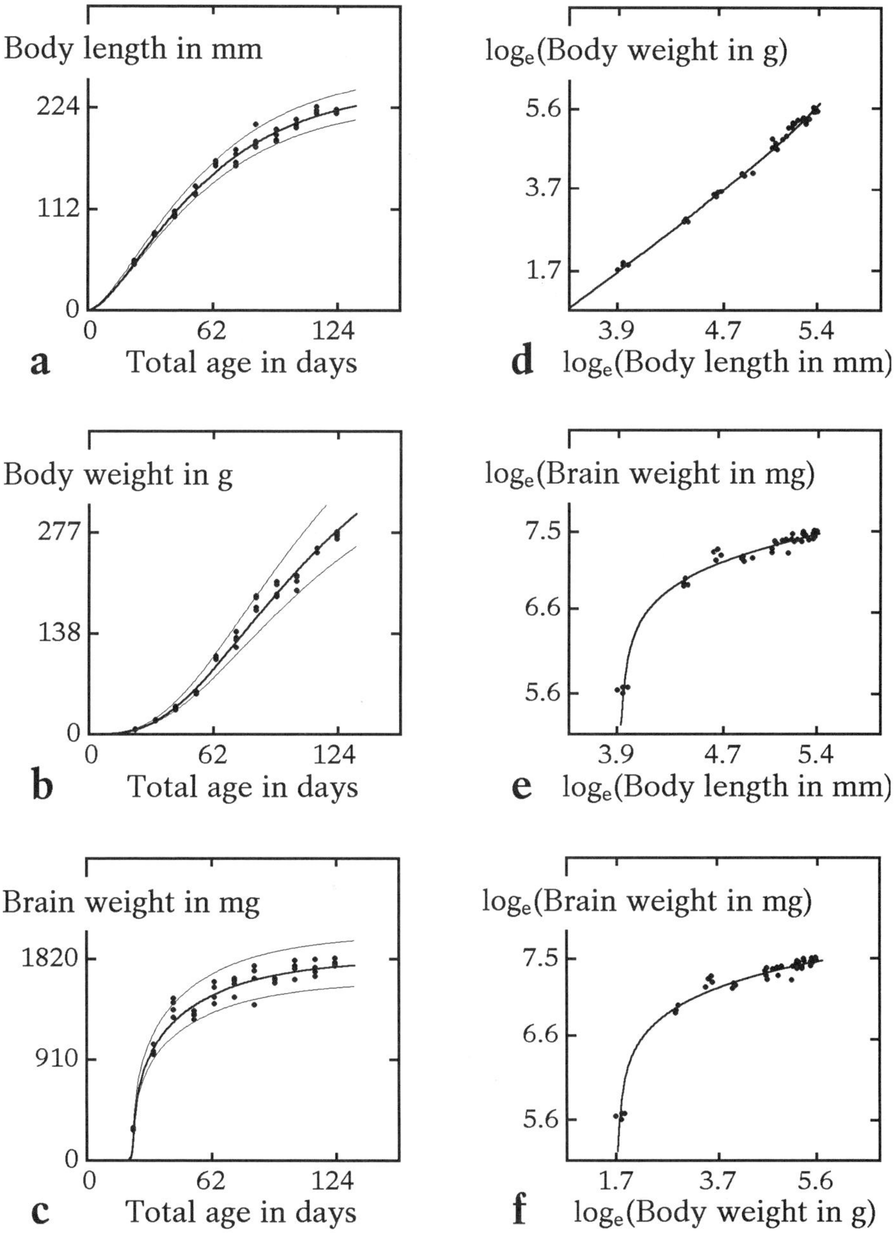

Figure 35.14.1
Chronological growth of the body length (a), body weight (b) and brain weight (c) of 44 male white rats, and bilogarithmic relationships, straight in d (simple allometry) but curved in e and in f (complex allometry); variation belts (thin lines) at level $\alpha = 0.05$; data from Jolicoeur and Pirlot (1988)

This would yield the modified version of Pütter's curve,

$$Y_i = A_i\{1-\exp[-(t-t_{0i})/C_i]\}^{B_i} = A_i\{1-\exp[-(t-D_i)/C_i]\}^{B_i},$$

often referred to as Bertalanffy's curve and which is frequently used in ichthyology and in fishery research. Instead of estimating the exponent B from the data as done in this chapter, however, Bertalanffy made it arbitrarily equal to $B=1$ in the case of a linear dimension (like fish length) and to $B=3$ in the case of a volume, a mass or a weight. Unfortunately, this modified version of Pütter's curve has negative or undefined initial ordinates when the delay $t_{0i}=D_i$ is positive, and thus yields an unsatisfactory description of early growth. This drawback could be avoided by replacing age t by the *piecewise-linear* delay function

$$\max(0, t-t_{0i}) = \max(0, t-D_i),$$

but, in the context of allometry, where the logarithmic transformation is often desirable, inserting the above delay function in the equation of Pütter's curve still has the disadvantage of yielding null initial ordinates. The delay function finally chosen was thus piecewise-linear in the limit (Prunty, 1983) but smooth and essentially positive,

$$(t^E + D_i^E)^{1/E} - D_i,$$

where the coefficient D is defined to be positive. Since the initial range covered by the curve,

$$Y_i = A_i\{1-\exp[[D_i-(t^E+D_i^E)^{1/E}]/C_i]\}^{B_i},$$

usually does not contain enough data to estimate exponent E, the latter was given the arbitrary value $E=100$.

The curves fitted to body length and weight and brain weight are illustrated in figure 35.14.1 (a, b and c) and multiplicative least-squares estimates of parameters, residual sums of squares and residual standard deviations are given in table 35.14.1. The present results differ slightly from those published by Jolicoeur and Pirlot (1988) because the

Table 35.14.1

Multiplicative least-squares estimates of parameters, residual sums of squares (RSS) and standard deviations for the curve $Y = A\{1-\exp[[D-(t^{100}+D^{100})^{1/100}]/C]\}^B$ fitted to the body length, body weight and brain weight of 44 male white rats, according to data from Jolicoeur and Pirlot (1988)

		body length in mm	body weight in g	brain weight in mg
Parameters	$\mathfrak{A}$ (A)	245.713 mm	444.295 g	1816.91 mg
	$\mathfrak{B}$ (B)	1.80660	4.12862	0.370571
	$\mathfrak{C}$ (C)	43.5002 days	55.2457 days	44.0217 days
	$\mathfrak{D}$ (D)	0 day	0 day	23.7944 days
RSS		0.0572125	0.275693	0.142824
Standard deviations		0.0378195	0.0830200	0.0597545

latter were obtained using a heteroscedastic model and a different delay function. Moreover, the confidence limits of parameters published by Jolicoeur and Pirlot (1988) were conditional, because an algorithm for computing unlinearized marginal confidence limits for the parameters of nonlinear models was not available at that time.

The major fact emerging from the analysis of those chronological growth curves is the presence of a considerable initial delay ($D = 23.7944$ days), almost equal to the average duration of pregnancy, for brain weight, and the absence of delays for the other two variates. Once the growth of brain weight has accelerated, however, it tends to progress faster ($C = 44.0217$ days) than that of body weight ($C = 55.2457$ days). To sum up, in the white rat, the growth of brain weight appears to accelerate almost exactly at the time of birth, when the individual begins having strong interactions with the external environment. The nature of allometric relationships between the three somatic dimensions is particularly evident in the bilogarithmic scatter diagrams (figure 35.14.1, d, e and f). The relationship between the logarithms of body length and body weight is practically straight (figure 35.14.1, d), which shows that these two variates are approximately related by a power function (simple allometry). The relationships between the logarithms of body dimensions and brain weight, however, are strongly curved (figure 35.14.1, e and f), showing that body weight is clearly not equal to a power function of body length or body weight. Substantially similar results have been obtained more recently by Jolicoeur, Cabana and Ducharme (1992) using a four-parameter generalization of the Gompertz curve,

$$Y_i = A_i \exp\{-\exp[(B_i - t)/C_i + (D_i/t)]\} .$$

The curves represented in the allometry diagrams (figure 35.14.1, d, e and f) describe the trajectories obtained by estimating both somatic dimensions from age through chronological growth curves, age being used as an *instrumental variate.*

It is generally more difficult to describe complex allometry satisfactorily when age is unknown or is not included in the analysis. In such a case, a simplified bivariate model of complex allometry may be derived (Jolicoeur, 1989) from the Gompertz curve (section 35.12):

$$Y_j = A\exp\{-B[\log_e(Y_{i\max}/Y_i)]^C\},$$

where $Y_{i\max}$ is the largest value of the somatic dimension Y_i observed in the sample. When the exponent C, which may be called the *exponent of complex allometry,* is equal to unity, the above function reduces to the power function

$$Y_j = A\,(Y_i/Y_{i\max})^B .$$

The hypothesis of simple allometry, which means that the two variates are related by a power function, can therefore be accepted when the confidence interval of parameter $\mathfrak{C}(C)$ covers unity. Strauss and Altig (1992) applied this approach to anuran tadpoles.

It would undoubtedly be difficult, and perhaps impossible, to develop a single general model which would always be satisfactory for complex allometry because, unlike simple allometry, which corresponds to a unique and well-defined kind of relationship, complex allometry is a heterogeneous category which lumps together very different entities.

Thus, the works of Dawood, Jolicoeur and Sharief (1988), Cabana, Jolicoeur and Baron (1990), Cabana, Jolicoeur and Michaud (1993) and of Jolicoeur and Pontier (1993) have shown that complex allometry relationships sometimes possess a single curvature which may be convex upwards or downwards, but may also have a sigmoid or an inverted sigmoid shape!

Even though complex allometry has been considered as somewhat exceptional until now, and has been little studied, it should be recognized to be the rule rather than the exception. Another example will now be given which concerns the reader more closely since it deals with the human species. In a reanalysis of cross-sectional autopsy data originally published by Marchand (1902), Jolicoeur, Baron and Cabana (1988) described the growth of stature and brain weight using a six-parameter model made up of two curvilinear segments. One of their main conclusions was that brain weight stops growing at approximately 7 years of age, much earlier than thought by many earlier workers, who felt that brain weight keeps increasing until 18 or 20 years of age. In order to check the conclusion of Jolicoeur, Baron and Cabana (1988) graphically, the scatter diagram of the logarithms of stature and brain weight in 699 males aged from 0 to 83 years is illustrated in figure 35.14.2. Individuals older than 7 years are represented by black dots (•), and younger individuals by white dots (○).

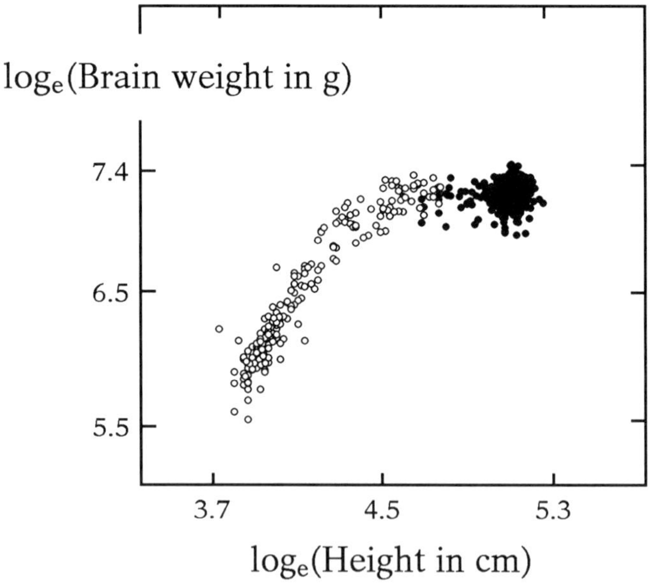

Figure 35.14.2
Scatter diagram of the logarithms of height and brain weight for 699 human males aged from 0 to 83 years; individuals are represented by white dots up to 7 years and by black dots above 7 years of age; after data from Marchand (1902) reanalyzed by Jolicoeur, Baron and Cabana (1988)

Since the relationship of logarithms shows a strong curvature, the hypothesis of simple allometry is clearly unacceptable. The dense accumulation of black dots at the upper right end of the relationship represents adult individuals whose growth has ended, and corresponds to the projection of the asymptotes of the chronological growth curves of stature and brain weight. In contrast with stature, brain weight does not appear to keep increasing clearly after 7 years of age.

Section 35.15: growth and decline of a biological population

When circumstances are unfavorable, as when individuals of one species must compete with those of another species, for instance, instead of fluctuating above and below an upper asymptote (section 35.10), the number of individuals in a population may decline after reaching its maximum. Thus figure 35.15.1 represents the growth and decline of an experimental population of the species *Paramecium caudatum* in a mixed culture with the related species *Paramecium aurelia,* according to data from Gause (1934).

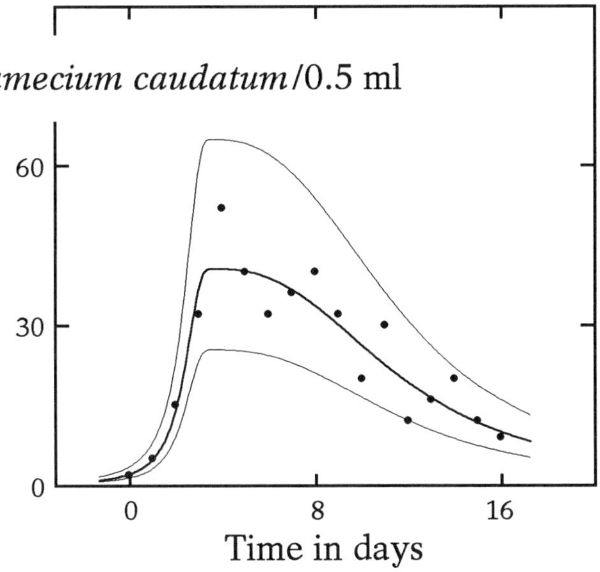

Figure 35.15.1
Growth and decline of the number of individuals of *Paramecium caudatum* /0.5 ml in a mixed culture with *Paramecium aurelia,* described using the segmented model
$Y = A/\{1 + \max[0, (B-t)/C]^E + \max[0, (t-B)/D]^E\}$;
variation belt (thin lines) at level $\alpha = 0.05$;
data from Gause (1934)

While we will not at this point go into Lotka's and Volterra's interesting theories, which are discussed in treatises of population ecology, let us illustrate a segmented empirical model which, in spite of its simplicity, allows maximum population size as well as rates of growth and decline to be estimated.

Since the initial rise of the population is obviously more rapid than its subsequent decline, the following model is used:

$$Y = A/\{1 + \max[0, (B-t)/C]^E + \max[0, (t-B)/D]^E\},$$

where Y denotes the number of individuals in the population at time t;
A is the average maximum number of individuals in the population;
B is the time at which the number Y of individuals in the population reaches its average maximum value A;
C determines the time $t=(B-C)$ at which the number Y of individuals has only reached $(A/2)$, half its average maximum value;
D determines the time $t=(B+D)$ at which the number Y of individuals, after reaching its average maximum value, has declined by one half, to $(A/2)$;
and E is an exponent which determines the degree of curvature and the shape of the relationship.

Multiplicative least-squares estimates, the residual sum of squares and the residual standard deviation are given in table 35.15.1. The average maximum number (40.5139 individuals/0.5 ml) of *Paramecium caudatum* is reached less than four days ($B=3.48408$ days) after the beginning of the experiment, and the half-decline time ($D=8.18793$ days) of the population is almost seven times longer than the time required initially by the population to reach half its maximum size ($C=1.16787$ days).

Table 35.15.1
Multiplicative least-squares estimates of parameters, residual sum of squares (RSS) and standard deviation for the segmented model of population growth and decline $Y = A/\{1 + \max[0, (B-t)/C]^E + \max[0, (t-B)/D]^E\}$ applied to data from Gause (1934) on *Paramecium caudatum*; the sample size $N=17$

Parameters	$\mathfrak{A}$ (A)	40.5139 individuals/0.5 ml
	$\mathfrak{B}$ (B)	3.48408 days
	$\mathfrak{C}$ (C)	1.16787 days
	$\mathfrak{D}$ (D)	8.18793 days
	$\mathfrak{E}$ (E)	2.64627 (dimensionless)
RSS		0.690069
Residual standard deviation		0.239804

Jolicoeur and Pontier (1989) have proposed an unsegmented growth and decline model related to exponential and logistic increase and decrease and having only four parameters,

$$Y = 1/\{\exp[(B-t)/C] + \exp[(t-A)/D)]\}.$$

While, in practice, this model is as satisfactory as the preceding model from a descriptive viewpoint (unweighted multiplicative least squares: RSS = 0.701419, residual standard deviation = 0.232283, A = 33.0944, B = − 0.547162, C = 0.863239, $D=7.23246$), its parameters are less readily interpretable.

Section 35.16: the rate of enzymatic reactions

From a historical viewpoint, the quantitative model of Michaelis and Menten (1913) has been an important step in the understanding of enzymatic reactions. According to Michaelis and Menten's model, the initial rate Y of an enzymatic reaction is related to the concentration X of the substrate through the equation

$$Y = A[X/(X+B)],$$

where A = the maximum speed which the reaction can reach, the amount of substrate available being taken into account;

and B = the substrate concentration at which the speed of the reaction is $(A/2)$, half its maximum value; the coefficient B is currently known as the *Michaelis constant* and denoted by the symbol K_M.

In biochemistry, Michaelis and Menten's curve is often linearized using unadvisable transformations, in which the fact that transformed variates have inconstant variances is neglected. Figure 35.16.1 illustrates the curve of Michaelis and Menten (1913) in the case of data published by Wilkinson (1961) and used as an example by Bliss and James (1966) in their detailed statistical study of the rectangular hyperbola. The concentration of the substrate (nicotinamide mononucleotide) is measured in millimoles per liter and the speed of the reaction is expressed in micromoles of nicotinamide-adenine dinucleotide formed per 3 minutes per milligram of enzyme protein. Additive least squares are used here in order to make results comparable to those of Bliss and James (1966).

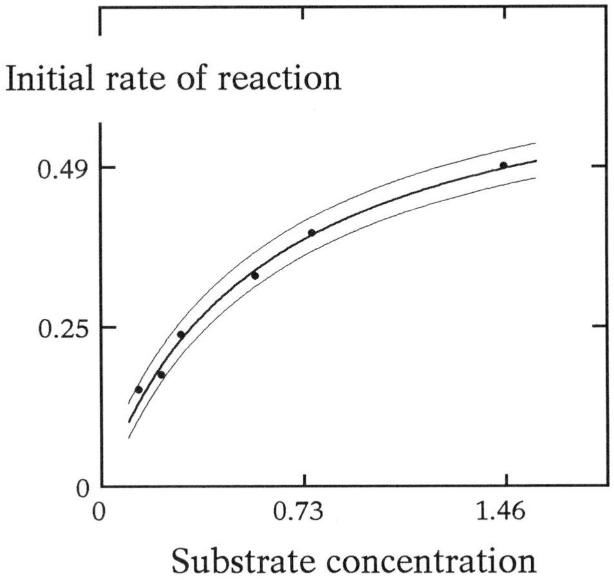

Figure 35.16.1

Relationship between the concentration X of the substrate and the initial rate Y of an enzymatic reaction, described using the curve of Michaelis and Menten (1913), $Y = A[X/(X+B)]$, fitted through additive least squares; data from Wilkinson (1961); variation belt at level $\alpha = 0.05$

Table 35.16.1
Multiplicative least-squares estimates of parameters, residual sum of squares (RSS) and standard deviation for the model of Michaelis and Menten (1913), $Y = A[X/(X+B)]$, fitted to data from Wilkinson (1961); $N=6$

Parameters	$\mathfrak{A}$ (A)	0.690399
	$\mathfrak{B}$ (B)	0.596548
RSS		0.000736018
Residual standard deviation		0.0135648

The means of computation available thirty years ago being taken into account, the present numerical results agree well enough with those published by Bliss and James (1966, bottom of page 581 and top of page 601), point estimates (table 35.16.1) having from 4 to 5 similar digits and confidence limits at significance level $\alpha = 0.05$ (table 35.16.2) having from 3 to 4 similar digits. For the record, the present calculations have been carried out with maximum accuracy (approximately 15 digits) and without transcribing any intermediate results from the computer.

Table 35.16.2
Marginal confidence limits of the parameters of the model $Y = A[X/(X+B)]$ (Michaelis and Menten, 1913) applied to data from Wilkinson (1961) on the rate of an enzymatic reaction

α	0.05		0.01	
Limit	lower	upper	lower	upper
$\mathfrak{A}$ (A)	0.598953	0.810749	0.549185	0.915077
$\mathfrak{B}$ (B)	0.433635	0.829533	0.351566	1.04382

Currie (1982) has studied the way in which the chosen values of the predictor variate X (the experimental design) affect the precision with which the parameters of Michaelis and Menten's curve can be estimated. If it were known a priori that the reaction studied agreed exactly with Michaelis and Menten's equation, one half of the observations should be made at the value $X = B$ and the other half at the highest possible value of X. However, such an experimental design would not allow the experimenter to check the goodness of fit of Michaelis and Menten's curve. Even though Michaelis and Menten's curve fits many sets of data satisfactorily, Jolivet (1982, chapter 5) noted that its derivation involves several simplifying assumptions, which may not always be fully justified. Surprisingly, the curve of Michaelis and Menten (1913) is cited much more often in the literature than the curve of Hill (1913),

$$Y = A/[1+(C/X)^B] = A[X^B/(X^B+C^B)],$$

or, if B and C are interchanged,

$$Y = A/[1+(B/X)^C] = A[X^C/(X^C+B^C)],$$

even though Hill's curve (section 35.11) is more general. In fact, Michaelis and Menten's curve is a particular case of Hill's curve when the exponent (denoted by B or by C) of the latter is equal to unity, these two models being nested (section 35.7).

Section 35.17: environmental temperature and biological processes

The influence of environmental temperature on biological processes was discussed by Belehradek (1935), whose excellent monograph is still well-worth reading. Moreover, Schwerdtfeger (1963, pages 94-170) made a substantial review of ecological aspects. The acceleration of biological processes with increasing environmental temperature was perceived empirically at first by the French physicist and naturalist Antoine Ferchault de Réaumur (1683-1757), who noted that fruits ripen faster when the average daily temperature is higher through the growing season. Analogous observations were later made not only in plant biology, on the duration of bud development for instance, but also on the duration of various developmental stages of poikilothermal animals (such as insects, fishes, amphibians, etc.). Such observations can be tentatively explained by postulating an inverse relationship between the duration Y of a biological process and the (average) temperature T to which this process is submitted:

$$Y = A/T,$$

where Y = the duration of the process in days,
T = the temperature in Celsius degrees to which the process is submitted,
and A = the expected duration of the process when the temperature equals 1° Celsius.

However, some biologists noted that the above equation implies that a biological process stops and that its duration becomes infinite when the environmental temperature $T = 0\,°C$, and that it would be more realistic to estimate from data the temperature threshold B above which a process can go on. This temperature threshold is sometimes called a *biological zero* or a *physiological zero,* and it may differ from the freezing point of fresh water, particularly in tropical or marine organisms. One then has the two-parameter model

$$Y = A/(T-B),$$

where B is the estimated temperature at which the speed of a biological process becomes null and its duration infinite. While the latter equation is often satisfactory and the *biological zero* concept is very interesting, the analysis of numerous data sets has shown that this two-parameter model is not always flexible enough, but the three-parameter generalization proposed by Belehradek (1935),

$$Y = A/(T-B)^C,$$

where C is an exponent which modifies the shape of the curve, does fit satisfactorily the relationship between temperature and the duration of most biological processes. However, the temperature must not be raised so much that the organism would be damaged and that the process would slow down and finally stop. As formulated above, Belehradek's model still leaves something to be desired because A denotes the estimated duration of the process at a very low temperature ($T = B + 1$), just above the biological zero, usually outside the range of the data, and at which observations cannot be made rapidly. It is therefore preferable to reparameterize the model of Belehradek (1935) in such a way that the coefficient A corresponds to a more moderate temperature T_1, approximately at the center of the range of the data.

The reparameterized version of Belehradek's curve (1935) has the following equation (Ratkowsky, 1990, pages 91 and 107; Jolicoeur, 1995, page 219):

$$Y = A\,[(T_1 - B)/(T - B)]^C ,$$

where Y = the duration of the process in days,
T = the temperature in Celsius degrees to which the process is submitted,
T_1 = the chosen (moderate) temperature corresponding to the duration A,
B = the temperature threshold above which the process can go on,
C = an exponent which determines the shape of the curve,
and A = the expected duration of the process when the temperature equals T_1.

While empirical studies were carried out, many research workers of the early 20th century tried analyzing biological processes with simplified theoretical formulas which had been originally developed to describe the influence of temperature on the speed of chemical reactions. In general, however, biological processes are more complex than chemical reactions, and their temperature threshold is higher that the absolute zero (0 °Kelvin = – 273.15 °Celsius). Therefore, it was gradually realized that the formulas derived from the research of Marcellin Berthelot (1827-1907), Jacobus Henricus Van't Hoff (1852-1911) and Svante Arrhenius (1859-1927) described biological relationships poorly and could not be interpreted clearly in biology. At the present time, the reparameterized version of Belehradek's curve (1935) is still the best descriptive model for the relationship between the duration of biological processes and environmental temperature. Nevertheless, the point estimate and the confidence limits of the coefficient B must be prevented from taking numerical values lower than the absolute zero (0 °Kelvin = – 273.15 °Celsius), something which could happen when the sample size N is small, when the residual variation is large, or when the curvature of the observed relationship between Y and T is weak. It is therefore better to use additive least squares rather than multiplicative least squares, because the latter would reduce the influence of long durations Y, which are critical for the determination of the temperature threshold. Moreover, all temperatures (T, T_1 and B) should be expressed in Kelvin degrees and the constraint $0 \leq B$ should be applied (see section 35.13) during the computations, at the end of which the results could be reexpressed in Celsius degrees.

Figure 35.17.1 illustrates the application of Belehradek's curve to the hatching time Y of eggs as related to water temperature T in the brown trout (*Salmo trutta,* also known as *Salmo fario*) and cod (*Gadus morrhua,* also known as *Gadus callarias*). The temperature $T_1 = 10\,°C = 283.15\,°K$ has been chosen here because it lies within the temperature range of both samples. Additive least-squares estimates of parameters are given in table 35.17.1 and confidence limits in tables 35.17 2 and 35.17.3. The hatching time of eggs at 10 °C is over four times longer in the brown trout than in cod. Like many other salmonoid fish species of temperate regions, the brown trout leaves the sea and migrates up freshwater streams where it spawns during the cold season. The long hatching time of eggs in this anadromous species, which may reach three or four months in cold water, is undoubtedly adaptive, since it allows hatching to coincide with the beginning of the next warm season, when environmental conditions are more propitious to fry development.

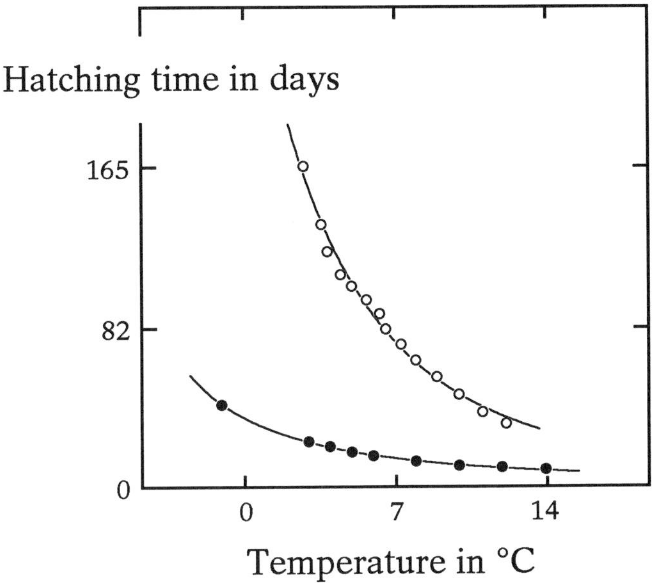

Figure 35.17.1
Relationship between the hatching time Y of eggs in days and water temperature $X = T$ in °C in the brown trout (○) and in cod (●), described using the reparameterized version $Y = A[(10 - B)/(T - B)]^C$ of Belehradek's curve (1935) fitted through additive least squares to data from Gray (1928) and Dannevig (1894; reproduced by Altman and Dittmer, 1966)

In the case of cod, however, which lives in cold seas where there is not much seasonal periodicity, and of which the larvae spend approximately two months and a half in plankton before moving down into the benthos, a long hatching time would not likely be an advantage.

Table 35.17.1
Additive least-squares estimates of parameters, residual sums of squares (RSS) and standard deviations for the curve $Y = A[(10 - B)/(T - B)]^C$ fitted to the hatching time Y of eggs according to water temperature in the brown trout and in cod

		brown trout	cod
Parameters	$\mathfrak{A}$ (A)	47.3408 days	10.9575 days
	$\mathfrak{B}$ (B)	– 13.0095 °C	– 10.1177 °C
	$\mathfrak{C}$ (C)	3.23611 (dimensionless)	1.69974 (dimensionless)
RSS		189.744	1.31265
Standard deviations		4.15325	0.467734

Table 35.17.2
Marginal confidence limits of the parameters of the model $Y = A[(10 - B)/(T - B)]^C$ applied to the hatching time Y of eggs in the brown trout according to data from Gray (1928); $N = 14$

α	0.05		0.01	
Limit	lower	upper	lower	upper
$\mathfrak{A}$ (A)	43.3218 days	51.3464 days	41.6815 days	52.9864 days
$\mathfrak{B}$ (B)	- 273.15 °C	- 2.49396 °C	- 273.15 °C	- 1.01948 °C
$\mathfrak{C}$ (C)	1.38807	50.1184	1.11441	52.1701

The fact that the point estimate of the *biological zero* B is lower than $-10\,°$C in both species (table 35.17.1) should be interpreted with caution. Whether the eggs of brown trout or cod would survive such low temperatures could only be ascertained by direct experiments. Even more, the fact that the lower marginal confidence limit of parameter $\mathfrak{B}$ (B) is equal to absolute zero (-273.15°C) in the brown trout (table 35.17.2) certainly does not mean that the eggs of this fish could survive such a temperature! – It only indicates that Gray's data (1928) do not confirm the existence of a *biological zero* distinct from *absolute zero,* and that they could be analyzed through a simpler nested model (section 35.7) involving only two parameters:

$$Y = A[(T_1 + 273.15)/(T + 273.15)]^C ,$$

where temperatures are expressed in Celsius degrees.

Table 35.17.3
Marginal confidence limits of the parameters of the model $Y = A[(10 - B)/(T - B)]^C$ applied to the hatching time Y of eggs in cod according to data from Dannevig (1894; reproduced by Altman and Dittmer, 1966); $N = 9$

α	0.05		0.01	
Limit	lower	upper	lower	upper
$\mathfrak{A}$ (A)	10.4378 days	11.4758 days	10.1694 days	11.7423 days
$\mathfrak{B}$ (B)	- 15.7641 °C	- 6.9888 °C	- 20.8364 °C	- 5.87766 °C
$\mathfrak{C}$ (C)	1.26703	2.45555	1.10833	3.1225

Section 35.18: theoretical versus empirical models

Some of the nonlinear models mentioned in this chapter or earlier (sections 20.6 and 20.7) possess or may be attributed a theoretical basis, while others are used empirically because they are similar in shape to the relationships observed on data about particular phenomena. Predicting a variate Y through a function $Y = f(t)$ may sometimes be justified by considering the derivative $Y' = dY/dt$ of that function with respect to time t or to another predictor variate, since the derivative Y' of a function Y corresponds to the rate of change of that function with respect to the predictor variate. Therefore, the terms of the derivative may reflect the effects of causes tending to increase or to decrease the value of the predicted variate Y. In an ideal case, the derivative of a prediction function might thus be interpreted as a causal explanation of the changes affecting a phenomenon.

For instance, the derivative of the exponential function $Y = A\,e^{Bt}$ (section 20.6),

$$Y' = dY/dt = AB\,e^{Bt} = BY,$$

is proportional to that function; consequently, the latter may reasonably be used to describe the increase in the number Y of individuals at a constant relative rate, $(Y'/Y) = B$, in a manner similar to compound interests, in a biological population which is unaffected by spatial or nutritional constraints. As for the logistic function (section 35.10),

$$Y = A/\{1 + \exp[(B - t)/C]\}$$

the first term (Y/C) of its derivative,

$$Y' = dY/dt = Y \times (1/C) \times [1 - (Y/A)] = (Y/C) - [Y^2/(AC)],$$

is proportional to the function Y and may reflect, similarly to the exponential function, the ability of living matter to self-reproduce, but the second term $-[Y^2/(AC)]$ is proportional to the square of function Y bearing a minus sign and may represent the negative interactions between individuals, which become stronger and stronger as a population increases in size while facing constraints. In the case of the Gompertz function (sections 35.11, 35.12 and 35.14),

$$Y = A \exp\{-\exp[(B - t)/C]\}$$

its relative rate of increase,

$$(Y'/Y) = (1/C) \exp[(B - t)/C],$$

is a decreasing exponential function of time, a condition which some research workers have considered to be a reasonable assumption for some biological tissues or organs.

As for the presence between two variates Y_1 and Y_2 of a *simple allometry* relationship or power function (sections 20.7, 22.5, 31.6, 35.12, and 35.14),

$$Y_2 = A\,Y_1^B,$$

if these two variates are functions of time, $Y_1 = Y_1(t)$ and $Y_2 = Y_2(t)$, and if the derivatives with respect to time t of both members of the logarithmic version of the allometry equation,

$$\log_e[Y_2(t)] = \log_e(A) + B \log_e[Y_1(t)],$$

are evaluated, it is found that

$$(1/Y_2) \times dY_2/dt = B \times (1/Y_1) \times dY_1/dt,$$

which shows that the ratio $(Y_2'/Y_2)/(Y_1'/Y_1)$ of the relative growth rate of the two variates Y_1 and Y_2 is constant and equal to the *allometry exponent* B. Finally, other prediction functions (sections 35.11 and 35.16), such as those of Hill (1913) and of Michaelis and Menten (1913), are based on theoretical considerations about the mechanism of chemical reactions (Jolivet, 1982, chapter 5).

Unfortunately, the theoretical "principles" on which the preceding models are based require many assumptions of which the legitimacy may often be debatable in practice. Thus, although exponential growth and early logistic growth imply that the number of new individuals should be strictly proportional to current population size, it is quite possible for the initial increase in population density to improve relative rates of reproduction in reality. As for the Gompertz curve, the decrease of the relative growth rate may seldom be exactly exponential in practice. Moreover, while theoretical considerations about the logistic and the Gompertz curves do not rule out their application to somatic growth, these curves are often not very suitable in that context, because they possess a lower asymptote and reach 0 only at $t = -\infty$ (section 35.11). Finally, it is noteworthy that, even if a model such as the logistic curve has some theoretical foundation, precise hypotheses concerning the numerical values of its parameters can seldom be made.

Yet, the present chapter shows that many empirical models can be used not only to provide satisfactory descriptions of observed biological relationships but also, contrary to common beliefs, to test important and meaningful hypotheses (sections 35.7, 35.8, 35.9, 35.11, 35.13, 35.14, 35.15 and 35.17). It may therefore be concluded that, even though theoretical knowledge is certainly more advanced in some biological fields than in others, the borderline between theoretical and empirical models is still fuzzy in much of biology. Good theoretical models are undoubtedly desirable, but it is better to have good empirical models than to have theoretical models which are overly simple or too remote from concrete reality, or not to have any models of any kind.

Appendix A

True BASIC™ computer program for the diagonalization of a square symmetric matrix following the method of Carl Gustav Jacobi (1804-1851)

(see section 31.3)

```
! Jacobi
! Diagonalization of a square symmetric matrix
!            following the method of Carl Gustav Jacobi (1846)
! P.J., May 17, 1996
OPTION ANGLE RADIANS
DIM S(3,3),U(3,3),D(3,3)
LET q=3                                ! q = the number of variates
MAT REDIM  S(q,q),U(q,q),D(q,q)
MAT READ S
CALL SubTrace(q,Trace,S(,))
CALL Jacobi(q,Trace,S(,),U(,),D(,))
PRINT "Matrix to be diagonalized"
PRINT
MAT PRINT S
PRINT "Trace =";Trace
PRINT
PRINT "Matrix of latent vectors"
PRINT
MAT PRINT U
PRINT "Diagonal matrix of latent roots"
PRINT
MAT PRINT D
DATA 1,0.5,0.5
DATA 0.5,1,0.5
DATA 0.5,0.5,1
PRINT "Type 'Yes' if you have finished recording the results"
INPUT YesNo$
END

SUB SubTrace(q,Trace,D(,))
    MAT REDIM D(q,q)
    LET Trace=0
    FOR i=1 TO q
        LET Trace=Trace+D(i,i)
    NEXT i
END SUB
```

```
SUB Jacobi(q,Trace,A(,),U(,),D(,))
    DIM T(3,3),V(3,3)
    MAT REDIM A(q,q),U(q,q),D(q,q),T(q,q),V(q,q)
    MAT U=IDN
    MAT D=A
    DO
       DO
      ! search for the largest off-diagonal element in absolute value, MaxEl
          LET MaxEl=0
          FOR i=1 TO q
              FOR j=i+1 TO q
                  IF ABS(D(i,j))>MaxEl THEN
                     LET MaxEl=ABS(D(i,j))
                     LET i0=i
                     LET j0=j
                  END IF
              NEXT j
          NEXT i
          IF  D(i0,i0)<=0  OR  D(j0,j0)<=0
                           OR  MaxEl<(1/10^11)*SQR(D(i0,i0)*D(j0,j0))
                           OR  MaxEl<(5/10^11)*Trace  THEN
! Nota Bene:  the 3 preceding lines must be typed as a single line
             LET D(i0,j0)=0
          ELSE
             EXIT DO
          END IF
       LOOP while MaxEl>0
       IF MaxEl>0 THEN
          MAT T=IDN
          IF D(i0,i0)=D(j0,j0) THEN
             LET F=Pi/4
          ELSE
             LET F=0.5*ATN(2*D(i0,j0)/(D(i0,i0)-D(j0,j0)))
          END IF
          LET T(i0,i0)=COS(F)
          LET T(i0,j0)=SIN(F)
          LET T(j0,i0)=-T(i0,j0)
          LET T(j0,j0)=T(i0,i0)
          MAT D=T*U
          MAT U=D
          MAT T=U*A
          MAT V=TRN(U)
          MAT D=T*V
       END IF
    LOOP WHILE MaxEl>0
```

```
    ! Latent vectors and roots arranged according to decreasing root magnitude
    MAT V=ZER
    FOR i=1 TO q
        LET F=-9*10^100
        FOR j=1 TO q
            IF D(j,j)>F THEN
               LET F=D(j,j)
               LET g=j
               FOR h=1 TO q
                   LET V(i,h)=U(j,h)
               NEXT h
            END IF
        NEXT j
        LET D(g,g)=-9*10^101
    NEXT i
    MAT U=V
    IF U(1,1)<0 THEN MAT U = (-1)*U
    ! the preceding line prevents element U(1,1) from being negative
    MAT T=TRN(U)
    MAT D=A*T
    MAT T=U*D
    MAT D=T
END SUB
```

Results of computations

```
Matrix to be diagonalized

 1              .5             .5
 .5             1              .5
 .5             .5             1

Trace = 3

Matrix of latent vectors

 .57735         .57735         .57735
-.707107        .707107        0
-.408248       -.408248        .816497

Diagonal matrix of latent roots

 2.             8.89429e-17   -2.5411e-17
-2.88398e-17    .5            -3.03577e-18
 2.6563e-17    -3.0443e-18     .5
```

Appendix B

True BASIC™ computer program for the minimization of the residual sum of squares with respect to a nonlinear model following the method of Nelder and Mead (1965)

(see sections 35.2 and 35.10; data from Gause (1934) reproduced with the permission of Dover Publications, Inc.)

```
! NelderMead
! Fitting of the logistic curve, Y=A/{1+exp[(B-t)/C]};
!  the residual sum of squares (RSS) is minimized by using the simplex
!   algorithm of Nelder and Mead (1965)
! P.J., May 22, 1996
DECLARE DEF Yestimate
DIM Mi(2),Ma(2),Observations(100,2)
DIM Increment(3),Par(3),Param(3)
DIM V(4,4),Vlow(4),Vhigh(4),Vmed(4),V1(4),V2(4)
LET q=3   ! q = the number of parameters of the fitted curve
! the dimensions of vectors and matrices are automatically readjusted
!     according to the number q of parameters of the fitted model
MAT REDIM Increment(q),Par(q),Param(q)
MAT REDIM V(q+1,q+1),Vlow(q+1),Vhigh(q+1),Vmed(q+1),V1(q+1),V2(q+1)
LET Display=1
! Iterations are not displayed if Display = 0,
!  but they are displayed at every 50th iteration if Display = 1,
!   and at every iteration if Display = 2.
LET Denominator=10
! The Denominator variable controls the size of the initial simplex:
!  the length of each side of the initial simplex is equal to the
!   corresponding coordinate of the starting point divided by Denominator;
!    the smaller Denominator, the larger the initial simplex.
LET MaxNoIter=2000 ! Maximum number of iterations, chosen by the user.

PRINT "Observations being read from DATA statements"
PRINT
READ N
MAT REDIM Observations(N,2)
FOR h=1 TO N
    FOR i=1 TO 2
        READ Observations(h,i)
    NEXT i
NEXT h

! Search for extreme values of variates
FOR i=1 TO 2
    LET Mi(i)=1.e+308
    LET Ma(i)=-1.e-308
NEXT i
FOR h=1 TO N
    FOR i=1 TO 2
        LET Mi(i)=min(Mi(i),Observations(h,i))
        LET Ma(i)=max(Ma(i),Observations(h,i))
    NEXT i
NEXT h
PRINT "Extreme values of variates"
PRINT
PRINT "Variate";TAB(10);"Minimum";TAB(20);"Maximum"
PRINT
FOR i=1 TO 2
    PRINT i;tab(10);Mi(i);tab(20);Ma(i)
    PRINT
NEXT i
```

```
! The minimization subroutine according to the method of Nelder and Mead
!  (1965)  is about to be called

! Starting point

LET Par(1)=Ma(1)
LET Par(2)=(Mi(2)+Ma(2))/2
LET Par(3)=(Ma(2)-Mi(2))/5

CALL NelderMead

PRINT "Results"
PRINT
PRINT "Number of iterations = ";Iteration
PRINT
PRINT "Residual sum of squares (RSS) = ";RSS
PRINT
PRINT "Estimates of parameters"
PRINT
FOR i=1 TO q
    PRINT "#";i; " : ";Param(i)
    PRINT
NEXT i

PRINT "Type ""Yes"" if you have finished recording the results"
INPUT YesNo$
IF Ucase$(YesNo$)="YES" THEN STOP


SUB NelderMead
    PRINT "The RSS will be minimized by the method of Nelder and Mead (1965)"
    PRINT
    LET Iteration=0

    ! Initial simplex: Starting point
    FOR i=1 TO q
        LET V(q+1,i)=Par(i)
    NEXT i

    ! Other points of the initial simplex

    ! The 'Increment' = the length of each side of the simplex
    FOR i=1 TO q
        LET Increment(i)=V(q+1,i)/Denominator
    NEXT i
    FOR i=1 TO q
        FOR j=1 TO q
            LET V(i,j)=V(q+1,j)
            IF i=j THEN LET V(i,i)=V(i,i)-Increment(i)
        NEXT j
    NEXT i

    CALL Heights
```

```
DO
    ! Iterative part of the algorithm

    ! Search for Vlow and Vhigh
    LET Vlow(q+1)=1.e+308
    LET Vhigh(q+1)=-1.e+308
    FOR i=1 TO q+1
        IF V(i,q+1)<Vlow(q+1) THEN
           FOR j=1 TO q+1
               LET Vlow(j)=V(i,j)
           NEXT j
        END IF
        IF V(i,q+1)>Vhigh(q+1) THEN
           LET NoVhigh=i
           FOR j=1 TO q+1
               LET Vhigh(j)=V(i,j)
           NEXT j
        END IF
    NEXT i

    ! Attempts to decrease the height of the highest point

    ! Calculation of Vmed
    MAT Vmed=ZER
    FOR i=1 TO q+1
        IF i<>NoVhigh THEN
           FOR j=1 TO q+1
               LET Vmed(j)=Vmed(j)+V(i,j)
           NEXT j
        END IF
    NEXT i
    MAT Vmed=(1/q)*Vmed

    ! Calculation of V1
    FOR i=1 TO q
        LET V1(i)=2*Vmed(i)-Vhigh(i)  !  reflection :Vmed + (Vmed - Vhigh)
        LET Par(i)=V1(i)
    NEXT i

    ! Evaluation of the RSS of V1
    CALL EvalRSS
    LET V1(q+1)=RSS
    ! RSS of V1 compared with those of Vlow and Vhigh
    IF V1(q+1)<Vlow(q+1) THEN
       ! Since V1(q+1)<Vlow(q+1), try extension
       ! Calculation of V2
       FOR i=1 TO q
           LET V2(i)=2*V1(i)-Vmed(i)  !  extension :  V1 + (V1 - Vmed)
           LET Par(i)=V2(i)
       NEXT i
       ! Evaluation of the RSS of V2
       CALL EvalRSS
       LET V2(q+1)=RSS
       ! RSS of V2 compared with that of Vlow
       IF V2(q+1)<Vlow(q+1) THEN
          ! Therefore, V2 is lower than Vlow,
          !   and replace Vhigh by V2
          FOR i=1 TO q+1
              LET V(NoVhigh,i)=V2(i)
          NEXT i
       ELSE
```

```
        ! Therefore, V2(q+1)>Vlow(q+1),
        !  and replace Vhigh by V1 since V1(q+1) < Vlow(q+1)
        FOR i=1 TO q+1
            LET V(NoVhigh,i)=V1(i)
        NEXT i
    END IF
ELSE
    ! Therefore, Vlow(q+1)<V1(q+1)
    !  Does V1(q+1) < V(i,q+1) for one of thevalues of i <> NoVhigh ?
    LET MaxRSSwithoutVhigh=0
    FOR i=1 TO q+1
         IF i<>NoVhigh THEN
            IF V(i,q+1)>MaxRSSwithoutVhigh THEN
               LET MaxRSSwithoutVhigh=V(i,q+1)
            END IF
         END IF
    NEXT i
    IF V1(q+1)<=MaxRSSwithoutVhigh THEN
        ! Therefore,  V1(q+1) <= RSSMax(i <> NoVhigh),
        !          and replace Vhigh by V1
        FOR i=1 TO q+1
            LET V(NoVhigh,i)=V1(i)
        NEXT i
    ELSE
        ! Therefore, V1(q+1) > RSSMax(i <> NoVhigh),
        !  and try contraction : calculation of V2
        IF V1(q+1) < Vhigh(q+1) THEN
           ! replace Vhigh by V1 before contracting
           FOR i=1 TO q+1
               LET V(NoVhigh,i)=V1(i)
           NEXT i
        END IF
        ! Contraction properly speaking
        FOR i=1 TO q
            LET V2(i)=(Vhigh(i)+Vmed(i))/2
            LET Par(i)=V2(i)
        NEXT i
        CALL EvalRSS
        LET V2(q+1)=RSS
        ! RSS of V2 compared to that of Vhigh
        IF V2(q+1)<Vhigh(q+1) THEN
           ! replace Vhigh by V2
           FOR i=1 TO q+1
               LET V(NoVhigh,i)=V2(i)
           NEXT i
        ELSE
           ! Contraction failed:
           !  bring all points of simplex closer to Vlow
           FOR i=1 TO q+1
               FOR j=1 TO q
                   LET V(i,j)=(V(i,j)+Vlow(j))/2
               NEXT j
           NEXT i
           CALL Heights
        END IF
    END IF
END IF
```

```
        ! Convergence criteria
        LET Criterion1=ABS((Vhigh(q+1)-Vlow(q+1))/Vlow(q+1))
        LET Criterion2=0
        FOR i=1 TO q
            LET Criterion2=Criterion2+ABS((Vhigh(i)-Vlow(i))/Vlow(i))
        NEXT i
        LET Criterion2=Criterion2/q
        LET Iteration=Iteration+1
        ! Optional display of iterations
        IF Display = 1 OR Display = 2 THEN
           IF Display = 2  OR  mod(Iteration,50)=0  OR  Criterion1<=1.e-14
                                    OR  Criterion2<=1.e-14 THEN
! Nota Bene:  the 2 lines above must be typed as a single line
              PRINT "Iteration ";Iteration
              PRINT "Criterion 1 (RSS): ";Criterion1;
              PRINT "Criterion 2 (parameters): ";Criterion2
              PRINT
              MAT PRINT V;
           END IF
        END IF
        ! Results recorded
        FOR i=1 TO q
            LET Param(i)=Vlow(i)
        NEXT i
        LET RSS=Vlow(q+1)
    LOOP WHILE Criterion1>1.e-14 AND Criterion2>1.e-14 AND Iteration<MaxNoIter
END SUB

SUB Heights
     ! RSS for the (q+1) points of the simplex
     FOR i=1 TO q+1
         FOR j=1 TO q
             LET Par(j)=V(i,j)
         NEXT j
         CALL EvalRSS
         LET V(i,q+1)=RSS
     NEXT i
END SUB

SUB EvalRSS
     ! Evaluation of the RSS corresponding to a vector of parameters Par()
     LET RSS=0
     FOR h=1 TO N
         LET Y=Observations(h,1)
         LET t=Observations(h,2)
         LET Yest=Yestimate(t,Par())
         LET RSS=RSS+(Y-Yest)^2
!        For multiplicative least squares, replace the preceding statement by
!        LET RSS=RSS+LOG(Y/Yest)^2
     NEXT h
END SUB

DEF Yestimate(t,Par())
     ! Evaluation of Yestimate at time t
     LET Yestimate=Par(1)/(1+exp((Par(2)-t)/Par(3)))
END DEF
```

```
! Observations

! Gause1934

! Number of Paramecium aurelia per 0.5 ml and age of the culture in days;
!  Table 4, page 145, column 2, in Gause (1934).

! Sample size
DATA 17

! Data
DATA 2,0
DATA 6,1
DATA 24,2
DATA 75,3
DATA 182,4
DATA 264,5
DATA 318,6
DATA 373,7
DATA 396,8
DATA 443,9
DATA 454,10
DATA 420,11
DATA 438,12
DATA 492,13
DATA 468,14
DATA 400,15
DATA 472,16

END
```

Results of computations

```
Number of iterations =  122

Residual sum of squares (RSS) =  8639.92

Estimates of parameters

# 1  :  448.183

# 2  :  4.74322

# 3  :  1.24261
```

Appendix C

Data of Adolph H. Schultz on the length in mm of the left and right limb bones of 117 male and 110 female white adult human skeletons

(published with the permission of Prof. Dr. Robert D. Martin, Adolph Schultz Foundation, Anthropologisches Institut und Museum der Universität Zürich-Irchel)

117 men								110 women							
X_1	X_2	X_3	X_4	X_5	X_6	X_7	X_8	X_1	X_2	X_3	X_4	X_5	X_6	X_7	X_8
357	360	260	262	461	453	379	381	328	325	240	241	455	450	389	382
355	360	259	260	473	473	399	394	322	319	229	227	433	433	368	370
354	360	261	266	476	476	390	389	321	327	234	236	445	445	385	386
354	356	258	257	460	455	387	389	318	325	239	244	449	448	373	369
351	346	268	267	471	470	413	408	317	320	239	239	444	438	383	380
348	357	266	265	465	465	389	391	317	320	232	229	418	422	343	345
347	358	265	262	452	449	392	394	316	317	205	205	410	408	336	334
344	342	236	235	441	442	370	366	313	317	224	227	428	424	362	358
343	348	255	260	473	473	395	401	312	310	207	205	409	411	323	325
341	343	250	251	463	461	362	363	311	303	225	222	416	415	351	349
341	339	255	255	457	456	383	379	309	317	229	234	427	424	375	378
340	343	243	243	458	463	375	374	308	306	218	221	424	422	350	345
340	338	250	249	449	447	381	378	307	311	236	239	419	421	364	361
338	340	246	249	468	462	389	391	306	314	228	234	414	414	343	347
338	339	239	246	444	444	367	370	306	310	228	234	418	416	345	357
337	338	250	250	463	464	375	378	306	310	215	217	409	405	340	337
337	336	244	246	433	433	358	363	306	306	216	215	410	414	343	343
335	337	263	265	463	467	401	398	306	299	220	218	428	428	362	358
335	335	247	245	444	446	376	375	305	310	214	224	428	428	353	347
334	338	245	247	449	443	387	392	305	309	223	227	428	429	360	362
334	332	251	247	437	433	367	368	305	305	231	235	436	435	374	374
333	336	255	260	448	443	374	367	304	309	224	229	426	422	355	352
333	336	243	246	465	465	378	373	303	307	228	233	396	396	348	344
333	332	258	262	453	450	384	382	303	305	226	226	420	417	358	360
333	329	238	236	428	430	366	370	303	300	232	232	405	407	345	345
332	338	241	249	461	461	385	387	302	308	217	226	419	420	355	348
332	336	238	239	427	430	357	352	302	306	233	237	444	442	386	383
331	343	250	251	453	451	375	379	302	306	216	218	425	428	343	345
331	330	250	251	461	457	393	394	302	305	213	217	424	423	347	352
330	335	249	252	452	448	375	375	301	310	219	220	412	412	350	356
330	335	238	238	423	425	345	347	301	305	221	221	424	427	346	347
330	334	248	250	437	439	369	365	301	300	229	232	415	411	357	360
330	332	249	250	451	447	376	376	300	311	225	231	419	417	332	333
330	331	246	250	453	445	373	369	300	298	212	217	403	401	345	349
329	344	254	268	433	431	362	373	299	306	219	221	423	418	343	342
329	336	239	242	437	437	367	369	299	304	225	226	395	392	337	336
329	333	245	248	438	438	373	371	299	300	233	232	405	398	341	339
329	330	257	257	462	460	396	393	298	304	218	222	406	406	344	353
329	330	247	249	450	445	383	385	298	300	216	220	381	378	320	319
329	325	249	246	443	442	375	376	298	296	212	218	409	405	337	338
328	328	240	249	435	433	369	368	297	304	219	223	396	399	334	337
327	330	241	245	425	425	354	354	297	301	224	226	415	411	355	356
326	329	249	250	432	432	366	369	297	297	226	228	417	414	351	350
326	329	240	242	437	438	358	360	297	297	220	220	411	410	347	336
326	321	231	230	410	409	342	341	297	297	220	215	420	420	358	353
325	327	245	248	429	424	374	369	297	296	213	216	421	415	345	344
325	321	245	243	402	399	351	350	296	305	209	213	405	398	340	339
324	328	236	240	430	433	350	350	296	300	211	215	376	377	314	320
324	328	238	238	431	434	358	359	295	299	223	228	406	407	344	347

117 men (continued)								110 women (continued)							
X_1	X_2	X_3	X_4	X_5	X_6	X_7	X_8	X_1	X_2	X_3	X_4	X_5	X_6	X_7	X_8
324	325	230	237	441	441	367	370	294	296	207	210	397	400	326	323
324	324	240	240	433	439	360	363	294	294	225	226	420	423	355	354
323	324	247	247	452	455	393	393	294	294	209	213	387	389	316	320
323	324	233	235	444	445	349	363	293	294	211	212	390	388	317	317
322	322	250	247	430	432	361	360	293	294	207	211	400	400	333	331
322	329	253	255	432	431	358	354	292	301	215	220	418	418	328	329
322	323	235	239	441	436	367	372	292	295	205	204	395	398	323	326
322	322	234	231	426	422	351	354	292	295	203	207	400	397	322	323
322	321	244	248	435	437	367	368	291	296	209	214	395	397	310	308
321	324	238	246	430	426	347	345	291	292	213	214	407	407	349	345
321	322	236	237	410	418	343	346	289	298	208	212	376	373	309	308
320	332	240	251	452	450	382	384	289	290	205	208	378	379	320	320
320	324	235	235	434	430	339	345	288	297	202	213	391	393	304	301
320	322	239	243	437	434	366	364	287	290	210	210	397	397	340	339
320	320	246	245	459	462	362	365	287	290	202	206	394	396	313	313
320	320	236	239	424	426	361	363	286	291	207	210	399	398	336	336
319	322	226	229	436	439	357	361	286	286	209	211	396	399	332	332
319	318	234	234	442	439	374	372	285	295	209	212	399	397	332	332
318	319	240	244	435	436	384	383	285	291	206	209	390	393	309	315
317	331	230	232	414	420	341	339	284	280	203	204	390	390	319	320
317	326	242	240	427	431	375	375	283	290	207	209	369	368	319	318
317	323	240	241	445	441	365	363	283	285	214	214	400	404	346	347
316	325	247	245	423	418	360	359	283	284	207	204	365	360	310	310
316	324	238	240	440	443	367	367	283	282	205	206	383	384	328	323
316	321	234	237	452	452	367	369	282	289	208	211	391	388	320	316
315	322	237	240	438	439	350	353	282	288	202	205	373	373	302	300
315	317	233	234	423	418	353	350	282	284	222	225	409	409	345	341
315	314	234	236	428	428	352	351	281	280	208	210	370	374	329	328
315	313	240	241	436	434	366	368	281	279	207	207	379	379	319	318
315	310	234	230	423	419	354	353	280	285	205	208	367	370	328	332
315	309	249	250	442	442	376	371	280	284	209	211	382	383	323	320
313	316	236	237	427	432	370	367	280	283	207	212	392	392	338	334
313	315	223	225	440	440	357	357	280	280	192	197	377	374	301	300
313	314	235	236	426	421	352	352	279	284	215	219	380	380	313	316
313	313	235	236	405	410	357	360	279	282	203	205	393	390	323	323
313	311	239	241	405	405	349	348	279	274	204	206	397	399	339	333
312	317	245	246	419	419	341	343	278	285	209	216	387	386	306	308
312	317	224	225	431	429	344	345	278	282	208	211	388	388	316	315
312	316	240	241	438	436	363	367	278	282	201	206	377	379	318	324
312	314	229	234	414	414	361	363	277	285	219	223	397	394	334	334
311	310	227	228	425	424	334	334	277	283	214	220	382	381	327	326
311	306	224	223	434	433	347	348	277	275	205	201	375	375	302	310
310	310	225	226	428	429	355	357	276	285	219	219	410	411	353	353
310	308	230	229	414	415	342	343	275	278	204	202	390	385	320	318
310	307	232	233	409	404	341	343	275	278	198	200	375	375	318	322
309	309	241	242	437	432	358	355	273	275	186	186	376	378	301	298
309	309	229	225	419	415	339	332	272	276	189	194	374	368	295	291
308	308	231	233	415	410	351	351	271	277	202	207	381	380	318	313
308	305	229	232	416	416	340	340	271	273	201	202	380	380	325	329
307	318	232	235	410	405	326	329	268	275	184	187	338	337	294	290
307	316	225	227	412	410	327	325	268	271	206	207	378	372	326	327
306	312	222	227	433	431	365	365	267	270	195	201	366	369	307	308
305	304	238	241	437	430	361	359	264	270	189	194	365	364	300	299
304	306	225	224	430	422	353	349	264	269	191	191	369	368	301	297
303	310	239	245	441	438	357	357	247	251	184	184	337	332	286	289
303	307	229	231	417	414	349	346	240	245	175	182	338	340	269	276
302	308	206	210	386	386	320	321	271	276	213	217	359	359	296	298
302	303	203	217	398	398	311	315	288	296	226	236	383	389	319	328

117 men (end)								110 women (end)							
X_1	X_2	X_3	X_4	X_5	X_6	X_7	X_8	X_1	X_2	X_3	X_4	X_5	X_6	X_7	X_8
301	303	233	234	403	400	334	324	291	295	219	221	367	367	307	310
301	296	226	225	412	414	350	347	305	306	223	224	400	400	345	346
300	307	230	236	423	424	338	339	313	311	237	234	404	397	335	339
299	304	220	221	403	399	325	320								
297	296	230	235	411	407	336	340								
293	298	223	228	388	384	326	328								
290	292	204	209	407	409	350	337								
287	294	227	231	413	410	341	342								
277	280	208	212	382	378	328	325								
276	282	215	219	377	374	329	328								

X_1 = left humerus length,
X_2 = right humerus length,
X_3 = left radius length,
X_4 = right radius length,
X_5 = left femur length,
X_6 = right femur length,
X_7 = left tibia length,
X_8 = right tibia length.

(see sections 29.4, 29.6, 30.2, and 31.5)

Appendix D

Data of Jolicoeur (1984) on the length in mm of the skull (X_1), humerus (X_2) and femur (X_3) of 68 male North American martens

(see sections 31.7 and 31.9)

X_1	X_2	X_3
79.2	61.9	69.4
79.5	66.0	72.4
77.9	62.8	70.3
79.4	64.7	71.5
79.0	63.0	69.7
78.5	64.6	71.4
78.4	62.8	69.6
80.1	65.6	70.9
78.6	63.2	71.1
79.7	64.3	70.8
78.7	63.0	69.5
79.5	62.4	68.2
79.7	66.8	73.5
77.3	65.2	71.9
79.7	65.4	72.6
81.9	65.1	71.5
79.5	62.6	69.5
79.0	62.7	70.2
77.7	61.7	68.3
80.1	66.4	73.1
81.1	66.2	75.0
79.5	63.8	70.4
76.8	60.8	65.5
78.7	64.0	69.7
79.5	66.0	73.3
77.8	63.7	68.4
79.0	64.8	72.1
75.2	61.6	67.7
78.5	63.7	71.3
81.1	66.1	71.9
79.0	65.2	71.8
80.5	66.2	72.8
79.6	63.8	70.4
77.7	64.5	71.8
78.2	64.6	71.8
81.0	67.2	72.7
81.5	66.7	72.5
78.2	64.8	72.1
79.1	67.6	73.2
78.7	63.8	70.6
78.6	62.3	67.6
80.6	64.9	71.0
82.3	66.1	73.5
80.4	64.7	73.1
80.7	63.2	71.2
76.5	61.2	68.5
76.3	62.7	70.2
81.4	65.2	71.5
82.3	65.2	71.7
76.8	63.2	69.0
79.1	64.8	70.3
80.8	65.7	71.9
80.7	64.0	72.3
77.4	62.2	69.5
79.2	63.2	70.4
77.5	67.1	73.2
80.6	64.6	71.9
73.2	59.3	65.3
81.5	64.1	71.0
75.7	61.6	68.7
78.3	62.9	69.9
81.3	64.3	71.8
76.7	63.4	68.4
78.1	64.3	71.2
75.0	61.8	68.7
78.3	63.2	69.4
80.9	64.6	70.4
77.7	63.5	71.5

Appendix E

Data (in mm) of the botanist Edgar Anderson, published by Fisher (1936), on the length (X_1) and breadth (X_2) of a sepal and on the length (X_3) and breadth (X_4) of a petal of 50 flowers of each of three iris species

(see section 33.8; reproduced with the permission of Cambridge University Press)

Iris setosa canadensis				*Iris versicolor*				*Iris virginica*			
X_1	X_2	X_3	X_4	X_1	X_2	X_3	X_4	X_1	X_2	X_3	X_4
5.1	3.5	1.4	0.2	7.0	3.2	4.7	1.4	6.3	3.3	6.0	2.5
4.9	3.0	1.4	0.2	6.4	3.2	4.5	1.5	5.8	2.7	5.1	1.9
4.7	3.2	1.3	0.2	6.9	3.1	4.9	1.5	7.1	3.0	5.9	2.1
4.6	3.1	1.5	0.2	5.5	2.3	4.0	1.3	6.3	2.9	5.6	1.8
5.0	3.6	1.4	0.2	6.5	2.8	4.6	1.5	6.5	3.0	5.8	2.2
5.4	3.9	1.7	0.4	5.7	2.8	4.5	1.3	7.6	3.0	6.6	2.1
4.6	3.4	1.4	0.3	6.3	3.3	4.7	1.6	4.9	2.5	4.5	1.7
5.0	3.4	1.5	0.2	4.9	2.4	3.3	1.0	7.3	2.9	6.3	1.8
4.4	2.9	1.4	0.2	6.6	2.9	4.6	1.3	6.7	2.5	5.8	1.8
4.9	3.1	1.5	0.1	5.2	2.7	3.9	1.4	7.2	3.6	6.1	2.5
5.4	3.7	1.5	0.2	5.0	2.0	3.5	1.0	6.5	3.2	5.1	2.0
4.8	3.4	1.6	0.2	5.9	3.0	4.2	1.5	6.4	2.7	5.3	1.9
4.8	3.0	1.4	0.1	6.0	2.2	4.0	1.0	6.8	3.0	5.5	2.1
4.3	3.0	1.1	0.1	6.1	2.9	4.7	1.4	5.7	2.5	5.0	2.0
5.8	4.0	1.2	0.2	5.6	2.9	3.6	1.3	5.8	2.8	5.1	2.4
5.7	4.4	1.5	0.4	6.7	3.1	4.4	1.4	6.4	3.2	5.3	2.3
5.4	3.9	1.3	0.4	5.6	3.0	4.5	1.5	6.5	3.0	5.5	1.8
5.1	3.5	1.4	0.3	5.8	2.7	4.1	1.0	7.7	3.8	6.7	2.2
5.7	3.8	1.7	0.3	6.2	2.2	4.5	1.5	7.7	2.6	6.9	2.3
5.1	3.8	1.5	0.3	5.6	2.5	3.9	1.1	6.0	2.2	5.0	1.5
5.4	3.4	1.7	0.2	5.9	3.2	4.8	1.8	6.9	3.2	5.7	2.3
5.1	3.7	1.5	0.4	6.1	2.8	4.0	1.3	5.6	2.8	4.9	2.0
4.6	3.6	1.0	0.2	6.3	2.5	4.9	1.5	7.7	2.8	6.7	2.0
5.1	3.3	1.7	0.5	6.1	2.8	4.7	1.2	6.3	2.7	4.9	1.8
4.8	3.4	1.9	0.2	6.4	2.9	4.3	1.3	6.7	3.3	5.7	2.1
5.0	3.0	1.6	0.2	6.6	3.0	4.4	1.4	7.2	3.2	6.0	1.8
5.0	3.4	1.6	0.4	6.8	2.8	4.8	1.4	6.2	2.8	4.8	1.8
5.2	3.5	1.5	0.2	6.7	3.0	5.0	1.7	6.1	3.0	4.9	1.8
5.2	3.4	1.4	0.2	6.0	2.9	4.5	1.5	6.4	2.8	5.6	2.1
4.7	3.2	1.6	0.2	5.7	2.6	3.5	1.0	7.2	3.0	5.8	1.6
4.8	3.1	1.6	0.2	5.5	2.4	3.8	1.1	7.4	2.8	6.1	1.9
5.4	3.4	1.5	0.4	5.5	2.4	3.7	1.0	7.9	3.8	6.4	2.0
5.2	4.1	1.5	0.1	5.8	2.7	3.9	1.2	6.4	2.8	5.6	2.2
5.5	4.2	1.4	0.2	6.0	2.7	5.1	1.6	6.3	2.8	5.1	1.5
4.9	3.1	1.5	0.2	5.4	3.0	4.5	1.5	6.1	2.6	5.6	1.4
5.0	3.2	1.2	0.2	6.0	3.4	4.5	1.6	7.7	3.0	6.1	2.3
5.5	3.5	1.3	0.2	6.7	3.1	4.7	1.5	6.3	3.4	5.6	2.4
4.9	3.6	1.4	0.1	6.3	2.3	4.4	1.3	6.4	3.1	5.5	1.8
4.4	3.0	1.3	0.2	5.6	3.0	4.1	1.3	6.0	3.0	4.8	1.8
5.1	3.4	1.5	0.2	5.5	2.5	4.0	1.3	6.9	3.1	5.4	2.1
5.0	3.5	1.3	0.3	5.5	2.6	4.4	1.2	6.7	3.1	5.6	2.4
4.5	2.3	1.3	0.3	6.1	3.0	4.6	1.4	6.9	3.1	5.1	2.3
4.4	3.2	1.3	0.2	5.8	2.6	4.0	1.2	5.8	2.7	5.1	1.9
5.0	3.5	1.6	0.6	5.0	2.3	3.3	1.0	6.8	3.2	5.9	2.3
5.1	3.8	1.9	0.4	5.6	2.7	4.2	1.3	6.7	3.3	5.7	2.5
4.8	3.0	1.4	0.3	5.7	3.0	4.2	1.2	6.7	3.0	5.2	2.3
5.1	3.8	1.6	0.2	5.7	2.9	4.2	1.3	6.3	2.5	5.0	1.9
4.6	3.2	1.4	0.2	6.2	2.9	4.3	1.3	6.5	3.0	5.2	2.0
5.3	3.7	1.5	0.2	5.1	2.5	3.0	1.1	6.2	3.4	5.4	2.3
5.0	3.3	1.4	0.2	5.7	2.8	4.1	1.3	5.9	3.0	5.1	1.8

Appendix F

Length and breadth in mm of the head of the first two adult sons in 25 Dutch families, extracted by Rao (1952) from the data originally published by Frets (1921)

(see section 34.4)

first adult son		second adult son	
length	breadth	length	breadth
191	155	179	145
195	149	201	152
181	148	185	149
183	153	188	149
176	144	171	142
208	157	192	152
189	150	190	149
197	159	189	152
188	152	197	159
192	150	187	151
179	158	186	148
183	147	174	147
174	150	185	152
190	159	195	157
188	151	187	158
163	137	161	130
195	155	183	158
186	153	173	148
181	145	182	146
175	140	165	137
192	154	185	152
174	143	178	147
176	139	176	143
197	167	200	158
190	163	187	150

Appendix G

Longitudinal data of Jolicoeur, Pontier, Pernin and Sempé (1988) on the postnatal age, height, and body weight of 13 boys and 14 girls from the French auxological survey

(see section 35.13; data kindly provided by Dr. Michel Sempé, pediatrician and research director, I.N.S.E.R.M., Lyon, France)

postnatal	boy No. 1		boy No. 2		boy No. 3		boy No. 4		boy No. 5	
age (years)	height (cm)	weight (kg)	height (cm)	weight (kg)	height (cm)	weight (kg)	height (cm)	weight (kg)	height (cm)	weight (kg)
4/52	52.4	4.00	53.9	4.02	53.6	4.02	50.7	3.49	54.6	4.65
0.25	57.5	5.09	61.7	5.87	60.3	5.50	59.0	5.46	62.1	6.47
0.50	64.4	6.33	69.2	7.81	65.5	7.55	64.2	7.23	67.2	7.50
0.75	69.4	7.50	74.0	8.88	69.7	8.94	68.8	8.37	70.9	8.52
1.0	71.7	8.05	77.0	9.43	72.5	9.38	72.9	8.85	75.2	9.70
1.5	77.9	9.62	84.2	11.62	78.4	10.72	79.5	10.56	79.8	11.35
2.0	82.3	10.70	88.8	12.20	82.6	11.50	85.0	11.54	85.6	12.72
2.5	86.5	11.60	93.0	13.94	87.3	11.86	89.0	12.74	90.8	14.45
3.0	90.2	12.87	98.1	14.82	90.5	12.42	93.3	13.87	95.7	14.94
3.5	94.2	13.65	102.0	15.93	93.7	12.97	97.5	15.61	98.8	15.83
4.0	97.8	14.72	106.5	16.44	96.2	13.80	101.0	16.00	102.5	16.74
4.5	100.5	15.32	110.0	17.88	99.3	14.70	104.9	17.15	106.8	18.20
5.0	103.2	15.93	113.0	19.15	102.5	15.60	108.8	18.35	110.9	19.70
5.5	106.0	16.75	115.9	20.25	106.3	16.53	111.4	19.25	113.7	20.85
6.0	108.8	17.60	118.5	21.30	110.2	17.44	114.1	20.20	116.5	22.00
6.5	111.6	18.45	122.7	22.60	113.4	18.10	117.8	21.65	120.0	22.50
7.0	113.7	19.15	126.0	24.75	116.2	19.10	120.8	22.45	122.9	24.95
7.5	116.5	20.20	129.8	26.40	119.2	20.20	123.4	23.80	125.8	26.75
8.0	119.0	21.10	134.0	27.90	122.1	21.30	125.9	25.05	128.7	28.60
8.5	121.7	22.00	136.2	28.90	124.9	22.80	128.7	26.85	132.3	29.80
9.0	124.2	23.00	138.2	29.85	127.7	24.10	131.5	28.65	135.8	31.05
9.5	126.4	24.10	141.0	30.90	130.3	25.15	134.0	29.80	138.3	33.30
10.0	128.6	25.20	144.0	32.00	132.8	26.25	136.5	30.90	140.8	35.53
10.5	130.8	26.40	146.7	34.10	135.3	27.70	139.2	32.56	143.0	37.25
11.0	133.3	27.50	149.4	36.30	137.9	29.11	142.0	34.23	145.2	39.00
11.5	135.6	28.70	152.2	38.50	140.4	30.70	144.4	35.30	148.3	43.70
12.0	138.0	29.95	156.3	44.00	142.9	32.25	146.7	36.30	151.4	48.42
12.5	140.1	31.90	161.4	46.30	145.6	33.60	149.3	39.65	155.2	52.55
13.0	143.4	33.70	166.5	49.80	148.6	35.45	154.1	44.40	161.5	58.35
13.5	146.6	34.15	170.0	52.70	153.4	37.75	159.9	49.20	165.0	59.45
14.0	149.6	36.75	174.0	56.60	157.8	41.10	165.3	52.55	167.4	62.40
14.5	152.5	39.40	175.9	59.40	161.9	43.00	167.9	56.30	169.3	64.20
15.0	157.9	42.60	178.7	61.20	164.8	47.75	170.7	57.00	170.7	63.55
15.5	163.0	46.00	178.9	60.40	167.2	48.70	172.2	58.30	171.4	63.65
16.0	167.1	49.20	180.0	59.65	168.3	50.80	172.5	58.00	171.5	64.90
16.5	170.0	49.40	180.6	60.25	169.3	50.95	173.2	61.30	171.8	68.00
17.0	171.5	50.85	181.1	60.85	169.8	52.55	173.6	61.40	172.0	72.60
17.5	172.7	52.30	181.6	62.05	170.3	53.00	173.9	62.65	172.2	77.15
18.0	173.5	53.50	182.0	63.25	171.0	52.55	174.1	59.10	173.0	78.20
18.5	174.3	54.50	182.0	63.50	171.3	54.30	174.4	59.80	173.5	78.40
19.0	175.1	55.40	182.0	63.70	171.6	54.15	174.7	60.50	173.9	78.50

Appendix G

(continued)

postnatal age (years)	boy No. 6 height (cm)	boy No. 6 weight (kg)	boy No. 7 height (cm)	boy No. 7 weight (kg)	boy No. 8 height (cm)	boy No. 8 weight (kg)	boy No. 9 height (cm)	boy No. 9 weight (kg)	boy No. 10 height (cm)	boy No. 10 weight (kg)
4/52	52.8	3.65	53.9	3.82	51.9	3.85	55.2	4.50	53.1	4.11
0.25	61.0	5.70	60.0	5.37	57.9	5.88	61.6	5.70	58.5	5.25
0.50	67.3	6.65	67.5	7.59	64.7	8.53	68.5	8.02	66.4	7.71
0.75	71.1	8.30	72.5	9.25	68.4	9.80	73.4	10.00	69.5	9.34
1.0	74.4	9.32	76.4	10.60	70.8	10.56	77.5	11.51	73.1	10.56
1.5	83.5	11.48	83.5	12.17	75.3	11.20	82.3	12.70	81.2	11.81
2.0	88.5	12.31	89.0	13.60	79.5	11.72	88.0	13.16	85.5	13.07
2.5	94.0	13.50	93.8	14.86	83.0	12.87	92.2	13.96	90.4	15.05
3.0	99.3	14.75	98.5	16.16	86.0	13.42	96.0	14.76	93.5	16.48
3.5	104.2	16.65	102.0	17.24	89.0	14.00	100.8	15.70	97.8	17.11
4.0	108.4	17.91	105.0	18.20	92.0	14.57	104.8	16.60	102.1	18.20
4.5	111.5	18.65	108.4	19.55	95.0	15.51	107.6	17.22	104.1	17.65
5.0	114.5	19.40	111.8	20.85	98.0	16.14	110.3	17.85	107.1	20.20
5.5	117.8	21.15	114.2	20.70	100.6	16.94	112.8	18.50	110.1	21.00
6.0	121.0	22.90	117.8	22.75	103.2	17.74	115.3	19.15	113.0	21.80
6.5	124.0	22.95	120.3	23.88	105.7	18.55	118.6	20.15	116.7	25.05
7.0	126.5	24.50	122.9	25.10	109.0	19.30	121.9	21.15	120.4	28.35
7.5	129.6	27.00	125.6	26.90	111.0	19.70	124.6	22.35	123.4	29.85
8.0	132.7	29.20	128.3	28.70	113.0	20.08	127.4	23.50	126.4	31.35
8.5	135.6	30.85	131.0	30.50	115.4	20.95	129.1	24.70	128.8	32.80
9.0	138.4	32.50	133.7	32.20	117.8	21.84	130.8	25.90	131.1	34.30
9.5	141.5	34.95	136.0	33.10	120.2	23.30	133.4	26.70	133.4	35.80
10.0	144.6	37.41	138.3	33.95	122.7	24.78	136.0	27.50	135.7	37.30
10.5	147.3	39.80	140.9	34.75	124.7	25.62	137.8	29.60	138.0	38.70
11.0	150.1	42.19	143.5	35.53	126.7	26.46	139.7	31.75	140.2	40.10
11.5	153.1	43.86	145.6	36.95	128.8	27.20	142.1	33.10	143.2	43.40
12.0	156.1	45.53	147.8	38.40	131.0	27.90	144.5	34.50	146.3	46.72
12.5	159.9	46.00	150.4	40.00	136.7	31.60	147.0	35.90	148.3	49.65
13.0	162.8	48.00	153.3	42.20	143.2	35.25	148.3	37.00	152.0	50.58
13.5	167.6	49.60	156.9	44.25	148.1	40.00	150.6	39.60	156.8	53.00
14.0	172.9	55.00	161.2	47.71	152.7	43.30	153.8	43.40	162.6	57.62
14.5	175.5	56.40	164.2	51.46	155.6	47.00	158.0	45.70	164.7	60.00
15.0	178.3	61.50	169.6	54.25	158.6	49.15	162.5	48.00	166.6	63.10
15.5	180.0	59.55	171.2	56.30	161.0	49.65	166.9	50.00	168.9	64.30
16.0	180.6	64.55	172.2	60.25	163.0	50.20	170.0	51.40	169.5	65.00
16.5	181.2	62.50	175.0	62.40	163.0	53.30	172.7	53.35	169.9	66.20
17.0	181.6	66.00	175.3	64.00	163.0	53.00	174.0	55.40	170.5	66.10
17.5	181.8	66.15	175.5	63.70	163.5	55.10	175.0	55.40	171.0	68.10
18.0	182.0	70.20	176.0	65.00	164.0	56.20	175.8	58.70	171.2	68.40
18.5	182.2	71.80	176.5	66.40	164.3	56.50	176.4	58.50	171.4	68.70
19.0	182.4	73.30	177.0	67.40	164.5	56.65	176.8	58.30	171.5	68.85

Appendix G

(continued)

postnatal age (years)	boy No. 11		boy No. 12		boy No. 13		girl No. 1		girl No. 2	
	height (cm)	weight (kg)	height (cm)	weight (kg)	height (cm)	weight (kg)	height (cm)	weight (kg)	height (cm)	weight (kg)
4/52	53.9	3.67	53.0	4.27	57.8	4.44	49.3	2.95	53.0	3.77
0.25	59.4	5.93	59.5	5.90	63.5	6.34	56.5	4.29	59.1	5.19
0.50	66.7	8.13	64.5	6.97	70.9	8.27	62.9	5.70	66.7	7.21
0.75	70.7	9.93	67.3	7.60	74.6	9.33	67.2	6.70	69.3	8.28
1.0	73.7	10.50	70.1	8.23	78.4	10.25	71.3	7.47	73.1	9.01
1.5	80.5	11.66	76.1	9.15	84.8	12.20	78.2	9.00	78.8	10.78
2.0	85.6	12.51	81.2	10.10	90.7	13.85	84.0	10.10	84.6	12.09
2.5	89.4	13.37	86.0	10.90	95.6	15.08	89.5	11.31	89.6	13.45
3.0	93.5	13.69	89.4	11.75	100.0	15.70	93.6	12.30	93.0	14.20
3.5	97.0	14.76	92.7	12.72	103.6	16.29	98.6	13.61	97.0	16.05
4.0	100.8	14.95	97.3	13.85	107.0	17.35	102.5	14.56	100.6	16.07
4.5	101.2	16.30	100.3	14.60	109.6	18.45	105.6	15.40	103.6	17.15
5.0	105.7	17.50	103.0	15.85	113.0	19.50	108.2	16.35	106.7	18.20
5.5	109.0	17.60	106.0	16.00	115.8	20.80	110.8	17.10	109.4	19.10
6.0	112.0	19.20	108.8	17.00	118.8	22.30	113.2	18.00	112.2	20.00
6.5	114.1	20.20	111.7	18.10	121.7	23.70	115.6	19.00	115.9	23.30
7.0	117.0	21.25	115.1	19.50	124.7	25.10	118.3	21.05	119.0	23.00
7.5	119.5	23.30	118.2	20.70	127.6	26.10	121.5	21.50	120.3	24.35
8.0	123.4	25.00	121.0	21.90	131.0	27.25	124.2	23.21	123.4	25.40
8.5	125.9	26.75	124.4	23.00	134.3	28.45	126.7	25.30	126.2	27.55
9.0	128.4	28.50	127.8	24.10	136.5	30.00	129.1	27.30	128.8	29.65
9.5	131.1	29.55	129.5	25.10	138.6	31.50	131.5	29.40	131.5	31.50
10.0	133.9	30.63	131.2	26.10	140.8	32.50	134.0	31.40	133.2	35.62
10.5	136.4	32.90	132.9	27.10	143.0	33.50	136.5	33.50	137.0	37.80
11.0	139.0	35.20	135.9	28.90	146.0	35.70	139.0	35.49	140.3	42.00
11.5	142.5	37.80	138.1	30.00	148.7	38.00	142.0	37.95	145.2	44.45
12.0	146.1	40.50	140.2	31.20	151.4	43.60	145.0	40.40	150.2	46.85
12.5	149.4	41.90	143.8	32.11	156.0	46.20	148.5	42.85	153.5	52.50
13.0	153.2	44.30	144.7	33.90	160.6	48.80	152.0	45.30	155.5	57.90
13.5	157.9	48.45	146.8	34.75	168.2	61.40	155.0	47.00	157.3	59.10
14.0	162.6	53.65	149.0	35.85	172.1	64.70	156.8	48.80	157.6	62.85
14.5	166.9	57.70	151.3	38.00	175.0	72.90	158.1	50.50	158.9	62.30
15.0	170.6	61.45	153.8	40.40	176.7	76.62	158.5	50.50	160.0	62.75
15.5	172.5	65.15	157.0	43.30	178.0	78.50	158.7	50.50	161.1	62.25
16.0	173.9	66.60	160.7	46.60	179.2	77.40	158.8	49.90	161.1	60.60
16.5	174.8	65.30	165.0	50.50	180.0	76.30	159.8	51.30	161.1	59.00
17.0	175.5	64.40	169.1	52.95	180.4	81.00	160.7	52.65	161.1	60.10
17.5	175.9	67.55	171.1	55.50	180.6	86.00	160.8	51.80	161.1	63.30
18.0	176.6	69.50	172.3	57.00	180.8	90.00	160.8	51.25	161.1	61.25
18.5	177.2	71.00	173.0	57.90	180.9	93.00	160.8	55.00	161.1	61.90
19.0	177.7	72.50	173.4	58.25	181.0	96.00	160.7	62.40	161.1	62.55

Appendix G

(continued)

postnatal age (years)	girl No. 3		girl No. 4		girl No. 5		girl No. 6		girl No. 7	
	height (cm)	weight (kg)	height (cm)	weight (kg)	height (cm)	weight (kg)	height (cm)	weight (kg)	height (cm)	weight (kg)
4/52	53.5	3.75	54.8	4.00	48.8	2.80	54.2	3.60	52.6	3.65
0.25	60.2	6.05	61.7	6.06	56.5	4.80	60.7	5.60	58.5	5.24
0.50	67.3	8.33	67.5	7.34	65.2	7.03	66.5	7.72	62.7	6.86
0.75	70.1	10.09	69.8	8.42	70.9	8.96	70.4	8.96	69.3	7.94
1.0	73.9	11.42	72.8	9.12	75.5	9.68	74.5	9.88	72.4	8.30
1.5	81.0	13.07	78.7	9.90	83.8	10.80	82.2	11.61	77.4	9.95
2.0	85.5	13.33	83.6	11.82	88.3	12.15	89.0	13.23	80.9	10.42
2.5	90.5	14.13	88.5	12.59	92.5	13.23	94.7	14.58	86.9	10.72
3.0	94.5	15.00	92.5	13.95	96.6	13.87	99.5	16.79	89.0	11.67
3.5	97.9	15.90	96.3	15.05	101.0	15.19	105.0	17.93	92.5	12.25
4.0	101.2	16.84	100.0	15.84	104.0	15.70	109.0	19.37	95.9	12.76
4.5	104.1	17.67	103.9	17.25	107.2	16.65	112.5	20.90	98.0	13.65
5.0	107.0	18.60	107.7	18.50	110.2	17.65	116.0	22.46	100.0	14.50
5.5	110.2	19.60	110.6	18.74	113.7	18.73	119.2	23.30	103.0	15.10
6.0	113.2	20.60	113.0	20.40	116.8	20.30	122.5	24.00	105.8	15.90
6.5	115.9	21.60	115.9	21.30	119.8	21.30	125.8	26.40	108.7	16.75
7.0	118.7	22.57	118.9	22.20	123.5	22.60	129.1	28.70	109.8	17.15
7.5	121.3	23.60	122.4	24.60	126.5	24.85	132.3	30.20	112.4	18.30
8.0	123.8	24.65	125.9	27.00	129.4	25.50	135.4	33.00	114.9	19.50
8.5	127.1	26.40	128.5	28.60	132.1	26.70	138.3	35.17	116.7	19.95
9.0	130.3	28.10	131.6	30.50	134.8	27.92	141.2	37.34	118.5	20.40
9.5	132.0	28.90	134.7	32.40	137.5	29.67	144.2	39.07	120.9	21.95
10.0	135.3	32.20	138.7	37.19	140.3	31.42	147.2	40.80	123.0	22.55
10.5	137.6	33.45	142.2	39.60	143.6	33.80	149.8	44.08	125.2	23.98
11.0	140.0	34.70	145.8	42.00	146.9	36.16	152.4	47.37	127.5	25.42
11.5	144.2	37.20	149.8	45.30	150.9	38.70	156.3	50.20	130.3	27.08
12.0	148.5	39.75	153.8	48.70	154.9	41.30	160.2	53.00	133.1	28.75
12.5	153.2	42.30	155.1	49.50	156.9	43.90	163.8	58.10	137.6	30.10
13.0	155.6	44.50	157.2	53.10	159.8	46.10	167.2	60.30	141.6	34.00
13.5	158.2	47.35	157.2	53.85	160.3	47.65	170.1	66.00	145.7	36.90
14.0	159.4	50.30	158.6	55.95	161.8	49.70	170.7	67.50	147.8	37.75
14.5	160.3	49.25	159.3	57.20	162.5	51.60	171.7	72.80	149.7	40.35
15.0	161.5	50.00	159.5	56.75	163.2	53.40	172.4	73.80	151.8	41.50
15.5	162.0	51.00	159.5	54.10	163.4	54.40	173.4	73.00	151.8	43.25
16.0	162.3	52.00	159.5	55.45	163.4	52.20	174.3	72.40	152.0	42.55
16.5	162.5	52.80	159.6	54.48	164.5	53.85	174.8	73.50	152.3	41.25
17.0	162.6	53.50	159.8	55.00	164.6	52.00	175.2	74.55	152.8	40.70
17.5	162.7	53.20	160.0	56.20	164.6	53.80	175.4	74.00	152.9	42.00
18.0	162.8	53.00	160.0	54.50	164.6	53.60	175.5	73.00	152.9	42.35
18.5	163.0	52.50	160.1	54.00	164.6	53.50	175.5	72.00	152.9	42.20
19.0	163.1	52.00	160.2	53.50	164.6	53.45	175.5	71.20	152.9	42.10

Appendix G

(continued)

postnatal age (years)	girl No. 8		girl No. 9		girl No. 10		girl No. 11		girl No. 12	
	height (cm)	weight (kg)	height (cm)	weight (kg)	height (cm)	weight (kg)	height (cm)	weight (kg)	height (cm)	weight (kg)
4/52	53.8	3.87	53.3	3.77	51.2	3.52	53.5	4.36	52.4	4.25
0.25	59.4	5.18	59.7	5.22	55.0	4.45	59.5	6.04	59.5	5.78
0.50	63.8	7.59	64.0	6.50	60.2	5.37	66.0	8.00	67.5	7.33
0.75	69.6	9.23	67.7	7.65	65.5	6.00	70.6	9.28	71.6	9.15
1.0	72.8	9.65	71.2	8.89	66.7	6.67	73.5	10.18	76.1	9.75
1.5	78.0	10.43	79.1	10.69	71.3	7.89	79.1	11.20	81.1	11.25
2.0	82.5	11.56	83.5	11.05	76.5	8.55	84.5	12.48	86.8	12.39
2.5	88.0	12.55	88.0	12.73	80.4	9.28	88.8	13.30	91.0	14.19
3.0	92.4	13.15	92.7	13.58	84.3	10.02	92.8	14.40	97.6	16.32
3.5	95.4	13.80	95.8	14.23	87.0	10.50	96.2	15.05	100.8	16.65
4.0	98.3	14.56	98.5	15.03	91.3	10.87	99.5	15.73	104.0	17.06
4.5	100.6	15.44	101.6	16.10	94.6	11.60	102.9	17.00	107.4	17.90
5.0	104.6	16.67	105.6	16.40	97.8	12.30	106.2	18.30	110.6	18.70
5.5	107.5	17.53	107.8	17.25	101.0	13.05	109.6	18.95	113.7	20.00
6.0	110.3	18.40	110.0	18.10	102.0	14.00	113.0	19.60	117.0	21.35
6.5	113.3	19.42	112.6	19.50	105.0	14.72	117.4	21.50	120.2	22.55
7.0	116.2	20.45	116.4	20.05	108.0	15.50	120.9	23.00	122.7	23.85
7.5	119.0	21.35	118.9	21.30	111.0	16.25	123.8	24.00	126.1	25.80
8.0	121.8	22.30	121.3	22.55	114.3	17.40	126.6	25.00	129.4	27.70
8.5	125.0	24.05	123.7	23.80	116.7	18.70	129.7	27.40	131.7	29.45
9.0	128.2	25.80	126.3	24.55	119.0	20.00	132.8	29.80	134.0	31.23
9.5	130.9	27.00	128.3	25.65	122.5	21.05	135.5	30.55	136.2	34.20
10.0	133.6	28.17	130.3	26.73	126.0	22.10	138.3	31.35	138.5	37.20
10.5	136.7	29.91	133.2	28.25	130.1	24.15	142.6	33.95	142.4	38.44
11.0	139.9	31.65	136.2	29.75	134.3	26.20	147.0	36.60	146.3	39.68
11.5	142.7	34.31	138.6	30.15	138.1	28.80	150.4	39.50	151.1	42.20
12.0	147.1	36.95	143.6	33.20	141.8	31.50	154.3	42.20	153.1	39.30
12.5	150.6	40.40	147.9	35.65	144.8	32.20	156.4	45.00	156.1	42.80
13.0	154.0	41.70	150.9	38.30	146.8	35.10	158.5	47.90	159.2	46.25
13.5	155.6	43.60	152.4	39.90	147.5	33.55	159.2	51.90	160.7	49.20
14.0	157.2	43.65	153.7	41.34	148.8	37.00	160.1	52.35	164.0	48.40
14.5	158.2	46.30	154.7	42.40	149.2	36.60	160.4	53.35	164.5	50.10
15.0	159.0	48.80	155.4	43.20	149.6	36.20	160.7	54.35	165.0	51.67
15.5	159.2	50.25	155.8	43.80	150.0	37.10	161.0	54.00	165.2	50.00
16.0	159.5	50.00	156.3	46.20	150.3	37.90	161.4	53.75	165.3	48.40
16.5	159.7	49.30	156.7	48.60	150.7	37.60	161.5	54.20	165.4	50.10
17.0	159.8	49.00	157.0	51.25	151.0	37.45	161.5	54.60	165.5	51.70
17.5	159.8	48.80	157.1	50.50	151.0	38.00	161.7	56.10	165.6	52.00
18.0	159.8	48.70	157.2	50.20	151.0	38.60	161.8	57.60	165.6	52.30
18.5	159.8	48.60	157.3	50.00	151.0	39.20	161.9	57.70	165.6	52.50
19.0	159.8	48.50	157.3	49.70	151.0	38.00	161.9	57.80	165.6	52.65

Appendix G

(continued)

postnatal age (years)	girl No. 13		girl No. 14	
	height (cm)	weight (kg)	height (cm)	weight (kg)
4/52	52.0	3.59	53.0	4.30
0.25	58.0	4.93	60.0	5.78
0.50	63.7	6.06	65.7	7.01
0.75	67.3	7.13	69.1	7.93
1.0	71.0	8.35	73.5	8.80
1.5	77.3	9.14	80.0	10.50
2.0	81.4	10.05	86.0	11.60
2.5	86.1	10.70	91.0	12.73
3.0	90.1	11.70	95.5	14.00
3.5	93.6	12.60	99.6	15.28
4.0	97.0	13.60	103.7	16.60
4.5	100.2	14.20	107.7	17.90
5.0	103.2	15.45	111.5	18.80
5.5	106.0	16.00	114.2	20.25
6.0	108.9	16.80	117.4	21.60
6.5	111.9	17.65	120.3	22.90
7.0	114.9	18.55	123.2	24.10
7.5	117.9	19.45	126.1	25.27
8.0	121.2	20.40	129.1	27.20
8.5	124.5	21.35	132.1	29.10
9.0	126.6	22.70	135.5	30.60
9.5	128.8	23.80	138.8	32.00
10.0	130.8	25.00	140.9	33.90
10.5	133.1	26.00	143.0	35.90
11.0	135.4	27.00	145.8	39.60
11.5	138.8	28.50	147.6	39.90
12.0	142.9	31.00	150.6	42.40
12.5	147.2	33.62	154.5	46.05
13.0	150.6	36.00	158.4	49.60
13.5	155.0	37.40	161.7	55.40
14.0	157.9	40.50	164.5	58.70
14.5	160.0	41.60	165.0	60.00
15.0	161.3	44.00	165.5	61.85
15.5	162.3	47.40	166.0	61.10
16.0	162.6	49.15	166.4	62.40
16.5	163.0	48.20	166.7	63.60
17.0	163.3	49.00	167.0	64.00
17.5	163.5	49.50	167.2	64.20
18.0	163.8	49.60	167.3	65.60
18.5	164.0	49.70	167.3	67.00
19.0	164.2	49.75	167.4	66.60

Appendix H

Cross-sectional data of Jolicoeur and Pirlot (1988) on the postnatal age, body length, body weight and brain weight of 44 male white rats

(see section 35.14; data reproduced with the permission of Dr. David E. Harrison, Editor-in-Chief, *Growth, Development & Aging*)

postnatal age (days)	body length (mm)	body weight (g)	brain weight (mg)
0	50	5.6	280
0	52	6.2	270
0	54	6.3	290
0	52	6.6	290
10	82	17.8	950
10	85	17.9	960
10	82	18.0	980
10	83	18.9	1040
20	105	32.3	1290
20	103	34.1	1420
20	109	37.0	1360
20	106	35.9	1460
30	137	57.5	1320
30	129	54.0	1270
30	127	57.3	1310
30	128	55.3	1350
40	159	103.9	1470
40	162	105.0	1610
40	165	101.6	1560
40	159	105.8	1410
50	172	128.6	1620
50	163	117.8	1590
50	177	139.8	1640
50	159	131.4	1470
60	180	169.5	1400
60	186	189.2	1750
60	205	187.2	1710
60	184	173.3	1640
70	199	208.8	1600
70	193	205.1	1640
70	186	191.8	1630
70	188	188.7	1600
80	205	196.5	1730
80	202	215.5	1800
80	210	209.5	1630
80	201	217.0	1750
90	220	254.2	1700
90	220	248.4	1740
90	224	254.4	1810
90	217	248.5	1660
100	218	270.6	1780
100	219	273.7	1820
100	217	276.7	1760
100	221	267.3	1750

Appendix I

Special symbols

symbol	example	meaning
$\equiv$	$X \equiv Y$	X is algebraically identical to Y
$\doteq$	$X \doteq Y$	X is approximately equal to Y
$\leftarrow$	$X \leftarrow \mathcal{N}(\mu, \sigma^2)$	X comes from (follows) a normal probability distribution of which the mean is μ and the variance is σ^2

Appendix J

Greek alphabet

upper case	lower case	name	upper case	lower case	name
Α	α	alpha	Ν	ν	nu
Β	β, ϐ	beta	Ξ	ξ	xi
Γ	γ	gamma	Ο	ο	omicron
Δ	δ	delta	Π	π	pi
Ε	ε	epsilon	Ρ	ρ	rho
Ζ	ζ	zeta	Σ	σ, ς	sigma
Η	η	eta	Τ	τ	tau
Θ	θ, ϑ	theta	Υ	υ	upsilon
Ι	ι	iota	Φ	φ, ϕ	phi
Κ	ϰ, κ	kappa	Χ	χ	chi
Λ	λ	lambda	Ψ	ψ	psi
Μ	μ	mu	Ω	ω	omega

Bibliography

(including some relevant references not cited in the text)

ABRAMOWITZ, M., STEGUN, I. A.
1968 *Handbook of Mathematical Functions, with Formulas, Graphs, and Mathematical Tables.* National Bureau of Standards, Applied Mathematics Series, No. 55. U. S. Government Printing Office, Washington, D. C.

AITKEN, A. C.
1956 *Determinants and Matrices.* 9th edition. Oliver and Boyd, Edinburgh

ALLARD, C., JOLICOEUR, P., GOULET, C.
1970 Cholesterol determinations: how many? *Canadian Medical Association Journal* **102,** 1398-1399.

ALTMAN, P. L., DITTMER, D. S.
1966 *Environmental Biology.* Biological Handbooks, Federation of American Societies for Experimental Biology, Bethesda, Maryland.

ANDERSON, E.
1954 Efficient and inefficient methods of measuring specific differences. Chapter 6, pages 93-106, in *Statistics and Mathematics in Biology* (O. Kempthorne, T. A. Bancroft, J. W. Gowen, J. L. Lush, ed.). Iowa State College Press, Ames. [reprinted in 1964 by Hafner, New York].

ANDERSON, R. L.
1946 Missing-plot techniques. *Biometrics* **2,** 41-47.

ANDERSON, T. W.
1958 *An Introduction to Multivariate Statistical Analysis.* Wiley, New York.

1963 Asymptotic theory for principal component analysis. *Annals of Mathematical Statistics* **34,** 122-148.

1984 *An Introduction to Multivariate Statistical Analysis.* 2nd edition. Wiley, New York.

ANDREWS, D. F., HERZBERG, A. M.
1985 *Data.* Springer-Verlag, New York.

ASHBY, E., OXLEY, T. A.
1935 The interaction of factors in the growth of *Lemna*. VI. *Annals of Botany* **49,** 309-336.

ATCHLEY, W. R., BRYANT, E. H.
1975 *Multivariate Statistical Methods, Among-Groups Covariation.* Benchmark Papers in Systematic and Evolutionary Biology, 1. Dowden, Hutchinson & Ross, Stroudsburg, Pennsylvania.

BARON, G., JOLICOEUR, P.
1980 Brain structure in Chiroptera: some multivariate trends. *Evolution* **34,** 386-393.

BARTLETT, M. S.
1954 A note on the multiplying factors for various χ^2 approximations. *Journal of the Royal Statistical Society, Series B* **16,** 296-298.

BATES, D. M., WATTS, D. G.
1988 *Nonlinear regression analysis and its applications.* Wiley, New York.

BATSCHELET, E.
1976 *Introduction to Mathematics for Life Scientists.* 2nd edition. Springer-Verlag, New York.

1981 *Circular Statistics in Biology.* Academic Press, London.

BEAUMONT, R. A.
1965 *Linear Algebra.* Harcourt, Brace & World, Inc., New York.

BEERSTECHER, E., SUTTON, H. E., BERRY, H. K., BROWN, W. D., REED, J., RICH, G. B., BERRY, L. J., WILLIAMS, R. J.
1950 Biochemical individuality. V. Explorations with respect to the metabolic patterns of compulsive drinkers. *Archives of Biochemistry* **29,** 27-40.

BELEHRADEK, J.
1935 *Temperature and Living Matter.* Borntraeger, Berlin.

BERTALANFFY, L. von
1951 *Theoretische Biologie.* Zweiter Band: *Stoffwechsel, Wachstum.* Francke, Bern.

BLACKITH, R. E., REYMENT, R. A.
1971 *Multivariate Morphometrics.* Academic Press, London.

BLISS, C. I., JAMES, A. T.
1966 Fitting the rectangular hyperbola. *Biometrics* **22,** 573-602.

BOCK, R. D., THISSEN, D.
1976 Fitting multi-component models for growth in stature. *Proceedings of the 9th International Biometrics Conference* 1, 431-442.

BOX, G. E. P., HUNTER, W. G., HUNTER, J. S.
1978 *Statistics for Experimenters.* Wiley, New York.

BRADSHAW, J. L., ROGERS, L. J.
1993 *The Evolution of Lateral Asymmetries, Language, Tool Use, and Intellect.* Academic Press, Inc., San Diego.

BREEDLOVE, D. E., EHRLICH, P. R.
1968 Plant-herbivore coevolution: lupines and lycaenids. *Science* **162,** 671-672.

BRIDGMAN, P. W.
1931 *Dimensional Analysis.* Yale University Press, New Haven.

BRODY, S.
1945 *Bioenergetics and Growth.* Reinhold Publishing Corporation, New York. [reprinted in 1964 by Hafner, New York].

BROWER, L. P.
1959 Speciation in butterflies of the *Papilio glaucus* group. I. Morphological relationships and hybridization. *Evolution* **13,** 40-63.

BROWN, J. D., BEERSTECHER, E.
1951 Metabolic patterns of underweight and overweight individuals. *Univ.Texas Publ., Biochem. Inst. Studies* **4,** 181-188.

BRYANT, E. H., ATCHLEY, W. R.
1975 *Multivariate Statistical Methods, Within-Groups Covariation.* Benchmark Papers in Systematic and Evolutionary Biology, 2. Dowden, Hutchinson & Ross, Stroudsburg, Pennsylvania.

BUTLER, G. E., McKIE, M., RATCLIFFE, S. G.
1990 The cyclical nature of prepubertal growth. *Annals of Human Biology* **17,** 177-198.

CABANA, T., JOLICOEUR, P., BARON, G.
1990 Brain and body growth and allometry in the Mongolian Gerbil (*Meriones unguiculatus*). *Growth, Development & Aging* **54,** 23-30.

CABANA, T., JOLICOEUR, P., MICHAUD, J.
1993 Prenatal and postnatal growth and allometry of stature, head circumference and brain weight in Québec children. *American Journal of Human Biology* **5,** 93-99.

CAUSTON, D. R., VENUS, J. C.
1981 *The Biometry of Plant Growth.* Arnold, London.

COCHRAN, W. G.
1954 Some methods for strengthening the common χ^2 tests. *Biometrics* **10,** 417-451.
1957 Analysis of covariance: its nature and uses. *Biometrics* **13,** 261-281.

COREN, S., PORAC, C.
1977 Fifty centuries of right-handedness: the historical record. *Science* **198,** 631-632.

CREASY, M. A.
1956 Confidence limits for the gradient in the linear functional relationship. *Journal of the Royal Statistical Society, Series B* **18,** 65-69.

CURRIE, D. J.
1982 Estimating Michaelis-Menten parameters: bias, variance and experimental design. *Biometrics* **38,** 907-919.

DAGNELIE, P.
1975 *Analyse statistique à plusieurs variables.* Les Presses agronomiques de Gembloux, Belgique.

DARROCH, J. N., MOSIMANN, J. E.
1985 Canonical and principal components of shape. *Biometrika* **72,** 241-252.

DAVID, F. N.
1938 *Tables of the Ordinates and Probability Integral of the Distribution of the Correlation Coefficient in Small Samples.* Cambridge University Press, London.

DAWES, E. A.
1967 *Quantitative Problems in Biochemistry.* 4th edition. Livingstone, Edinburgh.

DAWOOD, N., JOLICOEUR, P., SHARIEF, S. D.
1988 Postnatal brain growth and allometry in the rabbit *Oryctolagus cuniculus*. *Growth, Development & Aging* **52,** 169-175.

DEMPSTER, A. P.
1969 *Elements of Continuous Multivariate Analysis.* Addison-Wesley, Reading, Mass.

DONALDSON, J. R., SCHNABEL, R. B.
1987 Computational experience with confidence regions and confidence intervals for nonlinear least squares. *Technometrics* **29,** 67-82.

DRAPER, N. R., SMITH, H.
1981 *Applied Regression Analysis.* 2nd edition. Wiley, New York.

FELDMAN, H. A.
1995 On the allometric mass exponent, when it exists. *Journal of Theoretical Biology* **172,** 187-197.

FERRON, J.
1987 L'optimisation du comportement animal. Chapter 1, pages 2-36 in *Une approche mathématique de la biologie* (R. V. Jean, ed.). Gaëtan Morin, Chicoutimi.

FIELLER, E. C.
1954 Some problems in interval estimation. *Journal of the Royal Statistical Society, Series B* **16,** 175-185, 219-221.

FISHER, R. A.
1936 The use of multiple measurements in taxonomic problems. *Annals of Eugenics* **7,** 179-188.

1950 *Contributions to Mathematical Statistics.* Wiley, New York.

1960 *The Design of Experiments.* 7th edition. Oliver and Boyd, Edinburgh.

1970 *Statistical Methods for Research Workers.* 14th edition. Oliver and Boyd, Edinburgh.

FLURY, B.
1988 *Common Principal Components and Related Multivariate Models.* Wiley, New York.

FORD, P.
1958 Studies on the development of the kidney of the Pacific pink salmon (*Onchorynchus gorbuscha* (Walbaum)) II. Variation in glomerular count of the kidney of the Pacific pink salmon. *Canadian Journal of Zoology* **36,** 45-47.

FRETS, G. P.
1921 *Heredity of Headform in Man.* Martinus Nijhoff, The Hague.

GAUSE, G. F.
1934 *The Struggle for Existence.* Williams & Wilkins, Baltimore. [reprinted in 1971 by Dover Publications, Inc., New York].

GEISSLER, A.
1889 Beiträge zur Frage des Geschlechtsverhältnisses der Geborenen. *Zeitschrift des Königlich Sächsischen Statistischen Bureaus.*

GNANADESIKAN, R.
1977 *Methods for Statistical Data Analysis of Multivariate Observations.* Wiley, New York.

GRAY, J.
1928 The growth of fish. III. The effect of temperature on the development of the eggs of *Salmo fario*. *British Journal of Experimental Biology* **6**, 125-130.

GRAYBILL, F. A.
1969 *Introduction to Matrices with Applications in Statistics*. Wadsworth, Belmont.

GREEN, R. H.
1971 A multivariate statistical approach to the Hutchinsonian niche: Bivalve molluscs of central Canada. *Ecology* **52**, 543-556.

1979 *Sampling Design and Statistical Methods for Environmental Biologists*. Wiley, New York.

GREENWOOD, M., BROWN, J. W.
1913 A second study of the weight, variability and correlation of the human viscera. *Biometrika* **9**, 473-485.

HALDANE, J. S., PRIESTLEY, J. G.
1905 The regulation of the lung ventilation. *Journal of Physiology* **32**, 225-266.

HAUSPIE, R. C.
1989 Mathematical models for the study of individual growth patterns. *Revue d'Épidémiologie et de Santé Publique* **37**, 461-476.

HAYES, F. R., ANTHONY, E. H.
1958 Lake water and sediment. I. Characteristics and water chemistry of some Canadian east coast lakes. *Limnology and Oceanography* **3**, 299-307.

HEUSNER, A. A.
1991 Size and power in mammals. *Journal of Experimental Biology* **160**, 25-54.

HEUSNER, A. A., JOLICOEUR, P.
1972 Allometry of the standard oxygen consumption in the white rat. Pages 22-24 in *Proceedings of the International Symposium on Environmental Physiology: Bioenergetics*, Dublin, July 18-23, 1971. Federation of American Societies for Experimental Biology, Bethesda, Maryland.

HILL, A. V.
1913 The combinations of haemoglobin with oxygen and with carbon monoxide. *Biochemical Journal* **7**, 471-480.

HOTELLING, H.
1931 The generalization of Student's ratio. *Annals of Mathematical Statistics* **2**, 360-378.

1933 Analysis of a complex of statistical variables into principal components. *Journal of Educational Psychology* **24**, 417-441, 498-520.

1936 Relations between two sets of variates. *Biometrika* **28**, 321-377.

HUNT, R.
1982 *Plant Growth Curves*. Arnold, London.

HUXLEY, J. S.
1932 *Problems of Relative Growth*. Methuen, London.

JACKSON, J. E.
1991 *A User's Guide to Principal Components.* Wiley, New York.

JACOBI, C. G. J.
1846 Über ein leichtes Verfahren, die in der Theorie der Säkularstörungen vorkommenden Gleichungen numerisch aufzulösen.
Journal für die reine und angewandte Mathematik **30**, 51-95.

JAMES, A. T.
1977 Tests for a prescribed subspace of principal components. Pages 73-77 in *Multivariate Analysis – IV, Proceedings of the Fourth International Symposium on Multivariate Analysis* (P. R. Krishnaiah, ed.). North-Holland, Amsterdam.

JEAN, R. V.
1987 Croissance différentielle, formule allométrique de Huxley et croissance sigmoïde. Chapter 5, pages 128-155 in *Une approche mathématique de la biologie* (R. V. Jean, ed.). Gaëtan Morin, Chicoutimi.

JOLICOEUR, P.
1959 Multivariate geographical variation in the wolf *Canis lupus* L.
Evolution **13**, 283-299.

1963a The multivariate generalization of the allometry equation.
Biometrics **19**, 497-499.

1963b The degree of generality of robustness in *Martes americana.*
Growth **27**, 1-27.

1963c Bilateral symmetry and asymmetry in limb bones of *Martes americana* and man.
Revue Canadienne de Biologie **22**, 409-432.

1963d Les combinaisons multidimensionnelles de caractères anatomiques quantitatifs.
Proceedings of the XVI International Congress of Zoology, Washington, **1**, 183.

1965 Calcul d'un intervalle de confiance pour la pente de l'axe majeur de la distribution normale de deux variables. *Biométrie-Praximétrie* **6**, 31-35.

1968 Interval estimation of the slope of the major axis of a bivariate normal distribution in the case of a small sample. *Biometrics* **24**, 679-682.

1973 Imaginary confidence limits of the slope of the major axis of a bivariate normal distribution: a sampling experiment.
Journal of the American Statistical Association **68**, 866-871.

1975a Linear regressions in fishery research: some comments.
Journal of the Fisheries Research Board of Canada **32**, 1491-1494.

1975b Sexual dimorphism and geographical distance as factors of skull variation in the wolf *Canis lupus* L. Chapter 4, pages 54-61, in *The Wild Canids* (M. W. Fox, ed.). Van Nostrand Reinhold, New York.

1984 Principal components, factor analysis, and multivariate allometry: a small–sample direction test. *Biometrics* **40**, 685-690.

1985 A flexible 3-parameter curve for limited or unlimited somatic growth.
Growth **49**, 271-281.

JOLICOEUR, P.
1989 A simplified model for bivariate complex allometry. *Journal of Theoretical Biology* **140,** 41-49.

1990 Bivariate allometry: interval estimation of the slopes of the ordinary and standardized normal major axes and structural relationship. *Journal of Theoretical Biology* **144,** 275-285.

1995 Unlinearized confidence limits for the statistical recovery of non-linear relationships in biology. *Journal of Theoretical Biology* **172,** 217-224.

1996a Combinaisons linéaires *a priori* de logarithmes de variables lognormales. Communication présentée aux 28ièmes Journées de Statistique, 27-30 mai 1996, Université Laval, Québec.

1996b Biometry, a meeting ground between biologists and statisticians? *Biometric Bulletin* **13,** 2.

JOLICOEUR, P., ABIDI, H., PONTIER, J.
1991 Human stature: which growth model? *Growth, Development & Aging* **55,** 129-132.

JOLICOEUR, P., .BARON, G.
1980 Brain center correlations among Chiroptera. *Brain, Behavior & Evolution* **17,** 419-431.

JOLICOEUR, P., BARON, G., CABANA, T.
1988 Cross-sectional growth and decline of human stature and brain weight in 19th-century Germany. *Growth, Development & Aging* **52,** 201-206.

JOLICOEUR, P., BRUNEL, P.
1966 Application du diagramme hexagonal à l'étude de la sélection de ses proies par la morue. *Vie et Milieu* **17,** 419-433.

JOLICOEUR, P., CABANA, T., DUCHARME, G.
1992 A four-parameter generalization of the Gompertz curve suitable for somatic growth. *Growth, Development & Aging* **56,** 69-74.

JOLICOEUR, P., DUCHARME, G.
1992 Bivariate allometry: point estimation of the slope of the normal major axis. *Journal of Theoretical Biology* **154,** 35-41.

JOLICOEUR, P., HEUSNER, A. A.
1971 The allometry equation in the analysis of the standard oxygen consumption and body weight of the white rat. *Biometrics* **27,** 841-855.

1986 Log-normal variation belts for growth curves. *Biometrics* **42,** 785-794.

JOLICOEUR, P., MOSIMANN, J. E.
1960 Size and shape variation in the Painted Turtle. A principal component analysis. *Growth* **24,** 339-354.

1968 Intervalles de confiance pour la pente de l'axe majeur d'une distribution normale bidimensionnelle. *Biométrie-Praximétrie* **9,** 121-140.

JOLICOEUR, P., PIRLOT, P.
1988 Asymptotic growth and complex allometry of the brain and body in the white rat. *Growth, Development & Aging* **52,** 3-9.

JOLICOEUR, P., PIRLOT, P., BARON, G., STEPHAN, H.
1984 Brain structure and correlation patterns in Insectivora, Chiroptera, and Primates. *Systematic Zoology* **33,** 14-29.

JOLICOEUR, P., PONTIER, J.
1989 Population growth and decline: a four-parameter generalization of the logistic curve. *Journal of Theoretical Biology* **141,** 563-571.

1993 Peut-on utiliser des modèles de croissance longitudinale dans l'analyse de données transversales? *Biométrie-Praximétrie* **33,** 33-44.

JOLICOEUR, P., PONTIER, J., ABIDI, H.
1992 Asymptotic models for the longitudinal growth of human stature. *American Journal of Human Biology* **4,** 461-468.

JOLICOEUR, P., PONTIER, J., PERNIN, M.-O., SEMPÉ, M.
1988 A lifetime asymptotic growth curve for human height. *Biometrics* **44,** 995-1003.

JOLIVET, E.
1983 *Introduction aux modèles mathématiques en biologie.* Masson, Paris.

JUDAY, C., BIRGE, E. A., MELOCHE, V. W.
1941 Chemical analyses of the bottom deposits of Wisconsin lakes. II. Second Report. *Transactions of the Wisconsin Academy of Sciences, Arts and Letters* **33,** 99-114.

KANEFUJI, K., SHOHOJI, T.
1990 On a growth model of human height. *Growth, Development & Aging* 54, 155-165.

KARPINOS, B. D.
1958 Height and weight of selective service registrants processed for military service during world war II. *Human Biology* **30,** 292-321.

KAVANAGH, A. J., RICHARDS, O. W.
1942 Mathematical analysis of the relative growth of organisms. *Proceedings of the Rochester Academy of Science* **8,** 150-174.

KENDALL, M. G.
1957 *A Course in Multivariate Analysis.* Griffin, London.

KENDALL, M. G., STUART, A.
1963 *The Advanced Theory of Statistics.* Volume 1, 2nd edition. Griffin, London.

1966 *The Advanced Theory of Statistics.* Volume 3. Griffin, London.

1973 *The Advanced Theory of Statistics.* Volume 2, 3rd edition. Griffin, London.

KENNEDY, W. J., Jr., GENTLE, J. E.
1980 *Statistical Computing.* Dekker, New York.

KENNEY, J. F., KEEPING, E. S.
1951 *Mathematics of Statistics.* Part Two. 2nd edition. Van Nostrand, Princeton.

KENNEY, J. F., KEEPING, E. S.
1954 *Mathematics of Statistics*. Part One. 3rd edition. Van Nostrand, Princeton.

KERMACK, K. A., HALDANE, J. B. S.
1950 Organic correlation and allometry. *Biometrika* **37,** 30-41.

KIMURA, D. K.
1992 Symmetry and scale dependence in functional relationship regression. *Systematic Biology* **41,** 233-241.

KRUSKAL, W. H.
1953 On the uniqueness of the line of organic correlation. *Biometrics* **9,** 47-58.

KRZANOWSKI, W. J.
1989 On confidence regions in canonical variate analysis. *Biometrika* **76,** 107-116.

KSHIRSAGAR, A. M.
1961 The goodness-of-fit of a single (non-isotropic) hypothetical principal component. *Biometrika* **48,** 397-407.
1972 *Multivariate Analysis*. Dekker, New York.

LAWS, R. M.
1953 The elephant seal (*Mirounga leonina* Linn.). I. Growth and age. *Falkland Islands Dependencies Survey, Scientific Reports,* No. 8.

LEBART, L., MORINEAU, A., PIRON, M.
1995 *Statistique exploratoire multidimensionnelle*. Dunod, Paris.

LEBEAU, B.
1984 La reproduction du maskinongé *Esox masquinongy* Mitchill au lac Barrière et au lac St-Louis, Province de Québec. Mémoire de M. Sc., Département de Sciences biologiques, Université de Montréal.

LEBEAU, B., JOLICOEUR, P., PAGEAU, G., CROSSMAN, E. J.
1986 Asymptotic growth, egg production and trivariate allometry in *Esox masquinongy* Mitchill. *Growth* **50,** 185-200.

LEBRETON, J. D., MILLIER, C., JOLIVET, E., PAVÉ, A., VILA, J. P.
1982 *Modèles dynamiques déterministes en biologie*. Masson, Paris.

LEE, R. E., GILBERT, C. A.
1918 On the application of the mass law to the process of disinfection. *Journal of Physical Chemistry* **22,** 348-372.

LEFEBVRE, J., TASSENCOURT, L.
1968 Le problème de la précision dans le calcul des corrélations canoniques. *Biométrie-Praximétrie* **9,** 161-173.

LEGAY, J.-M.
1997 *L'expérience et le modèle, Un discours sur la méthode. "Sciences en questions".* Institut National de la Recherche Agronomique, Paris.

LEPRINCE, D. J., JOLICOEUR, P.
1986a Annual, intra-, and inter-specific variations in body size and potential fecundity of some *Tabanus* species (*Diptera: Tabanidae*).
Canadian Entomologist **118,** 1265-1272.

1986b Response to carbon dioxide of *Tabanus quinquevittatus* Wiedemann females (*Diptera, Tabanidae*) in relation to relative abundance, parity, follicle development, and sperm and fructose presence.
Canadian Entomologist **118,** 1273-1277.

LESLIE, P. H.
1945 On the use of matrices in certain population mathematics.
Biometrika **33,** 183-212.

LUBISCHEW, A. A.
1962 On the use of discriminant functions in taxonomy. *Biometrics* **18,** 455-477.

LUMER, H.
1937 The consequences of sigmoid growth for relative growth functions.
Growth **1,** 140-154.

MACKINTOSH, N. A., WHEELER, J. F. G.
1929 Southern blue and fin whales. *Discovery Reports* **I,** 257-540.

MALLOWS, C. L.
1961 Latent vectors of random symmetric matrices. *Biometrika* **48,** 133-149.

MARCHAND, F.
1902 Ueber das Hirngewicht des Menschen. *Abhandlungen der Königlich Säschsischen Gesellschaft der Wissenschaften, mathematisch-physikalische Classe* **27,** 391-482.

MARDIA, K. V.
1972 *Statistics of Directional Data.* Academic Press, London.

McALLISTER, D. E., JOLICOEUR, P., TSUYUKI, H.
1972 Morphological and myogen comparison of johnny and tessellated darters and their hybrids, genus *Etheostoma,* near Ottawa, Canada.
Journal of the Fisheries Research Board of Canada **29,** 1173-1180.

MICHAELIS, L., MENTEN, M. L.
1913 Die Kinetik der Invertinwirkung. *Biochemische Zeitschrift* **49,** 333-369.

MILLER, A. J.
1994 Bivariate confidence regions in non-linear regression.
Applied Statistics **43,** 275-284.

MILLER, R. G.
1966 *Simultaneous Statistical Inference.* McGraw-Hill, New York.

MILLIKAN, C. R.
1963 Effects of different levels of zinc and phosphorus on the growth of subterranean clover (*Trifolium subterraneum* L.).
Australian Journal of Agricultural Research **14,** 180-205.

MOOD, A. M., GRAYBILL, F. A.
1963 *Introduction to the Theory of Statistics.* 2nd edition. McGraw-Hill, New York.

MORRISON, D. F.
1967 *Multivariate Statistical Methods.* McGraw-Hill, New York.

1983 *Applied Linear Statistical Methods.* Prentice-Hall, Englewood Cliffs, New Jersey.

1990 *Multivariate Statistical Methods.* 3rd edition. McGraw-Hill, New York.

MOSIMANN, J. E.
1962 On the compound multinomial distribution, the multivariate β-distribution, and correlations among proportions. *Biometrika* **49,** 65-82.

1963 On the compound negative multinomial distribution and correlations among inversely sampled pollen counts. *Biometrika* **50,** 47-54.

1968 *Elementary Probability for the Biological Sciences.* Appleton-Century-Crofts, New York.

1970 Size allometry: size and shape variables with characterizations of the lognormal and generalized gamma distribution. *Journal of the American Statistical Association* **65,** 930-945.

MOSIMANN, J. E., CAMPBELL, G.
1988 Applications in Biology: Simple Growth Models. Chapter 11, pages 287-302, in *Lognormal Distributions: Theory and Applications* (E. L. Crow et K. Shimizu, ed.). Dekker, New York.

MÜNTER, A. H.
1936 A study of the lengths of the long bones of the arms and legs in man, with special reference to Anglo-Saxon skeletons. *Biometrika* **28,** 258-294.

MURDOCH, D. C.
1957 *Linear Algebra for Undergraduates.* Wiley, New York.

NELDER, J. A., MEAD, R.
1965 A simplex method for function minimization. *Computer Journal* **7,** 308-313.

NIKLAS, K. J.
1994 The allometry of safety-factors for plant height. *American Journal of Botany* **81,** 345-351.

OLSSON, D. M., NELSON, L. S.
1975 The Nelder-Mead simplex procedure for function minimization. *Technometrics* **17,** 45-51.

O'NEILL, R.
1971 Function minimization using a simplex procedure. *Applied Statistics* **20,** 338-345.

PAGEAU, G.
1967 Comportement, alimentation et croissance de l'achigan à petite bouche (*Micropterus dolomieui* Lacépède) dans la plaine de Montréal et dans les Laurentides. Thèse de Ph. D., Département de Sciences biologiques, Université de Montréal.

PEARSON, E. S., HARTLEY, H. O.
1966 *Biometrika Tables for Statisticians.* Volume 1. 3rd edition. Cambridge University Press, London.

PEARSON, K.
1900 On the criterion that a given system of deviations from the probable in the case of a correlated system of variables is such that it can be reasonably supposed to have arisen from random sampling. *London, Edinburgh and Dublin Philosophical Magazine and Journal of Science,* Series 5, **50,** 157-175.

1901 On lines and planes of closest fit to systems of points in space. *Philosophical Magazine,* Series 6, **2,** 559-572.

PEARSON, K., LEE, A.
1900 Mathematical contributions to the theory of evolution. VIII. On the inheritance of characters not capable of exact quantitative measurement. *Philosophical Transactions of the Royal Society of London (Series A)* **195,** 79-150.

PIRLOT, P., JOLICOEUR, P.
1982 Correlations between major brain regions in Chiroptera. *Brain, Behavior & Evolution* **20,** 172-181.

PONTIER, J., DUFOUR, A.-B., NORMAND, M.
1990 *Le modèle euclidien en analyse de données.* Éditions de l'Université de Bruxelles, Bruxelles.

PONTIER, J., JOLICOEUR, P.
1996 Pourquoi une modélisation mathématique de la courbe de croissance staturale? *Cahiers d'Anthropologie et Biométrie Humaine (Paris)* **14,** 45-53.

PONTIER, J., JOLICOEUR, P., PERNIN, M.-O.
1987 Analyse canonique complète. *Statistique et Analyse des Données* **12,** 124-148.

PONTIER, J., JOLICOEUR, P., ABIDI, H., SEMPÉ, M.
1988 Croissance staturale chez l'enfant: le modèle JPPS. *Biométrie-Praximétrie* **28,** 27-44.

PONTIER, J., JOLICOEUR, P., PERNIN, M.-O., ABIDI, H., SEMPÉ, M.
1988 Modélisation de la courbe de croissance staturale chez l'enfant: le modèle JPPS. *Cahiers d'Anthropologie et Biométrie Humaine (Paris)* **6,** 71-85.

PONTIER, J., NORMAND, M., JOLICOEUR, P.
1989 Modélisation JPPS d'une courbe de maturation osseuse. *Biométrie-Praximétrie* **29,** 49-62.

POPJAK, G., MUIR, H.
1950 In search of a phospholipin precursor. *Biochemical Journal* **46,** 103-113.

PREECE, M. A., BAINES, M. J.
1978 A new family of mathematical models describing the human growth curve. *Annals of Human Biology* **5,** 1-24.

PRUNTY, L.
1983 Curve fitting with smooth functions that are piecewise-linear in the limit. *Biometrics* **39,** 857-866.

PÜTTER, A.
1920 Studien über physiologische Ähnlichkeit. VI. Wachstumsähnlichkeiten. *Pfluegers Archiv für die gesamte Physiologie* **180,** 298-340.

RAO, C. R.
1952 *Advanced Statistical Methods in Biometric Research.* Wiley, New York. [reprinted with corrections in 1970 by Hafner, New York].

1965 *Linear Statistical Inference and Its Applications.* Wiley, New York.

RATKOWSKY, D. A.
1983 *Nonlinear Regression Modeling.* Dekker, New York.

1990 *Handbook of Nonlinear Regression Models.* Dekker, New York.

RAWSON, D. S.
1960 A limnological comparison of twelve large lakes in northern Saskatchewan. *Limnology and Oceanography* **5,** 195-211.

REMPE, U.
1962 Über einige statistische Hilfsmittel moderner zoologisch-systematischer Untersuchungen. *Zoologischer Anzeiger* **169,** 93-140.

RIGGS, D. S.
1970 *The Mathematical Approach to Physiological Problems.* M.I.T. Press, Cambridge.

RINGROSE, T. J., KRZANOWSKI, W. J.
1991 Simulation study of confidence regions for canonical variate analysis. *Statistics and Computing* **1,** 41-46.

ROSS, G. S.
1990 *Nonlinear Estimation.* Springer-Verlag, New York.

ROY, S. N.
1957 *Some Aspects of Multivariate Analysis.* Wiley, New York.

SCHEBESTA, P., LEBZELTER, V.
1933 *Anthropology of the Central African Pygmies in the Belgian Congo.* Czech Academy of Sciences and Arts, Prague.

SCHEFFÉ, H.
1959 *The Analysis of Variance.* Wiley, New York.

SCHREIER, O., SPERNER, E.
1959 *Modern Algebra and Matrix Theory.* Chelsea, New York.

SCHWERDTFEGER, F.
1963 *Ökologie der Tiere.* Band I: *Autökologie.* Verlag Paul Parey, Berlin.

SCOTT, J. H.
1957 Muscle growth and function in relation to skeletal morphology. *American Journal of Physical Anthropology* N. S., **15,** 197-234.

SEAL, H. L.
1964 *Multivariate Statistical Analysis for Biologists.* Methuen, London.

SEARLE, S. R.
1966 *Matrix Algebra for the Biological Sciences.* Wiley, New York.

SEBER, G. A. F.
1984 *Multivariate Observations.* Wiley, New York.

SEBER, G. A. F., WILD, C. J.
1989 *Nonlinear Regression.* Wiley, New York.

SHEA, B. T.
1985 Bivariate and multivariate growth allometry: statistical and biological considerations. *Journal of Zoology, London (A)* **206,** 367-390.

SHOHOJI, T., SASAKI, H.
1987 Individual growth of stature of Japanese. *Growth* **51,** 432-450.

SIMON, W.
1977 *Mathematical Techniques for Biology and Medicine.* M.I.T. Press, Cambridge.

SMITH, J. M.
1957 Temperature tolerance and acclimatization in *Drosophila subobscura. Journal of Experimental Biology* **34,** 85-96.

SNEDECOR, G. W., COCHRAN, W. G.
1967 *Statistical Methods.* 6th edition. Iowa State University Press, Ames, Iowa.

SPRENT, P.
1969 *Models in Regression and Related Topics.* Methuen, London.
1972 The mathematics of size and shape. *Biometrics* **28,** 23-37.
1993 *Applied Nonparametric Statistical Methods.* 2nd edition. Chapman & Hall, London.

STAMP, N. E.
1980 Egg deposition patterns in butterflies: why do some species cluster their eggs rather than deposit them singly? *American Naturalist* **115,** 367-380.

STIGLER, S. M.
1986 *The History of Statistics. The Measurement of Uncertainty before 1900.* Harvard University Press, Cambridge.

STRAUSS, R. E., ALTIG, R.
1992 Ontogenetic body form changes in three ecological morphotypes of anuran tadpoles. *Growth, Development & Aging* **56,** 3-16.

STUDENT
1908 The probable error of a mean. *Biometrika* **6,** 1-25.

TEISSIER, G.
1948 La relation d'allométrie: sa signification statistique et biologique. *Biometrics* **4,** 14-53.
1955 Sur la détermination de l'axe d'un nuage rectiligne de points. *Biometrics* **11,** 344-356.
1960 Relative growth. Chapter 16, pages 537-560, in *The Physiology of Crustacea, Vol.1, Metabolism and Growth* (T. H. Waterman, ed.). Academic Press, New York.
1961 Choix des paramètres définissant une relation structurale linéaire dans la pratique biométrique. *Biométrie-Praximétrie* **2,** 137-158.

TERROINE, É. F., TRAUTMANN, S.
1927 Influence de la température extérieure sur la production calorique des homéothermes et loi des surfaces.
Annales de Physiologie et de Physicochimie biologique **3,** 422-457.

TOMASSONE, R., AUDRAIN, S., LESQUOY-de TURCKHEIM, É., MILLIER, C.
1992 *La régression, nouveaux regards sur une ancienne méthode statistique.* 2nd edition. Masson, Paris.

TOMASSONE, R., DERVIN, C., MASSON, J.-P.
1993 *Biométrie: modélisation de phénomènes biologiques.* Masson, Paris.

TOUABTI-MIMOUNI, N.
1996 Contribution de méthodes biométriques à l'analyse de la morphotypologie des sportifs. Thèse de Doctorat, Faculté des Sciences du Sport et de l'Éducation Physique, Université Claude-Bernard – Lyon I.

TYLER, D. E.
1981 Asymptotic inference for eigenvectors. *Annals of Statistics* **9,** 725-736.

VAJDA, S.
1960 *An Introduction to Linear Programming and the Theory of Games.* Methuen, London.

VAUGHTON, M.
1970 Analysis of examination results. M. Sc. thesis, Department of Mathematics, University of Adelaide, Australia.

WADE, T. L.
1951 *The Algebra of Vectors and Matrices.* Addison-Wesley, Reading, Massachusetts.

WHITE, P. A.
1958 The computation of eigenvalues and eigenvectors of a matrix.
Journal of the Society for Industrial and Applied Mathematics **6,** 393-437.

WILKINSON, G. N.
1961 Statistical estimations in enzyme kinetics. *Biochemical Journal* **80,** 324-332.

WILSON, E. B., HILFERTY, M. M.
1931 The distribution of chi-square.
Proceedings of the National Academy of Sciences of the U. S. A. **17,** 684-688.

WINSOR, C. P.
1932 The Gompertz curve as a growth curve.
Proceedings of the National Academy of Sciences of the U. S. A. **18,** 1-8.

WORKING, H., HOTELLING, H.
1929 Applications of the theory of error to the interpretation of trends.
Journal of the American Statistical Association **24,** 73-89.

WRIGHT, S.
1968 *Evolution and the Genetics of Populations. Volume 1, Genetic and Biometric Foundations.* University of Chicago Press, Chicago.

YATES, F.
1934 Contingency tables involving small numbers and the χ^2 test.
Supplement to the Journal of the Royal Statistical Society **1,** 217-235.

The statistical tables most frequently used in biometry

The standardized normal distribution

Deviation	Cumulative probability	Deviation	Cumulative probability
0.00	0.500000	1.50	0.933193
0.05	0.519939	1.55	0.939429
0.10	0.539828	1.60	0.945201
0.15	0.559618	1.65	0.950529
0.20	0.579260	1.70	0.955435
0.25	0.598706	1.75	0.959941
0.30	0.617911	1.80	0.964070
0.35	0.636831	1.85	0.967843
0.40	0.655422	1.90	0.971283
0.45	0.673645	1.95	0.974412
0.50	0.691462	2.00	0.977250
0.55	0.708840	2.05	0.979818
0.60	0.725747	2.10	0.982136
0.65	0.742154	2.15	0.984222
0.70	0.758036	2.20	0.986097
0.75	0.773373	2.25	0.987776
0.80	0.788145	2.30	0.989276
0.85	0.802337	2.35	0.990613
0.90	0.815940	2.40	0.991802
0.95	0.828944	2.45	0.992857
1.00	0.841345	2.50	0.993790
1.05	0.853141	2.55	0.994614
1.10	0.864334	2.60	0.995339
1.15	0.874928	2.65	0.995975
1.20	0.884930	2.70	0.996533
1.25	0.894350	2.75	0.997020
1.30	0.903200	2.80	0.997445
1.35	0.911492	2.85	0.997814
1.40	0.919243	2.90	0.998134
1.45	0.926471	2.95	0.998411
1.50	0.933193	3.00	0.998650

The standardized normal distribution

Deviation	Cumulative probability	Deviation	Cumulative probability
3.00	0.998650	4.00	0.999968
3.05	0.998856	4.05	0.999974
3.10	0.999032	4.10	0.999979
3.15	0.999184	4.15	0.999983
3.20	0.999313	4.20	0.999987
3.25	0.999423	4.25	0.999989
3.30	0.999517	4.30	0.999991
3.35	0.999596	4.35	0.999993
3.40	0.999663	4.40	0.999995
3.45	0.999720	4.45	0.999996
3.50	0.999767	4.50	0.999997
3.55	0.999807	4.55	0.999997
3.60	0.999841	4.60	0.999998
3.65	0.999869	4.65	0.999998
3.70	0.999892	4.70	0.999999
3.75	0.999912	4.75	0.999999
3.80	0.999928	4.80	0.999999
3.85	0.999941	4.85	0.999999
3.90	0.999952	4.90	1.000000
3.95	0.999961	4.95	1.000000
4.00	0.999968	5.00	1.000000

NOTES:

The standardized normal distribution being symmetrical about the origin, the only deviations tabulated above are those corresponding to cumulative probabilities equal to or greater than 0.5. The t values corresponding to cumulative probabilities smaller than 0.5 may be obtained through the relationship

$$t_{(P)} = -t_{(1-P)} \,.$$

The values of standardized normal deviations having cumulative probabilities equal to the complements $(1-\alpha)$ of the usual significance levels, such as $\alpha = 0.05$, $\alpha = 0.01$ and $\alpha = 0.001$, or to the complements $(1-\alpha/2)$ of their halves, are given on the last line of the table of Student's t distribution, because the latter becomes more and more similar to the normal distribution when its number ν of degrees of freedom increases. Therefore, the last line of the table of Student's t, for which $\nu = \infty$, corresponds to the normal distribution. A standardized normal deviation having cumulative probability P may thus be denoted either by $t_{(P)}$ or by $t_{(P;\,\infty)}$.

The distribution of Student's t

ν	Cumulative probabilities					
	0.9500	0.9750	0.9900	0.9950	0.9990	0.9995
1	6.3138	12.7062	31.8205	63.6567	318.3088	636.6193
2	2.9200	4.3027	6.9646	9.9248	22.3271	31.5991
3	2.3534	3.1824	4.5407	5.8409	10.2145	12.9240
4	2.1318	2.7764	3.7469	4.6041	7.1732	8.6103
5	2.0150	2.5706	3.3649	4.0321	5.8934	6.8688
6	1.9432	2.4469	3.1427	3.7074	5.2076	5.9588
7	1.8946	2.3646	2.9980	3.4995	4.7853	5.4079
8	1.8595	2.3060	2.8965	3.3554	4.5008	5.0413
9	1.8331	2.2622	2.8214	3.2498	4.2968	4.7809
10	1.8125	2.2281	2.7638	3.1693	4.1437	4.5869
11	1.7959	2.2010	2.7181	3.1058	4.0247	4.4370
12	1.7823	2.1788	2.6810	3.0545	3.9296	4.3178
13	1.7709	2.1604	2.6503	3.0123	3.8520	4.2208
14	1.7613	2.1448	2.6245	2.9768	3.7874	4.1405
15	1.7531	2.1314	2.6025	2.9467	3.7328	4.0728
16	1.7459	2.1199	2.5835	2.9208	3.6862	4.0150
17	1.7396	2.1098	2.5669	2.8982	3.6458	3.9651
18	1.7341	2.1009	2.5524	2.8784	3.6105	3.9216
19	1.7291	2.0930	2.5395	2.8609	3.5794	3.8834
20	1.7247	2.0860	2.5280	2.8453	3.5518	3.8495
21	1.7207	2.0796	2.5176	2.8314	3.5272	3.8193
22	1.7171	2.0739	2.5083	2.8188	3.5050	3.7921
23	1.7139	2.0687	2.4999	2.8073	3.4850	3.7676
24	1.7109	2.0639	2.4922	2.7969	3.4668	3.7454
25	1.7081	2.0595	2.4851	2.7874	3.4502	3.7251
26	1.7056	2.0555	2.4786	2.7787	3.4350	3.7066
27	1.7033	2.0518	2.4727	2.7707	3.4210	3.6896
28	1.7011	2.0484	2.4671	2.7633	3.4082	3.6739
29	1.6991	2.0452	2.4620	2.7564	3.3962	3.6594
30	1.6973	2.0423	2.4573	2.7500	3.3852	3.6460

The distribution of Student's t

ν	Cumulative probabilities					
	0.9500	0.9750	0.9900	0.9950	0.9990	0.9995
31	1.6955	2.0395	2.4528	2.7440	3.3749	3.6335
32	1.6939	2.0369	2.4487	2.7385	3.3653	3.6218
33	1.6924	2.0345	2.4448	2.7333	3.3563	3.6109
34	1.6909	2.0322	2.4411	2.7284	3.3479	3.6007
35	1.6896	2.0301	2.4377	2.7238	3.3400	3.5911
36	1.6883	2.0281	2.4345	2.7195	3.3326	3.5821
37	1.6871	2.0262	2.4314	2.7154	3.3256	3.5737
38	1.6860	2.0244	2.4286	2.7116	3.3190	3.5657
39	1.6849	2.0227	2.4258	2.7079	3.3128	3.5581
40	1.6839	2.0211	2.4233	2.7045	3.3069	3.5510
41	1.6829	2.0195	2.4208	2.7012	3.3013	3.5442
42	1.6820	2.0181	2.4185	2.6981	3.2960	3.5377
43	1.6811	2.0167	2.4163	2.6951	3.2909	3.5316
44	1.6802	2.0154	2.4141	2.6923	3.2861	3.5258
45	1.6794	2.0141	2.4121	2.6896	3.2815	3.5203
46	1.6787	2.0129	2.4102	2.6870	3.2771	3.5150
47	1.6779	2.0117	2.4083	2.6846	3.2729	3.5099
48	1.6772	2.0106	2.4066	2.6822	3.2689	3.5051
49	1.6766	2.0096	2.4049	2.6800	3.2651	3.5004
50	1.6759	2.0086	2.4033	2.6778	3.2614	3.4960
51	1.6753	2.0076	2.4017	2.6757	3.2579	3.4918
52	1.6747	2.0066	2.4002	2.6737	3.2545	3.4877
53	1.6741	2.0057	2.3988	2.6718	3.2513	3.4838
54	1.6736	2.0049	2.3974	2.6700	3.2481	3.4800
55	1.6730	2.0040	2.3961	2.6682	3.2451	3.4764
56	1.6725	2.0032	2.3948	2.6665	3.2423	3.4729
57	1.6720	2.0025	2.3936	2.6649	3.2395	3.4696
58	1.6716	2.0017	2.3924	2.6633	3.2368	3.4663
59	1.6711	2.0010	2.3912	2.6618	3.2342	3.4632
60	1.6706	2.0003	2.3901	2.6603	3.2317	3.4602

The distribution of Student's t

ν	Cumulative probabilities					
	0.9500	0.9750	0.9900	0.9950	0.9990	0.9995
61	1.6702	1.9996	2.3890	2.6589	3.2293	3.4573
62	1.6698	1.9990	2.3880	2.6575	3.2270	3.4545
63	1.6694	1.9983	2.3870	2.6561	3.2247	3.4518
64	1.6690	1.9977	2.3860	2.6549	3.2225	3.4491
65	1.6686	1.9971	2.3851	2.6536	3.2204	3.4466
66	1.6683	1.9966	2.3842	2.6524	3.2184	3.4441
67	1.6679	1.9960	2.3833	2.6512	3.2164	3.4417
68	1.6676	1.9955	2.3824	2.6501	3.2145	3.4394
69	1.6672	1.9949	2.3816	2.6490	3.2126	3.4372
70	1.6669	1.9944	2.3808	2.6479	3.2108	3.4350
71	1.6666	1.9939	2.3800	2.6469	3.2090	3.4329
72	1.6663	1.9935	2.3793	2.6459	3.2073	3.4308
73	1.6660	1.9930	2.3785	2.6449	3.2057	3.4289
74	1.6657	1.9925	2.3778	2.6439	3.2041	3.4269
75	1.6654	1.9921	2.3771	2.6430	3.2025	3.4250
76	1.6652	1.9917	2.3764	2.6421	3.2010	3.4232
77	1.6649	1.9913	2.3758	2.6412	3.1995	3.4214
78	1.6646	1.9908	2.3751	2.6403	3.1980	3.4197
79	1.6644	1.9905	2.3745	2.6395	3.1966	3.4180
80	1.6641	1.9901	2.3739	2.6387	3.1953	3.4163
81	1.6639	1.9897	2.3733	2.6379	3.1939	3.4147
82	1.6636	1.9893	2.3727	2.6371	3.1926	3.4132
83	1.6634	1.9890	2.3721	2.6364	3.1913	3.4116
84	1.6632	1.9886	2.3716	2.6356	3.1901	3.4102
85	1.6630	1.9883	2.3710	2.6349	3.1889	3.4087
86	1.6628	1.9879	2.3705	2.6342	3.1877	3.4073
87	1.6626	1.9876	2.3700	2.6335	3.1866	3.4059
88	1.6624	1.9873	2.3695	2.6329	3.1854	3.4045
89	1.6622	1.9870	2.3690	2.6322	3.1843	3.4032
90	1.6620	1.9867	2.3685	2.6316	3.1833	3.4019

The distribution of Student's t

ν	Cumulative probabilities					
	0.9500	0.9750	0.9900	0.9950	0.9990	0.9995
91	1.6618	1.9864	2.3680	2.6309	3.1822	3.4007
92	1.6616	1.9861	2.3676	2.6303	3.1812	3.3994
93	1.6614	1.9858	2.3671	2.6297	3.1802	3.3982
94	1.6612	1.9855	2.3667	2.6291	3.1792	3.3971
95	1.6611	1.9853	2.3662	2.6286	3.1782	3.3959
96	1.6609	1.9850	2.3658	2.6280	3.1773	3.3948
97	1.6607	1.9847	2.3654	2.6275	3.1764	3.3937
98	1.6606	1.9845	2.3650	2.6269	3.1755	3.3926
99	1.6604	1.9842	2.3646	2.6264	3.1746	3.3915
100	1.6602	1.9840	2.3642	2.6259	3.1737	3.3905
102	1.6599	1.9835	2.3635	2.6249	3.1721	3.3885
104	1.6596	1.9830	2.3627	2.6239	3.1705	3.3865
106	1.6594	1.9826	2.3620	2.6230	3.1689	3.3847
108	1.6591	1.9822	2.3614	2.6221	3.1674	3.3829
110	1.6588	1.9818	2.3607	2.6213	3.1660	3.3812
112	1.6586	1.9814	2.3601	2.6204	3.1646	3.3795
114	1.6583	1.9810	2.3595	2.6196	3.1633	3.3779
116	1.6581	1.9806	2.3589	2.6189	3.1620	3.3764
118	1.6579	1.9803	2.3584	2.6181	3.1607	3.3749
120	1.6577	1.9799	2.3578	2.6174	3.1595	3.3735
122	1.6574	1.9796	2.3573	2.6167	3.1584	3.3721
124	1.6572	1.9793	2.3568	2.6161	3.1573	3.3707
126	1.6570	1.9790	2.3563	2.6154	3.1562	3.3694
128	1.6568	1.9787	2.3558	2.6148	3.1551	3.3682
130	1.6567	1.9784	2.3554	2.6142	3.1541	3.3669
132	1.6565	1.9781	2.3549	2.6136	3.1531	3.3658
134	1.6563	1.9778	2.3545	2.6130	3.1522	3.3646
136	1.6561	1.9776	2.3541	2.6125	3.1512	3.3635
138	1.6560	1.9773	2.3537	2.6119	3.1503	3.3624
140	1.6558	1.9771	2.3533	2.6114	3.1495	3.3614

The distribution of Student's t

ν	Cumulative probabilities					
	0.9500	0.9750	0.9900	0.9950	0.9990	0.9995
142	1.6557	1.9768	2.3529	2.6109	3.1486	3.3604
144	1.6555	1.9766	2.3525	2.6104	3.1478	3.3594
146	1.6554	1.9763	2.3522	2.6099	3.1470	3.3584
148	1.6552	1.9761	2.3518	2.6095	3.1462	3.3575
150	1.6551	1.9759	2.3515	2.6090	3.1455	3.3566
152	1.6549	1.9757	2.3511	2.6086	3.1447	3.3557
154	1.6548	1.9755	2.3508	2.6081	3.1440	3.3548
156	1.6547	1.9753	2.3505	2.6077	3.1433	3.3540
158	1.6546	1.9751	2.3502	2.6073	3.1426	3.3532
160	1.6544	1.9749	2.3499	2.6069	3.1419	3.3524
162	1.6543	1.9747	2.3496	2.6065	3.1413	3.3516
164	1.6542	1.9745	2.3493	2.6061	3.1407	3.3508
166	1.6541	1.9744	2.3490	2.6058	3.1401	3.3501
168	1.6540	1.9742	2.3487	2.6054	3.1395	3.3494
170	1.6539	1.9740	2.3485	2.6051	3.1389	3.3487
172	1.6538	1.9739	2.3482	2.6047	3.1383	3.3480
174	1.6537	1.9737	2.3480	2.6044	3.1377	3.3473
176	1.6536	1.9735	2.3477	2.6041	3.1372	3.3467
178	1.6535	1.9734	2.3475	2.6037	3.1366	3.3460
180	1.6534	1.9732	2.3472	2.6034	3.1361	3.3454
182	1.6533	1.9731	2.3470	2.6031	3.1356	3.3448
184	1.6532	1.9729	2.3468	2.6028	3.1351	3.3442
186	1.6531	1.9728	2.3466	2.6025	3.1346	3.3436
188	1.6530	1.9727	2.3463	2.6022	3.1341	3.3430
190	1.6529	1.9725	2.3461	2.6020	3.1337	3.3425
192	1.6528	1.9724	2.3459	2.6017	3.1332	3.3419
194	1.6527	1.9723	2.3457	2.6014	3.1328	3.3414
196	1.6527	1.9721	2.3455	2.6011	3.1323	3.3409
198	1.6526	1.9720	2.3453	2.6009	3.1319	3.3403
200	1.6525	1.9719	2.3451	2.6006	3.1315	3.3398

The distribution of Student's *t*

ν	Cumulative probabilities					
	0.9500	0.9750	0.9900	0.9950	0.9990	0.9995
250	1.6510	1.9695	2.3414	2.5956	3.1231	3.3299
300	1.6499	1.9679	2.3388	2.5923	3.1176	3.3233
400	1.6487	1.9659	2.3357	2.5882	3.1107	3.3150
500	1.6479	1.9647	2.3338	2.5857	3.1066	3.3101
600	1.6474	1.9639	2.3326	2.5840	3.1039	3.3068
700	1.6470	1.9634	2.3317	2.5829	3.1019	3.3045
800	1.6468	1.9629	2.3310	2.5820	3.1005	3.3027
900	1.6465	1.9626	2.3305	2.5813	3.0993	3.3014
1000	1.6464	1.9623	2.3301	2.5808	3.0984	3.3003
∞	1.644854	1.959964	2.326348	2.575829	3.090232	3.290527

NOTES:

The distribution of Student's t being symmetrical about the origin, the only standardized deviations tabulated here are those corresponding to cumulative probabilities equal to or greater than 0.5. The t values corresponding to cumulative probabilities smaller than 0.5 may be obtained through the relationship

$$t_{(P;\,\nu)} = -t_{(1-P;\,\nu)}\,.$$

In the case of Student's t distribution, like in the cases of the distributions of chi squared (χ^2) and of the variance ratio (F), interpolation is slightly more accurate if it is based on the inverse of the number of degrees of freedom rather than on the number of degrees of freedom itself. In the preceding table, however, all even numbers of degrees of freedom are given between $\nu = 100$ and $\nu = 200$. The t value corresponding to an odd number ν of degrees of freedom between 100 and 200 may thus be obtained with satisfactory accuracy by averaging the t values having one degree of freedom less and one degree of freedom more than desired:

$$t_{(P;\,\nu)} \doteq [\,t_{(P;\,\nu-1)} + t_{(P;\,\nu+1)}\,]/2\,.$$

The larger the number ν of degrees of freedom, the greater the similarity of Student's t distribution to the normal distribution. Therefore, the last line of the table of Student's t, for which $\nu = \infty$, corresponds to the normal distribution. A standardized normal deviation having cumulative probability P may thus be denoted either by $t_{(P)}$ or by $t_{(P;\,\infty)}$.

The distribution of χ^2 (chi squared)

ν	Cumulative probabilities					
	0.0005 0.9500	0.0010 0.9750	0.0050 0.9900	0.0100 0.9950	0.0250 0.9990	0.0500 0.9995
1	0.00000039 3.8415	0.00000157 5.0239	0.00003927 6.6349	0.00015709 7.8794	0.00098207 10.828	0.0039321 12.116
2	0.0010003 5.9915	0.0020010 7.3778	0.010025 9.2103	0.020101 10.597	0.050636 13.816	0.10259 15.202
3	0.015279 7.8147	0.024298 9.3484	0.071722 11.345	0.11483 12.838	0.21580 16.266	0.35185 17.730
4	0.063922 9.4877	0.090804 11.143	0.20699 13.277	0.29711 14.860	0.48442 18.467	0.71072 19.997
5	0.15814 11.070	0.21021 12.833	0.41174 15.086	0.55430 16.750	0.83121 20.515	1.1455 22.105
6	0.29941 12.592	0.38107 14.449	0.67573 16.812	0.87209 18.548	1.2373 22.458	1.6354 24.103
7	0.48488 14.067	0.59849 16.013	0.98926 18.475	1.2390 20.278	1.6899 24.322	2.1673 26.018
8	0.71038 15.507	0.85711 17.535	1.3444 20.090	1.6465 21.955	2.1797 26.124	2.7326 27.868
9	0.97170 16.919	1.1519 19.023	1.7349 21.666	2.0879 23.589	2.7004 27.877	3.3251 29.666
10	1.2650 18.307	1.4787 20.483	2.1559 23.209	2.5582 25.188	3.2470 29.588	3.9403 31.420
11	1.5868 19.675	1.8339 21.920	2.6032 24.725	3.0535 26.757	3.8157 31.264	4.5748 33.137
12	1.9344 21.026	2.2142 23.337	3.0738 26.217	3.5706 28.300	4.4038 32.909	5.2260 34.821

The distribution of χ^2 (chi squared)

ν	Cumulative probabilities					
	0.0005 0.9500	0.0010 0.9750	0.0050 0.9900	0.0100 0.9950	0.0250 0.9990	0.0500 0.9995
13	2.3051 22.362	2.6172 24.736	3.5650 27.688	4.1069 29.819	5.0088 34.528	5.8919 36.478
14	2.6967 23.685	3.0407 26.119	4.0747 29.141	4.6604 31.319	5.6287 36.123	6.5706 38.109
15	3.1075 24.996	3.4827 27.488	4.6009 30.578	5.2293 32.801	6.2621 37.697	7.2609 39.719
16	3.5358 26.296	3.9416 28.845	5.1422 32.000	5.8122 34.267	6.9077 39.252	7.9616 41.308
17	3.9802 27.587	4.4161 30.191	5.6972 33.409	6.4078 35.718	7.5642 40.790	8.6718 42.879
18	4.4394 28.869	4.9048 31.526	6.2648 34.805	7.0149 37.156	8.2307 42.312	9.3905 44.434
19	4.9123 30.144	5.4068 32.852	6.8440 36.191	7.6327 38.582	8.9065 43.820	10.117 45.973
20	5.3981 31.410	5.9210 34.170	7.4338 37.566	8.2604 39.997	9.5908 45.315	10.851 47.498
21	5.8957 32.671	6.4467 35.479	8.0337 38.932	8.8972 41.401	10.283 46.797	11.591 49.011
22	6.4045 33.924	6.9830 36.781	8.6427 40.289	9.5425 42.796	10.982 48.268	12.338 50.511
23	6.9237 35.172	7.5292 38.076	9.2604 41.638	10.196 44.181	11.689 49.728	13.091 52.000
24	7.4527 36.415	8.0849 39.364	9.8862 42.980	10.856 45.559	12.401 51.179	13.848 53.479

The distribution of χ^2 (chi squared)

ν	Cumulative probabilities					
	0.0005 0.9500	0.0010 0.9750	0.0050 0.9900	0.0100 0.9950	0.0250 0.9990	0.0500 0.9995
25	7.9910 37.652	8.6493 40.646	10.520 44.314	11.524 46.928	13.120 52.620	14.611 54.947
26	8.5379 38.885	9.2221 41.923	11.160 45.642	12.198 48.290	13.844 54.052	15.379 56.407
27	9.0932 40.113	9.8028 43.195	11.808 46.963	12.879 49.645	14.573 55.476	16.151 57.858
28	9.6563 41.337	10.391 44.461	12.461 48.278	13.565 50.993	15.308 56.892	16.928 59.300
29	10.227 42.557	10.986 45.722	13.121 49.588	14.256 52.336	16.047 58.301	17.708 60.735
30	10.804 43.773	11.588 46.979	13.787 50.892	14.953 53.672	16.791 59.703	18.493 62.162
31	11.389 44.985	12.196 48.232	14.458 52.191	15.655 55.003	17.539 61.098	19.281 63.582
32	11.979 46.194	12.811 49.480	15.134 53.486	16.362 56.328	18.291 62.487	20.072 64.995
33	12.576 47.400	13.431 50.725	15.815 54.776	17.074 57.648	19.047 63.870	20.867 66.403
34	13.179 48.602	14.057 51.966	16.501 56.061	17.789 58.964	19.806 65.247	21.664 67.803
35	13.787 49.802	14.688 53.203	17.192 57.342	18.509 60.275	20.569 66.619	22.465 69.199
36	14.401 50.998	15.324 54.437	17.887 58.619	19.233 61.581	21.336 67.985	23.269 70.588

The distribution of χ^2 (chi squared)

ν	Cumulative probabilities					
	0.0005 0.9500	0.0010 0.9750	0.0050 0.9900	0.0100 0.9950	0.0250 0.9990	0.0500 0.9995
37	15.020 52.192	15.965 55.668	18.586 59.893	19.960 62.883	22.106 69.346	24.075 71.972
38	15.644 53.384	16.611 56.896	19.289 61.162	20.691 64.181	22.878 70.703	24.884 73.351
39	16.273 54.572	17.262 58.120	19.996 62.428	21.426 65.476	23.654 72.055	25.695 74.725
40	16.906 55.758	17.916 59.342	20.707 63.691	22.164 66.766	24.433 73.402	26.509 76.095
41	17.544 56.942	18.575 60.561	21.421 64.950	22.906 68.053	25.215 74.745	27.326 77.459
42	18.186 58.124	19.239 61.777	22.138 66.206	23.650 69.336	25.999 76.084	28.144 78.820
43	18.832 59.304	19.906 62.990	22.859 67.459	24.398 70.616	26.785 77.419	28.965 80.176
44	19.483 60.481	20.576 64.201	23.584 68.710	25.148 71.893	27.575 78.750	29.787 81.528
45	20.137 61.656	21.251 65.410	24.311 69.957	25.901 73.166	28.366 80.077	30.612 82.876
46	20.794 62.830	21.929 66.617	25.041 71.201	26.657 74.437	29.160 81.400	31.439 84.220
47	21.456 64.001	22.610 67.821	25.775 72.443	27.416 75.704	29.956 82.720	32.268 85.560
48	22.121 65.171	23.295 69.023	26.511 73.683	28.177 76.969	30.755 84.037	33.098 86.897

The distribution of χ^2 (chi squared)

ν	Cumulative probabilities					
	0.0005 0.9500	0.0010 0.9750	0.0050 0.9900	0.0100 0.9950	0.0250 0.9990	0.0500 0.9995
49	22.789 66.339	23.983 70.222	27.249 74.919	28.941 78.231	31.555 85.351	33.930 88.231
50	23.461 67.505	24.674 71.420	27.991 76.154	29.707 79.490	32.357 86.661	34.764 89.561
51	24.136 68.669	25.368 72.616	28.735 77.386	30.475 80.747	33.162 87.968	35.600 90.887
52	24.814 69.832	26.065 73.810	29.481 78.616	31.246 82.001	33.968 89.272	36.437 92.211
53	25.495 70.993	26.765 75.002	30.230 79.843	32.018 83.253	34.776 90.573	37.276 93.531
54	26.179 72.153	27.468 76.192	30.981 81.069	32.793 84.502	35.586 91.872	38.116 94.849
55	26.866 73.311	28.173 77.380	31.735 82.292	33.570 85.749	36.398 93.168	38.958 96.163
56	27.555 74.468	28.881 78.567	32.490 83.513	34.350 86.994	37.212 94.461	39.801 97.475
57	28.248 75.624	29.592 79.752	33.248 84.733	35.131 88.236	38.027 95.751	40.646 98.784
58	28.943 76.778	30.305 80.936	34.008 85.950	35.913 89.477	38.844 97.039	41.492 100.090
59	29.640 77.931	31.020 82.117	34.770 87.166	36.698 90.715	39.662 98.324	42.339 101.394
60	30.340 79.082	31.738 83.298	35.534 88.379	37.485 91.952	40.482 99.607	43.188 102.695

The distribution of χ^2 (chi squared)

ν	Cumulative probabilities					
	0.0005 0.9500	0.0010 0.9750	0.0050 0.9900	0.0100 0.9950	0.0250 0.9990	0.0500 0.9995
61	31.043 80.232	32.459 84.476	36.301 89.591	38.273 93.186	41.303 100.888	44.038 103.993
62	31.748 81.381	33.181 85.654	37.068 90.802	39.063 94.419	42.126 102.166	44.889 105.289
63	32.455 82.529	33.906 86.830	37.838 92.010	39.855 95.649	42.950 103.442	45.741 106.583
64	33.165 83.675	34.633 88.004	38.610 93.217	40.649 96.878	43.776 104.716	46.595 107.875
65	33.877 84.821	35.362 89.177	39.383 94.422	41.444 98.105	44.603 105.988	47.450 109.164
66	34.591 85.965	36.093 90.349	40.158 95.626	42.240 99.330	45.431 107.258	48.305 110.451
67	35.307 87.108	36.826 91.519	40.935 96.828	43.038 100.554	46.261 108.526	49.162 111.736
68	36.025 88.250	37.561 92.689	41.713 98.028	43.838 101.776	47.092 109.791	50.020 113.018
69	36.745 89.391	38.298 93.856	42.494 99.228	44.639 102.996	47.924 111.055	50.879 114.299
70	37.467 90.531	39.036 95.023	43.275 100.425	45.442 104.215	48.758 112.317	51.739 115.578
71	38.192 91.670	39.777 96.189	44.058 101.621	46.246 105.432	49.592 113.577	52.600 116.854
72	38.918 92.808	40.519 97.353	44.843 102.816	47.051 106.648	50.428 114.835	53.462 118.129

The distribution of χ^2 (chi squared)

ν	Cumulative probabilities					
	0.0005 0.9500	0.0010 0.9750	0.0050 0.9900	0.0100 0.9950	0.0250 0.9990	0.0500 0.9995
73	39.646 93.945	41.264 98.516	45.629 104.010	47.858 107.862	51.265 116.092	54.325 119.402
74	40.376 95.081	42.010 99.678	46.417 105.202	48.666 109.074	52.103 117.346	55.189 120.673
75	41.107 96.217	42.757 100.839	47.206 106.393	49.475 110.286	52.942 118.599	56.054 121.942
76	41.841 97.351	43.507 101.999	47.997 107.583	50.286 111.495	53.782 119.850	56.920 123.209
77	42.576 98.484	44.258 103.158	48.788 108.771	51.097 112.704	54.623 121.100	57.786 124.475
78	43.312 99.617	45.010 104.316	49.582 109.958	51.910 113.911	55.466 122.348	58.654 125.739
79	44.051 100.749	45.764 105.473	50.376 111.144	52.725 115.117	56.309 123.594	59.522 127.001
80	44.791 101.879	46.520 106.629	51.172 112.329	53.540 116.321	57.153 124.839	60.391 128.261
81	45.533 103.010	47.277 107.783	51.969 113.512	54.357 117.524	57.998 126.083	61.261 129.520
82	46.276 104.139	48.036 108.937	52.767 114.695	55.174 118.726	58.845 127.324	62.132 130.778
83	47.021 105.267	48.796 110.090	53.567 115.876	55.993 119.927	59.692 128.565	63.004 132.033
84	47.767 106.395	49.557 111.242	54.368 117.057	56.813 121.126	60.540 129.804	63.876 133.288

The distribution of χ^2 (chi squared)

ν	Cumulative probabilities					
	0.0005 0.9500	0.0010 0.9750	0.0050 0.9900	0.0100 0.9950	0.0250 0.9990	0.0500 0.9995
85	48.515 107.522	50.320 112.393	55.170 118.236	57.634 122.325	61.389 131.041	64.749 134.540
86	49.264 108.648	51.085 113.544	55.973 119.414	58.456 123.522	62.239 132.277	65.623 135.792
87	50.015 109.773	51.850 114.693	56.777 120.591	59.279 124.718	63.089 133.512	66.498 137.041
88	50.767 110.898	52.617 115.841	57.582 121.767	60.103 125.913	63.941 134.745	67.373 138.290
89	51.521 112.022	53.386 116.989	58.389 122.942	60.928 127.106	64.793 135.978	68.249 139.537
90	52.276 113.145	54.155 118.136	59.196 124.116	61.754 128.299	65.647 137.208	69.126 140.782
91	53.032 114.268	54.926 119.282	60.005 125.289	62.581 129.491	66.501 138.438	70.003 142.027
92	53.790 115.390	55.698 120.427	60.815 126.462	63.409 130.681	67.356 139.666	70.882 143.269
93	54.549 116.511	56.472 121.571	61.625 127.633	64.238 131.871	68.211 140.893	71.760 144.511
94	55.309 117.632	57.246 122.715	62.437 128.803	65.068 133.059	69.068 142.119	72.640 145.751
95	56.070 118.752	58.022 123.858	63.250 129.973	65.898 134.247	69.925 143.344	73.520 146.990
96	56.833 119.871	58.799 125.000	64.063 131.141	66.730 135.433	70.783 144.567	74.401 148.228

The distribution of χ^2 (chi squared)

ν	Cumulative probabilities					
	0.0005 0.9500	0.0010 0.9750	0.0050 0.9900	0.0100 0.9950	0.0250 0.9990	0.0500 0.9995
97	57.597 120.990	59.577 126.141	64.878 132.309	67.562 136.619	71.642 145.789	75.282 149.465
98	58.362 122.108	60.356 127.282	65.694 133.476	68.396 137.803	72.501 147.010	76.164 150.700
99	59.128 123.225	61.137 128.422	66.510 134.642	69.230 138.987	73.361 148.230	77.046 151.934
100	59.896 124.342	61.918 129.561	67.328 135.807	70.065 140.169	74.222 149.449	77.929 153.167
102	61.434 126.574	63.484 131.838	68.965 138.134	71.737 142.532	75.946 151.884	79.697 155.629
104	62.977 128.804	65.054 134.111	70.606 140.459	73.413 144.891	77.672 154.314	81.468 158.087
106	64.524 131.031	66.629 136.382	72.251 142.780	75.092 147.247	79.401 156.740	83.240 160.541
108	66.075 133.257	68.207 138.651	73.899 145.099	76.774 149.599	81.133 159.162	85.015 162.990
110	67.631 135.480	69.789 140.917	75.550 147.414	78.458 151.948	82.867 161.581	86.792 165.435
112	69.191 137.701	71.375 143.180	77.204 149.727	80.146 154.294	84.604 163.995	88.570 167.876
114	70.754 139.921	72.965 145.441	78.862 152.037	81.836 156.637	86.342 166.406	90.351 170.314
116	72.321 142.138	74.558 147.700	80.522 154.344	83.529 158.977	88.084 168.813	92.134 172.747

The distribution of χ^2 (chi squared)

ν	Cumulative probabilities					
	0.0005 0.9500	0.0010 0.9750	0.0050 0.9900	0.0100 0.9950	0.0250 0.9990	0.0500 0.9995
118	73.892 144.354	76.155 149.957	82.185 156.648	85.225 161.314	89.827 171.217	93.918 175.177
120	75.467 146.567	77.755 152.211	83.852 158.950	86.923 163.648	91.573 173.617	95.705 177.603
122	77.045 148.779	79.359 154.464	85.520 161.250	88.624 165.980	93.320 176.014	97.493 180.025
124	78.626 150.989	80.965 156.714	87.192 163.546	90.327 168.308	95.070 178.408	99.283 182.445
126	80.211 153.198	82.575 158.962	88.866 165.841	92.033 170.634	96.822 180.799	101.074 184.860
128	81.799 155.405	84.188 161.209	90.543 168.133	93.741 172.957	98.576 183.186	102.867 187.273
130	83.390 157.610	85.804 163.453	92.222 170.423	95.451 175.278	100.331 185.571	104.662 189.682
132	84.984 159.814	87.422 165.696	93.904 172.711	97.163 177.597	102.089 187.953	106.459 192.088
134	86.582 162.016	89.044 167.936	95.588 174.996	98.878 179.913	103.848 190.331	108.257 194.491
136	88.182 164.216	90.669 170.175	97.275 177.280	100.595 182.226	105.609 192.707	110.056 196.891
138	89.785 166.415	92.296 172.412	98.964 179.561	102.314 184.538	107.372 195.080	111.857 199.289
140	91.391 168.613	93.926 174.648	100.655 181.840	104.034 186.847	109.137 197.451	113.659 201.683

The distribution of χ^2 (chi squared)

ν	Cumulative probabilities					
	0.0005 0.9500	0.0010 0.9750	0.0050 0.9900	0.0100 0.9950	0.0250 0.9990	0.0500 0.9995
142	93.000 170.809	95.558 176.882	102.348 184.118	105.757 189.154	110.903 199.819	115.463 204.074
144	94.612 173.004	97.193 179.114	104.044 186.393	107.482 191.458	112.671 202.184	117.268 206.463
146	96.226 175.198	98.831 181.344	105.741 188.666	109.209 193.761	114.441 204.547	119.075 208.849
148	97.843 177.390	100.471 183.573	107.441 190.938	110.937 196.062	116.212 206.907	120.883 211.233
150	99.463 179.581	102.113 185.800	109.142 193.208	112.668 198.360	117.985 209.265	122.692 213.613
152	101.085 181.770	103.758 188.026	110.846 195.476	114.400 200.657	119.759 211.620	124.502 215.992
154	102.709 183.959	105.405 190.251	112.551 197.742	116.134 202.951	121.534 213.973	126.314 218.368
156	104.336 186.146	107.055 192.474	114.259 200.006	117.869 205.244	123.312 216.324	128.127 220.741
158	105.966 188.332	108.706 194.695	115.968 202.269	119.607 207.535	125.090 218.673	129.941 223.112
160	107.597 190.516	110.360 196.915	117.679 204.530	121.346 209.824	126.870 221.019	131.756 225.481
162	109.231 192.700	112.016 199.134	119.392 206.790	123.086 212.111	128.651 223.363	133.572 227.847
164	110.867 194.883	113.674 201.351	121.107 209.047	124.828 214.396	130.434 225.705	135.390 230.211

The distribution of χ^2 (chi squared)

ν	Cumulative probabilities					
	0.0005 0.9500	0.0010 0.9750	0.0050 0.9900	0.0100 0.9950	0.0250 0.9990	0.0500 0.9995
166	112.506 197.064	115.335 203.567	122.823 211.304	126.572 216.680	132.218 228.045	137.209 232.573
168	114.146 199.244	116.997 205.782	124.541 213.558	128.318 218.962	134.003 230.383	139.028 234.933
170	115.789 201.423	118.661 207.995	126.261 215.812	130.064 221.242	135.790 232.719	140.849 237.291
172	117.434 203.602	120.327 210.208	127.983 218.063	131.813 223.521	137.578 235.053	142.671 239.646
174	119.080 205.779	121.995 212.419	129.706 220.314	133.563 225.798	139.367 237.385	144.494 242.000
176	120.729 207.955	123.665 214.628	131.430 222.563	135.314 228.074	141.157 239.716	146.318 244.351
178	122.380 210.130	125.337 216.837	133.157 224.810	137.066 230.347	142.949 242.044	148.143 246.701
180	124.033 212.304	127.011 219.044	134.884 227.056	138.820 232.620	144.741 244.370	149.969 249.048
182	125.687 214.477	128.687 221.251	136.614 229.301	140.576 234.891	146.535 246.695	151.796 251.394
184	127.344 216.649	130.364 223.456	138.344 231.544	142.332 237.160	148.330 249.018	153.623 253.737
186	129.002 218.820	132.043 225.660	140.077 233.786	144.090 239.428	150.126 251.339	155.452 256.079
188	130.662 220.991	133.724 227.863	141.810 236.027	145.850 241.694	151.923 253.659	157.282 258.419

The distribution of χ^2 (chi squared)

ν	Cumulative probabilities					
	0.0005 0.9500	0.0010 0.9750	0.0050 0.9900	0.0100 0.9950	0.0250 0.9990	0.0500 0.9995
190	132.324 223.160	135.406 230.064	143.545 238.266	147.610 243.959	153.721 255.976	159.113 260.758
192	133.988 225.329	137.090 232.265	145.282 240.505	149.372 246.223	155.521 258.292	160.944 263.094
194	135.654 227.496	138.776 234.465	147.020 242.742	151.135 248.485	157.321 260.607	162.776 265.429
196	137.321 229.663	140.463 236.664	148.759 244.977	152.900 250.746	159.122 262.920	164.610 267.762
198	138.990 231.829	142.152 238.861	150.499 247.212	154.665 253.006	160.925 265.231	166.444 270.093
200	140.660 233.994	143.843 241.058	152.241 249.445	156.432 255.264	162.728 267.541	168.279 272.423

NOTES:

In the case of the distribution of χ^2, like in the cases of the distributions of Student's t and of the variance ratio, F, interpolation is slightly more accurate if it is based on the inverse of the number of degrees of freedom rather than on the number of degrees of freedom itself. In the preceding table, however, all even numbers of degrees of freedom are given between $\nu = 100$ and $\nu = 200$. The value of χ^2 corresponding to an odd number ν of degrees of freedom between 100 and 200 may thus be obtained with satisfactory accuracy simply by averaging the χ^2 values having one degree of freedom less and one degree of freedom more than desired:

$$\chi^2_{(P;\,\nu)} \doteq [\,\chi^2_{(P;\,\nu-1)} + \chi^2_{(P;\,\nu+1)}\,]/2 \ .$$

When the number ν of degrees of freedom is larger than 200, the value of χ^2 may usually be obtained with an accuracy of at least 4 digits by using the excellent approximation of Wilson and Hilferty (1931):

$$\chi^2_{(P;\,\nu)} \doteq \nu\{1 - 2/(9\nu) + t_{(P;\,\infty)}\sqrt{[2/(9\nu)]}\}^3 \ .$$

The distribution of the variance ratio, $F = S_1^2/S_2^2$, when $\sigma_1^2 = \sigma_2^2$

ν_1	ν_2	Cumulative probabilities					
		0.9500	0.9750	0.9900	0.9950	0.9990	0.9995
1	1	161.4476	647.7890	4052.181	16210.72	405284.1	1621138
1	2	18.5128	38.5063	98.5025	198.5013	998.5003	1998.500
1	3	10.1280	17.4434	34.1162	55.5520	167.0292	266.5492
1	4	7.7086	12.2179	21.1977	31.3328	74.1373	106.2189
1	5	6.6079	10.0070	16.2582	22.7848	47.1808	63.6110
1	6	5.9874	8.8131	13.7450	18.6350	35.5075	46.0816
1	7	5.5914	8.0727	12.2464	16.2356	29.2452	36.9878
1	8	5.3177	7.5709	11.2586	14.6882	25.4148	31.5553
1	9	5.1174	7.2093	10.5614	13.6136	22.8571	27.9910
1	10	4.9646	6.9367	10.0443	12.8265	21.0396	25.4921
1	11	4.8443	6.7241	9.6460	12.2263	19.6868	23.6520
1	12	4.7472	6.5538	9.3302	11.7542	18.6433	22.2450
1	13	4.6672	6.4143	9.0738	11.3735	17.8154	21.1367
1	14	4.6001	6.2979	8.8616	11.0603	17.1434	20.2424
1	15	4.5431	6.1995	8.6831	10.7980	16.5874	19.5065
1	16	4.4940	6.1151	8.5310	10.5755	16.1202	18.8907
1	17	4.4513	6.0420	8.3997	10.3842	15.7222	18.3683
1	18	4.4139	5.9781	8.2854	10.2181	15.3793	17.9197
1	19	4.3807	5.9216	8.1849	10.0725	15.0808	17.5304
1	20	4.3512	5.8715	8.0960	9.9439	14.8188	17.1895
1	21	4.3248	5.8266	8.0166	9.8295	14.5869	16.8886
1	22	4.3009	5.7863	7.9454	9.7271	14.3803	16.6211
1	23	4.2793	5.7498	7.8811	9.6348	14.1950	16.3817
1	24	4.2597	5.7166	7.8229	9.5513	14.0280	16.1663
1	25	4.2417	5.6864	7.7698	9.4753	13.8767	15.9715
1	26	4.2252	5.6586	7.7213	9.4059	13.7390	15.7944
1	27	4.2100	5.6331	7.6767	9.3423	13.6131	15.6328
1	28	4.1960	5.6096	7.6356	9.2838	13.4976	15.4847
1	29	4.1830	5.5878	7.5977	9.2297	13.3912	15.3485
1	30	4.1709	5.5675	7.5625	9.1797	13.2930	15.2228
1	40	4.0847	5.4239	7.3141	8.8279	12.6094	14.3520
1	50	4.0343	5.3403	7.1706	8.6258	12.2221	13.8618
1	60	4.0012	5.2856	7.0771	8.4946	11.9730	13.5476
1	120	3.9201	5.1523	6.8509	8.1788	11.3802	12.8039
1	250	3.8789	5.0849	6.7373	8.0212	11.0880	12.4394
1	500	3.8601	5.0543	6.6858	7.9498	10.9567	12.2760
1	1000	3.8508	5.0391	6.6603	7.9145	10.8919	12.1955
1	∞	3.8415	5.0239	6.6349	7.8794	10.8276	12.1157

The distribution of the variance ratio, $F = S_1^2/S_2^2$, when $\sigma_1^2 = \sigma_2^2$

ν_1	ν_2	Cumulative probabilities					
		0.9500	0.9750	0.9900	0.9950	0.9990	0.9995
2	1	199.5000	799.5000	4999.500	19999.50	499999.5	2000000
2	2	19.0000	39.0000	99.0000	199.0000	999.0000	1999.000
2	3	9.5521	16.0441	30.8165	49.7993	148.5000	236.6102
2	4	6.9443	10.6491	18.0000	26.2843	61.2456	87.4427
2	5	5.7861	8.4336	13.2739	18.3138	37.1223	49.7820
2	6	5.1433	7.2599	10.9248	14.5441	27.0000	34.7976
2	7	4.7374	6.5415	9.5466	12.4040	21.6890	27.2057
2	8	4.4590	6.0595	8.6491	11.0424	18.4937	22.7496
2	9	4.2565	5.7147	8.0215	10.1067	16.3871	19.8655
2	10	4.1028	5.4564	7.5594	9.4270	14.9054	17.8653
2	11	3.9823	5.2559	7.2057	8.9122	13.8116	16.4053
2	12	3.8853	5.0959	6.9266	8.5096	12.9737	15.2972
2	13	3.8056	4.9653	6.7010	8.1865	12.3127	14.4298
2	14	3.7389	4.8567	6.5149	7.9216	11.7789	13.7336
2	15	3.6823	4.7650	6.3589	7.7008	11.3391	13.1632
2	16	3.6337	4.6867	6.2262	7.5138	10.9710	12.6880
2	17	3.5915	4.6189	6.1121	7.3536	10.6584	12.2862
2	18	3.5546	4.5597	6.0129	7.2148	10.3899	11.9423
2	19	3.5219	4.5075	5.9259	7.0935	10.1568	11.6446
2	20	3.4928	4.4613	5.8489	6.9865	9.9526	11.3847
2	21	3.4668	4.4199	5.7804	6.8914	9.7723	11.1557
2	22	3.4434	4.3828	5.7190	6.8064	9.6120	10.9526
2	23	3.4221	4.3492	5.6637	6.7300	9.4685	10.7712
2	24	3.4028	4.3187	5.6136	6.6609	9.3394	10.6083
2	25	3.3852	4.2909	5.5680	6.5982	9.2225	10.4611
2	26	3.3690	4.2655	5.5263	6.5409	9.1163	10.3275
2	27	3.3541	4.2421	5.4881	6.4885	9.0194	10.2058
2	28	3.3404	4.2205	5.4529	6.4403	8.9305	10.0944
2	29	3.3277	4.2006	5.4204	6.3958	8.8488	9.9921
2	30	3.3158	4.1821	5.3903	6.3547	8.7734	9.8977
2	40	3.2317	4.0510	5.1785	6.0664	8.2508	9.2470
2	50	3.1826	3.9749	5.0566	5.9016	7.9564	8.8829
2	60	3.1504	3.9253	4.9774	5.7950	7.7678	8.6505
2	120	3.0718	3.8046	4.7865	5.5393	7.3211	8.1033
2	250	3.0319	3.7439	4.6911	5.4122	7.1022	7.8368
2	500	3.0138	3.7162	4.6478	5.3549	7.0041	7.7176
2	1000	3.0047	3.7025	4.6264	5.3265	6.9557	7.6590
2	∞	2.9957	3.6889	4.6052	5.2983	6.9078	7.6009

The distribution of the variance ratio, $F = S_1^2/S_2^2$, when $\sigma_1^2 = \sigma_2^2$

ν_1	ν_2	Cumulative probabilities					
		0.9500	0.9750	0.9900	0.9950	0.9990	0.9995
3	1	215.7073	864.1630	5403.352	21614.74	540379.2	2161518
3	2	19.1643	39.1655	99.1662	199.1664	999.1666	1999.167
3	3	9.2766	15.4392	29.4567	47.4672	141.1085	224.7012
3	4	6.5914	9.9792	16.6944	24.2591	56.1772	80.0925
3	5	5.4095	7.7636	12.0600	16.5298	33.2025	44.4225
3	6	4.7571	6.5988	9.7795	12.9166	23.7033	30.4535
3	7	4.3468	5.8898	8.4513	10.8824	18.7723	23.4572
3	8	4.0662	5.4160	7.5910	9.5965	15.8295	19.3865
3	9	3.8625	5.0781	6.9919	8.7171	13.9018	16.7700
3	10	3.7083	4.8256	6.5523	8.0807	12.5527	14.9655
3	11	3.5874	4.6300	6.2167	7.6004	11.5611	13.6545
3	12	3.4903	4.4742	5.9525	7.2258	10.8042	12.6632
3	13	3.4105	4.3472	5.7394	6.9258	10.2089	11.8897
3	14	3.3439	4.2417	5.5639	6.6804	9.7294	11.2707
3	15	3.2874	4.1528	5.4170	6.4760	9.3353	10.7649
3	16	3.2389	4.0768	5.2922	6.3034	9.0059	10.3443
3	17	3.1968	4.0112	5.1850	6.1556	8.7269	9.9894
3	18	3.1599	3.9539	5.0919	6.0278	8.4875	9.6861
3	19	3.1274	3.9034	5.0103	5.9161	8.2799	9.4241
3	20	3.0984	3.8587	4.9382	5.8177	8.0984	9.1955
3	21	3.0725	3.8188	4.8740	5.7304	7.9383	8.9945
3	22	3.0491	3.7829	4.8166	5.6524	7.7960	8.8163
3	23	3.0280	3.7505	4.7649	5.5823	7.6688	8.6574
3	24	3.0088	3.7211	4.7181	5.5190	7.5545	8.5147
3	25	2.9912	3.6943	4.6755	5.4615	7.4511	8.3860
3	26	2.9752	3.6697	4.6366	5.4091	7.3572	8.2693
3	27	2.9604	3.6472	4.6009	5.3611	7.2715	8.1630
3	28	2.9467	3.6264	4.5681	5.3170	7.1931	8.0658
3	29	2.9340	3.6072	4.5378	5.2764	7.1210	7.9766
3	30	2.9223	3.5894	4.5097	5.2388	7.0545	7.8944
3	40	2.8387	3.4633	4.3126	4.9758	6.5945	7.3287
3	50	2.7900	3.3902	4.1993	4.8259	6.3364	7.0133
3	60	2.7581	3.3425	4.1259	4.7290	6.1712	6.8124
3	120	2.6802	3.2269	3.9491	4.4972	5.7814	6.3409
3	250	2.6407	3.1687	3.8609	4.3822	5.5909	6.1120
3	500	2.6227	3.1423	3.8210	4.3304	5.5057	6.0099
3	1000	2.6138	3.1292	3.8012	4.3048	5.4637	5.9597
3	∞	2.6049	3.1161	3.7816	4.2794	5.4221	5.9100

The distribution of the variance ratio, $F = S_1^2/S_2^2$, when $\sigma_1^2 = \sigma_2^2$

ν_1	ν_2	Cumulative probabilities					
		0.9500	0.9750	0.9900	0.9950	0.9990	0.9995
4	1	224.5832	899.5833	5624.583	22499.58	562499.6	2250000
4	2	19.2468	39.2484	99.2494	199.2497	999.2499	1999.250
4	3	9.1172	15.1010	28.7099	46.1946	137.1004	218.2508
4	4	6.3882	9.6045	15.9770	23.1545	53.4358	76.1242
4	5	5.1922	7.3879	11.3919	15.5561	31.0850	41.5344
4	6	4.5337	6.2272	9.1483	12.0275	21.9235	28.1152
4	7	4.1203	5.5226	7.8466	10.0505	17.1980	21.4408
4	8	3.8379	5.0526	7.0061	8.8051	14.3916	17.5782
4	9	3.6331	4.7181	6.4221	7.9559	12.5603	15.1060
4	10	3.4780	4.4683	5.9943	7.3428	11.2828	13.4068
4	11	3.3567	4.2751	5.6683	6.8809	10.3461	12.1759
4	12	3.2592	4.1212	5.4120	6.5211	9.6327	11.2474
4	13	3.1791	3.9959	5.2053	6.2335	9.0727	10.5244
4	14	3.1122	3.8919	5.0354	5.9984	8.6223	9.9469
4	15	3.0556	3.8043	4.8932	5.8029	8.2527	9.4757
4	16	3.0069	3.7294	4.7726	5.6378	7.9442	9.0844
4	17	2.9647	3.6648	4.6690	5.4967	7.6831	8.7547
4	18	2.9277	3.6083	4.5790	5.3746	7.4593	8.4732
4	19	2.8951	3.5587	4.5003	5.2681	7.2655	8.2302
4	20	2.8661	3.5147	4.4307	5.1743	7.0960	8.0185
4	21	2.8401	3.4754	4.3688	5.0911	6.9467	7.8324
4	22	2.8167	3.4401	4.3134	5.0168	6.8142	7.6676
4	23	2.7955	3.4083	4.2636	4.9500	6.6957	7.5207
4	24	2.7763	3.3794	4.2184	4.8898	6.5892	7.3889
4	25	2.7587	3.3530	4.1774	4.8351	6.4931	7.2701
4	26	2.7426	3.3289	4.1400	4.7852	6.4057	7.1624
4	27	2.7278	3.3067	4.1056	4.7396	6.3261	7.0643
4	28	2.7141	3.2863	4.0740	4.6977	6.2532	6.9747
4	29	2.7014	3.2674	4.0449	4.6591	6.1863	6.8925
4	30	2.6896	3.2499	4.0179	4.6234	6.1245	6.8168
4	40	2.6060	3.1261	3.8283	4.3738	5.6981	6.2966
4	50	2.5572	3.0544	3.7195	4.2316	5.4593	6.0072
4	60	2.5252	3.0077	3.6490	4.1399	5.3067	5.8232
4	120	2.4472	2.8943	3.4795	3.9207	4.9472	5.3921
4	250	2.4078	2.8373	3.3950	3.8121	4.7719	5.1833
4	500	2.3898	2.8114	3.3569	3.7632	4.6935	5.0903
4	1000	2.3808	2.7986	3.3380	3.7390	4.6549	5.0446
4	∞	2.3719	2.7858	3.3192	3.7151	4.6167	4.9993

The distribution of the variance ratio, $F = S_1^2/S_2^2$, when $\sigma_1^2 = \sigma_2^2$

ν_1	ν_2	Cumulative probabilities					
		0.9500	0.9750	0.9900	0.9950	0.9990	0.9995
5	1	230.1619	921.8479	5763.650	23055.80	576404.6	2305619
5	2	19.2964	39.2982	99.2993	199.2996	999.2999	1999.300
5	3	9.0135	14.8848	28.2371	45.3916	134.5800	214.1973
5	4	6.2561	9.3645	15.5219	22.4564	51.7116	73.6307
5	5	5.0503	7.1464	10.9670	14.9396	29.7524	39.7194
5	6	4.3874	5.9876	8.7459	11.4637	20.8027	26.6452
5	7	3.9715	5.2852	7.4604	9.5221	16.2058	20.1726
5	8	3.6875	4.8173	6.6318	8.3018	13.4847	16.4403
5	9	3.4817	4.4844	6.0569	7.4712	11.7137	14.0583
5	10	3.3258	4.2361	5.6363	6.8724	10.4807	12.4251
5	11	3.2039	4.0440	5.3160	6.4217	9.5784	11.2442
5	12	3.1059	3.8911	5.0643	6.0711	8.8921	10.3549
5	13	3.0254	3.7667	4.8616	5.7910	8.3541	9.6635
5	14	2.9582	3.6634	4.6950	5.5623	7.9218	9.1118
5	15	2.9013	3.5764	4.5556	5.3721	7.5674	8.6622
5	16	2.8524	3.5021	4.4374	5.2117	7.2719	8.2892
5	17	2.8100	3.4379	4.3359	5.0746	7.0219	7.9751
5	18	2.7729	3.3820	4.2479	4.9560	6.8078	7.7072
5	19	2.7401	3.3327	4.1708	4.8526	6.6225	7.4761
5	20	2.7109	3.2891	4.1027	4.7616	6.4606	7.2748
5	21	2.6848	3.2501	4.0421	4.6809	6.3179	7.0980
5	22	2.6613	3.2151	3.9880	4.6088	6.1914	6.9416
5	23	2.6400	3.1835	3.9392	4.5441	6.0783	6.8021
5	24	2.6207	3.1548	3.8951	4.4857	5.9768	6.6771
5	25	2.6030	3.1287	3.8550	4.4327	5.8851	6.5644
5	26	2.5868	3.1048	3.8183	4.3844	5.8018	6.4623
5	27	2.5719	3.0828	3.7848	4.3402	5.7259	6.3694
5	28	2.5581	3.0626	3.7539	4.2996	5.6565	6.2845
5	29	2.5454	3.0438	3.7254	4.2622	5.5927	6.2067
5	30	2.5336	3.0265	3.6990	4.2276	5.5339	6.1350
5	40	2.4495	2.9037	3.5138	3.9860	5.1283	5.6430
5	50	2.4004	2.8327	3.4077	3.8486	4.9013	5.3698
5	60	2.3683	2.7863	3.3389	3.7599	4.7565	5.1962
5	120	2.2899	2.6740	3.1735	3.5482	4.4157	4.7902
5	250	2.2501	2.6175	3.0912	3.4435	4.2497	4.5939
5	500	2.2320	2.5919	3.0540	3.3963	4.1756	4.5065
5	1000	2.2231	2.5792	3.0355	3.3730	4.1391	4.4635
5	∞	2.2141	2.5665	3.0173	3.3499	4.1030	4.4211

The distribution of the variance ratio, $F = S_1^2/S_2^2$, when $\sigma_1^2 = \sigma_2^2$

ν_1	ν_2	Cumulative probabilities					
		0.9500	0.9750	0.9900	0.9950	0.9990	0.9995
6	1	233.9860	937.1111	5858.986	23437.11	585937.1	2343750
6	2	19.3295	39.3315	99.3326	199.3330	999.3333	1999.333
6	3	8.9406	14.7347	27.9107	44.8385	132.8475	211.4118
6	4	6.1631	9.1973	15.2069	21.9746	50.5250	71.9160
6	5	4.9503	6.9777	10.6723	14.5133	28.8344	38.4703
6	6	4.2839	5.8198	8.4661	11.0730	20.0297	25.6326
6	7	3.8660	5.1186	7.1914	9.1553	15.5208	19.2984
6	8	3.5806	4.6517	6.3707	7.9520	12.8580	15.6553
6	9	3.3738	4.3197	5.8018	7.1339	11.1281	13.3351
6	10	3.2172	4.0721	5.3858	6.5446	9.9256	11.7469
6	11	3.0946	3.8807	5.0692	6.1016	9.0466	10.6002
6	12	2.9961	3.7283	4.8206	5.7570	8.3788	9.7377
6	13	2.9153	3.6043	4.6204	5.4819	7.8557	9.0678
6	14	2.8477	3.5014	4.4558	5.2574	7.4358	8.5337
6	15	2.7905	3.4147	4.3183	5.0708	7.0917	8.0988
6	16	2.7413	3.3406	4.2016	4.9134	6.8049	7.7383
6	17	2.6987	3.2767	4.1015	4.7789	6.5625	7.4348
6	18	2.6613	3.2209	4.0146	4.6627	6.3550	7.1761
6	19	2.6283	3.1718	3.9386	4.5614	6.1754	6.9531
6	20	2.5990	3.1283	3.8714	4.4721	6.0186	6.7590
6	21	2.5727	3.0895	3.8117	4.3931	5.8805	6.5885
6	22	2.5491	3.0546	3.7583	4.3225	5.7580	6.4377
6	23	2.5277	3.0232	3.7102	4.2591	5.6486	6.3033
6	24	2.5082	2.9946	3.6667	4.2019	5.5504	6.1829
6	25	2.4904	2.9685	3.6272	4.1500	5.4617	6.0744
6	26	2.4741	2.9447	3.5911	4.1027	5.3812	5.9761
6	27	2.4591	2.9228	3.5580	4.0594	5.3078	5.8867
6	28	2.4453	2.9027	3.5276	4.0197	5.2407	5.8050
6	29	2.4324	2.8840	3.4995	3.9831	5.1791	5.7301
6	30	2.4205	2.8667	3.4735	3.9492	5.1223	5.6611
6	40	2.3359	2.7444	3.2910	3.7129	4.7306	5.1882
6	50	2.2864	2.6736	3.1864	3.5785	4.5117	4.9259
6	60	2.2541	2.6274	3.1187	3.4918	4.3721	4.7594
6	120	2.1750	2.5154	2.9559	3.2849	4.0437	4.3704
6	250	2.1350	2.4591	2.8748	3.1826	3.8841	4.1824
6	500	2.1167	2.4335	2.8381	3.1366	3.8128	4.0988
6	1000	2.1076	2.4208	2.8200	3.1138	3.7777	4.0577
6	∞	2.0986	2.4082	2.8020	3.0913	3.7430	4.0171

The distribution of the variance ratio, $F = S_1^2/S_2^2$, when $\sigma_1^2 = \sigma_2^2$

ν_1	ν_2	Cumulative probabilities					
		0.9500	0.9750	0.9900	0.9950	0.9990	0.9995
7	1	236.7684	948.2169	5928.356	23714.57	592873.3	2371494
7	2	19.3532	39.3552	99.3564	199.3568	999.3571	1999.357
7	3	8.8867	14.6244	27.6717	44.4341	131.5829	209.3792
7	4	6.0942	9.0741	14.9758	21.6217	49.6579	70.6635
7	5	4.8759	6.8531	10.4555	14.2004	28.1626	37.5569
7	6	4.2067	5.6955	8.2600	10.7859	19.4634	24.8916
7	7	3.7870	4.9949	6.9928	8.8854	15.0186	18.6579
7	8	3.5005	4.5286	6.1776	7.6941	12.3980	15.0797
7	9	3.2927	4.1970	5.6129	6.8849	10.6979	12.8044
7	10	3.1355	3.9498	5.2001	6.3025	9.5175	11.2489
7	11	3.0123	3.7586	4.8861	5.8648	8.6553	10.1270
7	12	2.9134	3.6065	4.6395	5.5245	8.0009	9.2839
7	13	2.8321	3.4827	4.4410	5.2529	7.4886	8.6296
7	14	2.7642	3.3799	4.2779	5.0313	7.0775	8.1083
7	15	2.7066	3.2934	4.1415	4.8473	6.7408	7.6840
7	16	2.6572	3.2194	4.0259	4.6920	6.4604	7.3324
7	17	2.6143	3.1556	3.9267	4.5594	6.2234	7.0367
7	18	2.5767	3.0999	3.8406	4.4448	6.0206	6.7847
7	19	2.5435	3.0509	3.7653	4.3448	5.8452	6.5675
7	20	2.5140	3.0074	3.6987	4.2569	5.6920	6.3785
7	21	2.4876	2.9686	3.6396	4.1789	5.5571	6.2125
7	22	2.4638	2.9338	3.5867	4.1094	5.4376	6.0658
7	23	2.4422	2.9023	3.5390	4.0469	5.3308	5.9351
7	24	2.4226	2.8738	3.4959	3.9905	5.2349	5.8180
7	25	2.4047	2.8478	3.4568	3.9394	5.1484	5.7124
7	26	2.3883	2.8240	3.4210	3.8928	5.0698	5.6169
7	27	2.3732	2.8021	3.3882	3.8501	4.9983	5.5300
7	28	2.3593	2.7820	3.3581	3.8110	4.9328	5.4506
7	29	2.3463	2.7633	3.3303	3.7749	4.8727	5.3778
7	30	2.3343	2.7460	3.3045	3.7416	4.8173	5.3108
7	40	2.2490	2.6238	3.1238	3.5088	4.4355	4.8516
7	50	2.1992	2.5530	3.0202	3.3765	4.2224	4.5972
7	60	2.1665	2.5068	2.9530	3.2911	4.0864	4.4357
7	120	2.0868	2.3948	2.7918	3.0874	3.7670	4.0587
7	250	2.0463	2.3384	2.7114	2.9867	3.6117	3.8768
7	500	2.0279	2.3129	2.6751	2.9414	3.5424	3.7959
7	1000	2.0187	2.3002	2.6572	2.9190	3.5083	3.7561
7	∞	2.0096	2.2875	2.6393	2.8968	3.4746	3.7168

The distribution of the variance ratio, $F = S_1^2/S_2^2$, when $\sigma_1^2 = \sigma_2^2$

ν_1	ν_2	Cumulative probabilities					
		0.9500	0.9750	0.9900	0.9950	0.9990	0.9995
8	1	238.8827	956.6562	5981.070	23925.41	598144.2	2392578
8	2	19.3710	39.3730	99.3742	199.3746	999.3749	1999.375
8	3	8.8452	14.5399	27.4892	44.1256	130.6190	207.8303
8	4	6.0410	8.9796	14.7989	21.3520	48.9962	69.7080
8	5	4.8183	6.7572	10.2893	13.9610	27.6495	36.8595
8	6	4.1468	5.5996	8.1017	10.5658	19.0303	24.3251
8	7	3.7257	4.8993	6.8400	8.6781	14.6340	18.1680
8	8	3.4381	4.4333	6.0289	7.4959	12.0455	14.6391
8	9	3.2296	4.1020	5.4671	6.6933	10.3680	12.3978
8	10	3.0717	3.8549	5.0567	6.1159	9.2041	10.8671
8	11	2.9480	3.6638	4.7445	5.6821	8.3548	9.7639
8	12	2.8486	3.5118	4.4994	5.3451	7.7104	8.9355
8	13	2.7669	3.3880	4.3021	5.0761	7.2061	8.2930
8	14	2.6987	3.2853	4.1399	4.8566	6.8017	7.7813
8	15	2.6408	3.1987	4.0045	4.6744	6.4707	7.3650
8	16	2.5911	3.1248	3.8896	4.5207	6.1950	7.0203
8	17	2.5480	3.0610	3.7910	4.3894	5.9620	6.7303
8	18	2.5102	3.0053	3.7054	4.2759	5.7628	6.4833
8	19	2.4768	2.9563	3.6305	4.1770	5.5904	6.2705
8	20	2.4471	2.9128	3.5644	4.0900	5.4400	6.0853
8	21	2.4205	2.8740	3.5056	4.0128	5.3076	5.9229
8	22	2.3965	2.8392	3.4530	3.9440	5.1901	5.7791
8	23	2.3748	2.8077	3.4057	3.8822	5.0853	5.6512
8	24	2.3551	2.7791	3.3629	3.8264	4.9912	5.5366
8	25	2.3371	2.7531	3.3239	3.7758	4.9063	5.4333
8	26	2.3205	2.7293	3.2884	3.7297	4.8292	5.3398
8	27	2.3053	2.7074	3.2558	3.6875	4.7590	5.2547
8	28	2.2913	2.6872	3.2259	3.6487	4.6947	5.1771
8	29	2.2783	2.6686	3.1982	3.6131	4.6358	5.1059
8	30	2.2662	2.6513	3.1726	3.5801	4.5814	5.0403
8	40	2.1802	2.5289	2.9930	3.3498	4.2070	4.5914
8	50	2.1299	2.4579	2.8900	3.2189	3.9980	4.3428
8	60	2.0970	2.4117	2.8233	3.1344	3.8648	4.1851
8	120	2.0164	2.2994	2.6629	2.9330	3.5519	3.8170
8	250	1.9756	2.2429	2.5830	2.8333	3.3998	3.6395
8	500	1.9569	2.2172	2.5469	2.7885	3.3320	3.5606
8	1000	1.9476	2.2045	2.5290	2.7663	3.2986	3.5218
8	∞	1.9384	2.1918	2.5113	2.7444	3.2656	3.4835

The distribution of the variance ratio, $F = S_1^2/S_2^2$, when $\sigma_1^2 = \sigma_2^2$

ν_1	ν_2	Cumulative probabilities					
		0.9500	0.9750	0.9900	0.9950	0.9990	0.9995
9	1	240.5433	963.2846	6022.473	24091.00	602284.0	2409137
9	2	19.3848	39.3869	99.3881	199.3885	999.3888	1999.389
9	3	8.8123	14.4731	27.3452	43.8824	129.8600	206.6107
9	4	5.9988	8.9047	14.6591	21.1391	48.4745	68.9549
9	5	4.7725	6.6811	10.1578	13.7716	27.2445	36.3093
9	6	4.0990	5.5234	7.9761	10.3915	18.6882	23.8779
9	7	3.6767	4.8232	6.7188	8.5138	14.3299	17.7808
9	8	3.3881	4.3572	5.9106	7.3386	11.7665	14.2905
9	9	3.1789	4.0260	5.3511	6.5411	10.1066	12.0759
9	10	3.0204	3.7790	4.9424	5.9676	8.9558	10.5646
9	11	2.8962	3.5879	4.6315	5.5368	8.1163	9.4762
9	12	2.7964	3.4358	4.3875	5.2021	7.4797	8.6593
9	13	2.7144	3.3120	4.1911	4.9351	6.9818	8.0259
9	14	2.6458	3.2093	4.0297	4.7173	6.5826	7.5217
9	15	2.5876	3.1227	3.8948	4.5364	6.2559	7.1117
9	16	2.5377	3.0488	3.7804	4.3838	5.9839	6.7722
9	17	2.4943	2.9849	3.6822	4.2535	5.7541	6.4868
9	18	2.4563	2.9291	3.5971	4.1410	5.5575	6.2437
9	19	2.4227	2.8801	3.5225	4.0428	5.3876	6.0343
9	20	2.3928	2.8365	3.4567	3.9564	5.2392	5.8521
9	21	2.3660	2.7977	3.3981	3.8799	5.1087	5.6923
9	22	2.3419	2.7628	3.3458	3.8116	4.9929	5.5509
9	23	2.3201	2.7313	3.2986	3.7502	4.8896	5.4251
9	24	2.3002	2.7027	3.2560	3.6949	4.7968	5.3124
9	25	2.2821	2.6766	3.2172	3.6447	4.7131	5.2109
9	26	2.2655	2.6528	3.1818	3.5989	4.6372	5.1189
9	27	2.2501	2.6309	3.1494	3.5571	4.5680	5.0354
9	28	2.2360	2.6106	3.1195	3.5186	4.5047	4.9590
9	29	2.2229	2.5919	3.0920	3.4832	4.4466	4.8890
9	30	2.2107	2.5746	3.0665	3.4505	4.3930	4.8246
9	40	2.1240	2.4519	2.8876	3.2220	4.0243	4.3836
9	50	2.0734	2.3808	2.7850	3.0920	3.8185	4.1394
9	60	2.0401	2.3344	2.7185	3.0083	3.6873	3.9846
9	120	1.9588	2.2217	2.5586	2.8083	3.3792	3.6234
9	250	1.9174	2.1650	2.4789	2.7094	3.2296	3.4492
9	500	1.8986	2.1392	2.4429	2.6649	3.1628	3.3718
9	1000	1.8892	2.1264	2.4250	2.6429	3.1300	3.3338
9	∞	1.8799	2.1136	2.4073	2.6210	3.0975	3.2962

The distribution of the variance ratio, $F = S_1^2/S_2^2$, when $\sigma_1^2 = \sigma_2^2$

ν_1	ν_2	Cumulative probabilities					
		0.9500	0.9750	0.9900	0.9950	0.9990	0.9995
10	1	241.8817	968.6274	6055.847	24224.49	605621.0	2422485
10	2	19.3959	39.3980	99.3992	199.3996	999.3999	1999.400
10	3	8.7855	14.4189	27.2287	43.6858	129.2467	205.6255
10	4	5.9644	8.8439	14.5459	20.9667	48.0526	68.3460
10	5	4.7351	6.6192	10.0510	13.6182	26.9166	35.8639
10	6	4.0600	5.4613	7.8741	10.2500	18.4109	23.5156
10	7	3.6365	4.7611	6.6201	8.3803	14.0833	17.4669
10	8	3.3472	4.2951	5.8143	7.2106	11.5401	14.0078
10	9	3.1373	3.9639	5.2565	6.4172	9.8943	11.8146
10	10	2.9782	3.7168	4.8491	5.8467	8.7539	10.3189
10	11	2.8536	3.5257	4.5393	5.4183	7.9224	9.2423
10	12	2.7534	3.3736	4.2961	5.0855	7.2920	8.4346
10	13	2.6710	3.2497	4.1003	4.8199	6.7992	7.8086
10	14	2.6022	3.1469	3.9394	4.6034	6.4041	7.3104
10	15	2.5437	3.0602	3.8049	4.4235	6.0808	6.9054
10	16	2.4935	2.9862	3.6909	4.2719	5.8117	6.5701
10	17	2.4499	2.9222	3.5931	4.1424	5.5844	6.2883
10	18	2.4117	2.8664	3.5082	4.0305	5.3900	6.0483
10	19	2.3779	2.8172	3.4338	3.9329	5.2219	5.8416
10	20	2.3479	2.7737	3.3682	3.8470	5.0752	5.6618
10	21	2.3210	2.7348	3.3098	3.7709	4.9462	5.5041
10	22	2.2967	2.6998	3.2576	3.7030	4.8317	5.3646
10	23	2.2747	2.6682	3.2106	3.6420	4.7296	5.2405
10	24	2.2547	2.6396	3.1681	3.5870	4.6379	5.1293
10	25	2.2365	2.6135	3.1294	3.5370	4.5551	5.0292
10	26	2.2197	2.5896	3.0941	3.4916	4.4801	4.9385
10	27	2.2043	2.5676	3.0618	3.4499	4.4117	4.8560
10	28	2.1900	2.5473	3.0320	3.4117	4.3491	4.7808
10	29	2.1768	2.5286	3.0045	3.3765	4.2917	4.7117
10	30	2.1646	2.5112	2.9791	3.3440	4.2388	4.6482
10	40	2.0772	2.3882	2.8005	3.1167	3.8744	4.2134
10	50	2.0261	2.3168	2.6981	2.9875	3.6711	3.9727
10	60	1.9926	2.2702	2.6318	2.9042	3.5415	3.8201
10	120	1.9105	2.1570	2.4721	2.7052	3.2372	3.4643
10	250	1.8687	2.0999	2.3925	2.6068	3.0893	3.2927
10	500	1.8496	2.0740	2.3565	2.5625	3.0234	3.2164
10	1000	1.8402	2.0611	2.3386	2.5405	2.9909	3.1790
10	∞	1.8307	2.0483	2.3209	2.5188	2.9588	3.1420

The distribution of the variance ratio, $F = S_1^2/S_2^2$, when $\sigma_1^2 = \sigma_2^2$

ν_1	ν_2	Cumulative probabilities					
		0.9500	0.9750	0.9900	0.9950	0.9990	0.9995
11	1	242.9835	973.0252	6083.317	24334.36	608367.7	2433472
11	2	19.4050	39.4071	99.4083	199.4087	999.4090	1999.409
11	3	8.7633	14.3742	27.1326	43.5236	128.7408	204.8129
11	4	5.9358	8.7935	14.4523	20.8243	47.7043	67.8433
11	5	4.7040	6.5678	9.9626	13.4912	26.6456	35.4961
11	6	4.0274	5.4098	7.7896	10.1329	18.1816	23.2161
11	7	3.6030	4.7095	6.5382	8.2697	13.8791	17.2072
11	8	3.3130	4.2434	5.7343	7.1045	11.3525	13.7737
11	9	3.1025	3.9121	5.1779	6.3142	9.7183	11.5982
11	10	2.9430	3.6649	4.7715	5.7462	8.5864	10.1153
11	11	2.8179	3.4737	4.4624	5.3197	7.7614	9.0483
11	12	2.7173	3.3215	4.2198	4.9884	7.1362	8.2481
11	13	2.6347	3.1975	4.0245	4.7240	6.6474	7.6281
11	14	2.5655	3.0946	3.8640	4.5085	6.2556	7.1349
11	15	2.5068	3.0078	3.7299	4.3295	5.9352	6.7340
11	16	2.4564	2.9337	3.6162	4.1785	5.6684	6.4021
11	17	2.4126	2:8696	3.5185	4.0496	5.4431	6.1232
11	18	2.3742	2.8137	3.4338	3.9382	5.2505	5.8858
11	19	2.3402	2.7645	3.3596	3.8410	5.0840	5.6813
11	20	2.3100	2.7209	3.2941	3.7555	4.9386	5.5034
11	21	2.2829	2.6819	3.2359	3.6798	4.8107	5.3474
11	22	2.2585	2.6469	3.1837	3.6122	4.6973	5.2095
11	23	2.2364	2.6152	3.1368	3.5515	4.5962	5.0867
11	24	2.2163	2.5865	3.0944	3.4967	4.5053	4.9767
11	25	2.1979	2.5603	3.0558	3.4470	4.4233	4.8777
11	26	2.1811	2.5363	3.0205	3.4017	4.3490	4.7880
11	27	2.1655	2.5143	2.9882	3.3602	4.2812	4.7065
11	28	2.1512	2.4940	2.9585	3.3222	4.2193	4.6321
11	29	2.1379	2.4752	2.9311	3.2871	4.1624	4.5638
11	30	2.1256	2.4577	2.9057	3.2547	4.1100	4.5011
11	40	2.0376	2.3343	2.7274	3.0284	3.7490	4.0712
11	50	1.9861	2.2627	2.6250	2.8997	3.5476	3.8334
11	60	1.9522	2.2159	2.5587	2.8166	3.4193	3.6826
11	120	1.8693	2.1021	2.3990	2.6183	3.1179	3.3309
11	250	1.8271	2.0447	2.3193	2.5202	2.9715	3.1613
11	500	1.8078	2.0186	2.2833	2.4760	2.9061	3.0860
11	1000	1.7982	2.0056	2.2655	2.4541	2.8740	3.0490
11	∞	1.7886	1.9927	2.2477	2.4324	2.8422	3.0124

The distribution of the variance ratio, $F = S_1^2/S_2^2$, when $\sigma_1^2 = \sigma_2^2$

ν_1	ν_2	Cumulative probabilities					
		0.9500	0.9750	0.9900	0.9950	0.9990	0.9995
12	1	243.9060	976.7079	6106.321	24426.37	610667.8	2442672
12	2	19.4125	39.4146	99.4159	199.4163	999.4166	1999.417
12	3	8.7446	14.3366	27.0518	43.3874	128.3165	204.1312
12	4	5.9117	8.7512	14.3736	20.7047	47.4118	67.4214
12	5	4.6777	6.5245	9.8883	13.3845	26.4180	35.1870
12	6	3.9999	5.3662	7.7183	10.0343	17.9888	22.9643
12	7	3.5747	4.6658	6.4691	8.1764	13.7073	16.9888
12	8	3.2839	4.1997	5.6667	7.0149	11.1945	13.5766
12	9	3.0729	3.8682	5.1114	6.2274	9.5700	11.4159
12	10	2.9130	3.6209	4.7059	5.6613	8.4452	9.9436
12	11	2.7876	3.4296	4.3974	5.2363	7.6256	8.8847
12	12	2.6866	3.2773	4.1553	4.9062	7.0046	8.0908
12	13	2.6037	3.1532	3.9603	4.6429	6.5192	7.4758
12	14	2.5342	3.0502	3.8001	4.4281	6.1302	6.9867
12	15	2.4753	2.9633	3.6662	4.2497	5.8121	6.5892
12	16	2.4247	2.8890	3.5527	4.0994	5.5473	6.2602
12	17	2.3807	2.8249	3.4552	3.9709	5.3237	5.9837
12	18	2.3421	2.7689	3.3706	3.8599	5.1324	5.7483
12	19	2.3080	2.7196	3.2965	3.7631	4.9672	5.5457
12	20	2.2776	2.6758	3.2311	3.6779	4.8229	5.3694
12	21	2.2504	2.6368	3.1730	3.6024	4.6960	5.2148
12	22	2.2258	2.6017	3.1209	3.5350	4.5835	5.0781
12	23	2.2036	2.5699	3.0740	3.4745	4.4831	4.9565
12	24	2.1834	2.5411	3.0316	3.4199	4.3929	4.8475
12	25	2.1649	2.5149	2.9931	3.3704	4.3116	4.7494
12	26	2.1479	2.4908	2.9578	3.3252	4.2378	4.6606
12	27	2.1323	2.4688	2.9256	3.2839	4.1706	4.5798
12	28	2.1179	2.4484	2.8959	3.2460	4.1091	4.5061
12	29	2.1045	2.4295	2.8685	3.2110	4.0526	4.4385
12	30	2.0921	2.4120	2.8431	3.1787	4.0006	4.3763
12	40	2.0035	2.2882	2.6648	2.9531	3.6425	3.9505
12	50	1.9515	2.2162	2.5625	2.8247	3.4426	3.7149
12	60	1.9174	2.1692	2.4961	2.7419	3.3153	3.5655
12	120	1.8337	2.0548	2.3363	2.5439	3.0162	3.2172
12	250	1.7910	1.9971	2.2565	2.4459	2.8708	3.0493
12	500	1.7715	1.9708	2.2204	2.4018	2.8060	2.9747
12	1000	1.7618	1.9577	2.2025	2.3800	2.7740	2.9380
12	∞	1.7522	1.9447	2.1847	2.3583	2.7425	2.9018

The distribution of the variance ratio, $F = S_1^2/S_2^2$, when $\sigma_1^2 = \sigma_2^2$

ν_1	ν_2	Cumulative probabilities					
		0.9500	0.9750	0.9900	0.9950	0.9990	0.9995
13	1	244.6898	979.8368	6125.865	24504.54	612622.0	2450489
13	2	19.4189	39.4210	99.4223	199.4227	999.4230	1999.423
13	3	8.7287	14.3045	26.9831	43.2715	127.9553	203.5512
13	4	5.8911	8.7150	14.3065	20.6027	47.1627	67.0621
13	5	4.6552	6.4876	9.8248	13.2934	26.2240	34.9237
13	6	3.9764	5.3290	7.6575	9.9501	17.8244	22.7496
13	7	3.5503	4.6285	6.4100	8.0967	13.5607	16.8024
13	8	3.2590	4.1622	5.6089	6.9384	11.0596	13.4084
13	9	3.0475	3.8306	5.0545	6.1530	9.4433	11.2601
13	10	2.8872	3.5832	4.6496	5.5887	8.3245	9.7970
13	11	2.7614	3.3917	4.3416	5.1649	7.5094	8.7449
13	12	2.6602	3.2393	4.0999	4.8358	6.8920	7.9563
13	13	2.5769	3.1150	3.9052	4.5733	6.4094	7.3455
13	14	2.5073	3.0119	3.7452	4.3591	6.0228	6.8598
13	15	2.4481	2.9249	3.6115	4.1813	5.7066	6.4651
13	16	2.3973	2.8506	3.4981	4.0314	5.4434	6.1385
13	17	2.3531	2.7863	3.4007	3.9033	5.2212	5.8641
13	18	2.3143	2.7302	3.3162	3.7926	5.0312	5.6305
13	19	2.2800	2.6808	3.2422	3.6961	4.8669	5.4294
13	20	2.2495	2.6369	3.1769	3.6111	4.7236	5.2544
13	21	2.2222	2.5978	3.1187	3.5358	4.5975	5.1010
13	22	2.1975	2.5626	3.0667	3.4686	4.4857	4.9654
13	23	2.1752	2.5308	3.0199	3.4083	4.3859	4.8447
13	24	2.1548	2.5019	2.9775	3.3538	4.2963	4.7365
13	25	2.1362	2.4756	2.9389	3.3044	4.2155	4.6392
13	26	2.1192	2.4515	2.9038	3.2594	4.1422	4.5511
13	27	2.1035	2.4293	2.8715	3.2182	4.0754	4.4709
13	28	2.0889	2.4089	2.8418	3.1803	4.0143	4.3978
13	29	2.0755	2.3900	2.8144	3.1454	3.9582	4.3307
13	30	2.0630	2.3724	2.7890	3.1132	3.9065	4.2690
13	40	1.9738	2.2481	2.6107	2.8880	3.5507	3.8466
13	50	1.9214	2.1758	2.5083	2.7599	3.3521	3.6128
13	60	1.8870	2.1286	2.4419	2.6771	3.2255	3.4646
13	120	1.8026	2.0136	2.2818	2.4794	2.9282	3.1190
13	250	1.7595	1.9555	2.2018	2.3814	2.7837	2.9524
13	500	1.7398	1.9290	2.1656	2.3373	2.7192	2.8783
13	1000	1.7299	1.9158	2.1477	2.3155	2.6874	2.8419
13	∞	1.7202	1.9027	2.1299	2.2938	2.6560	2.8060

The distribution of the variance ratio, $F = S_1^2/S_2^2$, when $\sigma_1^2 = \sigma_2^2$

ν_1	ν_2	Cumulative probabilities					
		0.9500	0.9750	0.9900	0.9950	0.9990	0.9995
14	1	245.3640	982.5278	6142.674	24571.77	614302.8	2457212
14	2	19.4244	39.4265	99.4278	199.4282	999.4285	1999.429
14	3	8.7149	14.2768	26.9238	43.1716	127.6443	203.0517
14	4	5.8733	8.6838	14.2486	20.5148	46.9481	66.7524
14	5	4.6358	6.4556	9.7700	13.2148	26.0566	34.6967
14	6	3.9559	5.2968	7.6049	9.8774	17.6825	22.5643
14	7	3.5292	4.5961	6.3590	8.0279	13.4341	16.6415
14	8	3.2374	4.1297	5.5589	6.8721	10.9430	13.2631
14	9	3.0255	3.7980	5.0052	6.0887	9.3337	11.1256
14	10	2.8647	3.5504	4.6008	5.5257	8.2200	9.6702
14	11	2.7386	3.3588	4.2932	5.1031	7.4089	8.6239
14	12	2.6371	3.2062	4.0518	4.7748	6.7945	7.8399
14	13	2.5536	3.0819	3.8573	4.5129	6.3144	7.2327
14	14	2.4837	2.9786	3.6975	4.2993	5.9297	6.7500
14	15	2.4244	2.8915	3.5639	4.1219	5.6151	6.3577
14	16	2.3733	2.8170	3.4506	3.9723	5.3533	6.0331
14	17	2.3290	2.7526	3.3533	3.8445	5.1323	5.7604
14	18	2.2900	2.6964	3.2689	3.7341	4.9433	5.5283
14	19	2.2556	2.6469	3.1949	3.6378	4.7799	5.3285
14	20	2.2250	2.6030	3.1296	3.5530	4.6374	5.1547
14	21	2.1975	2.5638	3.0715	3.4779	4.5119	5.0022
14	22	2.1727	2.5285	3.0195	3.4108	4.4007	4.8675
14	23	2.1502	2.4966	2.9727	3.3506	4.3015	4.7476
14	24	2.1298	2.4677	2.9303	3.2962	4.2124	4.6402
14	25	2.1111	2.4413	2.8917	3.2469	4.1320	4.5435
14	26	2.0939	2.4171	2.8566	3.2020	4.0591	4.4559
14	27	2.0781	2.3949	2.8243	3.1608	3.9926	4.3763
14	28	2.0635	2.3743	2.7946	3.1231	3.9319	4.3036
14	29	2.0500	2.3554	2.7672	3.0882	3.8761	4.2370
14	30	2.0374	2.3378	2.7418	3.0560	3.8247	4.1757
14	40	1.9476	2.2130	2.5634	2.8312	3.4707	3.7561
14	50	1.8949	2.1404	2.4609	2.7032	3.2731	3.5239
14	60	1.8602	2.0929	2.3943	2.6205	3.1472	3.3766
14	120	1.7750	1.9773	2.2339	2.4228	2.8512	3.0332
14	250	1.7315	1.9187	2.1537	2.3248	2.7074	2.8676
14	500	1.7116	1.8921	2.1174	2.2806	2.6431	2.7940
14	1000	1.7017	1.8788	2.0994	2.2588	2.6115	2.7578
14	∞	1.6918	1.8656	2.0815	2.2371	2.5802	2.7221

The distribution of the variance ratio, $F = S_1^2/S_2^2$, when $\sigma_1^2 = \sigma_2^2$

ν_1	ν_2	Cumulative probabilities					
		0.9500	0.9750	0.9900	0.9950	0.9990	0.9995
15	1	245.9499	984.8668	6157.285	24630.21	615763.7	2463056
15	2	19.4291	39.4313	99.4325	199.4329	999.4333	1999.433
15	3	8.7029	14.2527	26.8722	43.0847	127.3736	202.6169
15	4	5.8578	8.6565	14.1982	20.4383	46.7612	66.4828
15	5	4.6188	6.4277	9.7222	13.1463	25.9108	34.4988
15	6	3.9381	5.2687	7.5590	9.8140	17.5587	22.4029
15	7	3.5107	4.5678	6.3143	7.9678	13.3237	16.5012
15	8	3.2184	4.1012	5.5151	6.8143	10.8413	13.1363
15	9	3.0061	3.7694	4.9621	6.0325	9.2381	11.0081
15	10	2.8450	3.5217	4.5581	5.4707	8.1288	9.5594
15	11	2.7186	3.3299	4.2509	5.0489	7.3210	8.5182
15	12	2.6169	3.1772	4.0096	4.7213	6.7092	7.7381
15	13	2.5331	3.0527	3.8154	4.4600	6.2312	7.1341
15	14	2.4630	2.9493	3.6557	4.2468	5.8483	6.6538
15	15	2.4034	2.8621	3.5222	4.0698	5.5351	6.2637
15	16	2.3522	2.7875	3.4089	3.9205	5.2745	5.9409
15	17	2.3077	2.7230	3.3117	3.7929	5.0544	5.6697
15	18	2.2686	2.6667	3.2273	3.6827	4.8663	5.4388
15	19	2.2341	2.6171	3.1533	3.5866	4.7037	5.2401
15	20	2.2033	2.5731	3.0880	3.5020	4.5618	5.0673
15	21	2.1757	2.5338	3.0300	3.4270	4.4369	4.9157
15	22	2.1508	2.4984	2.9779	3.3600	4.3262	4.7817
15	23	2.1282	2.4665	2.9311	3.2999	4.2274	4.6624
15	24	2.1077	2.4374	2.8887	3.2456	4.1387	4.5556
15	25	2.0889	2.4110	2.8502	3.1963	4.0587	4.4595
15	26	2.0716	2.3867	2.8150	3.1515	3.9861	4.3724
15	27	2.0558	2.3644	2.7827	3.1104	3.9200	4.2933
15	28	2.0411	2.3438	2.7530	3.0727	3.8595	4.2210
15	29	2.0275	2.3248	2.7256	3.0379	3.8039	4.1547
15	30	2.0148	2.3072	2.7002	3.0057	3.7527	4.0938
15	40	1.9245	2.1819	2.5216	2.7811	3.4003	3.6765
15	50	1.8714	2.1090	2.4190	2.6531	3.2035	3.4456
15	60	1.8364	2.0613	2.3523	2.5705	3.0781	3.2992
15	120	1.7505	1.9450	2.1915	2.3727	2.7833	2.9575
15	250	1.7065	1.8861	2.1110	2.2746	2.6399	2.7928
15	500	1.6864	1.8592	2.0746	2.2304	2.5759	2.7195
15	1000	1.6764	1.8459	2.0565	2.2085	2.5443	2.6835
15	∞	1.6664	1.8326	2.0385	2.1868	2.5132	2.6479

The distribution of the variance ratio, $F = S_1^2/S_2^2$, when $\sigma_1^2 = \sigma_2^2$

ν_1	ν_2	Cumulative probabilities					
		0.9500	0.9750	0.9900	0.9950	0.9990	0.9995
16	1	246.4639	986.9187	6170.101	24681.47	617045.2	2468182
16	2	19.4333	39.4354	99.4367	199.4371	999.4374	1999.437
16	3	8.6923	14.2315	26.8269	43.0083	127.1359	202.2352
16	4	5.8441	8.6326	14.1539	20.3710	46.5969	66.2460
16	5	4.6038	6.4032	9.6802	13.0861	25.7826	34.3249
16	6	3.9223	5.2439	7.5186	9.7582	17.4499	22.2609
16	7	3.4944	4.5428	6.2750	7.9148	13.2265	16.3777
16	8	3.2016	4.0761	5.4766	6.7633	10.7517	13.0247
16	9	2.9890	3.7441	4.9240	5.9829	9.1538	10.9046
16	10	2.8276	3.4963	4.5204	5.4221	8.0484	9.4618
16	11	2.7009	3.3044	4.2134	5.0011	7.2435	8.4250
16	12	2.5989	3.1515	3.9724	4.6741	6.6340	7.6483
16	13	2.5149	3.0269	3.7783	4.4132	6.1578	7.0470
16	14	2.4446	2.9234	3.6187	4.2005	5.7764	6.5690
16	15	2.3849	2.8360	3.4852	4.0237	5.4644	6.1807
16	16	2.3335	2.7614	3.3720	3.8747	5.2048	5.8594
16	17	2.2888	2.6968	3.2748	3.7473	4.9856	5.5895
16	18	2.2496	2.6404	3.1904	3.6373	4.7982	5.3597
16	19	2.2149	2.5907	3.1165	3.5412	4.6362	5.1620
16	20	2.1840	2.5465	3.0512	3.4568	4.4949	4.9900
16	21	2.1563	2.5071	2.9931	3.3818	4.3705	4.8391
16	22	2.1313	2.4717	2.9411	3.3150	4.2602	4.7058
16	23	2.1086	2.4396	2.8943	3.2549	4.1618	4.5871
16	24	2.0880	2.4105	2.8519	3.2007	4.0735	4.4808
16	25	2.0691	2.3840	2.8133	3.1515	3.9938	4.3851
16	26	2.0518	2.3597	2.7781	3.1067	3.9215	4.2985
16	27	2.0358	2.3373	2.7458	3.0656	3.8556	4.2197
16	28	2.0210	2.3167	2.7160	3.0279	3.7953	4.1478
16	29	2.0073	2.2976	2.6886	2.9932	3.7400	4.0819
16	30	1.9946	2.2799	2.6632	2.9611	3.6890	4.0212
16	40	1.9037	2.1542	2.4844	2.7365	3.3378	3.6060
16	50	1.8503	2.0810	2.3816	2.6086	3.1418	3.3761
16	60	1.8151	2.0330	2.3148	2.5259	3.0167	3.2303
16	120	1.7285	1.9161	2.1536	2.3280	2.7228	2.8902
16	250	1.6841	1.8567	2.0728	2.2297	2.5797	2.7261
16	500	1.6638	1.8297	2.0362	2.1854	2.5159	2.6531
16	1000	1.6536	1.8162	2.0180	2.1635	2.4844	2.6172
16	∞	1.6435	1.8028	2.0000	2.1417	2.4533	2.5818

The distribution of the variance ratio, $F = S_1^2/S_2^2$, when $\sigma_1^2 = \sigma_2^2$

ν_1	ν_2	Cumulative probabilities					
		0.9500	0.9750	0.9900	0.9950	0.9990	0.9995
17	1	246.9184	988.7331	6181.435	24726.80	618178.4	2472715
17	2	19.4370	39.4391	99.4404	199.4408	999.4411	1999.441
17	3	8.6829	14.2127	26.7867	42.9407	126.9254	201.8972
17	4	5.8320	8.6113	14.1146	20.3113	46.4515	66.0362
17	5	4.5904	6.3814	9.6429	13.0327	25.6691	34.1709
17	6	3.9083	5.2218	7.4827	9.7086	17.3534	22.1350
17	7	3.4799	4.5206	6.2401	7.8678	13.1403	16.2682
17	8	3.1867	4.0538	5.4423	6.7180	10.6722	12.9257
17	9	2.9737	3.7216	4.8902	5.9388	9.0790	10.8128
17	10	2.8120	3.4737	4.4869	5.3789	7.9770	9.3752
17	11	2.6851	3.2816	4.1801	4.9586	7.1747	8.3423
17	12	2.5828	3.1286	3.9392	4.6321	6.5672	7.5686
17	13	2.4987	3.0039	3.7452	4.3716	6.0926	6.9697
17	14	2.4282	2.9003	3.5857	4.1592	5.7124	6.4936
17	15	2.3683	2.8128	3.4523	3.9827	5.4015	6.1068
17	16	2.3167	2.7380	3.3391	3.8338	5.1428	5.7869
17	17	2.2719	2.6733	3.2419	3.7066	4.9244	5.5181
17	18	2.2325	2.6168	3.1575	3.5967	4.7376	5.2894
17	19	2.1977	2.5670	3.0836	3.5008	4.5762	5.0924
17	20	2.1667	2.5228	3.0183	3.4164	4.4353	4.9212
17	21	2.1389	2.4833	2.9602	3.3416	4.3114	4.7709
17	22	2.1138	2.4478	2.9082	3.2748	4.2015	4.6382
17	23	2.0910	2.4157	2.8613	3.2148	4.1034	4.5200
17	24	2.0703	2.3865	2.8189	3.1606	4.0154	4.4142
17	25	2.0513	2.3599	2.7803	3.1114	3.9359	4.3189
17	26	2.0339	2.3355	2.7451	3.0666	3.8638	4.2326
17	27	2.0179	2.3131	2.7127	3.0256	3.7981	4.1542
17	28	2.0030	2.2924	2.6830	2.9879	3.7381	4.0826
17	29	1.9893	2.2732	2.6555	2.9532	3.6829	4.0169
17	30	1.9765	2.2554	2.6301	2.9211	3.6321	3.9565
17	40	1.8851	2.1293	2.4511	2.6966	3.2820	3.5429
17	50	1.8313	2.0558	2.3481	2.5686	3.0865	3.3140
17	60	1.7959	2.0076	2.2811	2.4859	2.9618	3.1688
17	120	1.7085	1.8900	2.1194	2.2878	2.6685	2.8298
17	250	1.6638	1.8303	2.0384	2.1894	2.5257	2.6662
17	500	1.6432	1.8030	2.0016	2.1449	2.4619	2.5934
17	1000	1.6330	1.7895	1.9834	2.1229	2.4305	2.5577
17	∞	1.6228	1.7759	1.9652	2.1011	2.3994	2.5223

The distribution of the variance ratio, $F = S_1^2/S_2^2$, when $\sigma_1^2 = \sigma_2^2$

ν_1	ν_2	Cumulative probabilities					
		0.9500	0.9750	0.9900	0.9950	0.9990	0.9995
18	1	247.3232	990.3490	6191.529	24767.17	619187.7	2476752
18	2	19.4402	39.4424	99.4436	199.4440	999.4444	1999.444
18	3	8.6745	14.1960	26.7509	42.8804	126.7378	201.5960
18	4	5.8211	8.5924	14.0795	20.2581	46.3217	65.8491
18	5	4.5785	6.3619	9.6096	12.9850	25.5677	34.0334
18	6	3.8957	5.2021	7.4507	9.6644	17.2673	22.0227
18	7	3.4669	4.5008	6.2089	7.8258	13.0633	16.1704
18	8	3.1733	4.0338	5.4116	6.6775	10.6012	12.8373
18	9	2.9600	3.7015	4.8599	5.8994	9.0121	10.7307
18	10	2.7980	3.4534	4.4569	5.3403	7.9131	9.2977
18	11	2.6709	3.2612	4.1503	4.9205	7.1131	8.2683
18	12	2.5684	3.1081	3.9095	4.5945	6.5074	7.4973
18	13	2.4841	2.9832	3.7156	4.3344	6.0342	6.9005
18	14	2.4134	2.8795	3.5561	4.1221	5.6551	6.4261
18	15	2.3533	2.7919	3.4228	3.9459	5.3452	6.0408
18	16	2.3016	2.7170	3.3096	3.7972	5.0872	5.7220
18	17	2.2567	2.6522	3.2124	3.6701	4.8695	5.4542
18	18	2.2172	2.5956	3.1280	3.5603	4.6833	5.2263
18	19	2.1823	2.5457	3.0541	3.4645	4.5223	5.0301
18	20	2.1511	2.5014	2.9887	3.3802	4.3819	4.8595
18	21	2.1232	2.4618	2.9306	3.3054	4.2583	4.7098
18	22	2.0980	2.4262	2.8786	3.2387	4.1487	4.5775
18	23	2.0751	2.3940	2.8317	3.1787	4.0509	4.4598
18	24	2.0543	2.3648	2.7892	3.1246	3.9631	4.3544
18	25	2.0353	2.3381	2.7506	3.0754	3.8839	4.2594
18	26	2.0178	2.3137	2.7153	3.0306	3.8120	4.1735
18	27	2.0017	2.2912	2.6830	2.9896	3.7466	4.0953
18	28	1.9868	2.2704	2.6532	2.9520	3.6867	4.0240
18	29	1.9730	2.2512	2.6257	2.9173	3.6316	3.9586
18	30	1.9601	2.2334	2.6003	2.8852	3.5810	3.8984
18	40	1.8682	2.1068	2.4210	2.6607	3.2318	3.4863
18	50	1.8141	2.0330	2.3178	2.5326	3.0367	3.2581
18	60	1.7784	1.9846	2.2507	2.4498	2.9123	3.1133
18	120	1.6904	1.8663	2.0885	2.2514	2.6195	2.7754
18	250	1.6453	1.8062	2.0071	2.1528	2.4769	2.6122
18	500	1.6245	1.7787	1.9702	2.1082	2.4132	2.5395
18	1000	1.6142	1.7651	1.9519	2.0862	2.3818	2.5038
18	∞	1.6038	1.7515	1.9336	2.0642	2.3507	2.4685

The distribution of the variance ratio, $F = S_1^2/S_2^2$, when $\sigma_1^2 = \sigma_2^2$

ν_1	ν_2	Cumulative probabilities					
		0.9500	0.9750	0.9900	0.9950	0.9990	0.9995
19	1	247.6861	991.7973	6200.576	24803.35	620092.3	2480370
19	2	19.4431	39.4453	99.4465	199.4470	999.4473	1999.447
19	3	8.6670	14.1810	26.7188	42.8263	126.5696	201.3258
19	4	5.8114	8.5753	14.0480	20.2104	46.2053	65.6812
19	5	4.5678	6.3444	9.5797	12.9422	25.4768	33.9100
19	6	3.8844	5.1844	7.4219	9.6247	17.1900	21.9218
19	7	3.4551	4.4829	6.1808	7.7881	12.9941	16.0826
19	8	3.1613	4.0158	5.3840	6.6411	10.5374	12.7578
19	9	2.9477	3.6833	4.8327	5.8639	8.9520	10.6570
19	10	2.7854	3.4351	4.4299	5.3055	7.8557	9.2281
19	11	2.6581	3.2428	4.1234	4.8863	7.0577	8.2017
19	12	2.5554	3.0896	3.8827	4.5606	6.4535	7.4331
19	13	2.4709	2.9646	3.6888	4.3008	5.9816	6.8382
19	14	2.4000	2.8607	3.5294	4.0888	5.6035	6.3653
19	15	2.3398	2.7730	3.3961	3.9127	5.2944	5.9812
19	16	2.2880	2.6980	3.2829	3.7641	5.0372	5.6635
19	17	2.2429	2.6331	3.1857	3.6372	4.8200	5.3966
19	18	2.2033	2.5764	3.1013	3.5275	4.6343	5.1694
19	19	2.1683	2.5265	3.0274	3.4318	4.4738	4.9739
19	20	2.1370	2.4821	2.9620	3.3475	4.3337	4.8038
19	21	2.1090	2.4424	2.9039	3.2728	4.2104	4.6547
19	22	2.0837	2.4067	2.8518	3.2060	4.1011	4.5228
19	23	2.0608	2.3745	2.8049	3.1461	4.0036	4.4055
19	24	2.0399	2.3452	2.7624	3.0920	3.9160	4.3004
19	25	2.0207	2.3184	2.7238	3.0429	3.8370	4.2058
19	26	2.0032	2.2939	2.6885	2.9981	3.7653	4.1201
19	27	1.9870	2.2713	2.6561	2.9571	3.7000	4.0422
19	28	1.9720	2.2505	2.6263	2.9194	3.6402	3.9711
19	29	1.9581	2.2313	2.5987	2.8847	3.5853	3.9059
19	30	1.9452	2.2134	2.5732	2.8526	3.5348	3.8459
19	40	1.8529	2.0864	2.3937	2.6281	3.1863	3.4350
19	50	1.7985	2.0122	2.2903	2.4999	2.9916	3.2075
19	60	1.7625	1.9636	2.2230	2.4171	2.8674	3.0631
19	120	1.6739	1.8447	2.0604	2.2183	2.5750	2.7260
19	250	1.6283	1.7842	1.9786	2.1194	2.4325	2.5631
19	500	1.6074	1.7566	1.9415	2.0748	2.3688	2.4905
19	1000	1.5969	1.7428	1.9231	2.0526	2.3374	2.4549
19	∞	1.5865	1.7291	1.9048	2.0306	2.3063	2.4196

The distribution of the variance ratio, $F = S_1^2/S_2^2$, when $\sigma_1^2 = \sigma_2^2$

ν_1	ν_2	Cumulative probabilities					
		0.9500	0.9750	0.9900	0.9950	0.9990	0.9995
20	1	248.0131	993.1028	6208.730	24835.97	620907.7	2483632
20	2	19.4458	39.4479	99.4492	199.4496	999.4499	1999.450
20	3	8.6602	14.1674	26.6898	42.7775	126.4178	201.0820
20	4	5.8025	8.5599	14.0196	20.1673	46.1003	65.5298
20	5	4.5581	6.3286	9.5526	12.9035	25.3946	33.7987
20	6	3.8742	5.1684	7.3958	9.5888	17.1201	21.8307
20	7	3.4445	4.4667	6.1554	7.7540	12.9316	16.0033
20	8	3.1503	3.9995	5.3591	6.6082	10.4797	12.6860
20	9	2.9365	3.6669	4.8080	5.8318	8.8976	10.5903
20	10	2.7740	3.4185	4.4054	5.2740	7.8037	9.1651
20	11	2.6464	3.2261	4.0990	4.8552	7.0076	8.1415
20	12	2.5436	3.0728	3.8584	4.5299	6.4048	7.3750
20	13	2.4589	2.9477	3.6646	4.2703	5.9340	6.7818
20	14	2.3879	2.8437	3.5052	4.0585	5.5568	6.3103
20	15	2.3275	2.7559	3.3719	3.8826	5.2484	5.9273
20	16	2.2756	2.6808	3.2587	3.7342	4.9918	5.6105
20	17	2.2304	2.6158	3.1615	3.6073	4.7751	5.3444
20	18	2.1906	2.5590	3.0771	3.4977	4.5899	5.1179
20	19	2.1555	2.5089	3.0031	3.4020	4.4297	4.9229
20	20	2.1242	2.4645	2.9377	3.3178	4.2900	4.7534
20	21	2.0960	2.4247	2.8796	3.2431	4.1670	4.6047
20	22	2.0707	2.3890	2.8274	3.1764	4.0579	4.4732
20	23	2.0476	2.3567	2.7805	3.1165	3.9606	4.3562
20	24	2.0267	2.3273	2.7380	3.0624	3.8732	4.2514
20	25	2.0075	2.3005	2.6993	3.0133	3.7944	4.1571
20	26	1.9898	2.2759	2.6640	2.9685	3.7228	4.0716
20	27	1.9736	2.2533	2.6316	2.9275	3.6576	3.9940
20	28	1.9586	2.2324	2.6017	2.8899	3.5980	3.9230
20	29	1.9446	2.2131	2.5742	2.8551	3.5432	3.8580
20	30	1.9317	2.1952	2.5487	2.8230	3.4928	3.7982
20	40	1.8389	2.0677	2.3689	2.5984	3.1450	3.3884
20	50	1.7841	1.9933	2.2652	2.4702	2.9506	3.1614
20	60	1.7480	1.9445	2.1978	2.3872	2.8266	3.0174
20	120	1.6587	1.8249	2.0346	2.1881	2.5344	2.6809
20	250	1.6127	1.7640	1.9525	2.0889	2.3920	2.5182
20	500	1.5916	1.7362	1.9152	2.0441	2.3282	2.4458
20	1000	1.5811	1.7223	1.8967	2.0219	2.2968	2.4101
20	∞	1.5705	1.7085	1.8783	1.9998	2.2657	2.3749

The distribution of the variance ratio, $F = S_1^2/S_2^2$, when $\sigma_1^2 = \sigma_2^2$

ν_1	ν_2	Cumulative probabilities					
		0.9500	0.9750	0.9900	0.9950	0.9990	0.9995
21	1	248.3094	994.2856	6216.118	24865.52	621646.4	2486587
21	2	19.4481	39.4503	99.4516	199.4520	999.4523	1999.452
21	3	8.6540	14.1551	26.6635	42.7333	126.2801	200.8610
21	4	5.7945	8.5460	13.9938	20.1282	46.0050	65.3924
21	5	4.5493	6.3142	9.5281	12.8684	25.3201	33.6976
21	6	3.8649	5.1538	7.3722	9.5562	17.0567	21.7480
21	7	3.4349	4.4520	6.1324	7.7230	12.8749	15.9313
21	8	3.1404	3.9846	5.3364	6.5783	10.4273	12.6207
21	9	2.9263	3.6520	4.7856	5.8027	8.8482	10.5297
21	10	2.7636	3.4035	4.3831	5.2454	7.7565	9.1078
21	11	2.6358	3.2109	4.0769	4.8270	6.9620	8.0868
21	12	2.5328	3.0575	3.8363	4.5020	6.3605	7.3222
21	13	2.4479	2.9322	3.6425	4.2426	5.8907	6.7305
21	14	2.3768	2.8282	3.4832	4.0310	5.5143	6.2602
21	15	2.3163	2.7403	3.3498	3.8552	5.2066	5.8783
21	16	2.2642	2.6651	3.2367	3.7069	4.9505	5.5623
21	17	2.2189	2.6000	3.1394	3.5801	4.7343	5.2969
21	18	2.1791	2.5431	3.0550	3.4705	4.5494	5.0710
21	19	2.1438	2.4930	2.9810	3.3749	4.3896	4.8765
21	20	2.1124	2.4484	2.9156	3.2907	4.2501	4.7074
21	21	2.0842	2.4086	2.8574	3.2160	4.1274	4.5591
21	22	2.0587	2.3728	2.8052	3.1494	4.0186	4.4280
21	23	2.0356	2.3404	2.7583	3.0895	3.9214	4.3113
21	24	2.0146	2.3109	2.7157	3.0354	3.8342	4.2068
21	25	1.9953	2.2840	2.6770	2.9862	3.7555	4.1127
21	26	1.9776	2.2594	2.6416	2.9415	3.6841	4.0275
21	27	1.9613	2.2367	2.6092	2.9005	3.6190	3.9500
21	28	1.9462	2.2158	2.5793	2.8628	3.5595	3.8792
21	29	1.9322	2.1965	2.5517	2.8281	3.5048	3.8144
21	30	1.9192	2.1785	2.5262	2.7960	3.4545	3.7547
21	40	1.8260	2.0506	2.3461	2.5713	3.1072	3.3459
21	50	1.7709	1.9759	2.2423	2.4429	2.9131	3.1194
21	60	1.7346	1.9269	2.1747	2.3598	2.7892	2.9756
21	120	1.6447	1.8067	2.0109	2.1603	2.4972	2.6396
21	250	1.5983	1.7454	1.9285	2.0609	2.3547	2.4771
21	500	1.5770	1.7174	1.8910	2.0159	2.2910	2.4047
21	1000	1.5664	1.7034	1.8724	1.9936	2.2595	2.3691
21	∞	1.5557	1.6895	1.8539	1.9715	2.2284	2.3338

The distribution of the variance ratio, $F = S_1^2/S_2^2$, when $\sigma_1^2 = \sigma_2^2$

ν_1	ν_2	Cumulative probabilities					
		0.9500	0.9750	0.9900	0.9950	0.9990	0.9995
22	1	248.5791	995.3622	6222.843	24892.42	622318.8	2489276
22	2	19.4503	39.4525	99.4537	199.4541	999.4545	1999.455
22	3	8.6484	14.1438	26.6396	42.6929	126.1548	200.6598
22	4	5.7872	8.5332	13.9703	20.0925	45.9182	65.2672
22	5	4.5413	6.3011	9.5058	12.8364	25.2522	33.6055
22	6	3.8564	5.1406	7.3506	9.5264	16.9989	21.6726
22	7	3.4260	4.4386	6.1113	7.6947	12.8231	15.8656
22	8	3.1313	3.9711	5.3157	6.5510	10.3795	12.5612
22	9	2.9169	3.6383	4.7651	5.7760	8.8031	10.4744
22	10	2.7541	3.3897	4.3628	5.2192	7.7134	9.0556
22	11	2.6261	3.1970	4.0566	4.8012	6.9204	8.0368
22	12	2.5229	3.0434	3.8161	4.4765	6.3200	7.2740
22	13	2.4379	2.9181	3.6224	4.2173	5.8511	6.6836
22	14	2.3667	2.8139	3.4630	4.0058	5.4755	6.2145
22	15	2.3060	2.7260	3.3297	3.8301	5.1683	5.8334
22	16	2.2538	2.6507	3.2165	3.6819	4.9128	5.5182
22	17	2.2084	2.5855	3.1192	3.5552	4.6969	5.2535
22	18	2.1685	2.5285	3.0348	3.4456	4.5124	5.0281
22	19	2.1331	2.4783	2.9607	3.3500	4.3529	4.8341
22	20	2.1016	2.4337	2.8953	3.2659	4.2137	4.6654
22	21	2.0733	2.3938	2.8370	3.1912	4.0912	4.5174
22	22	2.0478	2.3579	2.7849	3.1246	3.9825	4.3866
22	23	2.0246	2.3254	2.7378	3.0647	3.8856	4.2702
22	24	2.0035	2.2959	2.6953	3.0106	3.7985	4.1659
22	25	1.9842	2.2690	2.6565	2.9615	3.7199	4.0720
22	26	1.9664	2.2443	2.6211	2.9167	3.6486	3.9870
22	27	1.9500	2.2216	2.5887	2.8757	3.5837	3.9097
22	28	1.9349	2.2006	2.5587	2.8380	3.5242	3.8391
22	29	1.9208	2.1812	2.5311	2.8033	3.4696	3.7744
22	30	1.9077	2.1631	2.5055	2.7712	3.4194	3.7148
22	40	1.8141	2.0349	2.3252	2.5463	3.0726	3.3068
22	50	1.7588	1.9599	2.2211	2.4178	2.8786	3.0807
22	60	1.7222	1.9106	2.1533	2.3346	2.7548	2.9372
22	120	1.6317	1.7899	1.9891	2.1347	2.4629	2.6016
22	250	1.5850	1.7282	1.9063	2.0350	2.3204	2.4392
22	500	1.5635	1.7000	1.8686	1.9899	2.2566	2.3668
22	1000	1.5528	1.6859	1.8500	1.9675	2.2252	2.3312
22	∞	1.5420	1.6719	1.8313	1.9453	2.1940	2.2960

The distribution of the variance ratio, $F = S_1^2/S_2^2$, when $\sigma_1^2 = \sigma_2^2$

ν_1	ν_2	Cumulative probabilities					
		0.9500	0.9750	0.9900	0.9950	0.9990	0.9995
23	1	248.8256	996.3462	6228.990	24917.00	622933.5	2491735
23	2	19.4523	39.4544	99.4557	199.4561	999.4564	1999.456
23	3	8.6432	14.1336	26.6176	42.6561	126.0401	200.4757
23	4	5.7805	8.5216	13.9488	20.0599	45.8387	65.1527
23	5	4.5339	6.2891	9.4853	12.8071	25.1900	33.5213
23	6	3.8486	5.1284	7.3309	9.4992	16.9460	21.6037
23	7	3.4179	4.4263	6.0921	7.6688	12.7758	15.8054
23	8	3.1229	3.9587	5.2967	6.5260	10.3357	12.5067
23	9	2.9084	3.6257	4.7463	5.7516	8.7618	10.4238
23	10	2.7453	3.3770	4.3441	5.1953	7.6739	9.0077
23	11	2.6172	3.1843	4.0380	4.7775	6.8823	7.9910
23	12	2.5139	3.0306	3.7976	4.4530	6.2829	7.2297
23	13	2.4287	2.9052	3.6038	4.1940	5.8148	6.6407
23	14	2.3573	2.8009	3.4445	3.9827	5.4399	6.1725
23	15	2.2966	2.7128	3.3111	3.8071	5.1332	5.7923
23	16	2.2443	2.6374	3.1979	3.6589	4.8781	5.4778
23	17	2.1987	2.5721	3.1006	3.5323	4.6627	5.2136
23	18	2.1587	2.5151	3.0161	3.4228	4.4784	4.9887
23	19	2.1233	2.4648	2.9421	3.3272	4.3192	4.7951
23	20	2.0917	2.4201	2.8766	3.2431	4.1802	4.6268
23	21	2.0633	2.3801	2.8183	3.1684	4.0579	4.4791
23	22	2.0377	2.3442	2.7661	3.1018	3.9494	4.3486
23	23	2.0144	2.3116	2.7191	3.0419	3.8526	4.2324
23	24	1.9932	2.2821	2.6765	2.9878	3.7657	4.1284
23	25	1.9738	2.2551	2.6377	2.9387	3.6872	4.0346
23	26	1.9560	2.2303	2.6022	2.8939	3.6160	3.9498
23	27	1.9396	2.2076	2.5697	2.8529	3.5511	3.8726
23	28	1.9244	2.1865	2.5398	2.8152	3.4918	3.8022
23	29	1.9103	2.1671	2.5121	2.7805	3.4373	3.7376
23	30	1.8972	2.1490	2.4865	2.7483	3.3870	3.6781
23	40	1.8031	2.0203	2.3059	2.5233	3.0406	3.2709
23	50	1.7475	1.9451	2.2016	2.3947	2.8469	3.0451
23	60	1.7108	1.8956	2.1336	2.3114	2.7231	2.9018
23	120	1.6197	1.7743	1.9688	2.1110	2.4313	2.5665
23	250	1.5726	1.7122	1.8857	2.0110	2.2886	2.4041
23	500	1.5509	1.6838	1.8479	1.9657	2.2248	2.3317
23	1000	1.5401	1.6696	1.8291	1.9433	2.1933	2.2961
23	∞	1.5292	1.6555	1.8104	1.9209	2.1621	2.2609

The distribution of the variance ratio, $F = S_1^2/S_2^2$, when $\sigma_1^2 = \sigma_2^2$

ν_1	ν_2	Cumulative probabilities					
		0.9500	0.9750	0.9900	0.9950	0.9990	0.9995
24	1	249.0518	997.2492	6234.631	24939.57	623497.5	2493991
24	2	19.4541	39.4562	99.4575	199.4579	999.4583	1999.458
24	3	8.6385	14.1241	26.5975	42.6222	125.9349	200.3067
24	4	5.7744	8.5109	13.9291	20.0300	45.7658	65.0475
24	5	4.5272	6.2780	9.4665	12.7802	25.1329	33.4438
24	6	3.8415	5.1172	7.3127	9.4742	16.8974	21.5403
24	7	3.4105	4.4150	6.0743	7.6450	12.7322	15.7502
24	8	3.1152	3.9472	5.2793	6.5029	10.2954	12.4566
24	9	2.9005	3.6142	4.7290	5.7292	8.7239	10.3772
24	10	2.7372	3.3654	4.3269	5.1732	7.6376	8.9637
24	11	2.6090	3.1725	4.0209	4.7557	6.8471	7.9489
24	12	2.5055	3.0187	3.7805	4.4314	6.2488	7.1891
24	13	2.4202	2.8932	3.5868	4.1726	5.7814	6.6011
24	14	2.3487	2.7888	3.4274	3.9614	5.4070	6.1339
24	15	2.2878	2.7006	3.2940	3.7859	5.1009	5.7544
24	16	2.2354	2.6252	3.1808	3.6378	4.8462	5.4405
24	17	2.1898	2.5598	3.0835	3.5112	4.6311	5.1769
24	18	2.1497	2.5027	2.9990	3.4017	4.4471	4.9524
24	19	2.1141	2.4523	2.9249	3.3062	4.2881	4.7592
24	20	2.0825	2.4076	2.8594	3.2220	4.1493	4.5912
24	21	2.0540	2.3675	2.8010	3.1474	4.0272	4.4438
24	22	2.0283	2.3315	2.7488	3.0807	3.9189	4.3136
24	23	2.0050	2.2989	2.7017	3.0208	3.8222	4.1976
24	24	1.9838	2.2693	2.6591	2.9667	3.7354	4.0937
24	25	1.9643	2.2422	2.6203	2.9176	3.6570	4.0002
24	26	1.9464	2.2174	2.5848	2.8728	3.5859	3.9155
24	27	1.9299	2.1946	2.5522	2.8318	3.5211	3.8385
24	28	1.9147	2.1735	2.5223	2.7941	3.4618	3.7681
24	29	1.9005	2.1540	2.4946	2.7594	3.4074	3.7036
24	30	1.8874	2.1359	2.4689	2.7272	3.3572	3.6443
24	40	1.7929	2.0069	2.2880	2.5020	3.0111	3.2377
24	50	1.7371	1.9313	2.1835	2.3732	2.8175	3.0122
24	60	1.7001	1.8817	2.1154	2.2898	2.6938	2.8690
24	120	1.6084	1.7597	1.9500	2.0890	2.4019	2.5339
24	250	1.5610	1.6973	1.8665	1.9887	2.2592	2.3716
24	500	1.5392	1.6687	1.8285	1.9432	2.1952	2.2992
24	1000	1.5282	1.6544	1.8096	1.9207	2.1637	2.2635
24	∞	1.5173	1.6402	1.7908	1.8983	2.1324	2.2283

The distribution of the variance ratio, $F = S_1^2/S_2^2$, when $\sigma_1^2 = \sigma_2^2$

ν_1	ν_2	Cumulative probabilities					
		0.9500	0.9750	0.9900	0.9950	0.9990	0.9995
25	1	249.2601	998.0808	6239.825	24960.34	624016.8	2496068
25	2	19.4558	39.4579	99.4592	199.4596	999.4599	1999.460
25	3	8.6341	14.1155	26.5790	42.5910	125.8379	200.1510
25	4	5.7687	8.5010	13.9109	20.0024	45.6986	64.9506
25	5	4.5209	6.2679	9.4491	12.7554	25.0803	33.3725
25	6	3.8348	5.1069	7.2960	9.4511	16.8525	21.4818
25	7	3.4036	4.4045	6.0580	7.6230	12.6920	15.6992
25	8	3.1081	3.9367	5.2631	6.4817	10.2583	12.4104
25	9	2.8932	3.6035	4.7130	5.7084	8.6888	10.3342
25	10	2.7298	3.3546	4.3111	5.1528	7.6041	8.9231
25	11	2.6014	3.1616	4.0051	4.7356	6.8147	7.9100
25	12	2.4977	3.0077	3.7647	4.4115	6.2172	7.1515
25	13	2.4123	2.8821	3.5710	4.1528	5.7505	6.5646
25	14	2.3407	2.7777	3.4116	3.9417	5.3767	6.0982
25	15	2.2797	2.6894	3.2782	3.7662	5.0710	5.7194
25	16	2.2272	2.6138	3.1650	3.6182	4.8167	5.4061
25	17	2.1815	2.5484	3.0676	3.4916	4.6019	5.1429
25	18	2.1413	2.4912	2.9831	3.3822	4.4182	4.9189
25	19	2.1057	2.4408	2.9089	3.2867	4.2594	4.7260
25	20	2.0739	2.3959	2.8434	3.2025	4.1208	4.5583
25	21	2.0454	2.3558	2.7850	3.1279	3.9988	4.4112
25	22	2.0196	2.3198	2.7328	3.0613	3.8906	4.2811
25	23	1.9963	2.2871	2.6856	3.0014	3.7940	4.1654
25	24	1.9750	2.2574	2.6430	2.9472	3.7073	4.0617
25	25	1.9554	2.2303	2.6041	2.8981	3.6291	3.9683
25	26	1.9375	2.2054	2.5686	2.8533	3.5580	3.8837
25	27	1.9210	2.1826	2.5360	2.8123	3.4933	3.8068
25	28	1.9057	2.1615	2.5060	2.7746	3.4341	3.7366
25	29	1.8915	2.1419	2.4783	2.7398	3.3797	3.6722
25	30	1.8782	2.1237	2.4526	2.7076	3.3296	3.6130
25	40	1.7835	1.9943	2.2714	2.4823	2.9838	3.2069
25	50	1.7273	1.9186	2.1667	2.3533	2.7902	2.9817
25	60	1.6902	1.8687	2.0984	2.2697	2.6665	2.8386
25	120	1.5980	1.7462	1.9325	2.0686	2.3746	2.5037
25	250	1.5502	1.6834	1.8487	1.9679	2.2317	2.3413
25	500	1.5282	1.6546	1.8105	1.9223	2.1677	2.2688
25	1000	1.5171	1.6402	1.7915	1.8996	2.1361	2.2332
25	∞	1.5061	1.6259	1.7726	1.8771	2.1048	2.1979

The distribution of the variance ratio, $F = S_1^2/S_2^2$, when $\sigma_1^2 = \sigma_2^2$

ν_1	ν_2	Cumulative probabilities					
		0.9500	0.9750	0.9900	0.9950	0.9990	0.9995
26	1	249.4525	998.8490	6244.624	24979.53	624496.7	2497988
26	2	19.4573	39.4594	99.4607	199.4611	999.4615	1999.461
26	3	8.6301	14.1074	26.5618	42.5622	125.7483	200.0071
26	4	5.7635	8.4919	13.8940	19.9769	45.6364	64.8610
26	5	4.5151	6.2584	9.4331	12.7327	25.0316	33.3065
26	6	3.8287	5.0973	7.2805	9.4298	16.8111	21.4278
26	7	3.3972	4.3949	6.0428	7.6027	12.6549	15.6520
26	8	3.1015	3.9269	5.2482	6.4620	10.2239	12.3677
26	9	2.8864	3.5936	4.6982	5.6892	8.6564	10.2945
26	10	2.7229	3.3446	4.2963	5.1339	7.5730	8.8854
26	11	2.5943	3.1516	3.9904	4.7170	6.7847	7.8739
26	12	2.4905	2.9976	3.7500	4.3930	6.1880	7.1167
26	13	2.4050	2.8719	3.5563	4.1344	5.7219	6.5308
26	14	2.3333	2.7673	3.3969	3.9234	5.3486	6.0652
26	15	2.2722	2.6790	3.2635	3.7480	5.0433	5.6870
26	16	2.2196	2.6033	3.1503	3.6000	4.7893	5.3742
26	17	2.1738	2.5378	3.0529	3.4735	4.5748	5.1114
26	18	2.1335	2.4806	2.9683	3.3641	4.3913	4.8878
26	19	2.0978	2.4300	2.8941	3.2686	4.2327	4.6952
26	20	2.0660	2.3851	2.8286	3.1845	4.0943	4.5278
26	21	2.0374	2.3450	2.7702	3.1098	3.9724	4.3809
26	22	2.0116	2.3088	2.7179	3.0432	3.8644	4.2511
26	23	1.9881	2.2761	2.6707	2.9833	3.7679	4.1355
26	24	1.9668	2.2464	2.6280	2.9291	3.6813	4.0319
26	25	1.9472	2.2192	2.5891	2.8800	3.6031	3.9387
26	26	1.9292	2.1943	2.5536	2.8352	3.5321	3.8542
26	27	1.9126	2.1714	2.5209	2.7941	3.4675	3.7774
26	28	1.8973	2.1502	2.4909	2.7564	3.4083	3.7073
26	29	1.8830	2.1306	2.4631	2.7216	3.3540	3.6430
26	30	1.8698	2.1124	2.4374	2.6894	3.3039	3.5839
26	40	1.7746	1.9827	2.2559	2.4639	2.9583	3.1783
26	50	1.7183	1.9066	2.1510	2.3348	2.7648	2.9533
26	60	1.6809	1.8566	2.0825	2.2511	2.6412	2.8103
26	120	1.5881	1.7335	1.9161	2.0494	2.3491	2.4755
26	250	1.5400	1.6704	1.8319	1.9484	2.2061	2.3130
26	500	1.5178	1.6414	1.7936	1.9026	2.1419	2.2405
26	1000	1.5067	1.6269	1.7745	1.8799	2.1103	2.2048
26	∞	1.4956	1.6124	1.7554	1.8573	2.0789	2.1695

The distribution of the variance ratio, $F = S_1^2/S_2^2$, when $\sigma_1^2 = \sigma_2^2$

ν_1	ν_2	Cumulative probabilities					
		0.9500	0.9750	0.9900	0.9950	0.9990	0.9995
27	1	249.6309	999.5609	6249.071	24997.32	624941.3	2499766
27	2	19.4587	39.4609	99.4621	199.4625	999.4629	1999.463
27	3	8.6263	14.1000	26.5460	42.5355	125.6652	199.8736
27	4	5.7586	8.4834	13.8784	19.9532	45.5788	64.7780
27	5	4.5097	6.2497	9.4182	12.7112	24.9865	33.2453
27	6	3.8230	5.0884	7.2661	9.4100	16.7726	21.3776
27	7	3.3913	4.3859	6.0287	7.5838	12.6204	15.6083
27	8	3.0954	3.9178	5.2344	6.4438	10.1920	12.3280
27	9	2.8801	3.5845	4.6845	5.6714	8.6262	10.2576
27	10	2.7164	3.3353	4.2827	5.1164	7.5442	8.8505
27	11	2.5877	3.1422	3.9768	4.6997	6.7568	7.8405
27	12	2.4838	2.9881	3.7364	4.3759	6.1608	7.0844
27	13	2.3982	2.8623	3.5427	4.1174	5.6954	6.4993
27	14	2.3264	2.7577	3.3833	3.9064	5.3225	6.0345
27	15	2.2652	2.6692	3.2499	3.7311	5.0176	5.6569
27	16	2.2125	2.5935	3.1366	3.5831	4.7638	5.3445
27	17	2.1666	2.5280	3.0392	3.4566	4.5496	5.0822
27	18	2.1262	2.4706	2.9546	3.3472	4.3663	4.8589
27	19	2.0905	2.4200	2.8804	3.2517	4.2079	4.6666
27	20	2.0586	2.3751	2.8148	3.1676	4.0696	4.4994
27	21	2.0299	2.3348	2.7563	3.0930	3.9479	4.3527
27	22	2.0040	2.2986	2.7040	3.0263	3.8400	4.2231
27	23	1.9805	2.2659	2.6568	2.9664	3.7436	4.1076
27	24	1.9591	2.2361	2.6140	2.9123	3.6571	4.0042
27	25	1.9395	2.2089	2.5751	2.8631	3.5789	3.9111
27	26	1.9215	2.1839	2.5395	2.8183	3.5080	3.8268
27	27	1.9048	2.1609	2.5069	2.7772	3.4434	3.7501
27	28	1.8894	2.1397	2.4768	2.7395	3.3843	3.6801
27	29	1.8751	2.1201	2.4490	2.7047	3.3300	3.6158
27	30	1.8618	2.1018	2.4233	2.6725	3.2800	3.5567
27	40	1.7663	1.9718	2.2415	2.4467	2.9346	3.1516
27	50	1.7097	1.8955	2.1363	2.3174	2.7411	2.9268
27	60	1.6722	1.8453	2.0677	2.2336	2.6175	2.7839
27	120	1.5789	1.7216	1.9008	2.0315	2.3253	2.4491
27	250	1.5305	1.6581	1.8163	1.9302	2.1821	2.2866
27	500	1.5081	1.6289	1.7777	1.8842	2.1178	2.2140
27	1000	1.4969	1.6144	1.7585	1.8614	2.0861	2.1783
27	∞	1.4857	1.5998	1.7394	1.8387	2.0547	2.1429

The distribution of the variance ratio, $F = S_1^2/S_2^2$, when $\sigma_1^2 = \sigma_2^2$

ν_1	ν_2	Cumulative probabilities					
		0.9500	0.9750	0.9900	0.9950	0.9990	0.9995
28	1	249.7966	1000.222	6253.203	25013.85	625354.5	2501419
28	2	19.4600	39.4622	99.4635	199.4639	999.4642	1999.464
28	3	8.6229	14.0930	26.5312	42.5106	125.5879	199.7496
28	4	5.7541	8.4755	13.8639	19.9312	45.5252	64.7007
28	5	4.5047	6.2416	9.4043	12.6914	24.9445	33.1884
28	6	3.8177	5.0802	7.2527	9.3915	16.7368	21.3309
28	7	3.3858	4.3775	6.0157	7.5662	12.5883	15.5676
28	8	3.0897	3.9093	5.2214	6.4268	10.1623	12.2910
28	9	2.8743	3.5759	4.6717	5.6548	8.5982	10.2232
28	10	2.7104	3.3267	4.2700	5.1001	7.5173	8.8180
28	11	2.5816	3.1334	3.9641	4.6835	6.7309	7.8093
28	12	2.4776	2.9793	3.7237	4.3599	6.1355	7.0543
28	13	2.3918	2.8534	3.5300	4.1015	5.6706	6.4701
28	14	2.3199	2.7487	3.3706	3.8906	5.2982	6.0058
28	15	2.2587	2.6602	3.2372	3.7153	4.9936	5.6288
28	16	2.2059	2.5844	3.1238	3.5674	4.7401	5.3169
28	17	2.1599	2.5187	3.0264	3.4409	4.5261	5.0549
28	18	2.1195	2.4613	2.9418	3.3315	4.3430	4.8319
28	19	2.0836	2.4107	2.8675	3.2360	4.1848	4.6399
28	20	2.0517	2.3657	2.8019	3.1519	4.0466	4.4729
28	21	2.0229	2.3254	2.7434	3.0773	3.9250	4.3264
28	22	1.9970	2.2891	2.6910	3.0106	3.8172	4.1969
28	23	1.9734	2.2563	2.6438	2.9507	3.7209	4.0817
28	24	1.9520	2.2265	2.6010	2.8965	3.6345	3.9784
28	25	1.9323	2.1992	2.5620	2.8473	3.5564	3.8854
28	26	1.9142	2.1742	2.5264	2.8025	3.4856	3.8012
28	27	1.8975	2.1512	2.4937	2.7614	3.4210	3.7246
28	28	1.8821	2.1299	2.4636	2.7236	3.3619	3.6546
28	29	1.8677	2.1102	2.4358	2.6888	3.3076	3.5904
28	30	1.8544	2.0919	2.4100	2.6566	3.2576	3.5314
28	40	1.7586	1.9615	2.2280	2.4307	2.9124	3.1266
28	50	1.7017	1.8850	2.1226	2.3012	2.7190	2.9020
28	60	1.6641	1.8346	2.0538	2.2172	2.5953	2.7592
28	120	1.5703	1.7104	1.8864	2.0147	2.3029	2.4244
28	250	1.5215	1.6466	1.8015	1.9131	2.1595	2.2617
28	500	1.4989	1.6172	1.7627	1.8669	2.0952	2.1891
28	1000	1.4876	1.6025	1.7435	1.8440	2.0634	2.1533
28	∞	1.4763	1.5879	1.7242	1.8212	2.0319	2.1179

The distribution of the variance ratio, $F = S_1^2/S_2^2$, when $\sigma_1^2 = \sigma_2^2$

ν_1	ν_2	Cumulative probabilities					
		0.9500	0.9750	0.9900	0.9950	0.9990	0.9995
29	1	249.9510	1000.839	6257.053	25029.25	625739.4	2502959
29	2	19.4613	39.4634	99.4647	199.4651	999.4654	1999.465
29	3	8.6196	14.0866	26.5174	42.4874	125.5159	199.6340
29	4	5.7498	8.4681	13.8503	19.9107	45.4753	64.6287
29	5	4.5001	6.2340	9.3914	12.6729	24.9054	33.1353
29	6	3.8128	5.0724	7.2402	9.3743	16.7034	21.2874
29	7	3.3806	4.3697	6.0034	7.5498	12.5583	15.5296
29	8	3.0844	3.9014	5.2094	6.4109	10.1346	12.2566
29	9	2.8688	3.5679	4.6598	5.6393	8.5720	10.1911
29	10	2.7048	3.3186	4.2581	5.0848	7.4922	8.7876
29	11	2.5759	3.1253	3.9522	4.6684	6.7066	7.7802
29	12	2.4718	2.9710	3.7119	4.3449	6.1119	7.0262
29	13	2.3859	2.8451	3.5182	4.0866	5.6475	6.4427
29	14	2.3139	2.7403	3.3587	3.8758	5.2754	5.9791
29	15	2.2525	2.6517	3.2253	3.7006	4.9712	5.6026
29	16	2.1997	2.5758	3.1119	3.5527	4.7180	5.2911
29	17	2.1536	2.5101	3.0145	3.4262	4.5042	5.0294
29	18	2.1131	2.4527	2.9298	3.3168	4.3213	4.8067
29	19	2.0772	2.4019	2.8555	3.2213	4.1632	4.6149
29	20	2.0452	2.3569	2.7898	3.1372	4.0251	4.4482
29	21	2.0164	2.3165	2.7313	3.0625	3.9036	4.3019
29	22	1.9904	2.2802	2.6789	2.9959	3.7959	4.1725
29	23	1.9668	2.2473	2.6316	2.9359	3.6997	4.0574
29	24	1.9453	2.2174	2.5888	2.8818	3.6133	3.9542
29	25	1.9255	2.1901	2.5498	2.8326	3.5353	3.8613
29	26	1.9074	2.1651	2.5141	2.7877	3.4645	3.7772
29	27	1.8907	2.1420	2.4814	2.7466	3.4000	3.7007
29	28	1.8752	2.1207	2.4513	2.7088	3.3409	3.6308
29	29	1.8608	2.1010	2.4234	2.6740	3.2867	3.5667
29	30	1.8474	2.0827	2.3976	2.6417	3.2367	3.5077
29	40	1.7513	1.9519	2.2153	2.4156	2.8916	3.1033
29	50	1.6942	1.8752	2.1097	2.2860	2.6982	2.8787
29	60	1.6564	1.8246	2.0408	2.2019	2.5745	2.7360
29	120	1.5621	1.6999	1.8728	1.9989	2.2819	2.4011
29	250	1.5130	1.6357	1.7876	1.8969	2.1383	2.2384
29	500	1.4903	1.6062	1.7486	1.8506	2.0738	2.1656
29	1000	1.4789	1.5914	1.7293	1.8276	2.0420	2.1298
29	∞	1.4675	1.5766	1.7099	1.8047	2.0104	2.0943

The distribution of the variance ratio, $F = S_1^2/S_2^2$, when $\sigma_1^2 = \sigma_2^2$

ν_1	ν_2	Cumulative probabilities					
		0.9500	0.9750	0.9900	0.9950	0.9990	0.9995
30	1	250.0951	1001.414	6260.649	25043.63	626099.0	2504397
30	2	19.4624	39.4646	99.4658	199.4663	999.4666	1999.467
30	3	8.6166	14.0805	26.5045	42.4658	125.4486	199.5260
30	4	5.7459	8.4613	13.8377	19.8915	45.4286	64.5614
30	5	4.4957	6.2269	9.3793	12.6556	24.8688	33.0857
30	6	3.8082	5.0652	7.2285	9.3582	16.6722	21.2468
30	7	3.3758	4.3624	5.9920	7.5345	12.5304	15.4941
30	8	3.0794	3.8940	5.1981	6.3961	10.1087	12.2244
30	9	2.8637	3.5604	4.6486	5.6248	8.5476	10.1611
30	10	2.6996	3.3110	4.2469	5.0706	7.4688	8.7592
30	11	2.5705	3.1176	3.9411	4.6543	6.6839	7.7530
30	12	2.4663	2.9633	3.7008	4.3309	6.0898	6.9999
30	13	2.3803	2.8372	3.5070	4.0727	5.6258	6.4171
30	14	2.3082	2.7324	3.3476	3.8619	5.2542	5.9541
30	15	2.2468	2.6437	3.2141	3.6867	4.9502	5.5780
30	16	2.1938	2.5678	3.1007	3.5389	4.6972	5.2669
30	17	2.1477	2.5020	3.0032	3.4124	4.4836	5.0056
30	18	2.1071	2.4445	2.9185	3.3030	4.3009	4.7831
30	19	2.0712	2.3937	2.8442	3.2075	4.1429	4.5916
30	20	2.0391	2.3486	2.7785	3.1234	4.0050	4.4250
30	21	2.0102	2.3082	2.7200	3.0488	3.8836	4.2788
30	22	1.9842	2.2718	2.6675	2.9821	3.7759	4.1496
30	23	1.9605	2.2389	2.6202	2.9221	3.6798	4.0346
30	24	1.9390	2.2090	2.5773	2.8679	3.5935	3.9316
30	25	1.9192	2.1816	2.5383	2.8187	3.5155	3.8388
30	26	1.9010	2.1565	2.5026	2.7738	3.4448	3.7547
30	27	1.8842	2.1334	2.4699	2.7327	3.3803	3.6783
30	28	1.8687	2.1121	2.4397	2.6949	3.3213	3.6084
30	29	1.8543	2.0923	2.4118	2.6600	3.2671	3.5444
30	30	1.8409	2.0739	2.3860	2.6278	3.2171	3.4855
30	40	1.7444	1.9429	2.2034	2.4015	2.8721	3.0814
30	50	1.6872	1.8659	2.0976	2.2717	2.6787	2.8569
30	60	1.6491	1.8152	2.0285	2.1874	2.5549	2.7142
30	120	1.5543	1.6899	1.8600	1.9840	2.2621	2.3793
30	250	1.5049	1.6254	1.7744	1.8817	2.1183	2.2164
30	500	1.4821	1.5957	1.7353	1.8352	2.0537	2.1435
30	1000	1.4706	1.5808	1.7158	1.8121	2.0218	2.1076
30	∞	1.4591	1.5660	1.6964	1.7891	1.9901	2.0721

The distribution of the variance ratio, $F = S_1^2/S_2^2$, when $\sigma_1^2 = \sigma_2^2$

ν_1	ν_2	Cumulative probabilities					
		0.9500	0.9750	0.9900	0.9950	0.9990	0.9995
40	1	251.1432	1005.598	6286.782	25148.15	628712.0	2514849
40	2	19.4707	39.4729	99.4742	199.4746	999.4749	1999.475
40	3	8.5944	14.0365	26.4108	42.3082	124.9590	198.7399
40	4	5.7170	8.4111	13.7454	19.7518	45.0886	64.0714
40	5	4.4638	6.1750	9.2912	12.5297	24.6020	32.7242
40	6	3.7743	5.0125	7.1432	9.2408	16.4445	20.9501
40	7	3.3404	4.3089	5.9084	7.4224	12.3260	15.2349
40	8	3.0428	3.8398	5.1156	6.2875	9.9194	11.9889
40	9	2.8259	3.5055	4.5666	5.5186	8.3685	9.9419
40	10	2.6609	3.2554	4.1653	4.9659	7.2971	8.5515
40	11	2.5309	3.0613	3.8596	4.5508	6.5178	7.5538
40	12	2.4259	2.9063	3.6192	4.2282	5.9278	6.8071
40	13	2.3392	2.7797	3.4253	3.9704	5.4670	6.2294
40	14	2.2664	2.6742	3.2656	3.7600	5.0979	5.7704
40	15	2.2043	2.5850	3.1319	3.5850	4.7959	5.3976
40	16	2.1507	2.5085	3.0182	3.4372	4.5446	5.0891
40	17	2.1040	2.4422	2.9205	3.3108	4.3323	4.8300
40	18	2.0629	2.3842	2.8354	3.2014	4.1507	4.6094
40	19	2.0264	2.3329	2.7608	3.1058	3.9936	4.4194
40	20	1.9938	2.2873	2.6947	3.0215	3.8564	4.2541
40	21	1.9645	2.2465	2.6359	2.9467	3.7357	4.1091
40	22	1.9380	2.2097	2.5831	2.8799	3.6285	3.9808
40	23	1.9139	2.1763	2.5355	2.8197	3.5328	3.8666
40	24	1.8920	2.1460	2.4923	2.7654	3.4468	3.7643
40	25	1.8718	2.1183	2.4530	2.7160	3.3692	3.6721
40	26	1.8533	2.0928	2.4170	2.6709	3.2987	3.5886
40	27	1.8361	2.0693	2.3840	2.6296	3.2344	3.5126
40	28	1.8203	2.0477	2.3535	2.5916	3.1755	3.4432
40	29	1.8055	2.0276	2.3253	2.5565	3.1215	3.3795
40	30	1.7918	2.0089	2.2992	2.5241	3.0716	3.3209
40	40	1.6928	1.8752	2.1142	2.2958	2.7268	2.9183
40	50	1.6337	1.7963	2.0066	2.1644	2.5329	2.6941
40	60	1.5943	1.7440	1.9360	2.0789	2.4086	2.5512
40	120	1.4952	1.6141	1.7628	1.8709	2.1128	2.2145
40	250	1.4430	1.5465	1.6740	1.7655	1.9665	2.0495
40	500	1.4186	1.5151	1.6332	1.7172	1.9004	1.9754
40	1000	1.4063	1.4993	1.6127	1.6932	1.8676	1.9387
40	∞	1.3940	1.4835	1.5923	1.6691	1.8350	1.9024

The distribution of the variance ratio, $F = S_1^2/S_2^2$, when $\sigma_1^2 = \sigma_2^2$

ν_1	ν_2	Cumulative probabilities					
		0.9500	0.9750	0.9900	0.9950	0.9990	0.9995
50	1	251.7742	1008.117	6302.517	25211.09	630285.4	2521143
50	2	19.4757	39.4779	99.4792	199.4796	999.4799	1999.480
50	3	8.5810	14.0099	26.3542	42.2131	124.6635	198.2656
50	4	5.6995	8.3808	13.6896	19.6673	44.8832	63.7754
50	5	4.4444	6.1436	9.2378	12.4535	24.4407	32.5056
50	6	3.7537	4.9804	7.0915	9.1697	16.3067	20.7705
50	7	3.3189	4.2763	5.8577	7.3544	12.2020	15.0778
50	8	3.0204	3.8067	5.0654	6.2215	9.8044	11.8461
50	9	2.8028	3.4719	4.5167	5.4539	8.2597	9.8086
50	10	2.6371	3.2214	4.1155	4.9022	7.1927	8.4251
50	11	2.5066	3.0268	3.8097	4.4876	6.4165	7.4325
50	12	2.4010	2.8714	3.5692	4.1653	5.8290	6.6896
50	13	2.3138	2.7443	3.3752	3.9078	5.3700	6.1149
50	14	2.2405	2.6384	3.2153	3.6975	5.0023	5.6582
50	15	2.1780	2.5488	3.0814	3.5225	4.7015	5.2872
50	16	2.1240	2.4719	2.9675	3.3747	4.4511	4.9803
50	17	2.0769	2.4053	2.8694	3.2482	4.2395	4.7224
50	18	2.0354	2.3468	2.7841	3.1387	4.0584	4.5028
50	19	1.9986	2.2952	2.7093	3.0430	3.9018	4.3136
50	20	1.9656	2.2493	2.6430	2.9586	3.7650	4.1490
50	21	1.9360	2.2081	2.5838	2.8837	3.6445	4.0046
50	22	1.9092	2.1710	2.5308	2.8167	3.5376	3.8768
50	23	1.8848	2.1374	2.4829	2.7564	3.4421	3.7630
50	24	1.8625	2.1067	2.4395	2.7018	3.3562	3.6611
50	25	1.8421	2.0787	2.3999	2.6522	3.2787	3.5691
50	26	1.8233	2.0530	2.3637	2.6070	3.2083	3.4859
50	27	1.8059	2.0293	2.3304	2.5655	3.1440	3.4101
50	28	1.7898	2.0073	2.2997	2.5273	3.0852	3.3408
50	29	1.7748	1.9870	2.2714	2.4921	3.0311	3.2773
50	30	1.7609	1.9681	2.2450	2.4594	2.9813	3.2188
50	40	1.6600	1.8324	2.0581	2.2295	2.6360	2.8165
50	50	1.5995	1.7520	1.9490	2.0967	2.4413	2.5920
50	60	1.5590	1.6985	1.8772	2.0100	2.3162	2.4485
50	120	1.4565	1.5649	1.7000	1.7981	2.0172	2.1091
50	250	1.4019	1.4945	1.6083	1.6897	1.8681	1.9417
50	500	1.3762	1.4616	1.5658	1.6398	1.8004	1.8660
50	1000	1.3632	1.4451	1.5445	1.6148	1.7668	1.8285
50	∞	1.3501	1.4284	1.5231	1.5898	1.7332	1.7912

The distribution of the variance ratio, $F = S_1^2/S_2^2$, when $\sigma_1^2 = \sigma_2^2$

ν_1	ν_2	Cumulative probabilities					
		0.9500	0.9750	0.9900	0.9950	0.9990	0.9995
60	1	252.1957	1009.800	6313.030	25253.14	631336.6	2525347
60	2	19.4791	39.4812	99.4825	199.4829	999.4833	1999.483
60	3	8.5720	13.9921	26.3164	42.1494	124.4658	197.9483
60	4	5.6877	8.3604	13.6522	19.6107	44.7457	63.5772
60	5	4.4314	6.1225	9.2020	12.4024	24.3326	32.3592
60	6	3.7398	4.9589	7.0567	9.1219	16.2143	20.6501
60	7	3.3043	4.2544	5.8236	7.3088	12.1189	14.9724
60	8	3.0053	3.7844	5.0316	6.1772	9.7272	11.7501
60	9	2.7872	3.4493	4.4831	5.4104	8.1865	9.7191
60	10	2.6211	3.1984	4.0819	4.8592	7.1224	8.3401
60	11	2.4901	3.0035	3.7761	4.4450	6.3483	7.3508
60	12	2.3842	2.8478	3.5355	4.1229	5.7623	6.6104
60	13	2.2966	2.7204	3.3413	3.8655	5.3046	6.0376
60	14	2.2229	2.6142	3.1813	3.6552	4.9378	5.5825
60	15	2.1601	2.5242	3.0471	3.4803	4.6377	5.2127
60	16	2.1058	2.4471	2.9330	3.3324	4.3878	4.9067
60	17	2.0584	2.3801	2.8348	3.2058	4.1767	4.6496
60	18	2.0166	2.3214	2.7493	3.0962	3.9960	4.4306
60	19	1.9795	2.2696	2.6742	3.0004	3.8396	4.2420
60	20	1.9464	2.2234	2.6077	2.9159	3.7030	4.0778
60	21	1.9165	2.1819	2.5484	2.8408	3.5827	3.9337
60	22	1.8894	2.1446	2.4951	2.7736	3.4759	3.8063
60	23	1.8648	2.1107	2.4471	2.7132	3.3804	3.6927
60	24	1.8424	2.0799	2.4035	2.6585	3.2946	3.5909
60	25	1.8217	2.0516	2.3637	2.6088	3.2171	3.4991
60	26	1.8027	2.0257	2.3273	2.5633	3.1467	3.4160
60	27	1.7851	2.0018	2.2938	2.5217	3.0825	3.3403
60	28	1.7689	1.9797	2.2629	2.4834	3.0236	3.2711
60	29	1.7537	1.9591	2.2344	2.4479	2.9695	3.2076
60	30	1.7396	1.9400	2.2079	2.4151	2.9196	3.1492
60	40	1.6373	1.8028	2.0194	2.1838	2.5737	2.7467
60	50	1.5757	1.7211	1.9090	2.0499	2.3782	2.5216
60	60	1.5343	1.6668	1.8363	1.9622	2.2523	2.3776
60	120	1.4290	1.5299	1.6557	1.7469	1.9502	2.0354
60	250	1.3723	1.4573	1.5614	1.6358	1.7985	1.8654
60	500	1.3455	1.4231	1.5174	1.5843	1.7292	1.7883
60	1000	1.3318	1.4058	1.4953	1.5585	1.6947	1.7499
60	∞	1.3180	1.3883	1.4730	1.5325	1.6601	1.7116

The distribution of the variance ratio, $F = S_1^2/S_2^2$, when $\sigma_1^2 = \sigma_2^2$

ν_1	ν_2	Cumulative probabilities					
		0.9500	0.9750	0.9900	0.9950	0.9990	0.9995
120	1	253.2529	1014.020	6339.391	25358.57	633972.4	2535891
120	2	19.4874	39.4896	99.4908	199.4912	999.4916	1999.492
120	3	8.5494	13.9473	26.2211	41.9895	123.9692	197.1511
120	4	5.6581	8.3092	13.5581	19.4684	44.3998	63.0788
120	5	4.3985	6.0693	9.1118	12.2737	24.0605	31.9905
120	6	3.7047	4.9044	6.9690	9.0015	15.9812	20.3467
120	7	3.2674	4.1989	5.7373	7.1933	11.9090	14.7064
120	8	2.9669	3.7279	4.9461	6.0649	9.5321	11.5078
120	9	2.7475	3.3918	4.3978	5.3001	8.0014	9.4926
120	10	2.5801	3.1399	3.9965	4.7501	6.9443	8.1248
120	11	2.4480	2.9441	3.6904	4.3367	6.1753	7.1437
120	12	2.3410	2.7874	3.4494	4.0149	5.5931	6.4094
120	13	2.2524	2.6590	3.2548	3.7577	5.1381	5.8413
120	14	2.1778	2.5519	3.0942	3.5473	4.7735	5.3897
120	15	2.1141	2.4611	2.9595	3.3722	4.4750	5.0228
120	16	2.0589	2.3831	2.8447	3.2240	4.2263	4.7190
120	17	2.0107	2.3153	2.7459	3.0971	4.0160	4.4636
120	18	1.9681	2.2558	2.6597	2.9871	3.8360	4.2461
120	19	1.9302	2.2032	2.5839	2.8908	3.6801	4.0585
120	20	1.8963	2.1562	2.5168	2.8058	3.5438	3.8952
120	21	1.8657	2.1141	2.4568	2.7302	3.4237	3.7518
120	22	1.8380	2.0760	2.4029	2.6625	3.3170	3.6249
120	23	1.8128	2.0415	2.3542	2.6015	3.2216	3.5117
120	24	1.7896	2.0099	2.3100	2.5463	3.1357	3.4102
120	25	1.7684	1.9811	2.2696	2.4961	3.0581	3.3186
120	26	1.7488	1.9545	2.2325	2.4501	2.9875	3.2355
120	27	1.7306	1.9299	2.1985	2.4079	2.9231	3.1599
120	28	1.7138	1.9072	2.1670	2.3690	2.8640	3.0907
120	29	1.6981	1.8861	2.1379	2.3331	2.8097	3.0272
120	30	1.6835	1.8664	2.1108	2.2998	2.7595	2.9686
120	40	1.5766	1.7242	1.9172	2.0636	2.4103	2.5640
120	50	1.5115	1.6386	1.8026	1.9254	2.2113	2.3361
120	60	1.4673	1.5810	1.7263	1.8341	2.0821	2.1891
120	120	1.3519	1.4327	1.5330	1.6055	1.7667	1.8341
120	250	1.2870	1.3506	1.4280	1.4831	1.6030	1.6520
120	500	1.2551	1.3105	1.3774	1.4245	1.5259	1.5669
120	1000	1.2385	1.2898	1.3513	1.3945	1.4867	1.5238
120	∞	1.2214	1.2684	1.3246	1.3637	1.4468	1.4800

The distribution of the variance ratio, $F = S_1^2/S_2^2$, when $\sigma_1^2 = \sigma_2^2$

ν_1	ν_2	Cumulative probabilities					
		0.9500	0.9750	0.9900	0.9950	0.9990	0.9995
250	1	253.8043	1016.222	6353.143	25413.58	635347.5	2541391
250	2	19.4917	39.4939	99.4952	199.4956	999.4959	1999.496
250	3	8.5375	13.9238	26.1713	41.9058	123.7096	196.7344
250	4	5.6425	8.2823	13.5088	19.3938	44.2188	62.8180
250	5	4.3811	6.0413	9.0644	12.2062	23.9178	31.7973
250	6	3.6861	4.8758	6.9229	8.9382	15.8589	20.1874
250	7	3.2479	4.1696	5.6918	7.1326	11.7986	14.5666
250	8	2.9466	3.6981	4.9009	6.0057	9.4294	11.3803
250	9	2.7264	3.3613	4.3527	5.2419	7.9038	9.3733
250	10	2.5583	3.1089	3.9513	4.6924	6.8502	8.0112
250	11	2.4256	2.9124	3.6450	4.2792	6.0838	7.0342
250	12	2.3179	2.7552	3.4037	3.9576	5.5034	6.3030
250	13	2.2287	2.6263	3.2087	3.7003	5.0498	5.7372
250	14	2.1536	2.5186	3.0477	3.4899	4.6861	5.2874
250	15	2.0893	2.4273	2.9126	3.3145	4.3883	4.9217
250	16	2.0336	2.3487	2.7974	3.1661	4.1402	4.6190
250	17	1.9849	2.2804	2.6981	3.0388	3.9303	4.3644
250	18	1.9418	2.2205	2.6115	2.9285	3.7504	4.1474
250	19	1.9035	2.1673	2.5353	2.8319	3.5947	3.9603
250	20	1.8691	2.1199	2.4677	2.7465	3.4584	3.7974
250	21	1.8381	2.0773	2.4073	2.6706	3.3383	3.6542
250	22	1.8099	2.0388	2.3530	2.6025	3.2315	3.5273
250	23	1.7843	2.0038	2.3038	2.5411	3.1359	3.4142
250	24	1.7608	1.9718	2.2591	2.4855	3.0499	3.3127
250	25	1.7391	1.9425	2.2183	2.4349	2.9721	3.2211
250	26	1.7191	1.9155	2.1809	2.3885	2.9013	3.1380
250	27	1.7006	1.8905	2.1464	2.3459	2.8367	3.0622
250	28	1.6834	1.8674	2.1145	2.3067	2.7774	2.9929
250	29	1.6674	1.8459	2.0850	2.2703	2.7228	2.9292
250	30	1.6524	1.8258	2.0575	2.2366	2.6723	2.8705
250	40	1.5425	1.6802	1.8602	1.9967	2.3200	2.4634
250	50	1.4748	1.5917	1.7425	1.8553	2.1179	2.2325
250	60	1.4285	1.5317	1.6635	1.7612	1.9859	2.0828
250	120	1.3047	1.3737	1.4592	1.5209	1.6579	1.7151
250	250	1.2318	1.2821	1.3432	1.3866	1.4807	1.5191
250	500	1.1940	1.2352	1.2846	1.3193	1.3936	1.4236
250	1000	1.1735	1.2098	1.2532	1.2835	1.3477	1.3734
250	∞	1.1515	1.1828	1.2198	1.2454	1.2993	1.3207

The distribution of the variance ratio, $F = S_1^2/S_2^2$, when $\sigma_1^2 = \sigma_2^2$

ν_1	ν_2	Cumulative probabilities					
		0.9500	0.9750	0.9900	0.9950	0.9990	0.9995
500	1	254.0593	1017.240	6359.501	25439.00	635983.1	2543934
500	2	19.4937	39.4959	99.4972	199.4976	999.4979	1999.498
500	3	8.5320	13.9130	26.1483	41.8671	123.5894	196.5415
500	4	5.6353	8.2698	13.4859	19.3593	44.1349	62.6972
500	5	4.3731	6.0283	9.0424	12.1749	23.8517	31.7078
500	6	3.6775	4.8625	6.9015	8.9088	15.8022	20.1136
500	7	3.2389	4.1560	5.6707	7.1044	11.7474	14.5018
500	8	2.9371	3.6842	4.8799	5.9782	9.3817	11.3210
500	9	2.7166	3.3471	4.3317	5.2148	7.8584	9.3178
500	10	2.5481	3.0944	3.9302	4.6656	6.8065	7.9584
500	11	2.4151	2.8977	3.6238	4.2525	6.0412	6.9832
500	12	2.3071	2.7401	3.3823	3.9309	5.4616	6.2534
500	13	2.2176	2.6109	3.1871	3.6735	5.0086	5.6886
500	14	2.1422	2.5030	3.0260	3.4630	4.6453	5.2396
500	15	2.0776	2.4114	2.8906	3.2875	4.3478	4.8746
500	16	2.0217	2.3326	2.7752	3.1389	4.0999	4.5723
500	17	1.9727	2.2640	2.6757	3.0115	3.8901	4.3180
500	18	1.9294	2.2038	2.5889	2.9010	3.7103	4.1013
500	19	1.8909	2.1504	2.5124	2.8042	3.5546	3.9143
500	20	1.8562	2.1027	2.4446	2.7186	3.4183	3.7515
500	21	1.8250	2.0599	2.3840	2.6425	3.2981	3.6083
500	22	1.7966	2.0211	2.3294	2.5742	3.1913	3.4815
500	23	1.7708	1.9859	2.2800	2.5126	3.0956	3.3684
500	24	1.7470	1.9537	2.2351	2.4568	3.0095	3.2668
500	25	1.7252	1.9242	2.1941	2.4059	2.9316	3.1751
500	26	1.7050	1.8970	2.1564	2.3594	2.8606	3.0920
500	27	1.6863	1.8718	2.1217	2.3166	2.7958	3.0161
500	28	1.6689	1.8485	2.0896	2.2771	2.7364	2.9467
500	29	1.6527	1.8268	2.0598	2.2405	2.6816	2.8828
500	30	1.6375	1.8065	2.0321	2.2066	2.6310	2.8240
500	40	1.5260	1.6590	1.8329	1.9647	2.2768	2.4153
500	50	1.4569	1.5689	1.7133	1.8214	2.0729	2.1826
500	60	1.4093	1.5075	1.6327	1.7256	1.9391	2.0312
500	120	1.2804	1.3434	1.4215	1.4778	1.6028	1.6549
500	250	1.2015	1.2448	1.2974	1.3347	1.4154	1.4483
500	500	1.1587	1.1918	1.2317	1.2596	1.3191	1.3431
500	1000	1.1342	1.1618	1.1947	1.2176	1.2661	1.2854
500	∞	1.1063	1.1277	1.1530	1.1704	1.2069	1.2213

The distribution of the variance ratio, $F = S_1^2/S_2^2$, when $\sigma_1^2 = \sigma_2^2$

ν_1	ν_2	Cumulative probabilities					
		0.9500	0.9750	0.9900	0.9950	0.9990	0.9995
1000	1	254.1868	1017.749	6362.682	25451.73	636301.2	2545206
1000	2	19.4947	39.4969	99.4982	199.4986	999.4989	1999.499
1000	3	8.5292	13.9075	26.1367	41.8477	123.5293	196.4450
1000	4	5.6317	8.2636	13.4745	19.3420	44.0929	62.6368
1000	5	4.3690	6.0218	9.0314	12.1592	23.8186	31.6629
1000	6	3.6732	4.8558	6.8908	8.8941	15.7737	20.0766
1000	7	3.2343	4.1492	5.6601	7.0902	11.7217	14.4693
1000	8	2.9324	3.6772	4.8694	5.9644	9.3577	11.2913
1000	9	2.7116	3.3400	4.3211	5.2012	7.8357	9.2900
1000	10	2.5430	3.0871	3.9196	4.6521	6.7845	7.9318
1000	11	2.4098	2.8902	3.6131	4.2390	6.0198	6.9576
1000	12	2.3017	2.7325	3.3716	3.9174	5.4406	6.2285
1000	13	2.2121	2.6032	3.1763	3.6601	4.9879	5.6642
1000	14	2.1365	2.4951	3.0150	3.4494	4.6248	5.2156
1000	15	2.0718	2.4034	2.8795	3.2739	4.3275	4.8508
1000	16	2.0157	2.3245	2.7641	3.1253	4.0796	4.5488
1000	17	1.9666	2.2558	2.6644	2.9978	3.8699	4.2946
1000	18	1.9232	2.1954	2.5775	2.8871	3.6901	4.0780
1000	19	1.8845	2.1419	2.5009	2.7902	3.5344	3.8911
1000	20	1.8497	2.0941	2.4329	2.7046	3.3981	3.7283
1000	21	1.8184	2.0511	2.3722	2.6283	3.2779	3.5852
1000	22	1.7899	2.0122	2.3175	2.5599	3.1710	3.4584
1000	23	1.7639	1.9769	2.2680	2.4982	3.0753	3.3452
1000	24	1.7401	1.9445	2.2230	2.4423	2.9891	3.2437
1000	25	1.7181	1.9149	2.1818	2.3913	2.9111	3.1519
1000	26	1.6978	1.8876	2.1440	2.3446	2.8401	3.0687
1000	27	1.6790	1.8623	2.1092	2.3017	2.7752	2.9928
1000	28	1.6615	1.8389	2.0769	2.2621	2.7156	2.9233
1000	29	1.6452	1.8170	2.0471	2.2254	2.6607	2.8594
1000	30	1.6299	1.7967	2.0192	2.1914	2.6100	2.8004
1000	40	1.5175	1.6481	1.8189	1.9483	2.2549	2.3908
1000	50	1.4477	1.5572	1.6984	1.8040	2.0499	2.1571
1000	60	1.3994	1.4950	1.6169	1.7073	1.9150	2.0046
1000	120	1.2675	1.3273	1.4015	1.4550	1.5737	1.6232
1000	250	1.1847	1.2241	1.2720	1.3059	1.3793	1.4093
1000	500	1.1378	1.1664	1.2007	1.2247	1.2759	1.2965
1000	1000	1.1097	1.1320	1.1586	1.1771	1.2161	1.2316
1000	∞	1.0747	1.0895	1.1070	1.1189	1.1439	1.1537

The distribution of the variance ratio, $F = S_1^2/S_2^2$, when $\sigma_1^2 = \sigma_2^2$

ν_1	ν_2	Cumulative probabilities					
		0.9500	0.9750	0.9900	0.9950	0.9990	0.9995
∞	1	254.3144	1018.258	6365.864	25464.46	636619.4	2546479
∞	2	19.4957	39.4979	99.4992	199.4996	999.4999	1999.500
∞	3	8.5264	13.9021	26.1252	41.8283	123.4691	196.3483
∞	4	5.6281	8.2573	13.4631	19.3247	44.0509	62.5762
∞	5	4.3650	6.0153	9.0204	12.1435	23.7854	31.6180
∞	6	3.6689	4.8491	6.8800	8.8793	15.7453	20.0396
∞	7	3.2298	4.1423	5.6495	7.0760	11.6960	14.4367
∞	8	2.9276	3.6702	4.8588	5.9506	9.3337	11.2616
∞	9	2.7067	3.3329	4.3105	5.1875	7.8128	9.2621
∞	10	2.5379	3.0798	3.9090	4.6385	6.7625	7.9053
∞	11	2.4045	2.8828	3.6024	4.2255	5.9983	6.9320
∞	12	2.2962	2.7249	3.3608	3.9039	5.4195	6.2035
∞	13	2.2064	2.5955	3.1654	3.6465	4.9671	5.6398
∞	14	2.1307	2.4872	3.0040	3.4359	4.6042	5.1915
∞	15	2.0658	2.3953	2.8684	3.2602	4.3070	4.8270
∞	16	2.0096	2.3163	2.7528	3.1115	4.0592	4.5251
∞	17	1.9604	2.2474	2.6530	2.9839	3.8496	4.2712
∞	18	1.9168	2.1869	2.5660	2.8732	3.6698	4.0546
∞	19	1.8780	2.1333	2.4893	2.7762	3.5141	3.8678
∞	20	1.8432	2.0853	2.4212	2.6904	3.3778	3.7050
∞	21	1.8117	2.0422	2.3603	2.6140	3.2575	3.5619
∞	22	1.7831	2.0032	2.3055	2.5455	3.1505	3.4351
∞	23	1.7570	1.9677	2.2558	2.4837	3.0548	3.3219
∞	24	1.7330	1.9353	2.2107	2.4276	2.9685	3.2203
∞	25	1.7110	1.9055	2.1694	2.3765	2.8904	3.1285
∞	26	1.6906	1.8781	2.1315	2.3297	2.8193	3.0452
∞	27	1.6717	1.8527	2.0965	2.2867	2.7543	2.9693
∞	28	1.6541	1.8291	2.0642	2.2470	2.6947	2.8997
∞	29	1.6376	1.8072	2.0342	2.2102	2.6397	2.8357
∞	30	1.6223	1.7867	2.0062	2.1760	2.5889	2.7767
∞	40	1.5089	1.6371	1.8047	1.9318	2.2326	2.3660
∞	50	1.4383	1.5452	1.6831	1.7863	2.0264	2.1312
∞	60	1.3893	1.4821	1.6006	1.6885	1.8905	1.9776
∞	120	1.2539	1.3104	1.3805	1.4311	1.5433	1.5901
∞	250	1.1661	1.2014	1.2442	1.2745	1.3401	1.3669
∞	500	1.1132	1.1365	1.1644	1.1840	1.2256	1.2424
∞	1000	1.0781	1.0938	1.1125	1.1254	1.1528	1.1637
∞	∞	1.0000	1.0000	1.0000	1.0000	1.0000	1.0000

The distribution of the variance ratio, $F = S_1^2/S_2^2$, when $\sigma_1^2 = \sigma_2^2$

NOTES:

In the case of the distribution of the variance ratio, F, like in the cases of the distributions of Student's t and of χ^2 (chi squared), interpolation is slightly more accurate if it is based on the inverse of the numbers of degrees of freedom rather than on the numbers of degrees of freedom themselves.

Even though the variance ratio distribution is not symmetrical, it possesses an analogous property:

$$F_{(1-P;\, \nu_2,\, \nu_1)} = 1/F_{(P;\, \nu_1,\, \nu_2)} \,.$$

The only F values listed in the present table are those corresponding to cumulative probabilities greater than 0.95, and the above property is used to evaluate the F values corresponding to complementary probabilities (0.05, 0.025, 0.01, 0.005, 0.001 and 0.0005).

The distribution of the variance ratio, F, is related to the distribution of Student's t as follows:

$$F_{(P;\, 1,\, \nu)} = t^2_{[(1-P)/2;\, \nu]} = t^2_{[(1+P)/2;\, \nu]} \,.$$

While the present table concerns primarily the case where the parametric variances σ_1^2 and σ_2^2 are equal, it may also be used when the parametric variances σ_1^2 and σ_2^2 are unequal provided their ratio $\Phi = \sigma_1^2/\sigma_2^2$ is either known or specified in a hypothesis. The ratio of observed variances should then be evaluated through the following expression:

$$F_{(\nu_1,\, \nu_2)} = (S_1^2/S_2^2) \big/ (\sigma_1^2/\sigma_2^2) \,.$$

The denominator (σ_1^2/σ_2^2) compensates, so to speak, for the inequality of parametric variances. In the particular case where parametric variances are equal, the above general expression reduces to its simpler usual form:

$$F_{(\nu_1,\, \nu_2)} = (S_1^2/S_2^2) \,.$$

The distribution of the correlation coefficient, r, when $\rho = 0$

ν	Cumulative probabilities					
	0.9500	0.9750	0.9900	0.9950	0.9990	0.9995
1	0.98769	0.99692	0.99951	0.99988	1.00000	1.00000
2	0.90000	0.95000	0.98000	0.99000	0.99800	0.99900
3	0.80538	0.87834	0.93433	0.95874	0.98593	0.99114
4	0.72930	0.81140	0.88219	0.91720	0.96326	0.97407
5	0.66944	0.75449	0.83287	0.87453	0.93496	0.95088
6	0.62149	0.70673	0.78872	0.83434	0.90490	0.92490
7	0.58221	0.66638	0.74978	0.79768	0.87514	0.89826
8	0.54936	0.63190	0.71546	0.76459	0.84669	0.87212
9	0.52140	0.60207	0.68510	0.73479	0.81993	0.84705
10	0.49726	0.57598	0.65807	0.70789	0.79495	0.82330
11	0.47616	0.55294	0.63386	0.68353	0.77173	0.80096
12	0.45750	0.53241	0.61205	0.66138	0.75014	0.78000
13	0.44086	0.51398	0.59227	0.64114	0.73007	0.76035
14	0.42590	0.49731	0.57425	0.62259	0.71139	0.74193
15	0.41236	0.48215	0.55774	0.60551	0.69396	0.72466
16	0.40003	0.46828	0.54255	0.58971	0.67767	0.70843
17	0.38873	0.45553	0.52852	0.57507	0.66241	0.69316
18	0.37834	0.44376	0.51550	0.56144	0.64809	0.67878
19	0.36874	0.43286	0.50340	0.54871	0.63462	0.66521
20	0.35983	0.42271	0.49209	0.53680	0.62193	0.65238
21	0.35153	0.41325	0.48151	0.52562	0.60994	0.64023
22	0.34378	0.40439	0.47158	0.51510	0.59860	0.62871
23	0.33652	0.39607	0.46223	0.50518	0.58785	0.61777
24	0.32970	0.38824	0.45341	0.49581	0.57765	0.60736
25	0.32328	0.38086	0.44508	0.48693	0.56795	0.59745
26	0.31722	0.37389	0.43718	0.47851	0.55871	0.58799
27	0.31149	0.36728	0.42969	0.47051	0.54990	0.57896
28	0.30606	0.36101	0.42257	0.46289	0.54149	0.57032
29	0.30090	0.35505	0.41579	0.45563	0.53344	0.56205
30	0.29599	0.34937	0.40933	0.44870	0.52574	0.55412

The distribution of the correlation coefficient, r, when $\rho = 0$

ν	Cumulative probabilities					
	0.9500	0.9750	0.9900	0.9950	0.9990	0.9995
31	0.29132	0.34396	0.40315	0.44207	0.51836	0.54651
32	0.28686	0.33879	0.39725	0.43573	0.51127	0.53920
33	0.28259	0.33384	0.39160	0.42965	0.50447	0.53218
34	0.27852	0.32911	0.38618	0.42381	0.49793	0.52541
35	0.27461	0.32457	0.38098	0.41821	0.49163	0.51890
36	0.27086	0.32022	0.37598	0.41282	0.48556	0.51262
37	0.26727	0.31603	0.37117	0.40764	0.47971	0.50655
38	0.26381	0.31201	0.36655	0.40264	0.47407	0.50070
39	0.26048	0.30813	0.36209	0.39782	0.46862	0.49504
40	0.25728	0.30440	0.35779	0.39317	0.46335	0.48957
41	0.25419	0.30079	0.35364	0.38868	0.45825	0.48427
42	0.25121	0.29732	0.34963	0.38434	0.45332	0.47914
43	0.24833	0.29396	0.34575	0.38014	0.44854	0.47417
44	0.24555	0.29071	0.34200	0.37608	0.44391	0.46935
45	0.24286	0.28756	0.33837	0.37214	0.43942	0.46467
46	0.24026	0.28452	0.33485	0.36833	0.43506	0.46013
47	0.23773	0.28157	0.33144	0.36462	0.43083	0.45572
48	0.23529	0.27871	0.32813	0.36103	0.42671	0.45143
49	0.23292	0.27594	0.32492	0.35754	0.42272	0.44726
50	0.23062	0.27324	0.32180	0.35415	0.41883	0.44320
51	0.22839	0.27063	0.31876	0.35086	0.41505	0.43925
52	0.22622	0.26809	0.31582	0.34765	0.41136	0.43540
53	0.22411	0.26561	0.31295	0.34453	0.40778	0.43166
54	0.22206	0.26321	0.31016	0.34150	0.40428	0.42800
55	0.22006	0.26087	0.30744	0.33854	0.40088	0.42444
56	0.21812	0.25859	0.30479	0.33566	0.39755	0.42096
57	0.21623	0.25637	0.30221	0.33284	0.39431	0.41757
58	0.21438	0.25420	0.29970	0.33010	0.39115	0.41426
59	0.21258	0.25209	0.29724	0.32743	0.38806	0.41103
60	0.21083	0.25003	0.29485	0.32482	0.38504	0.40787

The distribution of the correlation coefficient, r, when $\rho = 0$

ν	Cumulative probabilities					
	0.9500	0.9750	0.9900	0.9950	0.9990	0.9995
61	0.20912	0.24803	0.29251	0.32227	0.38210	0.40478
62	0.20745	0.24606	0.29022	0.31978	0.37921	0.40176
63	0.20582	0.24415	0.28799	0.31735	0.37640	0.39880
64	0.20423	0.24228	0.28581	0.31497	0.37364	0.39591
65	0.20267	0.24045	0.28368	0.31264	0.37094	0.39309
66	0.20115	0.23866	0.28160	0.31036	0.36831	0.39032
67	0.19967	0.23691	0.27956	0.30814	0.36572	0.38761
68	0.19821	0.23520	0.27756	0.30596	0.36319	0.38495
69	0.19679	0.23352	0.27561	0.30382	0.36071	0.38235
70	0.19539	0.23188	0.27370	0.30173	0.35829	0.37980
71	0.19403	0.23028	0.27182	0.29969	0.35591	0.37730
72	0.19269	0.22871	0.26999	0.29768	0.35357	0.37485
73	0.19139	0.22716	0.26819	0.29571	0.35128	0.37244
74	0.19010	0.22565	0.26642	0.29379	0.34904	0.37009
75	0.18885	0.22417	0.26469	0.29189	0.34684	0.36777
76	0.18761	0.22272	0.26300	0.29004	0.34468	0.36550
77	0.18641	0.22130	0.26133	0.28822	0.34255	0.36327
78	0.18522	0.21990	0.25970	0.28643	0.34047	0.36108
79	0.18406	0.21853	0.25810	0.28468	0.33843	0.35893
80	0.18292	0.21718	0.25653	0.28296	0.33642	0.35682
81	0.18180	0.21586	0.25498	0.28127	0.33444	0.35474
82	0.18070	0.21457	0.25346	0.27961	0.33251	0.35270
83	0.17961	0.21329	0.25197	0.27797	0.33060	0.35069
84	0.17855	0.21204	0.25051	0.27637	0.32873	0.34872
85	0.17751	0.21081	0.24907	0.27479	0.32688	0.34678
86	0.17649	0.20960	0.24765	0.27324	0.32507	0.34488
87	0.17548	0.20841	0.24626	0.27172	0.32329	0.34300
88	0.17449	0.20725	0.24490	0.27022	0.32154	0.34115
89	0.17352	0.20610	0.24355	0.26875	0.31981	0.33934
90	0.17256	0.20497	0.24223	0.26730	0.31812	0.33755

The distribution of the correlation coefficient, r, when $\rho = 0$

ν	Cumulative probabilities					
	0.9500	0.9750	0.9900	0.9950	0.9990	0.9995
91	0.17162	0.20386	0.24092	0.26587	0.31644	0.33579
92	0.17069	0.20276	0.23964	0.26447	0.31480	0.33406
93	0.16978	0.20169	0.23838	0.26308	0.31318	0.33235
94	0.16888	0.20063	0.23714	0.26172	0.31159	0.33067
95	0.16800	0.19958	0.23592	0.26038	0.31002	0.32901
96	0.16713	0.19856	0.23472	0.25906	0.30847	0.32738
97	0.16627	0.19755	0.23353	0.25776	0.30694	0.32578
98	0.16543	0.19655	0.23236	0.25648	0.30544	0.32419
99	0.16460	0.19557	0.23121	0.25522	0.30396	0.32263
100	0.16378	0.19460	0.23008	0.25398	0.30250	0.32110
102	0.16218	0.19271	0.22786	0.25155	0.29965	0.31808
104	0.16063	0.19088	0.22571	0.24918	0.29687	0.31516
106	0.15912	0.18909	0.22361	0.24688	0.29417	0.31231
108	0.15765	0.18736	0.22158	0.24465	0.29154	0.30953
110	0.15622	0.18567	0.21959	0.24247	0.28898	0.30683
112	0.15483	0.18402	0.21766	0.24035	0.28649	0.30420
114	0.15348	0.18242	0.21578	0.23828	0.28406	0.30164
116	0.15216	0.18086	0.21395	0.23627	0.28169	0.29913
118	0.15087	0.17934	0.21216	0.23431	0.27938	0.29669
120	0.14962	0.17786	0.21042	0.23240	0.27713	0.29431
122	0.14840	0.17641	0.20872	0.23053	0.27493	0.29199
124	0.14720	0.17500	0.20706	0.22870	0.27278	0.28972
126	0.14604	0.17362	0.20544	0.22692	0.27068	0.28750
128	0.14490	0.17228	0.20386	0.22518	0.26863	0.28533
130	0.14379	0.17096	0.20231	0.22348	0.26662	0.28321
132	0.14270	0.16967	0.20080	0.22182	0.26466	0.28114
134	0.14164	0.16842	0.19932	0.22019	0.26274	0.27911
136	0.14060	0.16719	0.19787	0.21860	0.26086	0.27712
138	0.13959	0.16598	0.19645	0.21704	0.25902	0.27518
140	0.13859	0.16481	0.19507	0.21552	0.25722	0.27327

The distribution of the correlation coefficient, r, when $\rho = 0$

ν	Cumulative probabilities					
	0.9500	0.9750	0.9900	0.9950	0.9990	0.9995
142	0.13762	0.16365	0.19371	0.21402	0.25546	0.27141
144	0.13666	0.16252	0.19238	0.21256	0.25373	0.26958
146	0.13573	0.16142	0.19108	0.21113	0.25204	0.26779
148	0.13482	0.16033	0.18980	0.20973	0.25038	0.26604
150	0.13392	0.15927	0.18855	0.20835	0.24875	0.26432
152	0.13304	0.15823	0.18733	0.20700	0.24716	0.26263
154	0.13218	0.15721	0.18612	0.20568	0.24559	0.26097
156	0.13133	0.15621	0.18494	0.20438	0.24406	0.25935
158	0.13050	0.15523	0.18379	0.20310	0.24255	0.25775
160	0.12969	0.15426	0.18265	0.20185	0.24107	0.25618
162	0.12889	0.15331	0.18153	0.20062	0.23961	0.25465
164	0.12811	0.15238	0.18044	0.19942	0.23819	0.25313
166	0.12734	0.15147	0.17936	0.19823	0.23678	0.25165
168	0.12658	0.15058	0.17831	0.19707	0.23541	0.25019
170	0.12584	0.14969	0.17727	0.19593	0.23405	0.24876
172	0.12511	0.14883	0.17625	0.19480	0.23272	0.24735
174	0.12439	0.14798	0.17524	0.19370	0.23141	0.24596
176	0.12368	0.14714	0.17426	0.19261	0.23013	0.24460
178	0.12299	0.14632	0.17329	0.19154	0.22886	0.24326
180	0.12231	0.14551	0.17234	0.19049	0.22762	0.24194
182	0.12164	0.14472	0.17140	0.18946	0.22639	0.24065
184	0.12098	0.14393	0.17047	0.18844	0.22519	0.23937
186	0.12033	0.14316	0.16957	0.18744	0.22400	0.23811
188	0.11969	0.14240	0.16867	0.18646	0.22283	0.23688
190	0.11906	0.14166	0.16779	0.18549	0.22168	0.23566
192	0.11844	0.14092	0.16693	0.18454	0.22055	0.23446
194	0.11783	0.14020	0.16607	0.18360	0.21944	0.23328
196	0.11723	0.13949	0.16523	0.18267	0.21834	0.23212
198	0.11664	0.13879	0.16441	0.18176	0.21726	0.23097
200	0.11606	0.13810	0.16359	0.18086	0.21619	0.22984

The distribution of the correlation coefficient, r, when $\rho = 0$

ν	Cumulative probabilities					
	0.9500	0.9750	0.9900	0.9950	0.9990	0.9995
250	0.10385	0.12361	0.14648	0.16199	0.19378	0.20608
300	0.09483	0.11289	0.13382	0.14802	0.17715	0.18843
400	0.08215	0.09782	0.11600	0.12834	0.15369	0.16352
500	0.07350	0.08753	0.10381	0.11487	0.13761	0.14644
600	0.06710	0.07992	0.09480	0.10491	0.12571	0.13379
700	0.06213	0.07400	0.08779	0.09716	0.11644	0.12393
800	0.05812	0.06923	0.08214	0.09091	0.10896	0.11598
900	0.05480	0.06528	0.07745	0.08573	0.10276	0.10939
1000	0.05199	0.06194	0.07348	0.08134	0.09751	0.10380
∞	0.00000	0.00000	0.00000	0.00000	0.00000	0.00000

NOTES:

Since the distribution of the observed correlation coefficient r is symmetrical when the parametric correlation coefficient $\rho = 0$, the values of r having a cumulative probability P smaller than 0.5 may be evaluated by using the following relationship

$$r_{(P;\,\nu)} = -r_{(1-P;\,\nu)} \,.$$

The number of degrees of freedom of an ordinary correlation coefficient (also called a *total correlation coefficient*) is $\nu = (N-2)$, where N is sample size. In the case of a *partial correlation coefficient,* the number of degrees of freedom is $\nu = (N-2-k)$, where k denotes the number of predictor variates which are held constant.

A total or partial correlation coefficient may be transformed into a Student's t by using the equation

$$t_{(\nu)} = r\sqrt{[\nu/(1-r^2)]} \,.$$

Conversely, the value of r having a cumulative probability P may be obtained from the equation

$$r_{(P;\,\nu)} = t_{(P;\,\nu)}/\sqrt{[t^2_{(P;\,\nu)} + \nu]} \,.$$

The present table must be used neither when $\rho \neq 0$ (see section 19.6) nor in the case of a multiple correlation coefficient (see section 23.5).

Detailed table of contents

Author index

Subject index